Textbook of Biodiversity

Textbook of Biodiversity

Edited by Jason Hendon

SYRAWOOD
PUBLISHING HOUSE

New York

Published by Syrawood Publishing House,
750 Third Avenue, 9th Floor,
New York, NY 10017, USA
www.syrawoodpublishinghouse.com

Textbook of Biodiversity
Edited by Jason Hendon

International Standard Book Number: 978-1-68286-423-4 (Hardback)

Cataloging-in-publication Data

Textbook of biodiversity / edited by Jason Hendon.
      p. cm.
Includes bibliographical references and index.
ISBN 978-1-68286-423-4
1. Biodiversity. 2. Biodiversity conservation. 3. Ecology. I. Hendon, Jason.
QH541.15.B56 T49 2017
577--dc23

Printed in the United States of America.

TABLE OF CONTENTS

PREFACE

Biodiversity refers to the vast variety that is present in the flora and fauna of the earth's environment. Wildlife and forests are unevenly distributed around the earth and changes in environment as well as human activity can drastically alter the habits and habitat of both. This book on biodiversity, discusses the system of classification and management that has been practiced by experts in this field in the past few decades. This book is a complete source of knowledge on the present status of this important field. It traces the progress of this field and highlights some of its key concepts and applications and it elucidates the multidisciplinary aspects of biodiversity. This book is a vital tool for all researching or studying this field as it provides incredible insights into emerging trends and concepts.

Various studies have approached the subject by analyzing it with a single perspective, but the present book provides diverse methodologies and techniques to address this field. This book contains theories and applications needed for understanding the subject from different perspectives. The aim is to keep the readers informed about the progress in the field; therefore, the contributions were carefully examined to compile novel researches by specialists from across the globe.

Indeed, the job of the editor is the most crucial and challenging in compiling all chapters into a single book. In the end, I would extend my sincere thanks to the chapter authors for their profound work. I am also thankful for the support provided by my family and colleagues during the compilation of this book.

Editor

How Can We Identify and Communicate the Ecological Value of Deep-Sea Ecosystem Services?

Niels Jobstvogt[1,2]*, Michael Townsend[3], Ursula Witte[1], Nick Hanley[4]

1 Oceanlab, University of Aberdeen, Aberdeen, United Kingdom, **2** Aberdeen Centre for Environmental Sustainability (ACES), University of Aberdeen, Aberdeen, United Kingdom, **3** National Institute of Water and Atmospheric Research (NIWA), Hamilton, New Zealand, **4** Department of Geography and Sustainable Development, University of St Andrews, St Andrews, United Kingdom

Abstract

Submarine canyons are considered biodiversity hotspots which have been identified for their important roles in connecting the deep sea with shallower waters. To date, a huge gap exists between the high importance that scientists associate with deep-sea ecosystem services and the communication of this knowledge to decision makers and to the wider public, who remain largely ignorant of the importance of these services. The connectivity and complexity of marine ecosystems makes knowledge transfer very challenging, and new communication tools are necessary to increase understanding of ecological values beyond the science community. We show how the Ecosystem Principles Approach, a method that explains the importance of ocean processes via easily understandable ecological principles, might overcome this challenge for deep-sea ecosystem services. Scientists were asked to help develop a list of clear and concise ecosystem principles for the functioning of submarine canyons through a Delphi process to facilitate future transfers of ecological knowledge. These ecosystem principles describe ecosystem processes, link such processes to ecosystem services, and provide spatial and temporal information on the connectivity between deep and shallow waters. They also elucidate unique characteristics of submarine canyons. Our Ecosystem Principles Approach was successful in integrating ecological information into the ecosystem services assessment process. It therefore has a high potential to be the next step towards a wider implementation of ecological values in marine planning. We believe that successful communication of ecological knowledge is the key to a wider public support for ocean conservation, and that this endeavour has to be driven by scientists in their own interest as major deep-sea stakeholders.

Editor: James P. Meador, Northwest Fisheries Science Center, NOAA Fisheries, United States of America

Funding: This research project was funded by MASTS (Marine Alliance for Science and Technology for Scotland; URL: www.masts.ac.uk). MASTS is funded by the Scottish Funding Council (grant reference HR09011) and contributing institutions. Townsend's involvement was funded by NIWA (National Institute of Water and AtmosphericResearch; URL: www.niwa.co.nz) under the Coasts and OceansResearch Programme 3 (2013/14 SCI). The funders had no role in study design, data collection and analysis, decision to publish, or preparation of the manuscript.

Competing Interests: The authors have declared that no competing interests exist.

* Email: niels.jobstvogt@abdn.ac.uk

Introduction

The concept of ecosystem services (ES) has inspired a movement away from conservation for the sake of nature's inherent value to one that explicitly identifies, links and communicates the benefits of conservation to human wellbeing [1–3]. The endeavour of describing, quantifying and valuing the economic benefits that nature provides to society through ES has been identified as a powerful tool to make ecosystems count in cost-benefit analysis for environmental decision making [1,4]. Throughout this paper, however, the term 'value' is used in a broader sense, as a holistic concept which can include social, ecological and economic values. This broadening of the concept of value is needed because for the remotest places on earth like the deep sea, it is particularly challenging to make direct links between changes in system functioning and effects on the delivery of final ES (and thus on human well-being) [5,6].

The deep sea accounts for nearly 91% of the world's oceans with depths ranging from 200 m to almost 11,000 m. Despite its remoteness and size, its ecosystems are far from being unaffected by anthropogenic impacts such as fishing, climate change, and pollution [7–10]. To date many knowledge gaps remain around the functioning of deep-sea ecosystems. This is partially explained through the high costs, difficulties, and risks that are associated with deep-sea research. This lack of ecological knowledge means that we also know very little about the social and economic value of protecting the deep sea. By identifying and quantifying the ES benefits provided by the deep sea it is likely that appreciation for these benefits will change. This should lead to a larger emphasis on mitigating anthropogenic impacts in the oceans.

The major challenges of accounting for deep-sea ES stem from most people's lack of awareness about the deep-sea environment, and from the prevalence of intermediate services relative to easier-to-appreciate final services. Intermediate services in this paper refer to the indirect services that the ecosystem provides, such as habitat provision and nutrient cycling (the Millennium Ecosystem Assessment [1] refers to this category as supporting services). Intermediate services are the functional basis of the final services supplied by the system (Figure 1). The final services are considered as the ecosystem's contribution to human well-being [11] and include the ES categories of provisioning (e.g. commercial fish species), regulating (e.g. waste absorption and detoxification) and

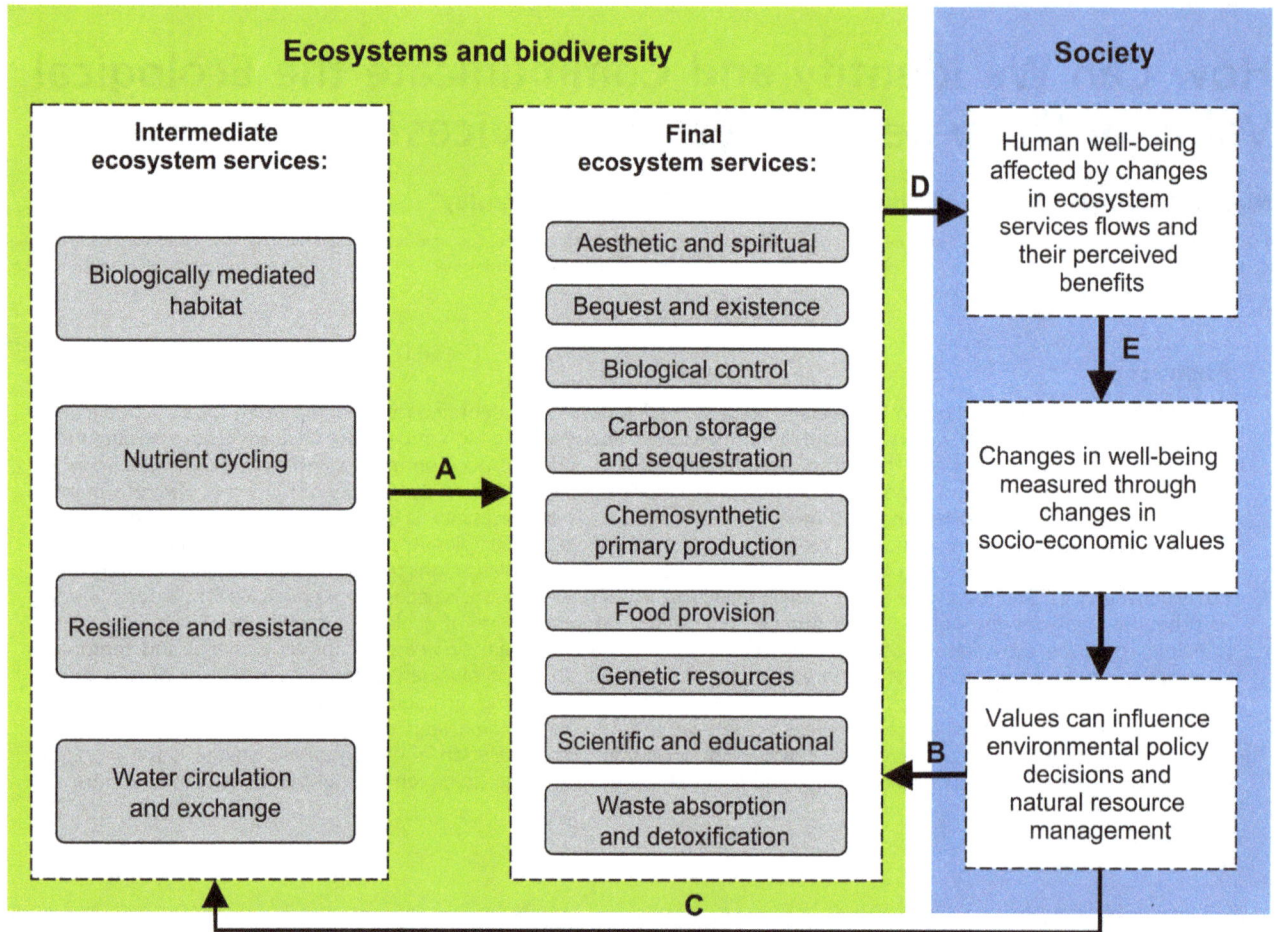

Figure 1. The ecosystem services framework for the example of deep-sea ecosystem services (ES). Environmental policies can either influence the management of final ES directly (arrow B) or indirectly via the intermediate ES (arrow C). The latter requires a sufficient understanding of the dependencies between intermediate and final ES (arrow A). Our understanding for the benefits provided by deep-sea ES (arrow D) and the values associated with them (arrow E) is currently very limited. The framework was simplified from [53] and adapted to the ES used for this expert consultation.

cultural (e.g. aesthetic values) [1]. There is a need to improve the integration of intermediate services and the processes that sustain them into the way in which we assess ES [12], since this ecological understanding is essential for demonstrating how human well-being ultimately depends on ecological processes and biodiversity [13,14].

In this paper we analyse the links between ES and the underlying processes that sustain them. This analysis was undertaken with the help of an expert panel of deep-sea scientists, focussing on submarine canyons. Within the deep sea, submarine canyons are an important ecosystem, which host unique reservoirs of biodiversity [15–18]. Along with much of the deep sea, they remain mostly underexplored and unprotected [9,19,20]. We selected this deep-sea ecosystem to test the Ecosystem Principles Approach (EPA) [21,22]. The EPA has recently been developed as a way of incorporating and translating ecological knowledge into easily understandable 'units' of information ('ecosystem principles') suitable for a wide range of audiences and thus for use in an ecosystem management context. The focus is on known and broadly-accepted information, with scientists from a wide field of expertise condensing this knowledge into principles that explain the linkages between ES, and their dependencies on underlying processes. The ecosystem principles also provide marine managers

with qualitative information on temporal, spatial, and causal dependencies of ES flows. In the New Zealand case study by Townsend and colleagues [22], the EPA highlighted the importance of accounting for intermediate services in marine management, which were often provided by different geographical areas relative to the location at which final services were taken into account in the ES assessment. Economic theory suggests that intermediate ES can and should be valued through the final services that they support and the resultant direct benefits to people [40]. Indeed, ES are often perceived as a purely economic concept [23], but they also have social and ecological values, which when integrated with economic values produce a more holistic ecosystem assessment that can better inform natural resource management decisions [24]. This dominance of economic approaches and monetary valuation of ES stems from the often-felt pressure among the nature conservation sector to "speak the same language" as business and policy sectors in order to make conservation count [25–29].

In contrast, the EPA's advantage lies where economic valuation reaches its limits, by offering a more holistic picture which might help non-experts to better understand the high ecological value that scientists associate with the deep sea. Links between ecosystem processes and ES are less well known for the deep sea. For this reason it may be more helpful to highlight the links between deep-sea ES and processes themselves, rather than presenting decision makers with a set of economic values that are likely to underestimate the ecosystems' holistic value due to the omission of important ecological aspects of the deep sea [23]. In this paper we test the applicability of the EPA to little known and remote deep-sea ecosystems, such as submarine canyons, and demonstrate how the approach can provide decision makers with an accessible knowledge base for conservation decisions, despite some deficiencies in scientific data and associated uncertainty. We further discuss the approach's utility for expert consultation and cross-disciplinary knowledge transfer.

Methodology

2.1 Case study area: The Nazaré Canyon

The Nazaré Canyon on the Portuguese continental margin (Figure 2), also described as "Europe's Grand Canyon" [16], was chosen as the case study area of this paper to test the applicability of the EPA for deep-sea ecosystems. The Nazaré Canyon has attracted scientific interest due to its habitat heterogeneity and is considered to be a biodiversity hotspot [16]. Like other submarine canyon ecosystems it plays an important role in transportation processes at the continental margins [30,31]. The canyon is shallowest at 1 km off the Portuguese coast (50 m depth) and with a total length of 210 km it extends into the Iberian abyssal plain where it reaches depths of over 4,900 m [32].

2.2.1 Sampling and survey structure. Twenty-three researchers with knowledge of the Nazaré Canyon, covering a broad range of disciplines such as ecology, biology, microbiology, biogeochemistry, geography, geomorphology, geology, sedimentology and oceanography, were invited to participate in a HERMIONE (Hotspot Ecosystem Research and Man's Impact On European Seas project; URL: www.eu-hermione.net; last access July 2013) workshop in September 2012 and in an online pre-workshop survey (Figure 3). These opportunities were used to gather ideas and feedback for the two main surveys that followed. We used an email-based Delphi process to gather structured information by consensus from the invited expert panel in two consecutive rounds of surveying (post-workshop survey I and II; Figure 3). The Delphi process was originally developed as an interactive forecasting technique, where an expert panel goes through iterative survey rounds. The group results of each round are fed back to participants who are able to adapt their responses in the next survey round. The main idea of the Delphi process is to lead the group towards a consensus through the indirect exchange of information via a process coordinator. This process allowed us to subsequently include experts' requests for changes and additional information into the post-workshop surveys (Figure 3).

The Delphi process makes it less likely that some researchers dominate the discussion and the outcomes, by maintaining anonymity throughout the communication process, thus avoiding the potential peer pressure of an expert workshop setting (further detail on the Delphi process in [33]). The survey questionnaires are available on request from the corresponding author.

2.2.2 Ethics statement. The nature of this research did not require ethical approval according to the University of Aberdeen Research Ethics Framework (*University of Aberdeen Research Ethical Review Checklist, Appendix A,* pp. 26–28; URL: www.abdn.ac.uk/documents/research-governance-framework-appendix4.pdf; last access March 2013). All study participants were recruited using an opt-in strategy and therefore consent was not explicitly recorded. Workshop and surveys did not include any sensitive personal questions. We asked participants to state their age, years of research experience, and field of studies; answering these questions was optional. Participants were provided with information on study objectives, sponsors, the participants' role, survey and workshop durations, potential benefits to the participants, summarised methodology, destination of gathered data and research results, the potential science impact and a contact address for further questions. Throughout the post-workshop surveys participants were identifiable via their email addresses. However, data on stated opinions and personal information was stored anonymously and kept confidential at all times. We chose email as the preferred communication method to facilitate the exchange of information during the Delphi phase of the expert consultation.

2.3 Submarine canyon ecosystem services

The experts helped to identify ES that were either perceived as less important or not relevant for the submarine canyon based on Table 1. Subsequently, the ES 'genetic resources', 'biological control', 'aesthetic and spiritual', 'scientific and educational', and 'chemosynthetic primary production' were excluded from Table 1 as less important relative to the other deep-sea ES. Accordingly eight ES (Table 1) were taken forward as a focus for the development of ecosystem principles.

2.4 Ecosystem Principles Approach

One of the main goals of the expert consultation was to develop a list of submarine canyon ecosystem principles, which could then be linked to ES. As an initial step, a review on the submarine canyon literature identified relatively well-explored ecosystem processes and relationships. The review findings were then discussed in the expert workshop and principles added or refined according to experts' suggestions. The following paragraph shows how the concept of ecosystem principles was described to canyon experts: "[An ecosystem principle] explicitly defines a key element

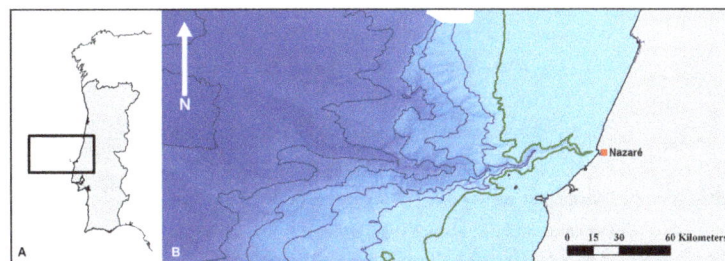

Figure 2. Nazaré Canyon. (A) Overview map of Portugal and the Nazaré Canyon area. **(B)** Nazaré Canyon bathymetry map with the Portuguese coastline to the east. Contour lines (blue) at 1000 m intervals; the 200 m depth contour, indicating the shelf edge, is marked in green. Data courtesy of Instituto Hidrografico, Lisbon and National Oceanography Centre, Southampton.

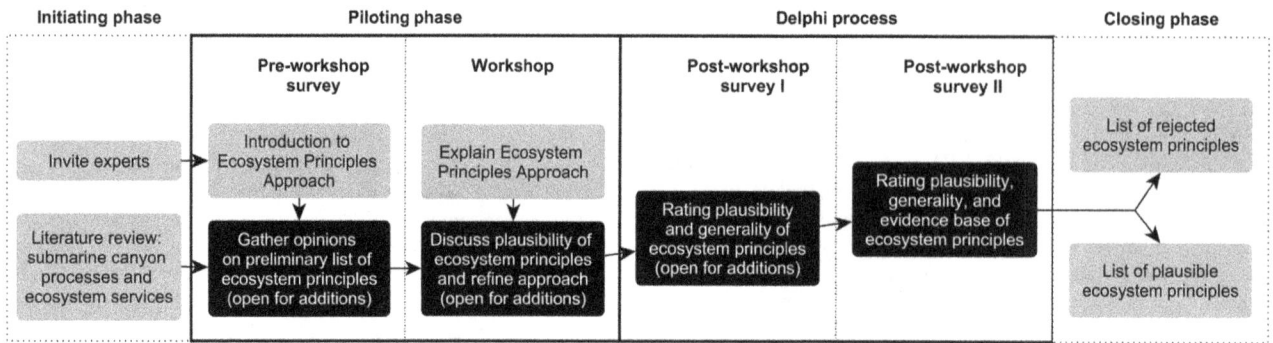

Figure 3. The survey phases of the submarine canyon expert consultation. Survey steps where experts were directly involved are highlighted as black boxes.

of how we expect the ecological system to operate" [22]. The workshop invitation summarised the EPA and included the methodology paper by Townsend and colleagues [22] as preparation for the workshop. During the workshop we provided further information in the form of a presentation that helped to increase participants' familiarity with the EPA. Principles were excluded from the initial list after each of the two consecutive post-workshop surveys (Delphi process; Figure 3) when fewer than 50% of the experts agreed with the plausibility of the principle. Experts were also able to propose new principles or suggest changes to the list of principles that was identified from the literature and subsequently refined throughout the piloting and Delphi phases (Figure 3).

We also asked the expert panel to categorise ecosystem principles according to their level of generality. The following categories were available: (i) general deep-sea principle, (ii) general

canyon principle, (iii) shelf-incising canyon specific principles, to (iv) Nazaré Canyon-specific principle. The option with the highest frequency was then presented as the group vote in the subsequent survey.

Only in the second stage of the Delphi process were experts asked to distinguish their rating based on evidence on the one side and their expert view (as individuals) on the other side. The evidence base was rated on a five-point scale from 'very poor' to 'very good'. For the presentation of the group result, this was then divided into three categories of good, intermediate or poor evidence, according to the average group scores. These evidence scores had no influence on the decision to include or exclude any principle, but were introduced to separate personal opinion from the levels of evidence that existed in support of the principle.

During the workshop experts stressed the importance of the connectivity function of submarine canyons at the continental

Table 1. Submarine canyon ecosystem services.

Ecosystem services	Descriptions
Provisioning services:	
Carbon sequestration and storage	The uptake, storage, and burial of organic material within the canyon.
Food provision	The provision of marine organisms for human consumption.
Genetic resources and chemical compounds*	The use of canyon organisms in biotechnological, pharmaceutical, or industrial applications.
Regulating services:	
Biological control*	The control of diseases and invasive species.
Waste absorption and detoxification	The burial, decomposition and transformation of waste within the canyon ecosystem.
Cultural services:	
Aesthetic and spiritual*	The canyon ecosystem aesthetic and spiritual or inspirational source for religion, arts, movies, documentaries, books and folklore.
Bequest and existence	Safeguarding the canyon ecosystem for future generations and for the existence of marine species.
Scientific and educational*	The cognitive use of the canyon ecosystem for science and education.
Intermediate services:	
Biologically mediated habitat	Canyon habitats formed by marine organisms that provide nursery and refuge sites for other marine life.
Nutrient cycling	The storage and recycling of nutrients by canyon organisms.
Chemosynthetic primary production*	Primary productivity that is not dependent on energy from the sun.
Resilience and resistance	The amount of disturbance that the canyon ecosystem can cope with and its ability to regenerate after disturbance.
Water circulation and exchange	The currents, such as up-and down-welling, dense shelf water cascading, and mixing of water masses.

Services are grouped into four categories: provisioning, regulating, cultural and intermediate.
Listed items taken from [5,7,36,52] with alterations.
*Deep-sea ES that were not taken forward for the development of submarine canyon principles.

margin and we therefore chose 'water circulation and exchange' as an example to demonstrate how ecosystem principles link the supply of ES with their underlying processes. In the final results section, those ecosystem principles which in their description indicated a relationship between the ES 'water circulation and exchange' and other services were linked. The ecosystem principles relevant to this particular ES were assigned by the authors of this paper.

Results

3.1 Sample characteristics

The workshop was attended by 14 deep-sea researchers, 11 of whom had completed the pre-workshop pilot survey (Figure 3). All 14 workshop attendees were invited to participate in the two stages of the Delphi process (Figure 3), which 11 did for the first round and 10 for the second round. The average survey participant had 21 years of research experience and the survey covered academics from senior professors to PhD researchers, with male and female researchers equally represented.

3.2 Ecosystem principles

Over the course of the Delphi process, 21 ecosystem principles were identified from the literature and were then assessed and refined by the expert group (Tables 2 and 3). To highlight the nature of ecosystem principles, we present principle P1 as an example, which was rated to be plausible by all experts: 'canyons host a large number of different habitats and as a result increase species diversity at a regional scale' (for further principle descriptions we refer the reader to Tables 2 and 3). Four principles were discarded, whereas 17 principles were rated as plausible. Ten ecosystem principles fell into the category 'general submarine canyon principles', five into category 'general deep-sea principles', and two into 'shelf-incising canyon specific principles' (Tables 2 and 3). This indicated that the majority of ecosystem principles were at an appropriate level to describe processes and linkages between ES for submarine canyons in general and that they can be readily transferred to other canyons. The Delphi process had the expected effect of driving opinions closer towards consensus. Seven of the ten experts who participated in Delphi-rounds I and II (Figure 3) were closer to the group rating after the second Delphi-round.

Comparing the ratings of evidence and plausibility, we recognised that the existence of supporting evidence was not necessarily a requisite for an ecosystem principle to be plausible. Seven principles obtained intermediate evidence scores and for P16 and P17 evidence was rated as poor. However, this lack of evidence did not translate into a lack of plausibility. It was therefore an advantage to separate the two ratings from each other to distinguish between the experts' opinions and their evidence-based judgments. However, plausibility was clearly lower overall when evidence ratings were poor (Tables 2 and 3).

Developing principles to link food availability with biodiversity was challenging, and none were rated as plausible (P18 and P20). Experts had strong concerns of oversimplification when it came to the type, quality and amount of organic matter as a source of food and how changes of those parameters affected biodiversity. For biodiversity there was again a concern of over-simplification by omitting information on the spatial scale of biodiversity. In the same way, geographical scale mattered to experts, and lack of information on depth ranges and exact geographical position was criticised. The rating on generality provided a preliminary solution for implementing information on the geographical transferability of principles. Giving experts the chance to express uncertainty

about the generality of principles as well as disentangling opinions about the rating of evidence allowed them to express their expectations for submarine canyons based on their research experience, and to transfer widely accepted knowledge from other ecosystems.

3.3 Linking principles and services

Many of the principles in Tables 2 and 3 have the capacity to provide information on *where* and *when* principles are likely to operate: for the principles included in Figure 4, particularly P8 and P7 reflect these spatio-temporal components. Other principles such as P3 and P12 explain *how* certain ES are provided and go into more detail on the processes involved. Principles like P17 and P16 that address effects of high biodiversity on ecosystem processes are capable of linking a broader set of ES such as 'carbon storage', 'food provision', 'bequest and existence' and 'waste absorption'. The ecosystem principles associated with 'biologically mediated habitat' were mainly thought to have an effect on biodiversity (e.g. P1 and P13) and to indirectly affect final ES such as 'food provision' and 'bequest and existence values' (P5).

For sustainable ecosystem management it can be equally important to understand the processes and principles that are involved in the provision of ES, as it is to understand the social and economic benefits of those services. We used 'water circulation and exchange' as an example to showcase how ecosystem principles explain links between ecological processes and ES (Figure 4; see also Table S1 for further details). 'Water circulation and exchange' has an important connectivity function in the submarine canyon (P11) and upwelling effects can lead to enhanced 'nutrient cycling' and as a result enhance productivity (P7 and P6; Figure 4). Further, 'nutrient cycling' might be important as an intermediate service for two different final ES, 'food provision' and 'bequest and existence values', because it can enhance fish abundance (P7). The 'bequest and existence' value can arise through the value that people tend to hold for iconic species (including fish), whereas 'food provision' relies on the abundance of commercially important fish as a consumptive resource. Trophic relationships, enhanced biomass, maintenance of deep-sea organisms (including non-iconic and non-commercial species) are important processes that sustain 'bequest and existence' as well as 'food provision' (P3, P4, P9; Figure 4) and should therefore be considered for management purposes. For 'carbon storage' and 'waste absorption' ecosystem management might be more concerned with other processes such as the transportation of organic and inorganic material, means of transportation, sedimentation rates, storage time, and burial processes that are important in parts of the submarine canyon (P3, P8, P12; Figure 4). How 'water circulation and exchange' is linked to 'resilience and resistance' as well as 'biologically mediated habitats' could not be resolved through the ecosystem principles developed in our workshop. This might be an indication that either too little evidence exists to support any ecosystem principles or that the links with processes that sustain these two ES are too complex to be described in the simplified form of ecosystem principles.

Discussion

The deep-sea case study for the Nazaré Canyon resulted in new insights on how to address the difficulties of assessing marine ES for ecosystem management purposes, especially when uncertainty is high due to lack of scientific data. In times where the demands on deep-sea resources are increasing, and scientific data on the potential impacts on marine biodiversity is scarce [5,34–36],

Table 2. Submarine canyon ecosystem principles with expert ratings on their plausibility and evidence base.

ID	Ecosystem principles	Plausibility	Evidence (mean score ± SE)
	General submarine canyon principles:		
P1	Canyons host a large number of different habitats and as a result increase species diversity at a regional scale.	100%	GOOD (3.8±0.2)
P2	The canyon topography tends to have a focusing or channelling effect for sediment and organic material.	100%	GOOD (3.8±0.2)
P3	The strength of large scale transportation events varies and occurrence ranges from a yearly to decadal pattern. They can be triggered by storms, high sediment load in the water column, cooling and increasing salinity of surface waters, or slope failures.	100%	GOOD (3.9±0.3)
P4	The transport of organic material from shallower waters to the deep seabed, which is mainly driven by large scale transportation events, is an important source of food for deep-sea organisms.	100%	MEDIUM (3.3±0.3)
P5	Canyons can serve as fish feeding ground, refuge and nursery area and therefore often show higher abundance of fish than their surroundings.	90%	MEDIUM (2.8±0.3)
P6	Canyons can enhance the mixing of water masses and as a result influence the exchange of nutrients, heat and salt between the shelf and the deep sea.	90%	MEDIUM (3.4±0.4)
P7	The canyon topography affects up- and down-welling of water masses at the continental margin. Upwelling events around the canyon head enhances productivity locally; as a result fish abundance can be higher.	90%	MEDIUM (3.3±0.3)
P8	By transporting large amounts of organic material from the shelf into deeper waters, canyons act as temporary stores of sediment and carbon. It can take decades or even centuries until the transported material reaches the abyssal plain, where it is then deposited on geological time scales.	80%	GOOD (3.8±0.2)
P9	Food quantity and quality tends to be higher within some canyon areas compared to the surrounding slope. This can enhance the biomass of the benthic and pelagic fauna.	80%	MEDIUM (3.3±0.4)
P10	Many species that are found in canyons are not found on the slope. They are therefore contributing to regional diversity.	80%	MEDIUM (3.1±0.4)

ID = principle identification number. The plausibility rating: ten experts participated in the full rating process (i.e. 100% = 10 experts). The evidence rating (1–5 from 'very good' to 'very poor'): poor (mean score <2.5), medium (2.5≤ mean score <3.5) and good (mean score ≥3.5); SE = standard error.

approaches such as the Ecosystem Principles Approach (EPA) are crucial to draw the link between the ecological and socio-economic dimensions of the ecosystem. Currently this linkage is poorly understood, contributing to the under-valuation of deep-sea ecosystems which is likely to undermine conservation efforts. We briefly outline the utility of expert consultation under these circumstances and reflect on the ability of the EPA to integrate more ecology into the assessment of ES and into the decision-making process. We first discuss how the EPA can help to communicate the overall importance of deep-sea ecosystems for the provision of ecosystem services. Further, we explain how the EPA can improve marine ecosystem-based management by promoting the inclusion of information on ecosystem processes into an ecosystem services assessment.

4.1 Communicating ecological values

The EPA has the ability to broaden access to ecological knowledge so that decision makers are not dependent on science advisors alone, but can take informed decisions on the basis of simplified ecological knowledge made available to them [37,38].

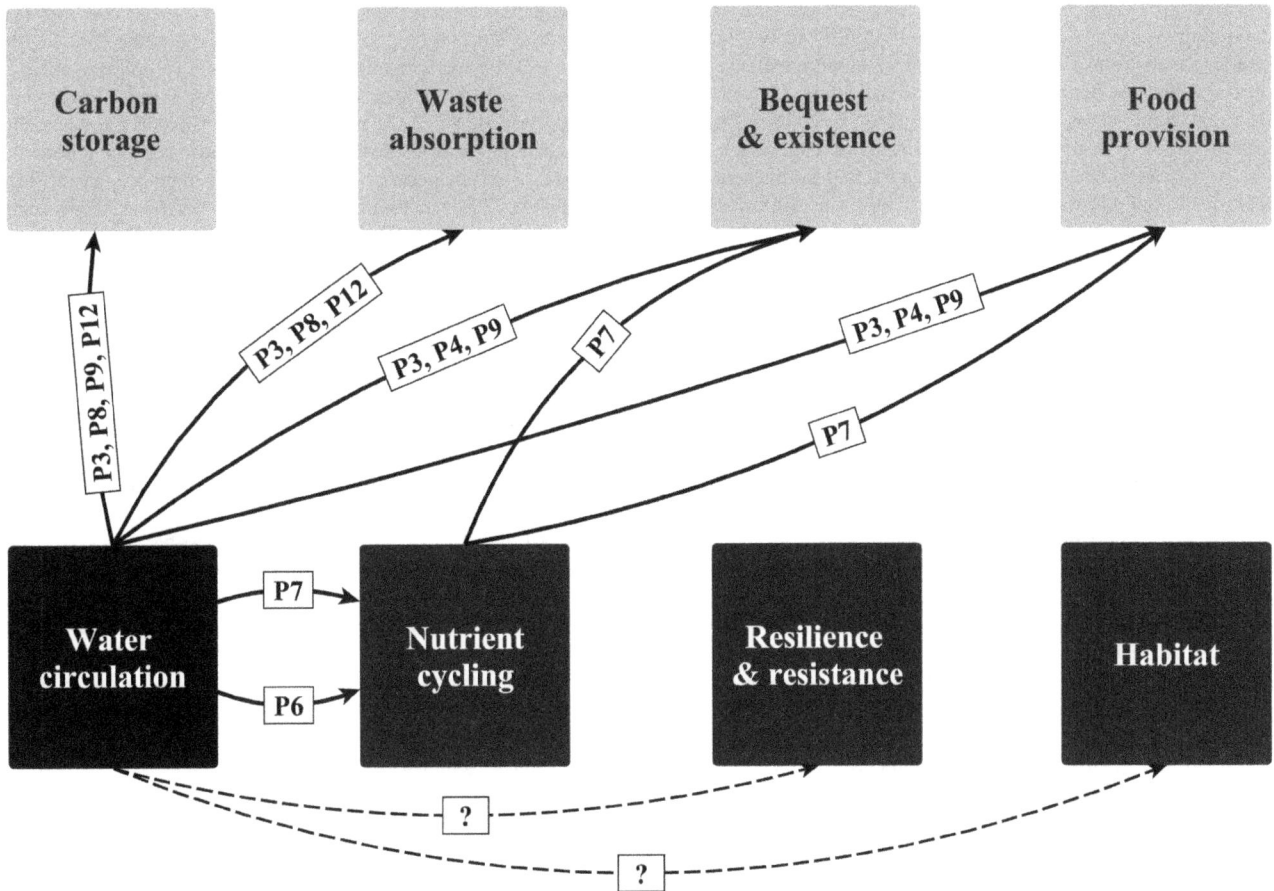

Figure 4. Links between 'water circulation' (black box) and other canyon ecosystem services explained through ecosystem principles. The intermediate services are in the lower half (dark grey and black boxes) and final services in the upper half (light grey boxes) of the diagram. Principles are indicated as arrows with their respective ID (cf. explanation in Table S1). Research gaps highlighted as question marks with dotted lines. Principles unrelated to 'water circulation' were omitted from this figure.

This broadening of access could also assist in increasing information about deep-sea ES amongst the general public. The process of developing ecosystem principles for the submarine canyon environment made clear that experts were able and willing to make predictions on connections between ES, generalisations on important canyon processes, and to link canyon characteristics to effects on ES. Also, while there remain research gaps and uncertainties, the list of ecosystem principles presented in Tables 2 and 3 provides the best available science knowledge to date, presented in easily understandable units of information. The list includes spatial and temporal information, as well as information on how the principles influence the supply of ES. We found that it was very important for the participating scientists to distinguish between their opinions (i.e. plausibility of the principles) and the (less subjective) rating of the existing scientific evidence base to inform principles. It was not imperative for principles to have a good evidence base, but rather to be generally accepted as plausible (cf. P5, P17 and P16). Herein lies a predictive strength of the EPA, in backing up the uncertainty associated with deep-sea science with a consensus-based approach, thereby decreasing uncertainty about ES linkages. The generality of principles was equally important, accounting for concerns that some ecosystem principles were valid on a larger scale than others i.e. 'general deep-sea principles' or 'shelf-incising canyon specific principles'. This additional type of information is crucial to highlight the

ability of the EPA to transfer principles to other submarine canyons or even other deep-sea ecosystems. The majority of experts rated the principles as either very broadly applicable to the deep sea or to submarine canyons in general, irrespective of the type of canyon. The broad applicability of principles was thought to be an effect of reducing the complexity of ecological information.

We share the view of one workshop participant who stated that it will be difficult to determine when the list of ecosystem principles is complete. New evidence, the inclusion of researchers with different academic backgrounds, and assigning more time to the task might increase the number of principles on the list. Thus including a broad range of disciplines into the principle development process is crucial. Nonetheless, there exists an asymptotic relationship between effort expended and the number of principles, where spending more time on identifying and reviewing submarine canyon principles might increase the detail of principles, but not their utility for management decisions. The EPA's utility lies not in providing large amounts of detail, but in providing meaningful, concise information to better understand the overall functioning of the ecosystem in conjunction with the ES it provides. The EPA is based on what we know today and the ecosystem principles in this paper cover a broad range of topics, with further workshops or surveys being likely to provide diminishing returns of new principles to our established list.

There are many aspects of the deep sea that the science community remains uncertain about, but the EPA helped to focus and assimilate known information with the underlying ecosystem processes that are better understood and agreed on. However, the fact that some ecosystem principles were discarded (Table 3) indicated that there remain gaps in understanding on how canyon biodiversity is influenced by current regimes, and by different types and quality of organic material, and also the importance of recruitment processes between deeper and shallower waters. The high specificity and complexity of these and other processes might not allow us to develop these processes as ecosystem principles at the current time.

4.2 Utility of the Ecosystem Principles Approach

The deep sea is hard to sample and poorly understood, yet we were able to draw on experts' knowledge and condense what they know about submarine canyons, one of the deep-sea's biodiversity hotspots. We showed that the EPA, in combination with a Delphi process, can be a useful tool for working at the fringes of our current knowledge, using collective expert opinions to evaluate and arbitrate on the content of ecological understanding. Through this process we can also highlight where knowledge and hence research gaps lie, and where funding is needed. The EPA might be seen as a balancing act between the precautionary principle on the one side and economic reasoning on the other. The precautionary principle as framed in Rio in 1992 states that lack of scientific certainty shall not be used as an excuse to postpone actions that might prevent environmental degradation [39]. Lubchenco [37] lists guidance for decision-making under uncertainty as one of the roles that science should play in society. This might include reliance on more subjective approaches such as the EPA to support more holistic decision-making in marine resource management until we have a greater body of scientific evidence to prove or disprove what researchers have outlined as ecosystem principles.

The principles for 'water circulation and exchange' (Figure 4) demonstrate the EPA's ability to provide information on the ecological value of the ecosystem, and how these are linked to the kinds of final ES which economists are likely to value in monetary terms. The approach does not focus on final ES alone, but provides information at multiple levels without losing sight of the indirect impacts on ecosystems through intermediate services and underlying processes, and of the multiple connections between ecosystems. By linking processes with services through ecosystem principles, we draw the attention towards the network character of ecosystems. The highly interlinked nature of this network means that it is actually far from straightforward to categorise and separate services for ES valuation, especially given the multitude of connections within the marine environment. This is so irrespective of whether social, ecological, or economic definitions of 'value' are assessed. Presenting the information on final and intermediate services together with their underlying processes in a network style can better inform future management scenarios with more realistic ecological information than assessments that are limited to final ES alone.

Current ES valuation frameworks suggest that intermediate services should be valued only in terms of the final services that they support and not be included directly in a valuation of ES flows to avoid double-counting of their social or economic value [25,40–42]. The resulting requirement for effective management is that underlying processes and linkages are sufficiently well understood [23] and the ES they support are provided within the managed area [5]. However, the spatial and temporal distances between marine intermediate and final services can span millennia and act on a global scale, as is the case with the ocean nutrient cycle [35,43]. Marine ecosystems are highly connected systems with many processes being important for the provision of intermediate and final ES and crossing ecosystem boundaries [5]. Hence, if intermediate and final ES are spatially separated, the chances are that recommendations which focus only on final ES will be poor for marine resource management [22,44]. The field of ES valuation has its roots in terrestrial ecosystems where ecosystems with their services and processes are less open than in marine ecosystems [45,46]; valuation approaches might have to be adapted for the marine environment to account for its higher connectivity. Also, to capture the holistic value of intermediate services, we would have to successfully value all final ES (including cultural ES values and other non-marketed ES), which is still more of a research aspiration rather than a currently-achievable outcome. A failure to recognise the contribution of intermediate services for the ES valuation in ecosystems where they dominate, such as the deep sea, will likely lead to misguided policy decisions [47].

The EPA should be seen as an addition to baseline ecological research and economic ES valuation, rather than as a substitute for either. The EPA's advantage lies where monetary valuation reaches its limits, in highlighting links between ES and their underlying processes, and in linking intermediate services with final ES. While economic ES values can help set marine management priorities that are socially and economically desirable (Daily et al., 2009), the EPA focusses on the ecological 'value' of the ecosystem and can provide important information on how such management priorities can be achieved. Where economic values require empirical links to well-being and monetary quantification, ecological 'value' is more focussed on the importance of the ecosystem processes and characteristics that lead to such economically valued benefits being produced.

4.3 Future research opportunities and lessons learnt

The EPA might not only enhance the availability of ecological information and its uptake by decision makers, but can also improve how research results are shared across disciplines. Inter- as well as trans-disciplinary collaborations are complicated by the existence of language barriers. The use of different key terms or jargon restricts access to the pool of knowledge to only a small number of experts. The lack of frameworks that translate research findings into understandable and meaningful formats has been described as one of the major reasons why information might not reach beyond disciplinary boundaries [48,49]. Different methodologies, attitudes and perceptions between disciplines might further decrease the flow of scientific evidence [38,48]. Thus to allow economists, geologists, biologists, oceanographers and other disciplines involved in marine science to share information it would be beneficial to work on a global matrix of ecosystem principles similar to the ES valuation databases provided by the Marine Ecosystem Services Partnership (MESP; URL: www.marineecosystemservices.org; last access August 2013) which gathers studies on monetary ES values. In contrast to the MESP database, the EPA would be able to add to the evidence base not just on economic and social, but also on ecological values. Also, a more extensive dataset on ecosystem principles for marine ecosystems would increase the chances that more complex management scenarios could be developed, such as in Bayesian belief networks (BBNs), which depend heavily on the availability of baseline information on ecosystem processes, even though BBNs are able to deal with knowledge gaps when expert knowledge is available [50]. Other fields that are using approaches like habitat mapping as well as biological value mapping might benefit from

Table 3. Submarine canyon ecosystem principles continued from Table 2.

ID	Ecosystem principles	Plausibility	Evidence (mean score ± SE)
	Shelf-incising canyon specific principles:		
P11	Canyons function as major transport pathways between the shelf and the deep sea.	100%	GOOD (4.0±0.3)
P12	Sediment, organic material, and pollutants that are transported alongshore get trapped by the canyon and transported down the canyon slope.	90%	MEDIUM (3.4±0.3)
	General deep-sea principles:		
P13	Areas with reef forming or habitat creating organisms can support higher diversity than their surroundings. These habitats are most common on hard substrates, such as areas with steep slopes, rocks, boulders, vertical walls, or overhangs.	100%	GOOD (4.1±0.3)
P14	The biomass of invertebrates living in and on the seafloor can constitute an important food source for commercially important deep-sea species.	100%	GOOD (3.7±0.3)
P15	The organisms inhabiting soft substrates play a major role for the recycling of nutrients. The process is largely dominated by bacteria, but is to a smaller extent also attributed to the animals living in and on the sediment.	70%	GOOD (3.5±0.3)
P16	Higher biodiversity can support higher rates of ecosystem processes.	70%	POOR (2.4±0.5)
P17	Higher biodiversity increases the insurance value of an ecosystem by increasing the likelihood that the ecosystem is able to provide the same ecosystem functions after an ecosystem impact occurred.	60%	POOR (2.0±0.4)
	Rejected ecosystem principles*:		
P18	Diversity tends to be lower in areas with high food availability.	10%	Not assessed
P19	Space and resource occupancy by native species can decrease invasion risk.	30%	POOR (1.4±0.2)
P20	Where strong bottom currents are common, food availability and substrate characteristics become less important and current speed becomes the main driver for species abundance and diversity.	40%	Not assessed
P21	The disturbances caused by strong bottom currents keep species diversity and abundance at low levels.	40%	POOR (2.3±0.4)

ID = principle identification number. The plausibility rating: ten experts participated in the full rating process (i.e. 100% = 10 experts). The evidence rating (1–5 from 'very good' to 'very poor'): poor (mean score <2.5), medium (2.5≤mean score <3.5) and good (mean score ≥3.5); SE = standard error.
*Principles P18–P21 were rejected by the majority of experts, i.e. their plausibility was below 50%.

the EPA as well, given that spatial ecosystem principles were developed [51].

Using the EPA it should be possible to provide more precise temporal and spatial information on ecosystem principles, and to develop management strategies based on the list of principles and evidence on social and economic values and resulting management priorities. Showing the EPA's potential to improve people's understanding of ecosystem functioning was beyond the scope of this study, but future research involving the wider public and

decision makers would be beneficial to test the effect of such simplified ecological knowledge on their decisions.

Three insights in particular emerge from our interdisciplinary workshop, which might improve future marine conservation initiatives and their acceptance:

(1) To further the field of marine ES valuation it would be beneficial to acknowledge that the traditional approaches to ES valuation, which have their roots in terrestrial research, might not be easily transferable to a highly linked marine

environment. Marine ecosystem boundaries are much less clearly defined than in terrestrial environments, and ES flows are less easily traceable. We might therefore need different approaches to valuing ES flows in the marine environment. The EPA is but one potential approach to improve integration of ecological values with social and economic values.

(2) The precautionary principle demands that we are cautious with our exploitation of the environment, but in the same time that management recommendations are made on a timely basis to the best of our knowledge, without postponing decisions for indefinite time until more certainty has been gained. The marine science community should more willingly embrace its important societal role in providing recommendations for nature conservation management with the support of social science approaches.

(3) We propose greater transparency in decisions on the conservation importance of marine areas. It should be possible to enhance understanding of the social, ecological as well as the economic values of certain areas, and to justify their protection, by providing easy understandable information on marine ES and how they relate to underlying ecosystem processes.

Acknowledgments

We are deeply thankful to the workshop and survey participants for their time and input, and to Dr Rene van der Wal, Dr Nicolas Krucien, and Prof Graham Pierce (University of Aberdeen) for constructive feedback on early drafts of this paper. We also like to thank Dr Veerle Huvenne (National Oceanography Centre) and the Instituto Hidrografico for the Nazaré Canyon bathymetry data, and the conference organizers for the workshop opportunity during the HERMIONE Annual Meeting 2012. We also thank two anonymous reviewers for their comments.

Author Contributions

Conceived and designed the experiments: NJ NH UW MT. Performed the experiments: NJ. Analyzed the data: NJ MT. Contributed reagents/materials/analysis tools: MT. Wrote the paper: NJ MT NH UW.

References

1. MEA (2005) Millenium Ecosystem Assessment: ecosystems and human well-being synthesis report. Washington DC: Island Press, 141.
2. Salles J-M (2011) Valuing biodiversity and ecosystem services: Why put economic values on Nature? Comptes rendus biologies 334: 469–482.
3. Lele S, Springate-baginski O, Lakerveld R, Deb D, Dash P (n.d.) Ecosystem services: origins, contributions, pitfalls, and alternatives. Conservation & Society: 1–46.
4. Hanley N, Barbier E (2009) Pricing Nature: Cost-Benefit Analysis and Environmental Policy. Cheltenham: Edward Elgar, 353.
5. Armstrong CW, Foley NS, Tinch R, Van den Hove S (2012) Services from the deep: Steps towards valuation of deep sea goods and services. Ecosystem Services 2: 2–13.
6. Jobstvogt N, Hanley N, Hynes S, Kenter J, Witte U (2014) Twenty thousand sterling under the sea: Estimating the value of protecting deep-sea biodiversity. Ecological Economics 97: 10–19.
7. Van den Hove S, Moreau V (2007) Deep-sea biodiversity and ecosystems: A scoping report on their socio-economy, management and governance. Cambridge: UNEP-WCMC, 84.
8. Benn AR, Weaver PP, Billet DSM, Hove S Van Den, Murdock AP, et al. (2010) Human activities on the deep seafloor in the North East Atlantic: an assessment of spatial extent. PLoS one 5(9): 1–15.
9. Ramirez-Llodra E, Tyler PA, Baker MC, Bergstad OA, Clark MR, et al. (2011) Man and the Last Great Wilderness: Human Impact on the Deep Sea. PLoS ONE 6(7): 1–25.
10. Jones DOB, Yool A, Wei C-L, Henson SA, Ruhl HA, et al. (2013) Global reductions in seafloor biomass in response to climate change. Global change biology: 1–12.
11. Haines-Young R, Potschin M (2013) Common International Classification of Ecosystem Services (CICES): Consultation on Version 4, August–December 2012. Nottingham, 34.
12. Turner RK, Paavola J, Cooper P, Farber S, Jessamy V, et al. (2003) Valuing nature: lessons learned and future research directions. Ecological Economics 46: 493–510.
13. Cardinale BJ, Duffy JE, Gonzalez A, Hooper DU, Perrings C, et al. (2012) Biodiversity loss and its impact on humanity. Nature 486: 59–67.
14. Mace GM, Norris K, Fitter AH (2012) Biodiversity and ecosystem services: a multilayered relationship. Trends in ecology & evolution 27: 19–26.
15. Stiles ML, Ylitalo-Ward H, Faure P, Hirshfield MF (2007) There's no place like home: Deep seafloor ecosystems of New England and the mid-Atlantic. Oceana, 40.
16. Tyler P, Amaro T, Arzola R, Cunha M, De Stigter H, et al. (2009) Europe's Grand Canyon: Nazaré Submarine Canyon. Oceanography 22: 46–57.
17. Danovaro R, Company JB, Corinaldesi C, D'Onghia G, Galil B, et al. (2010) Deep-sea biodiversity in the Mediterranean Sea: the known, the unknown, and the unknowable. PLoS one 5(8): 1–25.
18. Auster PJ, Moore J, Heinonen KB, Watling L (2005) A habitat classification scheme for seamount landscapes: assessing the functional role of deep-water corals as fish habitat. In: Freiwald A, Roberts JM, editors. Cold-Water corals and ecosystems. Berlin Heidelberg: Springer, 761–769.

19. Glover AG, Smith CR (2003) The deep-sea floor ecosystem: current status and prospects of anthropogenic change by the year 2025. Environmental Conservation 30: 219–241.
20. Würtz M (2012) Mediterranean submarine canyons: Ecology and governance. Gland, Switzerland and Malaga, Spain: IUCN, 216.
21. Townsend M, Thrush S (2010) Ecosystem functioning, goods and services in the coastal environment. Prepared by the National Institute of Water and Atmospheric Research for Auckland Regional Council. Auckland Regional Council Technical Report 2010/033.
22. Townsend M, Thrush S, Carbines M (2011) Simplifying the complex: an "Ecosystem Principles Approach" to goods and services management in marine coastal ecosystems. Marine Ecology Progress Series 434: 291–301.
23. Cornell S (2010) Valuing ecosystem benefits in a dynamic world. Climate Research 45: 261–272.
24. De Groot RS, Wilson M a, Boumans RM. (2002) A typology for the classification, description and valuation of ecosystem functions, goods and services. Ecological Economics 41: 393–408.
25. Fisher B, Turner K, Zylstra M, Brouwer R, De Groot R, et al. (2008) Ecosystem services and economic theory: integration for policy-relevant research. Ecological Applications 18: 2050–2067.
26. Brink P, Mazza L, Badura T, Kettunen M, Withana S (2012) Nature and its role in the transition to a green economy. The Economics of Ecosystems & Biodiversity (TEEB), 72.
27. Kallis G, Gómez-Baggethun E, Zografos C (2013) To value or not to value? That is not the question. Ecological Economics 94: 97–105.
28. Balmford A, Fisher B, Green RE, Naidoo R, Strassburg B, et al. (2010) Bringing Ecosystem Services into the Real World: An Operational Framework for Assessing the Economic Consequences of Losing Wild Nature. Environmental and Resource Economics 48: 161–175.
29. Peterson MJ, Hall DM, Feldpausch-Parker AM, Peterson TR (2010) Obscuring ecosystem function with application of the ecosystem services concept. Conservation biology: the journal of the Society for Conservation Biology 24: 113–119.
30. De Leo FC, Smith CR, Rowden AA, Bowden DA, Clark MR (2010) Submarine canyons: hotspots of benthic biomass and productivity in the deep sea. Proceedings Biological sciences/The Royal Society 277: 2783–2792.
31. Harris PT, Whiteway T (2011) Global distribution of large submarine canyons: Geomorphic differences between active and passive continental margins. Marine Geology 285: 69–86.
32. Lastras G, Arzola RG, Masson DG, Wynn RB, Huvenne V a I, et al. (2009) Geomorphology and sedimentary features in the Central Portuguese submarine canyons, Western Iberian margin. Geomorphology 103: 310–329.
33. Linstone HA, Turoff M (2002) The Delphi Method: Techniques and Applications. New Jersey: New Jersey Institute of Technology, 616.
34. Mengerink KJ, Van Dover CL, Ardron J, Baker M, Escobar-Briones E, et al. (2014) A Call for Deep-Ocean Stewardship. Science 344: 696–698.
35. Thurber AR, Sweetman AK, Narayanaswamy BE, Jones DOB, Ingels J, et al. (2013) Ecosystem function and services provided by the deep sea. Biogeosciences Discussions 10: 18193–18240.
36. Armstrong CW, Foley N, Tinch R, Van den Hove S (2010) Ecosystem goods and services of the deep sea. Deliverable D6.2 HERMIONE Project, 68.

37. Lubchenco J (1998) Entering the Century of the Environment: A New Social Contract for Science. Science 279: 491–497.

38. Rudd MA, Lawton RN (2013) Scientists' prioritization of global coastal research questions. Marine Policy 39: 101–111.

39. Myers N (2002) The Precautionary Principle Puts Values First. Bulletin of Science, Technology & Society 22: 210–219.

40. UK NEA (2011) UK NEA: Synthesis of the key findings. UK National Ecosystem Assessment, Cambridge, UNEP-WCMC, 87.

41. Boyd J, Banzhaf S (2007) What are ecosystem services? The need for standardized environmental accounting units. Ecological Economics 63: 616–626.

42. Wallace KJ (2007) Classification of ecosystem services: Problems and solutions. Biological Conservation 139: 235–246.

43. Roberts CM, Andelman S, Branch G, Bustamante RH, Carlos Castilla J, et al. (2003) Ecological criteria for evaluating candidate sites for marine reserves. Ecological Applications 13: 199–214.

44. McLeod K, Leslie H (2009) Ecosystem-based management for the oceans. Washington, DC: Island Press, 368.

45. Hawkins SJ (2004) Scaling up: the role of species and habitat patches in functioning of coastal ecosystems. Aquatic Conservation: Marine and Freshwater Ecosystems 14: 217–219.

46. Raffaelli D (2006) Biodiversity and ecosystem functioning: issues of scale and trophic complexity. Marine Ecology Progress Series 311: 285–294.

47. Johnston RJ, Russell M (2011) An operational structure for clarity in ecosystem service values. Ecological Economics 70: 2243–2249.

48. Davies AL, Colombo S, Hanley N (2013) Improving the application of long-term ecology in conservation and land management. Journal of Applied Ecology 51: 63–70.

49. Peters DPC (2010) Accessible ecology: synthesis of the long, deep, and broad. Trends in Ecology & Evolution 25: 592–601.

50. Campbell D, Chilton S, Clark S, Fitzsimmons C, Gazzola P, et al. (2012) Bayesian Belief Networks as an Interdisciplinary Marine Governance and Policy Tool. Valuing Nature Network Report, 27.

51. Townsend M, Thrush SF, Lohrer AM, Hewitt JE, Lundquist CJ, et al. (2014) Overcoming the challenges of data scarcity in mapping marine ecosystem service potential. Ecosystem Services 8: 44–55.

52. Beaumont N, Townsend M, Mangi S, Austen MC (2006) Marine biodiversity: an economic valuation. Building the evidence base for the Marine Bill. Department for Environment Food and Rural Affairs, 64.

53. Mace GM, Bateman I, Albon S, Balmford A, Brown C, et al. (2011) UK NEA Technical Report: Conceptual Framework and Methodology. UK National Ecosystem Assessment: Technical Report. Cambridge: UK National Ecosystem Assessment, UNEP-WCMC, 11–26.

2

Optimized Spatial Priorities for Biodiversity Conservation in China: A Systematic Conservation Planning Perspective

Ruidong Wu[1,2]*, Yongcheng Long[3], George P. Malanson[4], Paul A. Garber[5], Shuang Zhang[3], Diqiang Li[6], Peng Zhao[3], Longzhu Wang[3], Hairui Duo[7]

1 Institute of International Rivers and Eco-security, Yunnan University, Kunming, Yunnan, China, 2 Yunnan Key Laboratory of International Rivers and Transboundary Eco-security, Yunnan University, Kunming, Yunnan, China, 3 The Nature Conservancy China Program, Kunming, Yunnan, China, 4 Department of Geography, The University of Iowa, Iowa City, Iowa, United States of America, 5 Department of Anthropology, University of Illinois, Urbana, Illinois, United States of America, 6 Institute of Forest Ecology, Environment, and Protection, Chinese Academy of Forestry, Beijing, China, 7 School of Nature Reserve, Beijing Forestry University, Beijing, China

Abstract

By addressing several key features overlooked in previous studies, i.e. human disturbance, integration of ecosystem- and species-level conservation features, and principles of complementarity and representativeness, we present the first national-scale systematic conservation planning for China to determine the optimized spatial priorities for biodiversity conservation. We compiled a spatial database on the distributions of ecosystem- and species-level conservation features, and modeled a human disturbance index (HDI) by aggregating information using several socioeconomic proxies. We ran Marxan with two scenarios (HDI-ignored and HDI-considered) to investigate the effects of human disturbance, and explored the geographic patterns of the optimized spatial conservation priorities. Compared to when HDI was ignored, the HDI-considered scenario resulted in (1) a marked reduction ($\sim$9%) in the total HDI score and a slight increase ($\sim$7%) in the total area of the portfolio of priority units, (2) a significant increase ($\sim$43%) in the total irreplaceable area and (3) more irreplaceable units being identified in almost all environmental zones and highly-disturbed provinces. Thus the inclusion of human disturbance is essential for cost-effective priority-setting. Attention should be targeted to the areas that are characterized as moderately-disturbed, $<$2,000 m in altitude, and/or intermediately- to extremely-rugged in terrain to identify potentially important regions for implementing cost-effective conservation. We delineated 23 primary large-scale priority areas that are significant for conserving China's biodiversity, but those isolated priority units in disturbed regions are in more urgent need of conservation actions so as to prevent immediate and severe biodiversity loss. This study presents a spatially optimized national-scale portfolio of conservation priorities – effectively representing the overall biodiversity of China while minimizing conflicts with economic development. Our results offer critical insights for current conservation and strategic land-use planning in China. The approach is transferable and easy to implement by end-users, and applicable for national- and local-scale systematic conservation prioritization practices.

Editor: Duccio Rocchini, Fondazione Edmund Mach, Research and Innovation Centre, Italy

Funding: This study was funded by the National Natural Science Foundation of China (No. 31260148) and the National Key Technologies R&D Program of China (No. 2011BAC09B07). The funders had no role in study design, data collection and analysis, decision to publish, or preparation of the manuscript.

Competing Interests: The authors have declared that no competing interests exist.

* Email: rdwu@ynu.edu.cn

Introduction

Anthropogenic effects have resulted in the loss of biodiversity at an unprecedented rate, while resources for biodiversity conservation remain constrained in terms of both human and financial capacity [1]. That is why the systematic planning of priority areas is crucial to achieve the most cost-effective conservation, such as identifying large-scale biodiversity hotspots or assembling fine-resolution portfolios of conservation priorities [2–4]. In the last two decades, systematic conservation planning (SCP) has emerged as an effective approach for identifying conservation priorities [3–6]. SCP aims to identify a network of priority areas so as to effectively achieve explicit conservation goals in terms of representing the full range of biodiversity and sustaining their long-term survival [5]. Efficient conservation priorities can be identified through an optimized planning algorithm for meeting conservation goals at

the minimum land area or other costs (e.g., land prices, management and opportunity costs [7]). SCP provides an operational framework for minimizing land-use conflicts between conserving natural environments and economic development, and thus increase the likelihood of implementing the proposed conservation actions [3,6]. Here, we present the first national-scale SCP study for China to determine the optimized spatial priorities for biodiversity conservation.

China – one of the world's "megadiversity countries" – is home to many globally valued conservation priorities [2]. However, China's biodiversity is under severe threat due to the increasing pressure resulting from the country's historically unprecedented economic growth [8]. Meanwhile, China's conservation investment is considerably lower compared to developed and other developing countries [9]. Thus, the systematic conservation priority-setting has been emphasized in China during last two decades [8,10–11].

During this period, China has developed several templates of national-scale conservation priorities, which were based principally on the species (e.g., endemic, threatened, and/or other indicator species) richness patterns as well as expert judgments (e.g., [10,12–15]). These templates are crucial in guiding China's national-level conservation decisions; however, we think there are several critical limitations in previous priority-setting studies.

First, the effects of human disturbance are not incorporated in previous studies, whereas we believe explicit inclusion of human disturbance in priority-setting can minimize land-use conflicts and lower costs for meeting conservation goals [3,6]. Second, the scoring procedure in these studies is inefficient for achieving the goal of full representation of all biodiversity targets [16], i.e. the goal for representativeness – one of the core principles for designing an efficient reserve system [5]. The current scoring procedure requires a greater amount of land (and increases other costs) to achieve the same conservation goals and these greater demands are unlikely to get support from local authorities. Third, a study designed to systematically integrate conservation features at both ecosystem- and species-level is still lacking, as the conservation features used in previous studies are either species or ecosystem based. By incorporating biodiversity features from multiple organization levels, the resulting portfolio of conservation priorities is more efficient in representing the full range of biodiversity concerns and in maintaining the ecological integrity of ecosystems [5,16]. SCP can overcome this inefficiency in scoring procedure by employing the principle of between-site complementarity that serves to boost the efficient representation of all biodiversity targets, and provide mechanisms for integrating human disturbance and conservation features at multiple organization levels [17].

This study aims to determine the optimized national-scale spatial priorities in China and to ensure effectively fulfilling biodiversity conservation goals given the constraints of human disturbance by implementing a SCP approach. Taiwan, Hong Kong and Macao are not included in our analysis due to lack of required information. Specifically, we are trying to address two questions: (1) How will the inclusion of human disturbance affect the result of conservation priority-setting? (2) What are the geographical patterns of the optimized conservation priorities in China? In this analysis, we integrated human disturbance, conservation features at both the ecosystem- and species-level, and the principles of complementarity and representativeness. We used the software Marxan [17] to determine each unit's conservation value and to identify priorities with regular hexagons (100 km^2 per cell) as the planning units. We investigated the effects of human disturbance using two Marxan scenarios – a disturbance ignored scenario and a disturbance considered scenario. For the second scenario, human disturbance was included as a penalty function by aggregating information on several socioeconomic proxies as an index layer. We then explored the spatial patterns of the priority units, irreplaceable areas, and primary large-scale priority areas (i.e., the large clustered regions of high-conservation-value units). The analysis is limited to the data available at national-scale and applicable resolution, and misses some variability within the range of human disturbances. We believe this study is applicable to national- and local-scale conservation and other sustainable land-use planning for systematically evaluating each site's conservation value and identifying spatially optimized priority areas.

Methods

Conservation Features Mapping

Considering the complexity of biodiversity and severe lack of detailed spatial distribution data, surrogates (e.g., endangered/endemic species, key habitat types and environmental features) are often used in conservation planning [3,7,16,18]. Integrating conservation features from multiple levels can ensure the efficient representation of biodiversity [5] and compensate for limitations in the data [16]. In this analysis, we used both ecosystem- and species-level features as the surrogates.

The ecosystem-level features included were: (1) priority natural ecosystems as defined by Li, Song & Ouyang [13], including 129 natural ecosystems of forests, grasslands, meadows, deserts and wetlands, and (2) natural vegetation types derived from the national 1: 1,000,000 vegetation map, including 559 natural vegetation formations [19]. This study considered wetlands and lakes (in the priority natural ecosystems and natural vegetation types), but data on aquatic systems and species was lacking. We expected that China's key ecological elements, processes and services were covered with priority natural ecosystems and that basic habitat types were represented by finer-scale classifications of natural vegetation types. The species-level features were endangered species of plants, mammals, and birds. Endangered mammals and birds were identified according to China's "National List of Key Protected Wildlife" and the IUCN Red List Categories of critically endangered, endangered and vulnerable species [20]. Endangered plants were defined in the "China Plant Red Data Book: Endangered and Rare Plants" [21].

Previous studies often use county-level species distribution data derived from the published literatures [14–15], while our analysis was performed using a finer-scale resolution. For plants and mammals, we mapped each species' geographic range by combining its distribution data for counties, preferred habitat types and elevation range. For a bird species, the range was derived by intersecting only counties and habitat types, because knowledge of the altitude distribution of most avian species is lacking. This mapping process included: (1) collecting each species' attribute information, i.e. species name, taxonomy, endangered category, distribution across counties, preferred habitat types, and elevation range, (2) mapping each species' distributions across counties, habitat types and elevation range, respectively, and (3) identifying the overlap region among these distribution layers as each species' current range.

We collected the attribute information using the following resources. For plants, we used "National Key Protected Wild Plant Resources Survey" [22] as the primary source and other supplementary sources including "Subject Database of China Plants" [23], "China Species Information Services" [24] and "China Plant Red Data Book: Endangered and Rare Plants" [21]. For mammals and birds, we used "National Key Terrestrial Wildlife Resources Survey" [25] as the primary source and other supplementary sources including "Database of Fauna Sinica" [26], "Distributions of China Mammal Species" [27] and "China Red Data Book of Endangered Animals: Mammals" [28].

The datasets on county boundaries and habitat types were derived from the national 1: 1,000,000 geographic databases and the national 1: 1,000,000 vegetation map [19], respectively. The elevation range for each species was extracted from the Shuttle Radar Topography Mission (SRTM) 90 m Digital Elevation Model (DEM) [29]. We mapped the species' geographic ranges for 373 plant, 115 mammal and 81 bird species.

Human Disturbance Index Mapping

We used several socioeconomic proxies, including proportion of land converted by human use, human population density, gross domestic product (GDP) and road density to calculate the human disturbance index (HDI) or human footprint [4,30]. The basic planning units were regular hexagons, each sized 100 km^2. The analysis included three steps. First, we calculated an individual HDI (IHDI) for each of these proxies. For proportion of converted land, we calculated the IHDI as the percent area of human-developed-land use – including croplands, plantations, rural settlements and urban/industrial areas – within each hexagon. For human population density and GDP, we calculated the IHDIs as their mean values per square kilometer within each unit. For road density, we considered four transportation levels (i.e., railway, expressway, national-provincial road and other-level roads), and calculated an IHDI for each level as the total road length within each unit. Second, we normalized the data ranges of all IHDIs on a scale of from 0.00 to 1.00, and then summed them to get the total HDI. Finally, we empirically transformed the data range of the total HDI on a scale of from 10.00 to 300.00 (Figure 1) so as to clearly demonstrate the overall human disturbance pattern.

We obtained datasets on land uses, human population density and GDP from the Data Center for Resources and Environmental Sciences of the Chinese Academy of Sciences [31], and all are 1 km×1 km resolution grid files. The road networks were derived from the national 1: 1,000,000 geographic database.

Conservation Priority-setting

We used the software Marxan (v2.0.2) to implement the conservation priority-setting process. Marxan was developed to cost-effectively solve an optimization problem of representing a suite of biodiversity targets [17]. To ensure that all conservation features were captured across their ranges of environmental and genetic variations [32], we first stratified their ranges with China's 53 terrestrial ecoregions [33], and then defined a quantitative conservation target for each feature per ecoregion. Due to limited data available for setting up appropriate conservation targets [34], we defined the target for each conservation feature as a uniform percentage area of its distribution range as suggested in previous studies (e.g., [3,32]). Specifically, the quantitative targets were selected based on expert opinions as follows: 30% for endangered species, 20% for priority natural ecosystems and 10% for natural vegetation types. An internationally recognized lowest target of 10% was set for natural vegetation types because they were assumed to represent the variety of basic habitat types.

We ran Marxan with two scenarios – a HDI-considered scenario and a HDI-ignored scenario. For the HDI-considered scenario, we integrated HDI values as a penalty function in Marxan analysis, i.e. a unit having a higher degree of disturbance would receive a greater penalty. For the HDI-ignored scenario, we used a uniform penalty of 1.0 per unit. The units with greater HDI values exhibit a more highly degraded ecological condition and should offer less potential from a conservation perspective [4]. Therefore, Marxan's algorithm sought to identify the optimized priority areas by minimizing the total HDI score in the HDI-considered scenario or the total land area in the HDI-ignored scenario. For the Marxan configurations, we: (1) generated 1000 solutions; (2) included a boundary length file and a modifier factor to control the compactness of priority areas; (3) implemented Simulated Annealing followed by Iterative Improvement; and (4)

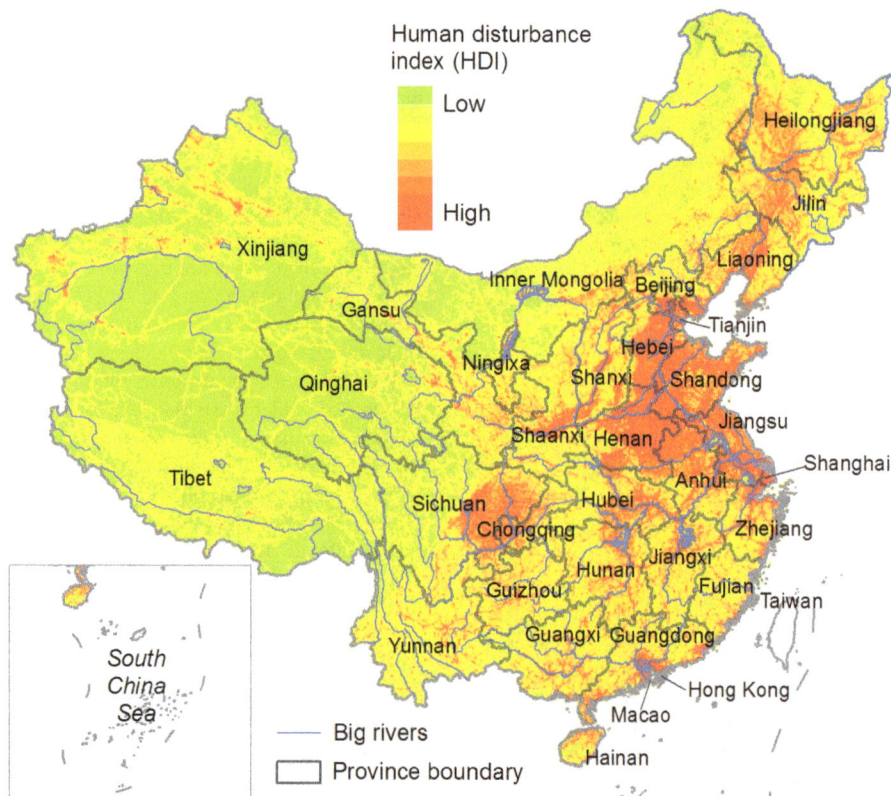

Figure 1. Human disturbance index (HDI). HDI was modeled by aggregating information on several socioeconomic proxies, including proportion of land converted by human use, human population density, gross domestic product, and road density.

used the default values for Number of Iterations (1,000,000) and Temperature Decreases (10,000).

We derived the conservation value that reflects the relative priority or irreplaceability of each planning unit [3] from the frequency of solutions selected, and used the best of the 1,000 solutions as the most cost-effective portfolio of priority units. We then identified the irreplaceable units as those selected in more than 800 solutions; 80% is often used for accuracy assessment for spatial data (e.g., [35]).

Effects of Human Disturbance

We compared the total HDI score and total area of the two portfolios of priority units generated by the HDI-ignored and HDI-considered scenarios, respectively. The changes in irreplaceable areas between the two scenarios were assessed in terms of the total area and the proportional area changes by province.

To assess the effects of human disturbance at a finer-scale, we further investigated the distributions of the priority units and irreplaceable areas on different environmental zones of HDI, elevation, and terrain ruggedness. We derived seven zones for each variable as follows: (1) We classified HDI zones by applying the Quantile Classification Scheme on HDI values; (2) We derived elevation zones from the SRTM 90 m DEM according to studies on geomorphology [36] (the elevation classification schemes were <200, 200–500, 500–1,000, 1,000–1,500, 1,500–2,000, 2,000–4,000, and >4,000 m); (3) We calculated a terrain ruggedness index (TRI) as the average difference in elevation between a center cell and its eight neighboring cells using the SRTM DEM, and the Quantile Classification Scheme was then used to break the TRI values into seven terrain categories, i.e. level, near-level, slightly-rugged, intermediately-rugged, moderately-rugged, highly-rugged and extremely-rugged [37–38].

Spatial Patterns of Conservation Priorities

Using the outputs from the HDI-considered scenario, we analyzed the spatial distributions of priority units and irreplaceable areas on HDI, elevation, TRI zones and provinces. We then delineated the primary large-scale priority areas as the large clusters of high-conservation-value planning units through an expert-based visual interpretation process.

Results

Effects of Human Disturbance

We presented the conservation value (based on a scale of from 0 to 1,000) of individual 100 km² hexagon units distributed throughout China (Figure 2). The portfolio of priority units in the HDI-ignored scenario (Figure 3) covered 24.6% of China's land area. By explicitly including the HDI as an additional penalty, we achieved the same conservation targets with a small increase (~7%) in the total area of priority units compared to when HDI was ignored, meanwhile a clear reduction of ~9% in the total HDI score was observed. The overlapping region (Figure 3) covered 46.3% and 43.2% of the priority units in the HDI-ignored and HDI-considered scenarios, respectively. A strong and positive spatial correlation exists (Spearman's rank correlation, $r = 0.871$, $p << 0.001$) between the two conservation value layers.

The irreplaceable units in the HDI-ignored scenario (Figure 2A) covered 2.8% of China's landmass, while an increase of ~43% in the total irreplaceable area was observed in the HDI-considered scenario (Figure 2B). The overlapping region occupied 82.7% and 57.7% of the irreplaceable areas in the HDI-ignored and HDI-considered scenarios, respectively. High proportional increases in irreplaceable area occurred principally in provinces located in the eastern coastal region, middle-lower Yangtze River Basin and northeastern China, whereas provinces in western and southwestern China had the fewest changes (Figure 4). Several provinces in the eastern highly-disturbed regions (Figure 1), including Guangdong, Jiangxi, Henan and Hebei, also were found to have small changes in their irreplaceable areas (Figure 4).

Compared to the results in the HDI-ignored scenario, the portfolio of priority units in the HDI-considered scenario contained: (1) fewer units in the three highest HDI zones and more units in the four lower HDI zones, (2) fewer units only in the lowest (<200 m) elevation zone and more units in the other six zones, and (3) fewer units in the level TRI zone and more units in each of the other TRI zones (Figure 5). The HDI-considered scenario identified a greater number of irreplaceable units in almost all environmental zones than did the HDI-ignored scenario, with the sole exception of the highest HDI zone (Figure 6).

Spatial Patterns of Conservation Priorities

We analyzed the spatial patterns of conservation priorities using the outcomes from the HDI-considered scenario. The priority units consistently decreased with increasing HDI value (Figure 5A), with the majority (~76%) located in the four lower HDI zones. The <200 m elevation zone included only 5.8% of all priority units, and the zones of 200–1,000, 1,000–2,000 and >2,000 m contained 29.6%, 25.8% and 38.8% of the priority units, respectively. The priority units generally had an increasing distribution trend on TRI zones from level to extremely-rugged terrain (Figure 5C), with the vast majority located in slightly- to extremely-rugged zones, and only 3.2% identified in level zone and 11.5% in near-level zone. All provinces included some units that were required for meeting the conservation targets (Figure 3), with the greatest proportion occurring in Xinjiang followed by Tibet, Inner Mongolia, Qinghai, Sichuan and Yunnan. These six western provinces contained 72.5% of the total priority units.

The irreplaceable units had a normal-like distribution on the HDI zones that peaked in the fourth zone (Figure 6A). Compared to the distribution of priority units, greater proportions of irreplaceable units were selected in lower elevation zones, with 10.7%, 39.3%, 26.7% and 23.3% of the total irreplaceable area located at <200, 200–1,000, 1,000–2,000 and >2,000 m zones, respectively. In particular, the highest zone (>4,000 m) contained the smallest proportion of irreplaceable areas (Figure 6B) although the greatest number of priority units occurred there (Figure 5B). In addition, over 75% of the irreplaceable areas were located in intermediately- to extremely-rugged TRI zones (Figure 6C). Provinces with the greatest number of irreplaceable areas were Yunnan followed by Guangxi, Tibet, Xinjiang, Inner Mongolia and Sichuan, and they contained 51.5% of the total irreplaceable area.

Overall, many more units in western China were assigned higher conservation values compared to eastern and southern regions, where the distributions of high-value units were severely fragmented (Figure 7). Based on the conservation value data and expert knowledge, we visually delineated the boundaries of 23 primary large-scale priority areas and excluded many small isolated areas (Figure 7). These large-scale priority areas covered ~28% of China's landmass and were mainly distributed in remote regions at high elevation and/or rugged terrain. Regions that have experienced high-intensity disturbances, e.g. Northeast China Plain, North China Plain, South Huaihe and Middle-lower Yangtze River Plain, Sichuan Basin and Pearl River Delta Area, did not contain any large-scale priority areas (Figure 7).

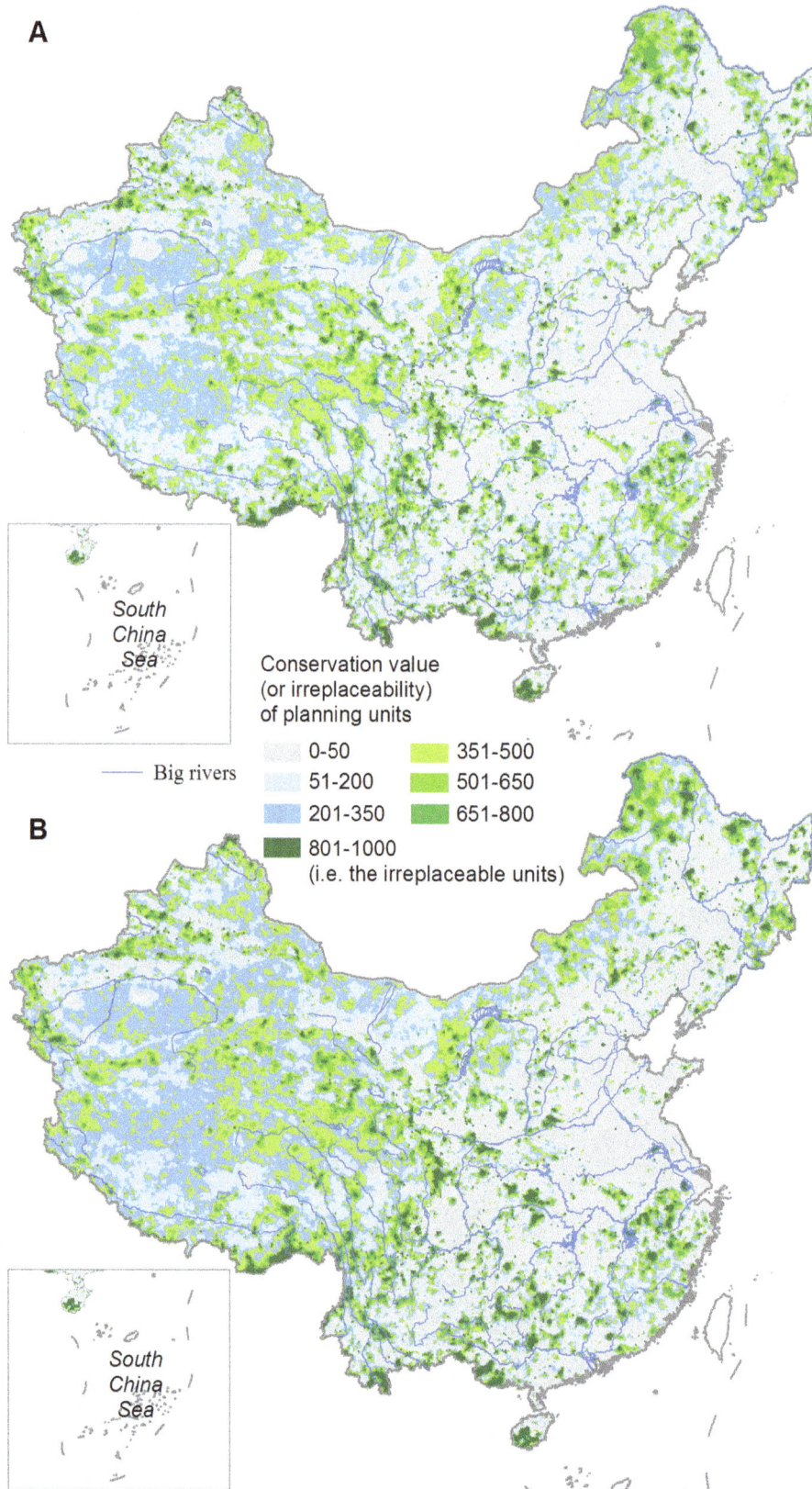

Figure 2. The conservation value of 100 km^2 hexagon units for achieving the defined conservation targets. (A) HDI-ignored scenario and (B) HDI-considered scenario.

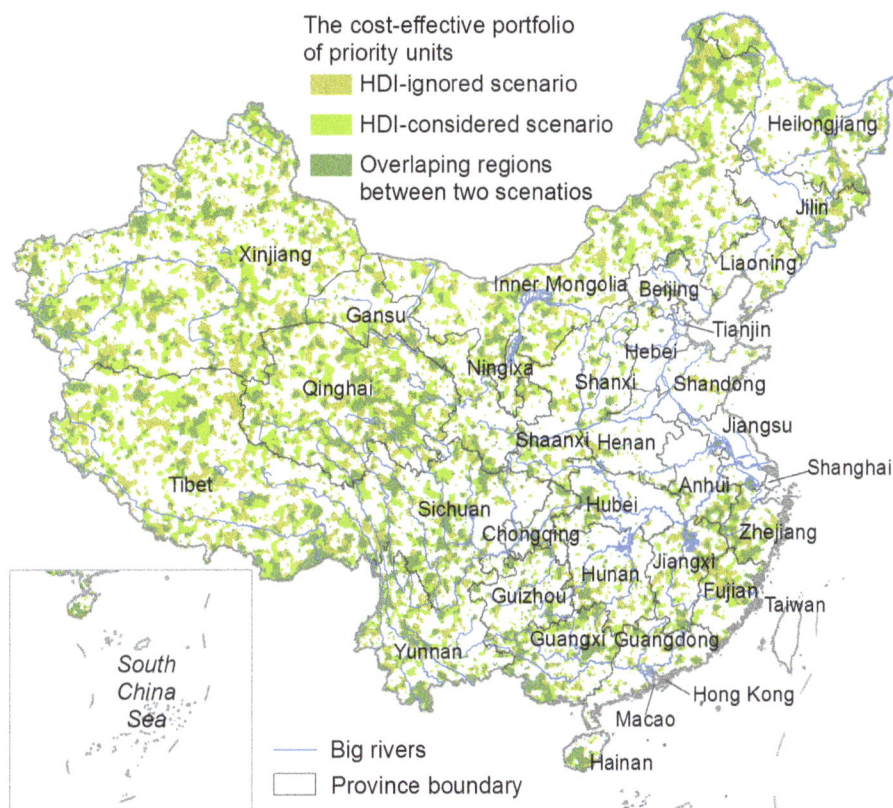

Figure 3. The cost-effective portfolios of priority units identified by the HDI-ignored and HDI-considered scenarios, respectively.

Discussion

In this study we implement a rigorous planning framework to identify the optimized national-scale conservation priorities in China. Our framework addresses several key features overlooked in previous studies, i.e. human disturbance, integration of ecosystem- and species-level conservation features, and principles of complementarity and representativeness.

Effects of Human Disturbance

Due to a lack of site-specific data on the ecological integrity of most biodiversity features [16], a HDI (or suitability index) is often modeled by aggregating human disturbance data to provide an indirect measure of ecological condition [4,30]. By explicitly considering HDI, our goal is to direct conservation towards the least-disturbed regions while still fully meeting conservation goals. We feel that this approach will promote conservation success and more efficiently achieve conservation goals [3,6]. Moreover, areas with higher disturbances offer less conservation potential as they have lower habitat suitability for sustaining conservation features [4].

Our result indicates that the portfolio of priority units in the HDI-considered scenario is characterized by a marked reduction in the total HDI score and a slight increase in the total area, and in addition, more priority units are identified at less-disturbed, higher and/or rugged regions (Figure 5). Such effects are derived from implementing Marxan's algorithm for identifying an optimized portfolio of priority areas that has the minimum total penalty score [17]. Therefore, many priority units identified in the HDI-ignored scenario, especially those distributed as fragments on highly-disturbed lands, were excluded or devalued in the HDI-considered

scenario so as to minimize the total HDI score of the portfolio. This requires the HDI-considered scenario to select a greater number of priority units with lower HDI values to achieve the same conservation goals, because each of these units contains relatively fewer conservation features and/or covers smaller areas within their distribution ranges.

The total irreplaceable area in the HDI-considered scenario increased significantly (~43%) and more irreplaceable units were selected in almost all HDI, elevation, and TRI zones except the highest HDI zone (Figure 6). We think the increase results from the fact that Marxan solutions favor those units with relatively lower penalty scores, which also was reported by Carwardine et al. [3]. This indicates that human disturbance can partly degrade the potential options available for implementing cost-effective conservation. Our result, that the most highly-developed provinces had the greatest proportional increases in irreplaceable area while western less-disturbed provinces had smaller changes (Figure 4), also supports this perspective. However, we also found that several highly-developed provinces had only small changes in irreplaceable area. We think this is because those provinces contain relatively fewer conservation features and limited overlap exists between the distributions of conservation features and areas of human disturbances.

A fundamental concern in including human disturbance is that priority areas may be biased to remote, higher and more rugged places. Such a biased distribution has been a severe problem resulting in the existing reserve networks failing to adequately represent the overall biodiversity [34,38]. Does our analysis further increase the existing biases in the location of established reserves? We feel it does not, because our framework implements 'representativeness' as a core principle in identifying priority areas

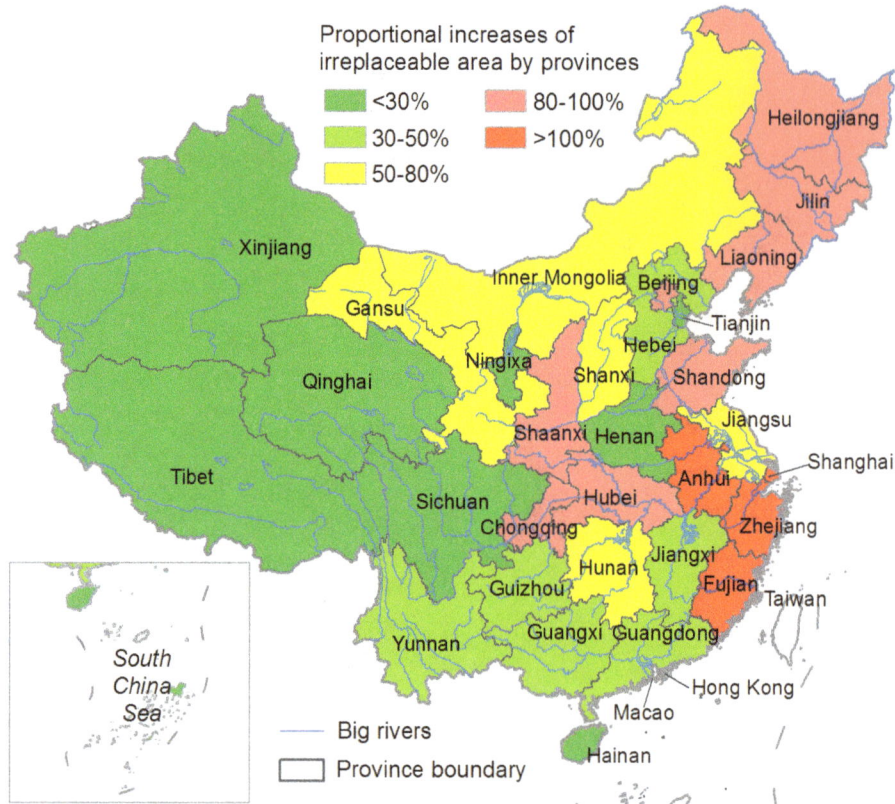

Figure 4. Proportional changes in irreplaceable area between the HDI-ignored and HDI-considered scenarios by province.

and defines explicit conservation targets for all selected conservation features. The goal for representing the full range of biodiversity requires that the priority-setting process also focuses on disturbed landscapes of high biodiversity conservation significance [5]. Similar to Linke et al. [4], we integrated human disturbance as a discounting factor for ecological condition so as to ensure that the resulting portfolio was optimized for maximizing conservation achievements.

Although apparent shifts of priority units towards less-disturbed zones were observed (Figure 5A), the HDI-considered scenario only selected fewer priority units in the <200 m elevation zone (Figure 5B) and level zone (Figure 5C), and identified more irreplaceable units in almost all HDI, elevation, and TRI zones except the highest HDI zone (Figure 6). The lowest/level zone may provide less conservation potential because of limited current biodiversity in response to long-term human disturbance [33]. We also found considerable overlap, and strong and positive pairwise associations between the portfolios of priority units and the portfolios of irreplaceable areas identified by the HDI-ignored and HDI-considered scenarios, respectively. These results demonstrate that our analysis is conservation target based, and the inclusion of human disturbance did not result in the biased distribution of conservation priorities.

Spatial Patterns of Conservation Priorities

Recognizing the advantages of including human disturbance in priority-setting, we analyzed the spatial patterns of conservation priorities using the results from the HDI-considered scenario. Human disturbance has caused severe degradation of natural ecosystems and many species extinctions, which can greatly

diminish the conservation value of a region that was historically rich in biodiversity [14,33]. Therefore, the higher the disturbance intensity, the lower the proportion of priority units was allocated in a region (Figure 5A). Rugged terrain often serves as a natural barrier for human development, and these mountainous areas have become refuges for many endangered species; These areas also are preferred as conservation priorities because they maintain more diverse habitats and higher animal and plant biodiversity [14].

We found higher percentages of irreplaceable area occurred in lower elevation zones (Figure 6B) compared to the distribution of priority units (Figure 5B). This implies that there are relatively fewer cost-effective options for fulfilling conservation targets in lowland regions, whereas the highland areas have greater flexibility in priority-setting. As moderately-disturbed and/or intermediately- to extremely-rugged zones contain the majority of irreplaceable areas (Figure 6), these habitats should be targeted to identify potentially important areas for implementing cost-effective conservation. These habitats are mainly found in western provinces, which include the vast majority of both priority units and irreplaceable areas, and therefore we consider those provinces to be of great significance in conserving China's biodiversity.

Previous researches have revealed that the remaining natural landscapes in eastern and southern China are highly fragmented, and western China supports more intact natural ecosystems and endangered species [14,33]. This study similarly found that western China contains more high-value units clustered in relatively larger patches, while the high-value units in eastern and southern regions are severely fragmented and principally located in mountainous areas (Figure 7). Our result shows that the

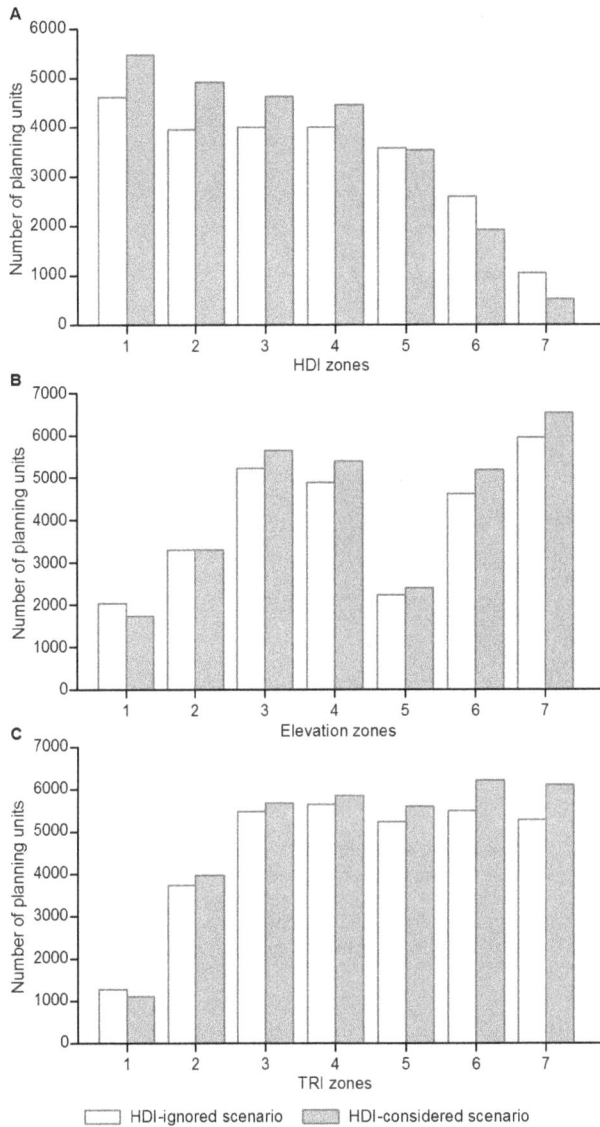

Figure 5. The distribution of priority units on (A) HDI, (B) elevation, and (C) TRI zones. The numbers 1 to 7 on the horizontal axes represent (A) low to high HDI value classifications, (B) elevation zones of <200, 200–500, 500–1,000, 1,000–1,500, 1,500–2,000, 2,000–4,000, and >4,000 m, and (C) terrain categories of level, near-level, slightly-rugged, intermediately-rugged, moderately-rugged, highly-rugged, and extremely-rugged.

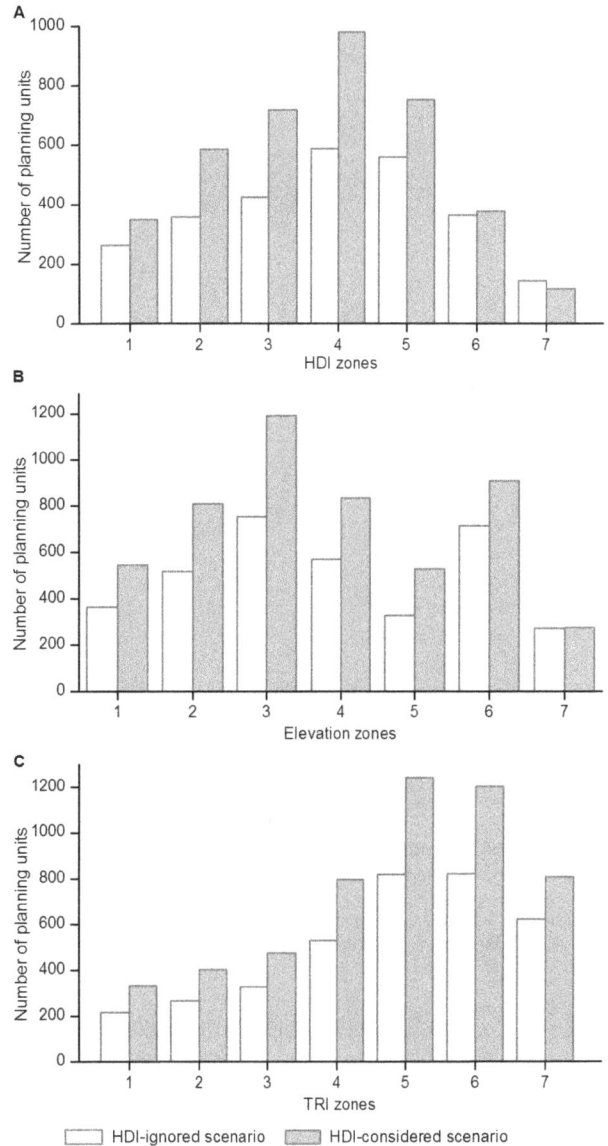

Figure 6. The distribution of irreplaceable units on (A) HDI, (B) elevation, and (C) TRI zones. See Figure 5 for the explanation of numbers 1 to 7 on the horizontal axes.

primary large-scale priority areas are mainly distributed in remote places with high elevation and rugged terrain (Figure 7). This finding is generally consistent with the results in previous studies (e.g., [10,12–15]).

However, we also identified several priority areas that were rarely considered before, including the Hulunbuir Grassland, Xilingol Grassland, Alashan-Ordos Region, Altai Mountain, and Pamirs Plateau (Figure 7). All are located in Inner Mongolia and Xinjiang, covering grassland, semi-desert, alpine and tundra biomes. These areas are not rich in species diversity, but they are valued for maintaining several important ecosystems that sustain many endemic species and critical ecosystem services [39]. Our result agrees with the limited number of studies that have considered goals for ecosystem conservation. For example, the

Alashan-Ordos Region and Altai Mountain are recognized as the key areas for protecting priority terrestrial ecosystems [39], and each of these five areas exhibits some overlap with the global 200 priority ecoregions, including Daurian/Mongolian Steppe, Altai-Sayan Montane Forests, and Middle Asian Montane Woodlands and Steppe [40].

The primary large-scale priority areas are the centers of biodiversity and evolution as they provide refuges for many species and sustain important ecosystem services [8,12], and in these areas it is usually simpler and less expensive to implement conservation actions. In addition to establishing reserves, these areas should be subject to a variety of sustainable management approaches that seek to balance extractive uses with the retention of natural resources and ecosystem functions, such as the various ecosystem service policies currently implemented in China [9]. However, these large-scale priority areas are not sufficient to fulfill China's overall conservation goals [15], because many species,

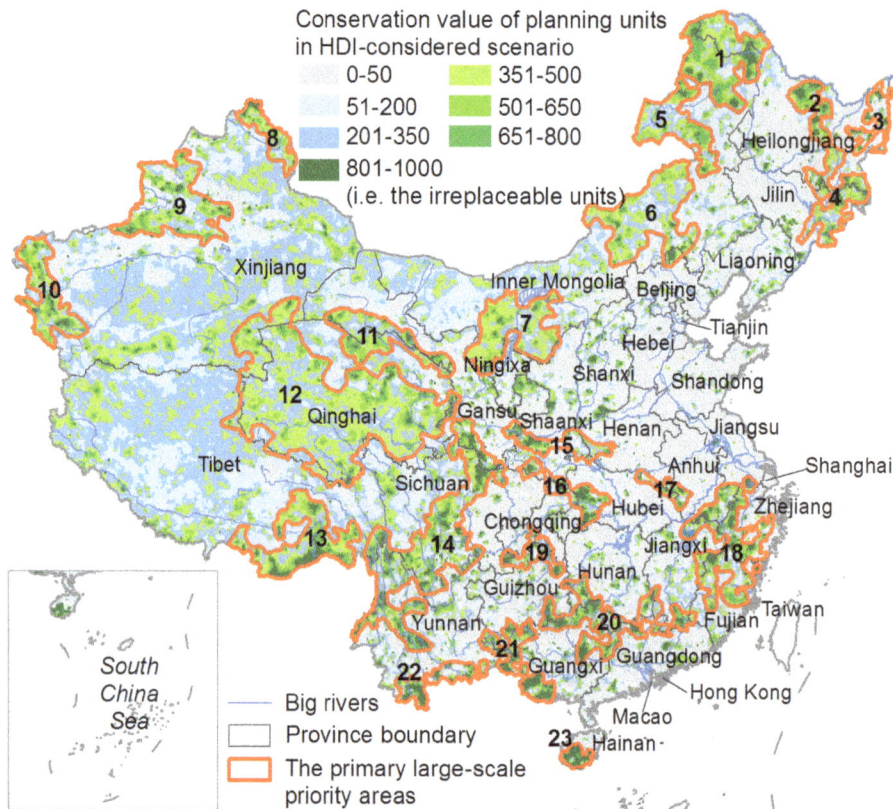

Figure 7. The distribution of the 23 primary large-scale priority areas. 1 – Daxing'anling Mountain, 2 – Xiaoxing'anling Mountain, 3 – Sanjiang Plain, 4 – Changbai Mountain, 5 – Hulunbuir Grassland, 6 – Xilingol Grassland, 7 – Alashan-Ordos Region, 8 – Altai Mountain, 9 – Tianshan Mountain, 10 – Pamirs Plateau, 11 – Qilian Mountain, 12 – Sanjiangyuan-Qiangtang Region, 13 – Southeast Himalaya Mountain, 14 – Hengduan Mountain, 15 – Qinling Mountain, 16 – Daba Mountain, 17 – Dabieshan Mountain, 18 – Mountain Region connecting Fujian-Zhejiang-Jiangxi-Anhui, 19 – Wuling Mountain, 20 – Nanling Mountain, 21 – Mountain Region in western Guangxi, 22 – Xishuangbanna, and 23 – Southern Hainan Island.

particularly those that exploit special microhabitats, may only occur in places close to developed landscapes and are already highly threatened [5]. Therefore, the priority units distributed in highly-disturbed regions that are not included in the large-scale priority areas should be included in local conservation actions. This may be even more urgent in order to prevent the immediate loss of biodiversity [41].

The Priority-setting Framework

Using this priority-setting framework, we are trying to ensure the identification of a comprehensive and cost-effective portfolio of conservation priorities for China. The process is driven by explicitly delineating spatial distributions and quantitatively defined targets for representative conservation features. We believe the analysis is rigorous, objective, transparent, and replicable.

We acknowledge that the availability and accuracy of spatial data on biodiversity and disturbances are a primary constraint for national-scale priority-setting. Therefore, our results can be further refined as more comprehensive data become available. This study has used the most up-to-date national survey data on key protected plant and animal species [22,25], as well as highly recognized information sources that have been used in previous studies (e.g., [14]). The ecosystem-level features represent meaningful biodiversity surrogates because they are the emergent entities of unique species assemblages and easily mapped; moreover, they are useful indicators of ecological processes and ecosystem services [16]. The integration of conservation features from different levels and multiple taxa can improve the effectiveness of priority areas in representing the overall biodiversity [5]. The HDI, a coarse simplification of current ecological condition, could be improved when better disturbance data and modeling methods are developed.

Hence we suggest that China increase its budget for improving the GIS-based conservation decision-making platform and enhance data sharing mechanisms. Moreover, the integration of ecological processes, ecosystem services, socioeconomic objectives and climate projections represents future research priorities in SCP [4,6,7].

Application

Systematic conservation priority-setting has significant implications in assisting China in achieving its cost-effective conservation goals as a megadiversity country. For instance, this approach has been applied in conservation priority-setting for China nationwide, and the work is a key component for developing the National Biodiversity Strategies and Action Plans (NBSAPs) [8]. China's Ministry of Environmental Protection requires all provinces, major river watersheds, and counties develop their Local Biodiversity Strategies and Action Plans (LBSAPs) [8]. Thus, we applied this priority-setting approach to come up with the first provincial LBSAPs for Sichuan [42], and now this approach is in high demand in China.

Not only is this approach useful in its direct application to conservation planning, but it also has important applicability for strategic land-use planning and sustainable development practices; e.g. the Ecological Function Regionalization and Major Function Oriented Zoning [43–44] could be further refined using our approach. Such planning seeks to optimize the spatial patterns of economic development and environment protection by investigating the synergies and trade-offs between their distributions [45]. Areas recognized as conservation priorities should be primarily preserved for sustaining biodiversity and ecosystem services. As China is now adopting a new paradigm of sustainable development by undertaking a transition from conventional industrialization to ecological civilization [46] numerous redlines on natural resources and environment management (e.g., the Key Ecological Function Regions and Development Prohibited/Restricted Zones) have been established to ensure the country's ecological security [47]. This priority-setting approach is of great significance for determining the spatially optimized conservation network or redlines for strategic land-use planning.

Conclusions

This study presents optimized national-scale spatial priorities for biodiversity conservation in China by implementing a systematic priority-setting approach with the integration of human disturbances, ecosystem- and species-level conservation features, and principles of complementarity and representativeness. Inclusion of human disturbance is essential for a cost-effective priority-setting – maximizing conservation achievement while minimizing conflicts with economic development. Such an approach will ensure the optimal spatial distribution of priority areas and reduce biases in conservation investment and/or land-use planning. The majority of priority units we identified are located in relatively remote, high and/or rugged places, however, areas that are moderately-disturbed, <2,000 m in altitude, and/or intermediately- to extremely-rugged in terrain should be targeted to identify potentially important regions for implementing cost-effective conservation. To achieve the overall biodiversity conservation goal in China, we delineate 23 primary large-scale priority areas, as well as recognize many isolated priority units in disturbed regions that need even more urgent conservation so as to prevent the immediate loss of biodiversity.

While requiring further refinement, our results provide valuable insights for current conservation and strategic land-use planning in China. This approach uses publicly available information, and is transferable and easy to implement by end-users, and applicable for national- and local-scale systematic conservation prioritization practices. Improved data, especially in the details of human disturbance and for aquatic systems at national-scale, will further enhance its applicability.

Acknowledgments

We thank Siobhan Kenney and Bastian Bertzky at UNEP-WCMC for valuable comments and English editing. We thank Bei Huang, Rui Zhang and Shan Sun, master students in Southwest Forestry University for their valuable help in building the spatial database, and the Data center for Resources and Environmental Sciences of the Chinese Academy of Sciences for providing data on land uses, human population density and GDP. PAG acknowledges the support of Chrissie, Sara, and Jenni.

Author Contributions

Conceived and designed the experiments: RDW YCL GPM PAG SZ. Performed the experiments: RDW SZ DQL PZ LZW HRD. Analyzed the data: RDW YCL LZW HRD. Contributed reagents/materials/analysis tools: RDW DQL LZW HRD. Wrote the paper: RDW YCL GPM PAG SZ.

References

1. McCarthy DP, Donald PF, Scharlemann JPW, Buchanan GM, Balmford A, et al. (2012) Financial costs of meeting global biodiversity conservation targets: current spending and unmet needs. Science 338: 946–949.
2. Brooks TM, Mittermeier RA, da Fonseca GAB, Gerlach J, Hoffmann M, et al. (2006) Global biodiversity conservation priorities. Science 313: 58–61.
3. Carwardine J, Wilson KA, Ceballos G, Ehrlich PR, Naidoo R, et al. (2008) Cost-effective priorities for global mammal conservation. Proceedings of the National Academy of Sciences of USA 105: 11446–11450.
4. Linke S, Kennard MJ, Hermoso V, Olden JD, Stein J, et al. (2012) Merging connectivity rules and large-scale condition assessment improves conservation adequacy in river systems. Journal of Applied Ecology 49: 1036–1045.
5. Margules CR, Pressey RL (2000) Systematic conservation planning. Nature 405: 243–253.
6. Venter O, Possingham HP, Hovani L, Dewi S, Griscom B, et al. (2013) Using systematic conservation planning to minimize REDD+ conflict with agriculture and logging in the tropics. Conservation Letters 6: 116–124.
7. Withey JC, Lawler JJ, Polasky S, Plantinga AJ, Nelson EJ, et al. (2012) Maximising return on conservation investment in the conterminous USA. Ecology Letters 15: 1249–1256.
8. The Ministry of Environmental Protection of China (2011) China national biodiversity conservation strategy and action plans 2011–2030. Beijing: Chinese Environmental Science Press.
9. Liu JG, Ouyang ZY, Yang W, Xu WH, Li SX (2013) Evaluation of ecosystem service policies from biophysical and social perspectives: the case of China. In: Levin SA, editor. Encyclopedia of biodiversity, second edition, volume 3. pp. 372–384. Waltham: Academic Press.
10. Chen LZ (1993) China's biodiversity and conservation strategy. Beijing: Science Press.
11. Ma KP (2001) Hotspots assessment and conservation priorities identification of biodiversity in China should be emphasized. Acta Phytoecologica Sinica 25: 125.
12. Chen CD (1998) Biodiversity of China: a country study. Beijing: Chinese Environmental Science Press.
13. Li DQ, Song YL, Ouyang ZY (2003) Research on the national forestry nature reserve system plan. Beijing: China Land Press.
14. Tang ZY, Wang ZH, Zheng CY, Fang JY (2006) Biodiversity in China's mountains. Frontiers in Ecology and the Environment 4: 347–352.
15. Zhang YB, Ma KP (2008) Geographic distribution patterns and status assessment of threatened plants in China. Biodiversity and Conservation 17: 1783–1798.
16. Groves CR (2003) Drafting a conservation blueprint: a practitioner's guide to planning for biodiversity. Washington DC: Island Press.
17. Ardon JA, Possingham HP, Klein CJ (2010) Marxan good practices handbook, Version 2. Victoria: Pacific Marine Analysis and Research Association.
18. Crous CJ, Samways MJ, Pryke JS (2013) Exploring the mesofilter as a novel operational scale in conservation planning. Journal of Applied Ecology 50: 205–214.
19. Zhang XS (2007) Vegetation map of People's Republic of China (1: 1,000,000). Beijing: Geology Press.
20. IUCN (2012) The IUCN red list of threatened species. Available: http://www.iucnredlist.org. Accessed 10 November 2012.
21. Fu LG (1992) China plant red data book: endangered and rare plants. Beijing: Science Press.
22. State Forestry Administration of China (2009) National key protected wild plant resources survey. Beijing: China Forestry Publishing House.
23. Institute of Botany (2009) Subject database of China plants. Available: http://www.plant.csdb.cn. Accessed 25 October 2012.
24. WCS (2005) China species information services. Available: http://www.baohu.org. Accessed 20 May 2012.
25. State Forestry Administration of China (2009) National key terrestrial wildlife resources survey. Beijing: China Forestry Publishing House.
26. Institute of Zoology (2009) Database of Fauna Sinica. Available: http://www.zoology.nsdc.cn. Accessed 15 September 2012.
27. Zhang RZ (1997) Distributions of China mammal species. Beijing: China Forestry Publishing House.
28. Wang S (1998) China red data book of endangered animals: mammals. Beijing: Science Press.
29. USGS (2004) Shuttle Radar Topography Mission DEM. Available: http://glcf.umiacs.umd.edu/data/srtm. Accessed 16 July 2012.
30. Sanderson EW, Jaiteh M, Levy MA, Redford KH, Wannebo AV, et al. (2002) The human footprint and the last of the wild. BioScience 52: 891–904.
31. Yang XH, Ma HQ (2009) Natural environment suitability of China and its relationship with population distributions. International Journal of Environmental Research and Public Health 6: 3025–3039.

32. Chan KMA, Shaw MR, Cameron DR, Underwood EC, Daily GC (2006) Conservation planning for ecosystem services. PLoS Biology 4: e379.

33. Wu RD, Zhang S, Yu DW, Zhao P, Li XH, et al. (2011) Effectiveness of China's nature reserves in representing ecological diversity. Frontiers in Ecology and the Environment 9: 383–389.

34. Rodrigues ASL, Andelman SJ, Bakarr MI, Boitani L, Brooks TM, et al. (2004) Effectiveness of the global protected area network in representing species diversity. Nature 428: 640–643.

35. Lea C, Curtis AC (2010) Thematic accuracy assessment procedures: National Park Service vegetation inventory, version 2.0. Natural resource report NPS/2010/NRR-2010/204. Fort Collins: National Park Service, US Department of the Interior.

36. Li BY, Pan BT, Han JF (2008) Basic terrestrial geomorphological types in China and their circumscriptions. Quaternary Sciences 28: 535–543.

37. Riley S, DeGloria SD, Elliot R (1999) A terrain ruggedness index that quantifies topographic heterogeneity. Intermountain Journal of Sciences 5: 23–27.

38. Wu RD, Ma GZ, Long YC, Yu JH, Li SN, et al. (2011) The performance of nature reserves in capturing the biological diversity on Hainan Island, China. Environmental Science and Pollution Research 18: 800–810.

39. Xu WH, Ouyang ZY, Huang H, Wang XK, Miao H, et al. (2006) Priority analysis on conserving China's terrestrial ecosystems. Acta Ecologica Sinica 26: 271–280.

40. Olson DM, Dinerstein E (2002) The global 200: priority ecoregions for global conservation. Annals of the Missouri Botanical garden 89: 199–224.

41. Joppa LN, Pfaff A (2009) High and far: biases in the location of protected areas. PLoS One 4: e8273.

42. The Ministry of Environmental Protection of China (2011) Issue of Sichuan biodiversity conservation strategy and action plans. Available: http://www.zhb.gov.cn/zhxx/gzdt/201112/t20111215_221379.htm. Accessed 15 July 2013.

43. The Ministry of Environmental Protection of China, Chinese Academy of Sciences (2008) Announcement on the issue of "National Ecological Function Regionalization". Available: http://www.mep.gov.cn/info/bgw/bgg/200808/t20080801_126867.htm. Accessed 15 May 2013.

44. The State Council of China (2011) Circular of the State Council on the issue of "Major Function Oriented Zoning". http://www.gov.cn/zwgk/2011-06/08/content_1879180.htm. Accessed 20 May 2013.

45. Fan J, Li P (2009) The scientific foundation of Major Function Oriented Zoning in China. Journal of Geographical Sciences 19: 515–531.

46. Pan J (2012) From industrial toward ecological in China. Science 336: 1397.

47. Lü Y, Ma Z, Zhang L, Fu B, Gao G (2013) Redlines for the greening of China. Environmental Science & Policy 33: 346–353.

3

The Costs of Evaluating Species Densities and Composition of Snakes to Assess Development Impacts in Amazonia

Rafael de Fraga[1]*, **Adam J. Stow**[2], **William E. Magnusson**[3], **Albertina P. Lima**[3]

1 Instituto Nacional de Pesquisas da Amazônia – Programa de Pós-graduação em Ecologia, Manaus, Amazonas, Brazil, 2 Department of Biological Sciences, Macquarie University, Sydney, New South Wales, Australia, 3 Instituto Nacional de Pesquisas da Amazônia – Coordenação de Biodiversidade, Manaus, Amazonas, Brazil

Abstract

Studies leading to decision-making for environmental licensing often fail to provide accurate estimates of diversity. Measures of snake diversity are regularly obtained to assess development impacts in the rainforests of the Amazon Basin, but this taxonomic group may be subject to poor detection probabilities. Recently, the Brazilian government tried to standardize sampling designs by the implementation of a system (RAPELD) to quantify biological diversity using spatially-standardized sampling units. Consistency in sampling design allows the detection probabilities to be compared among taxa, and sampling effort and associated cost to be evaluated. The cost effectiveness of detecting snakes has received no attention in Amazonia. Here we tested the effects of reducing sampling effort on estimates of species densities and assemblage composition. We identified snakes in seven plot systems, each standardised with 14 plots. The 250 m long centre line of each plot followed an altitudinal contour. Surveys were repeated four times in each plot and detection probabilities were estimated for the 41 species encountered. Reducing the number of observations, or the size of the sampling modules, caused significant loss of information on species densities and local patterns of variation in assemblage composition. We estimated the cost to find a snake as $ 120 U.S., but general linear models indicated the possibility of identifying differences in assemblage composition for half the overall survey costs. Decisions to reduce sampling effort depend on the importance of lost information to target-issues, and may not be the preferred option if there is the potential for identifying individual snake species requiring specific conservation actions. However, in most studies of human disturbance on species assemblages, it is likely to be more cost-effective to focus on other groups of organisms with higher detection probabilities.

Editor: Maura (Gee) Geraldine Chapman, University of Sydney, Australia

Funding: Programa de Pesquisa em Biodiversidade (PPBio), the National Institute for Science, Technology and Innovation for Amazonian Biodiversity (INCT-CENBAM) and the Instituto Nacional de Pesquisas da Amazônia (INPA) have funded field work. The Coordenação de Aperfeiçoamento de Pessoal de Nível Superior - CAPES provided a scholarship to R. de Fraga. The funders played a role in financing the construction of plot systems and data collection, but they had no direct involvement in those and other tasks (e.g. analysis, decision to publish, preparation of the manuscript).

* Email: r.defraga@gmail.com

Introduction

Obtaining environmental licenses to build and operate infrastructure or industrial facilities typically requires environmental assessment, and this process usually evaluates abiotic features, such as soil and water, and biotic features, such as fauna and flora. In Brazil, taxonomic groups to be included in environmental-impact assessments are defined by environmental agencies, and these usually include "herpetofauna", which includes snakes. Compared with other surveyed vertebrate groups, snakes are often rarely encountered, and can be more difficult to detect because they are secretive and cryptic [1], [2]. The problem of detecting snakes may be exacerbated by dense vegetation in rainforest areas, such as those in the Amazon Basin. This raises the question of whether collecting data on snakes is a cost-effective means of evaluating the impacts of environmental change in the rainforests of the Amazon Basin.

Describing human impacts on wildlife assemblages and developing conservation strategies can be based on monitoring that repeatedly measures the biotic response to disturbance [3] and provides data with direct application to setting priorities for research and conservation [4], [5]. Complementary phylogenetic and functional diversities have been shown to be more suitable meaures to assess asemblage changes, in comparison to species densities and composition [6], because disturbance is usually not acting at the level of species alone, but changing a network of abiotic and biotic factors interacting to filter species assemblies [7]. However defining functional groups depends on accurate estimates of multivariate niche overlap, which has been a challenge in tropical forests. Also, phylogenetic diversity alone does not necessarily reflect functional distance, because evolutionary traits may converge and diverge rapidly [6]. Because this study aimed to test only the effects of sampling design on ecological patterns, we represent snake assemblages using the number of species detected per plot system and species composition.

Recently, the Brazilian environmental agencies have recognized the value of standardizing sampling designs, and require or

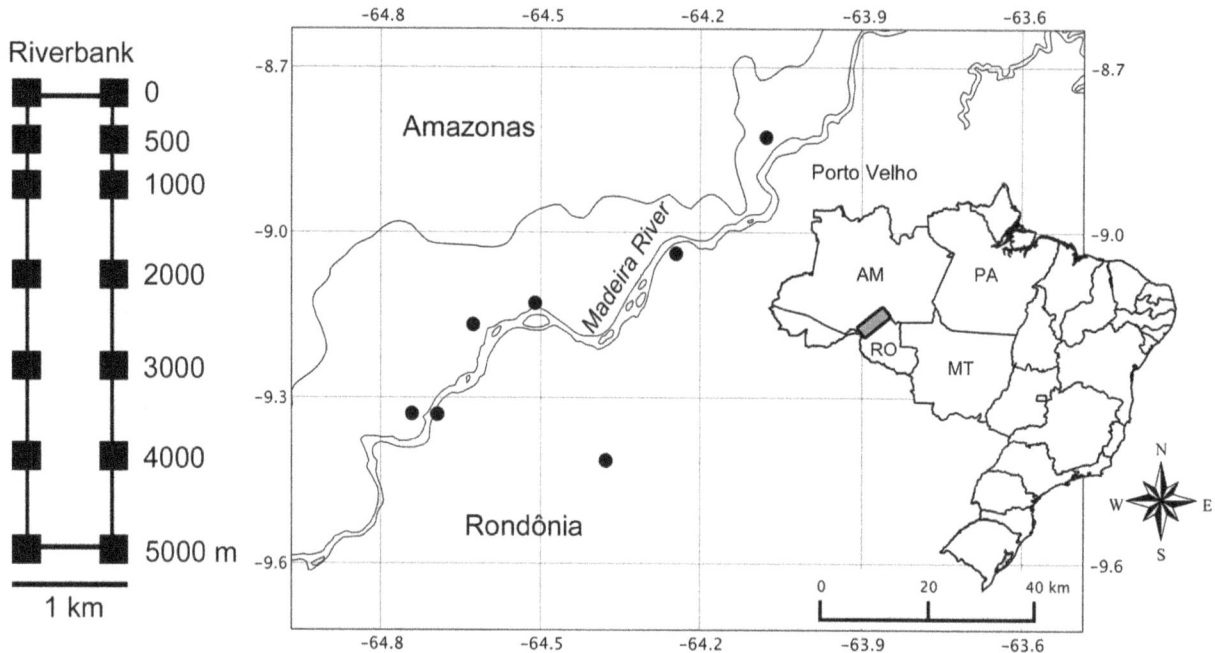

Figure 1. Plot systems in southwestern Amazonia. Sampling systems of 5 km² (black circles) located near the banks of the Madeira River in southwestern Brazilian Amazonia (Rondônia state). In detail on the left side, standard configuration of each system, with 14 plots (black squares).

recommend the use of the RAPELD system [8], [9]. RAPELD is an acronym for rapid assessments (RAP) combined with long term ecological research (PELD; in Portuguese). This method was modified from the 0.1 ha survey method developed by Gentry [10], differing primarily in that the direction of the long axis of each plot is along the altitudinal contour; use of different widths of plot for different taxa; and regular distribution of the plots across the landscape to be sampled [8]. Surveying along the contour line reduces the effects of change in altitude along the plot. Altitude probably does not directly affect organisms in lowland Amazonia, but it is related to other factors influencing plant and animal assemblages, such as edaphic characteristics [8].

The RAPELD system was designed to assess ecological parameters, such as species densities and assemblage composition, across spatially standardized sampling units [8]. Compared to individual or species-based sampling, it offers at least four main advantages: (1) the spatialization of sampling units has been useful both for rapid assessments of biological diversity and long-term monitoring; (2) due to its modular design, data from sites with different sampling intensity can be compared; (3) the sample design allows sampling for taxa of different sizes and mobility in the same sampling units; and (4) because the sampling plots follow the altitudinal contours, more precise measures of habitat factors, such as altitude, vegetation and soil characteristics, can be used as predictor variables in ecological models. The RAPELD system has been used to assess ecological and biogeographic processes that generate patterns of animal (e.g. [11]–[13]) and plant (e.g. [14]–[16]) distributions. However, being a relatively recent approach, its application has to be adjusted, especially in relation to the amount of sampling effort required to answer questions relevant to quantifying disturbance from human activities, and also to improve our knowledge about patterns of species distribution at regional scales.

Sampling effort differs among taxa as a function of detectability, and adjustments are important to generate useful results with the

least possible financial investment. In fact, high cost has been identified as a major barrier to the maintenance of biodiversity monitoring programs [17], and exceeding the limits of budgets is typical in multi-taxa studies [18], [19]. Recommendations for adjustments of standardized sampling in order to reduce costs have been provided in the Amazon for ants [20], [21] and mites [22], and some high-performance taxa for biological monitoring have been identified [23].

Although monitoring of snakes is mandatory in most impact assessments associated with major infrastructure projects in the Amazon, there has been no evaluation of the cost-effectiveness of targeting snakes. We used data from RAPELD monitoring of snakes in an area of about 1,500 km² covered by primary and secondary tropical rainforest to quantify the loss of information on differences in species densities and composition under reduced sampling effort.

Materials and Methods

Study Area

We obtained data on snake species-assemblage composition with the support of the Wildlife Conservation Program from Santo Antonio Energia, the concessionaire responsible for building and operating the Santo Antônio Hydroelectric Plant, in the Madeira River in southwestern Brazilian Amazonia (Rondônia State). This dam, which began operations in 2012, flooded about 210 km² of mainly primary rainforest, but here we use data collected prior to dam construction.

The study area is covered by "terra-firme" (not seasonally flooded) forest, seasonally flooded forests and "campinaranas" (white-sand forests). In the "terra-firme" areas, the canopy is up to 30 m in height and density of the understory varies according to altitude. In areas covered by "campinarana" the canopy is up to 20 m in height, and the understory is rich in ground bromeliads. The flooded forests are restricted to lowland areas near the banks of the Madeira River. Further details on the physiognomic classification can be found in [24].

The climate is consistently warm with average monthly temperatures between 20° and 30°C, and minima and maxima between 18° and 33°C in July. The dry season extends from June to September, with rainfall less than 30 mm per month in June and July, and a rainy season from October to May, with up to 330 mm of rain per month in December and January [25]. Tributaries of varying sizes are present in the study area, and the smallest of these often dry completely in the dry season.

Sampling design

We sampled seven plot systems (Figure 1), each consisting of two parallel 5 km trails, separated by 1 km. Seven 250 m long by 10 m wide plots with centre lines following the altitudinal contours were installed along each trail (14 per system). Plots were established at distances of 0, 500, 1000, 2000, 3000, 4000 and 5000 m from the river bank. Plot systems are called modules hereafter. All modules were installed perpendicular to the river. Four modules were on the left bank of the Madeira River (Madeira - Purus interfluve), two were on the right bank (Madeira - Tapajós interfluve), and one was on the right bank of the Jací-Paraná River, a tributary of the right bank of the Madeira River.

Snake sampling and sampling effort

We found snakes by visually searching at night, limited by space, with two observers per plot. We undertook four sampling campaigns (March-April 2010, November 2010, January 2011 and May-June 2011). Each campaign lasted about 30 days, and all plots were sampled in each campaign. We standardized the search time to one hour per plot (14 hours per module), but we had an average variation of 15 minutes in total time due to differences in the number of snakes encountered. As we did not search for other snakes while processing captured snakes, the effective search time was about one hour in each plot.

We held a maximum of six voucher specimens per species, per module. Voucher specimens were killed by overdose of a topical benzocaine-based anesthesic, which was applied in the oral mucosa. Voucher specimens were fixed by injecting a solution of 10% formalin and they were preserved in a 70% ethanol solution. All voucher specimens were deposited in the herpetology section of the zoological collections of the Instituto Nacional de Pesquisas da Amazônia, Manaus, Amazonas, Brazil. Further details on ethics will be shown below.

Data analyses

To test spatial autocorrelation among the modules, we used a Moran's correlogram of geographical distance between pairs of modules and Bray-Curtis dissimilarities in species composition between pairs of modules. We used the correlog function of the Pgirmess package [26] in R V.3.1.0.

We used the number of species detected as an index of the species density in modules. The species-accumulation curves based on rarefaction did not approach asymptotes, so we did not attempt to estimate the total number of species vulnerable to our sampling techniques that use each area. Such methods rarely produce useful information for decisions about megadiverse taxa suring short sampling periods [27]. If the methods are not useful to detect differences in observed species density, they are unlikely to be useful to compare estimates of total number of species, which have far greater standard errors. It was not necessary to use rarefaction to account for differences in search effort between modules because the same temporal and spatial efforts were expended in each module.

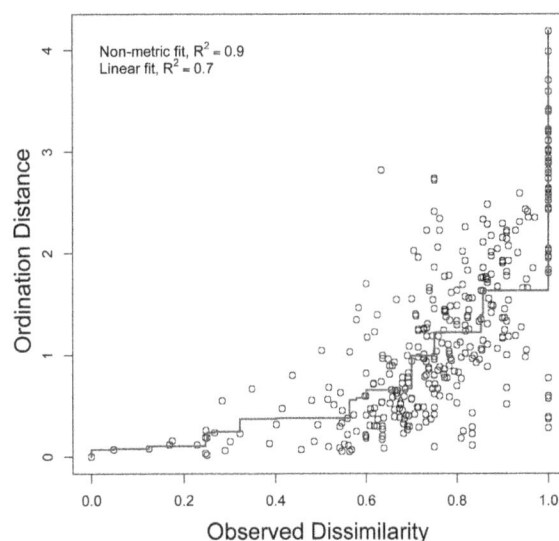

Figure 2. Shepard diagram. Relationship between NMDS ordination distance and original observed distance. NMDS ordination was undertaken on an abundance per species matrix.

We evaluated variation in assemblage composition using a dissimilarity matrix among modules. Dissimilarities were calculated by applying the Bray-Curtis index to the number of individuals per species, per module. We were interested in obtaining one-dimensional proportions of multivariate Bray-Curtis dissimilarities among modules, in order to use t-tests and general linear models. We reduced the dimensionality of the dissimilarity matrix and reordered the modules using Non-Metric Multidimensional Scaling (NMDS) in one dimension. We have chosen this method rather than alternative ordinations such as PCoA, because NMDS is apparently less sensitive to arch effects generated by heterogeneity in species distributions [28]. Moreover, NMDS has been recognized as the most efficient method to recover original multivariate ecological distances [29]. However, we also ran the analyses using PCoA ordinations and they produced qualitatively similar results not reported here. We used the metaMDS function (arguments k = 1, distance = "bray", trymax = 1000) of the Vegan package [30] in R v2.15.2. The NMDS axis captured 60% (P< 0.0001) of the variation in species composition among plots (Stress = 0.15). The reduction in dimensionality often causes distortion in relation to the observed dissimilarities [31], although distance can be reduced by rearranging the placement of points along the NMDS axis [32]. We reordered 1,000 times and we used the Shepard diagram drawn by the function stressplot in the Vegan Package in R to show that the observed dissimilarities and the ordination distances were up to 90% correlated (P<0.0001, Figure 2).

We used paired t-tests to investigate the differences among three different sampling intensities on the number of species recorded and on assemblage composition (NMDS scores). In the first test, we used complete modules (5 km^2) to show the changes in number of species and assemblage composition due to increasing the number of observations (campaigns). In the second test, we used the maximum number of observations to pair complete modules with modules constituted by only one 5 km^2 trail (1.25 km^2). In the third test, we used the maximum number of observations to pair complete modules with modules constituted by two 3 km trails (3 km^2 modules). The NMDS ordination produced some negative scores, and they directed the vectors (modules) in opposite

Table 1. Snake species found in seven 5 km^2 sampling systems in the southwestern Brazilian Amazon.

Taxon	N	O.M.	P
Boidae			
Boa constrictor Linnaeus, 1758	4	43	0.18 (0.03–0.63)
Corallus batesi (Gray, 1860)	2	28.6	0.07 (0.01–0.24)
Corallus hortulanus (Linnaeus, 1758)	10	57.1	0.3 (0.11–0.6)
Eunectes murinus (Linnaeus, 1758)	1	14.3	0.03 (0.005–0.21)
Colubridae (Colubrinae)			
Chironius fuscus (Linnaeus, 1758)	1	14.3	0.03 (0.005–0.21)
Chironius multiventris Schmidt & Walker, 1943	3	42.8	0.1 (0.03–0.28)
Dendrophidion dendrophis (Schlegel, 1837)	1	14.3	0.03 (0.005–0.21)
Drymoluber dichrous (Peters, 1863)	7	42.8	0.22 (0.0009–0.63)
Mastigodryas boddaerti (Sentzen, 1796)	2	14.3	0.07 (0.01–0.24)
Oxybelis aeneus (Wagler, 1824)	3	28.6	0.1 (0.03–0.28)
Pseustes poecilonotus (Günther, 1858)	2	28.6	0.07 (0.01–0.24)
Pseustes sulphureus (Wagler, 1824)	1	14.3	0.03 (0.005–0.21)
Rhinobothryum lentiginosum (Scopoli, 1785)	4	42.8	0.18 (0.03–0.63)
Spilotes pullatus (Linnaeus, 1758)	1	14.3	0.03 (0.005–0.21)
Colubridae (Dipsadinae)			
Apostolepis nigrolineata (Peters, 1896)	1	14.3	0.03 (0.005–0.21)
Dipsas catesbyi (Sentzen, 1796)	11	85.7	0.32 (0.17–0.51)
Dipsas indica Laurenti, 1768	3	42.8	0.1 (0.03–0.28)
Drepanoides anomalus (Jan, 1863)	4	42.8	0.18 (0.03–0.63)
Helicops angulatus (Linnaeus, 1758)	2	28.6	0.07 (0.01–0.24)
Imantodes cenchoa (Linnaeus, 1758)	9	71.4	0.3 (0.11–0.6)
Leptodeira annulata (Linnaeus, 1758)	19	100	0.42 (0.26–0.61)
Liophis reginae (Linnaeus, 1758)	2	28.6	0.07 (0.01–0.24)
Liophis typhlus (Linnaeus, 1758)	2	28.6	0.07 (0.01–0.24)
Oxyrhopus melanogenys (Tschudi, 1845)	5	71.4	0.19 (0.07–0.36)
Oxyrhopus occipitalis (Wied-Neuwied, 1824)	2	14.3	0.07 (0.01–0.24)
Oxyrhopus petolarius (Linnaeus, 1758)	1	14.3	0.03 (0.005–0.21)
Philodryas argentea (Daudin, 1803)	6	42.8	0.19 (0.07–0.36)
Philodryas georgeboulengeri (Grazziotin et al., 2012)	13	57.1	0.36 (0.14–0.66)
Pseudoboa coronata Schneider, 1801	2	28.6	0.07 (0.01–0.24)
Pseudoboa martinsi Zaher, Oliveira & Franco, 2008	1	14.3	0.03 (0.005–0.21)
Siphlophis compressus (Daudin, 1803)	10	71.4	0.3 (0.11–0.6)
Siphlophis worontzowi (Prado, 1940)	2	28.6	0.07 (0.01–0.24)
Taeniophallus sp.	5	57.1	0.19 (0.07–0.36)
Thamnodynastes pallidus (Linnaeus, 1758)	1	14.3	0.03 (0.005–0.21)
Xenopholis scalaris (Wucherer, 1861)	5	42.8	0.19 (0.07–0.36)
Elapidae			
Micrurus hemprichii (Jan, 1858)	4	42.8	0.18 (0.03–0.63)
Micrurus lemniscatus (Linnaeus, 1758)	5	42.8	0.19 (0.07–0.36)
Micrurus remotus Roze, 1987	4	28.6	0.18 (0.03–0.63)
Micrurus surinamensis (Cuvier, 1817)	1	14.3	0.03 (0.005–0.21)
Viperidae			
Bothrops atrox (Linnaeus, 1758)	18	85.7	0.39 (0.2–0.5)
Bothrops bilineatus smaragdinus Hoge, 1966	1	14.3	0.03 (0.005–0.21)

N = Number of individuals recorded in the whole study, O.M. = Proportion of modules estimated to be occupied (%), P = species detection probability and confidence intervals (95%) for a single survey of a module.

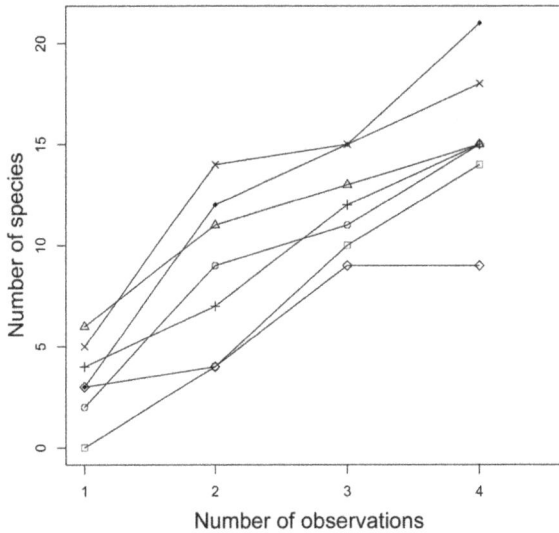

Figure 3. Cumulative number of snake species. Cumulative number of snake species in standardized modules in southwestern Brazilian Amazonia. Modules were sampled 4 times. Different symbols represent different modules.

Figure 4. Representation of snake assemblage. Multivariate representation of variation in snake assemblages among modules, based on NMDS scores. Lines connect data for the same module based on different levels of sampling. Black circles = one observation, triangles = two observations, open circles = three observations and crosses = four observations.

directions in paired t-tests, canceling each other. Therefore we added 1 to the NMDS scores to avoid negative numbers. To quantify the similarity of the representation of assemblage composition among modules with different sizes and sampling efforts, we tested for associations among NMDS scores using general linear models.

The major rivers in the Amazon basin are associated with the limits of species distributions of several taxonomic groups (e.g. [33]). Although rivers have not been identified as resulting in vicariance for snakes, we compared number of species per module and assemblage composition on the opposing river banks of the Madeira River using an analysis of similiarities (ANOSIM) with assemblage composition represented by NMDS scores. We used the function anosim of the Vegan package in R.

Detection probabilities

We quantified detection probabilities for each species using single-season models based on presence-absence data in the Program Presence v.5.2, with 100 bootstrap randomizations [34]. A single-season model provides probabilities of occupancy when detection of the target species is not guaranteed, even in places where they are present. The estimated occupancy and detection probabilities describe a history of detecting species over a series of surveys in the same locations [34]. Although the Amazon rainforest is apparently homogeneous on satellite images, subtle changes in habitat features across the landscape at a scale of a few kilometers can influence co-occurrence of species in some taxonomic groups, such as frogs [13], understory birds [12] and snakes [11]. Detection probabilities possibly vary slightly among areas within each module as a function of change in habitat features, such as vegetation density along the trails. We expected higher detectability in more open plots and the number of trees was quantified for all plots during the impact assessment for the hydroelectric dam, but differences in the number of trees among modules were negligible (ANOVA $F_{6-82} = 1.836$, $P = 0.1$). Other environmental factors, such as distance from the streams, can directly affect the composition of snake species in Amazonia [11].

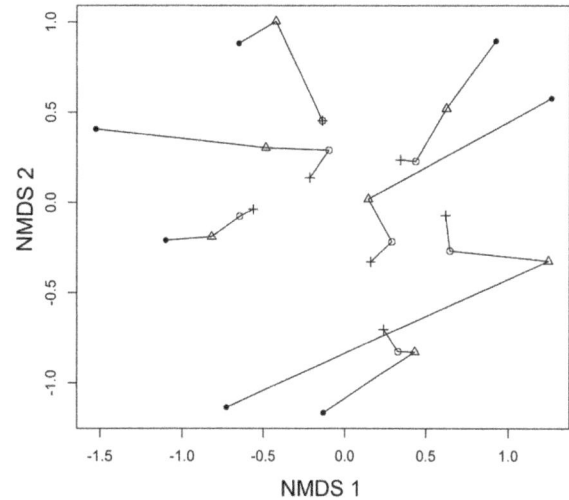

However most of the species recorded here are widely distributed in the study area, and similarity in assemblage composition of up to 60% among modules is expected at scales of tens of kilometers (see [35]). Therefore, we expected the occurrence of all species in all modules, with differences in co-occurrence resulting from the variation in habitat use over a few kilometers. As we were interested in estimating average detectability per species scaled to tens of kilometers, we assumed the same probabilities of occupancy and detection across all modules, and across all campaigns. The average detectability, rather than detectability on any given occasion is what is needed to compare costs of detection of different species in general surveys.

Cost estimates

To estimate the cost of sampling snakes, we considered fuel for transport among modules, batteries for headlamps, food and field-assistant salaries. We did not include the costs of construction and maintenance of the modules because the same field infrastructure was used for sampling many other taxa. For fuel, we calculated the cost considering the average consumption per km for a diesel-powered pickup truck for modules accessible by road, and the average consumption per kilometer of a boat powered by a 60 hp gasoline outboard motor for modules accessible by river. To estimate the number of headlight batteries, we considered eight people searching for snakes simultaneously (two per plot), each carrying a headlamp powered by three AA batteries. The batteries were changed every second night. For food, we used $ 8.68 United States Dollars (USD) per day, per person. This is an average value on the local market. For payment of field assistants, we used a daily value of $ 21.71 USD, a work-contract stipulated value.

Ethics and data availability

Snakes were collected under IBAMA/SISBIO (Ministry of Environment, Government of Brazil) permit n° 02001.000508/2008-99. This permit was subject to approval of all procedures for catching and collecting snakes, and it was allowing us to collect

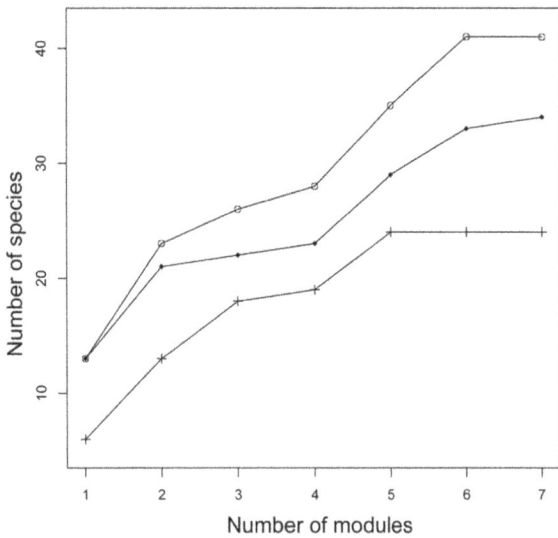

Figure 5. Number of snake species per sample design. Cumulative number of snake species with increasing number of standardized sample modules surveyed in southwestern Brazilian Amazon. Open circles = modules with two 5 km trails (5 km^2), diamonds = modules with one 5 km trail (1.25 km^2) and crosses = modules with two 3 km trails (3 km^2).

eight specimens per species, per module. However the limit has not been reached for any of the species found.

All data are available for free download on the website of the Programa de Pesquisa em Biodiversidade (PPBio) http://ppbio.inpa.gov.br/knb/metacat?action = read&qformat = ppbio&sessionid = 0&docid = naman.594.1

Results

Species densities and assemblage composition

We found 41 species of snakes (Table 1), but the number of species included in each test depended on the module size and number of observations. Neither the variation in number of species detected nor the assemblage composition among modules was spatially autocorrelated (P>0.25 in all cases). The number of species detected per module varied from nine to 21, and did not differ when the opposite banks of the Madeira River were compared (ANOVA $F_{1-6} = 0.762$, P=0.83) and assemblage composition based on NMDS scores was only 4% (P=0.01) different when comparing either side of the river.

Using complete modules, species density increased with the number of observations (Figure 3), even increasing between the third and fourth sampling occasions ($t_{1-6} = 4.459$, P=0.004). The assemblage composition (Figure 4) differed depending on the number of sampling occasions per module ($t_{1-6} = 2.497$, P=0.04). However, a general linear model indicated that NMDS scores were correlated between two and three observations ($r^2 = 0.85$, P=0.001), and highly correlated between three and four observations ($r^2 = 0.97$, P=0.00001). The assemblage composition resulting from four surveys per module was based on more species but provided a representation similar to that with only two surveys per module.

Removal of one trail per module (1.25 km^2 modules) significantly reduced ($t_{1-6} = 7.262$, P=0.0003) the number of species encountered (Figure 5). Assemblage composition (Figure 6) also differed between complete and partial sampling of modules

($t_{1-6} = 2.404$, P=0.05), but NMDS scores for complete and partial sampling were correlated in a general linear model ($r^2 = 0.62$, P=0.02). Reduction of 2 km in each module (3 km^2) significantly reduced the number of species detected ($t_{1-6} = 4.289$, P=0.005). However, assemblage composition did not differ significantly ($t_{1-6} = 1.347$, P=0.21), and NMDS scores were correlated ($r^2 = 0.48$, P=0.04). Although the size of the modules influenced the number of species and assemblage composition, conclusions based on similarity in assemblage composition among modules were similar with up to a 50% reduction in the size of the modules.

Detection probabilities

Detection probabilities of species per expedition per module ranged between 0.03 (SE = 0.03) and 0.42 (SE = 0.09), and were below 10% for almost half of the species detected (Figure 7). Confidence intervals for detection probabilities were very wide for most species (Table 1).

Costs

Each full survey of all modules cost \$ 5,450 USD, and the full study cost \$ 21,799 USD (Table 2). The highest costs were for field assistants, followed by food, headlamp batteries and fuel. We had an encounter rate of 0.89 snakes per hour (total number of individuals/total search time). Considering only the costs of food, field assistants, headlamp batteries and fuel, we calculated the cost to find a snake as \$ 120 USD (total cost/total number of snakes found).

Discussion

Sample reduction and decision-making

The number of species per module increased with each additional survey, and there was no tendency for the rate of species accumulation to lessen with the maximum sampling effort. About 95 snake species occur in the region of Porto Velho, state of Rondônia (literature compilation in [36]), more than twice the number of species found in this study. However, this is an estimate based on decades of herpetological collection, an effort generally not viable for biological monitoring applied to assess the impact of human disturbance. Impacts of human activities have been assessed using secondary data, but this method is not appropriate for detecting the influence of habitat factors on the regional distribution of species, because these data usually are based on specimens, and not sites, as sampling units. In addition, secondary data on snakes are generally not comparable due to the lack of sampling standardization and the sampling of snakes in this manner is rare in Brazil (for exceptions, see [11], [37], [38]).

Although there was variation in assemblage composition with increasing number of observations, the general linear models indicated that two surveys were sufficient for multivariate techniques to capture patterns of species dissimilarities between 5 km^2 modules. Removal of an entire trail from each module (1.25 km^2 modules) caused differences in the assemblage composition between pairs of modules, but not in a general linear model. Similar multivariate patterns were detected even with reduced sampling effort in time and space. However, paired tests are strongly influenced by the spatial configuration of the sampling units and the number of observations and this has traditionally varied among studies. The sampling configuration has often varied in relation to the answers required for the land-use management or improving knowledge about regional patterns of species distribution. A standardized sampling system overcomes some of these issues and allows data to be compared among studies.

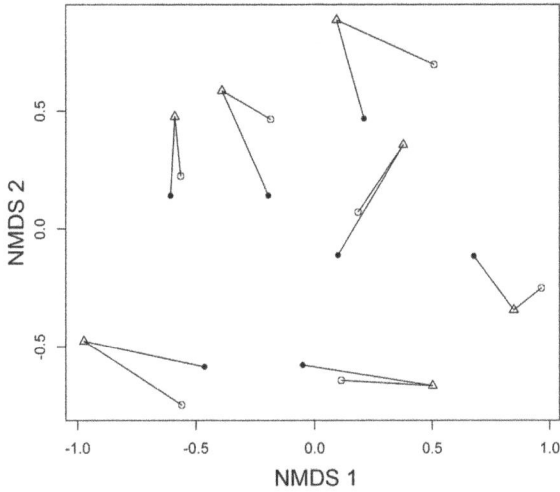

Figure 6. Representation of snake assemblage per sample design. Multivariate representation of variation in snake-assemblage composition based on NMDS scores from data obtained in standardized sampling modules in southwestern Brazilian Amazonia. Lines connect the same module sampled at different intensities. Black circles = modules with two 5 km trails (5 km²), triangles = modules with one 5 km trail (1.25 km²) and open circles = modules with two 3 km trails (3 km²).

We found that assemblage composition varied by up to 40% among modules. Differences in assemblage composition should be expected as a response to environmental gradients that subtly change on a scale of a few kilometers (e.g. [11]). We could have found the same pattern of differences in assemblage composition with a 50% reduction in the costs of food, fuel for transport, batteries for headlamps and field assistants, or a 40% reduction in construction costs of the modules (by reducing of 2 km on each trail). However, these decisions would imply the loss of about 24%

of the detected species, and make it difficult to detect local variations in assemblage composition. Deciding on sampling effort should be guided by the target-issues, but such decisions are usually made subjectively by the researchers, rather than in consultation with the regulatory agencies that will make management decisions [39].

Species detectability

All of the tests were limited by the low detectability of most species and associated wide confidence intervals, despite the relatively high sampling effort in time (980 observers*hours of searching) and space (3,500 ha of plots). We can not discount an effect of variation in the probability of detection of species among the modules because different types of habitat occur in patches throughout each module. However, we do not expect much more than the maximum detectability found in this study (42%), because detection probability for snakes have generally been found to be below 40%, even in areas known to be occupied (e.g. [37], [40]), and species that are not considered rare can be virtually undetectable, showing detection probabilities below 1% [2]. Failure to detect species in occupied habitats can generate erroneous predictions of species responses to natural variation in habitat factors [41], and generating reliable models of habitat use depends on very high sampling effort, and consequently very high financial costs. Thus, conservation programs based on species that are difficult to detect usually prioritize areas where habitat factors favor the detection, and not necessarily the responses of organisms to habitat change [42]. Overcoming these biases for snakes would come at a high monetary cost. In assessments of disturbance, it is generally advantageous to focus on sampling a limited set of high-detectability taxa that reflect the broader patterns of diversity [43], [44].

Cost-benefit

We calculated the cost to find a snake as $ 120 USD. Mesquita et al. [45] spent $ 0.49 USD per snake found by visual search in a semiarid region of Brazil, more than 230 times cheaper than this

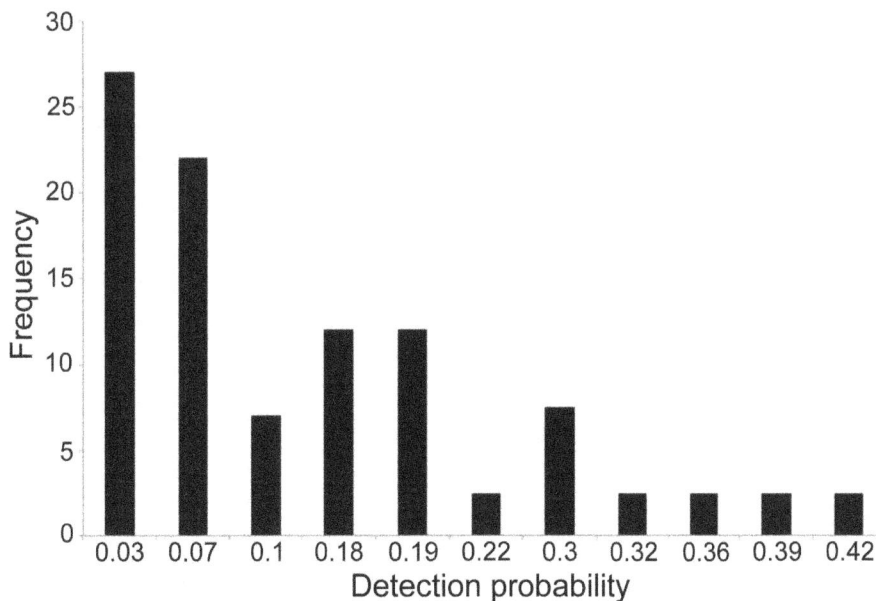

Figure 7. Detection probabilities of Amazonian snakes. Frequency of species of snakes with different probabilities of detection in visual surveys of RAPELD modules.

Table 2. Costs for sampling snakes in standardized modules in southwestern Brazilian Amazon.

	Each survey	Full study
Field assistances	2,605.00	10,420.00
Food	2,084.00	8,336.00
Batteries	625.00	2,500.00
Fuel	136.00	544.00
Total per observation	**5,450.00**	
Total for the full study		**21,800.00**

The values are in United States Dollars.

study. Differences in costs should be more evident among regions with different climate, terrain, vegetation and logistics, but those authors did not present a detailed description of the spatial distribution of the sites observed, and therefore the independence of sampling units was not clear. Furthermore, they did not estimate costs for food, fuel and field assistants, without which data collection would be impossible in the wide-scale sampling used in this study.

Although snakes are highly diverse in species and lifestyles in the Amazon, ecological models based on snake data are frequently influenced by false absences, resulting from low detectability and lack of specific methods for efficiently catching specimens (see [46]). Complementary methods for detecting snakes, such as passive traps, can increase species lists, especially because they optimize the catching of small litter and fossorial snakes (e.g. [47], [48], but the results usually do not justify the high cost and physical effort (e.g. [45]).

In some cases, a snake species with a limited distribution may be the primary target for impact assessment (e.g. *Bothrops alcatraz* and *Bothrops insularis*, species endemic to islands in southeastern Brazil), and general snake sampling during environmental-impact studies may be useful to generate landscape-scale models of species distribution and to obtain natural-history data, such as those on diet and reproduction. Futhermore, snakes can be found during visual search for other taxa, such as lizards and frogs. However, our simulations show that snakes are generally not good models for impact assessments, because detection of strong ecological patterns requires high financial costs, and cost reduction depends on subjective decisions about the importance of the lost information for management decisions. Also, if visual identification is not sufficient, collecting snakes may involve risks to personnel, and such risks might not be covered under general work insurance.

Limiting studies on human disturbance to a few taxa may be a necessary pragmatic decision in biodiverse regions, because weak cross-taxon congruence in assemblage composition is expected among higher-taxa [49]. Different groups of organisms may respond to habitat changes in different ways, and thus represent only small fractions of the total ecological funcionality of an area (see [50], [23]). Therefore we do not expect that there are suitable ecological surrogates for snakes, which makes it a difficult decision to not include them in multi-taxa studies aiming to understand human disturbance on the overall functioning of biodiversity networks. Combining snake sampling with surveys on other taxa, such as frogs and lizards, could potentially add value to measurements of biological diversity. However, due to the very low probability of detection, the data may not be useful for impact assessments. In any case, no snake species or assemblages are considered endangered in the Brazilian Amazon [51]. Therefore, due to the practical restrictions imposed by limitations of time and money, we recommend that reports on environmental impacts in the Amazon should focus primarily on high-detectability taxa, for which these limitations usually are less of a barrier for decision-making. Strategically appropriate taxa are those that simultaneously reflect useful measures of ecological patterns and are feasibly sampled [23], such as birds [52] and dung beetles [53]. We do not have detailed detectability probabilities for these species. However, standard methods obviously collect more than the mean of 6.5 individuals and 5.75 species of snakes per module per sampling occasion that we found, so it can be expected that they will cost much less than the U.S. $120 per individual that we found for snakes. If the number of taxa is reduced, the resources saved can be used to increase the spatial scale of sampling, which is usually one of the most important restrictions on decision making in the context of environmental impacts [9].

Acknowledgments

We thank E. Farias, M.C. Araújo and P.I. Simões for assistance in fieldwork. D.B. Provete and an anonymous reviewer provided us comments that greatly improved the manuscript. Overall coordination and management of data benefited from the support of the Programa de Pesquisa em Biodiversidade (PPBio), the National Institute for Science, Technology and Innovation for Amazonian Biodiversity (INCT-CEN-BAM) and the Instituto Nacional de Pesquisas da Amazônia (INPA). H. Espírito-Santo helped with R-program scripts.

Author Contributions

Conceived and designed the experiments: RF WEM APL. Performed the experiments: RF AJS. Analyzed the data: RF AJS WEM. Contributed reagents/materials/analysis tools: RF APL. Wrote the paper: RF AJS WEM APL.

References

1. Kéry M (2002) Inferring the absence of a species - a case study of snakes. J Wildl Manage 66: 330–338.
2. Steen D (2010) Snakes in the grass: secretive natural histories defy both conventional and progressive statistics. Herpetol Conserv and Biol 5(2): 183–188.
3. Hinds WT (1984) Towards monitoring of long term trends in terrestrial ecosystems. Environ Conserv 11: 11–17.
4. Burbidge AA (1991) Cost constraints on surveys for nature conservation. In: Margules CR, Austin MP, editors. Nature conservation: cost effective biological surveys and data analysis. Melbourne: CSIRO. pp. 3–6.
5. Stork NE, Samways MJ (1995) Inventorying and monitoring: executive summary. In Heywood VH, Watson RT, editors. Global biodiversity assessment. Cambridge: Cambridge University Press. pp. 457–458.

6. Bässler C, Ernst R, Cadotte M, Heibl C, Müller J (2014) Near-to-nature logging influences fungal community assembly processes in a temperate forest. J App Ecol 51: 1–9.

7. Ernst R, Linsenmairs KE, Thomas R, Rödol MO (2007). Amphibian communities in disturbed forests: lessons from the Neo- and Afrotropics. In Tscharntke T, Leuschner C, Zeller M, Guhardja E, Bidin A, editors. The stability of tropical rainforest margins, linking ecological, economic and social constraints of land use and conservation. Berlin: Springer Verlag. pp. 61–87.

8. Magnusson WE, Lima AP, Luizão R, Luizão F, Costa FRC, et al. (2005) RAPELD: a modification of the Gentry method of floristic survey for biodiversity surveys in long-term ecological research sites. Biota Neotropica 2: 1–6.

9. Magnusson WE, Braga-Neto R, Pezzini F, Baccaro F, Bergallo H, et al. (2013) Biodiversity and Integrated Environmental Monitoring. Manaus: Áttema. 356 p.

10. Gentry AH (1982) Patterns of Neotropical plant species. Evol Biol 15: 1–84.

11. Fraga R de, Lima AP, Magnusson WE (2011) Mesoscale spatial ecology in a tropical snake assemblage: the width of riparian corridors in central Amazon. Herpetol J 21: 51–57.

12. Bueno AS, Bruno RS, Pimentel TP, Sanaiotti TM, Magnusson WE (2012) The width of riparian habitats for understory birds in an Amazonian forest. Ecol Appl 22: 722–734.

13. Rojas-Ahumada DP, Landeiro VM, Menin M (2012) Role of environmental and spatial processes in structuring anuran communities across a tropical rainforest. Austral Ecol 37: 865–873.

14. Freitas CG, Costa FRC, Svenning JC, Balslev H (2012) Topographic separation of two sympatric palms in the central Amazon – does dispersal play a role? Acta Oecol 39: 128–135.

15. Pansonato MP, Costa FRC, Castilho CV, Carvalho FA, Zuquim G (2013) Spatial scale or amplitude of predictors as determinants of the relative importance of environmental factors to plant community structure. Biotropica 45(3): 299–307.

16. Schietti J, Emilio T, Rennó CD, Drucker DP, Costa FRC, et al. (2013) Vertical distance from drainage drives floristic composition changes in an Amazonian rainforest. Plant Ecol Divers 6: 1–13.

17. Danielsen F, Mendonza MM, Alviola P, Balete DS, Enghoff M, et al. (2003) Biodiversity monitoring in developing countries: what are we trying to archieve? Oryx 37: 407–409.

18. Margules CR, Austin MP (1991) Nature conservation: cost effective biological surveys and data analysis. Melbourne: CSIRO. 207 p.

19. Lawton JH, Bignell DE, Bolton B, Bloemers GF, Eggleton P (1998) Biodiversity inventories, indicator taxa and effects of habitat modification in tropical forest. Nature 391: 72–76.

20. Souza JLP de, Moura CAR de, Harada AY, Franklin E (2007) Diversidade de espécies dos gêneros Crematogaster, Gnamptogenys e Pachycondyla (Hymenoptera: Formicidae) e complementaridade dos métodos de coleta durante a estação seca numa estação ecológica do estado do Pará, Brasil. Acta Amazonica 37(4): 649–656.

21. Souza JLP de, Moura CAR de, Franklin E (2009) Efficiency in inventories of ants in a forest reserve in Central Amazonia. Pesqu Agropec Bras 44(8): 940–948.

22. Santos EMR, Franklin E, Magnusson WE (2008) Cost-efficiency of subsampling protocols to evaluate Oribatid-Mite communities in an Amazonian Savanna. Biotropica 40(6): 728–735.

23. Gardner TA, Barlow J, Araujo IS, Ávila-Pires TC, Bonaldo AB, et al. (2008) The cost-effectiveness of biodiversity surveys in tropical forests. Ecol Lett 11: 139–150.

24. Ribeiro JELS, Hopkins MJG, Vincenti A, Sothers CA, Costa MAS, et al. (1999) Flora da Reserva Ducke: Guia de identificação das plantas vasculares de uma Floresta de Terra Firme na Amazônia Central. Manaus: INPA. 800pp.

25. Horbe AMC, Queiroz MMA, Moura CAV, TORO MAG (2013). Geoquímica das águas do médio e baixo Rio Madeira e seus principais tributários – Amazonas – Brasil. Acta Amazonica 43(4) 489–504.

26. Giradoux P (2013). Miscellaneous functions for analusis and display of ecological and spatial data. Available: http://cran.r-project.org. Accessed June 11 2014.

27. Gotelli NJ, Colwell RK (2010). Estimating species richness. In Magurran AE, McGill, BJ, editors. Biological Diversity: Frontiers In Measurement and Assessment. Oxford: Oxford University Press. pp 39–54.

28. Ruokolainen L, Salo K (2006). Differences in performance of four ordination methods on a complex vegetation dataset. Ann Bot Fennici 43: 269–275.

29. Austin MP (2013). Inconsistences between theory and methodology: a recurrent problem in ordination studies. J Veg Sci 24: 251–268.

30. Oksanen J, Blanchet FG, Kindt R, Legendre P, Minchin PR, et al. (2010) Vegan: Community Ecology Package. Available: http://cran.r-project.org. Accessed October 25 2013.

31. James FC, McCulloch CE (1990) Multivariate analysis in ecology and systematics Panacea or Pandora's box? Ann Rev Ecol Syst 21: 129–166.

32. Magnusson WE, Mourão G (2005) Estatística sem matemática – A ligação entre as questões e a análise. Londrina: Planta. pp. 103–122.

33. Colwell RK (2000) A barrier runs through it...or maybe just a river. Proc Nat Acad of Sci 97(25): 13470–13472.

34. MacKenzie DI, Nichols JD, Lachman GB, Droege S, Royle JA (2002) Estimating site occupancy rates when detection probabilities are less than one. Ecology 83(8): 2248–2255.

35. da Silva NJ, Sites-Jr JW (1995) Patterns od diversity of Neotropical squamate reptile species with emphasis on the Brazilian Amazon and the conservation potential of indigenous reserves. Cons Biol 9: 873–901.

36. Bernarde PS, Albuquerque S de, Barros TO, Turci LCBT (2012) Serpentes do estado de Rondônia, Brasil. Biota Neotropica 12(3): 1–29.

37. Abrahão C (2007) Influência do tamanho do riacho, chuva, e disponibilidade de alimento na distribuição de Bothrops atrox na Reserva Florestal Adolpho Ducke. Masters dissertation. INPA/UFAM, Manaus, Amazonas. Available: http://ppbio.inpa.gov.br/sites/default/files/Dissertacao_Abraao_Carlos.pdf. Accessed 25 October 2013.

38. Fraga R de, Magnusson WE, Abrahão CR, Sanaiotti T, Lima AP (2013) Habitat selection by Bothrops atrox (Serpentes: Viperidae) in Central Amazonia, Brazil. Copeia 2013(4): 684–690.

39. Stow AJ, Magnusson WE (2012) Genetically defining populations is of limited use for evaluating and managing human impacts of gene flow. Wildlife Research 39: 290–294.

40. Tibor H, Öllerer K, Farcźády L, Moga CI, Băancilă R (2009) Using species detectability to infer distribution, habitat use and absence of a cryptic species: the smooth snake (Coronella austriaca) in Saxon Transylvania. Acta Sci Transylv 17(1): 61–76.

41. Gu W, Swihart RK (2003) Absent or undetected? Effects of non-detection of species occurance on wildlife-habitats models. Biol Conserv 116: 195–203.

42. Pulliam HR (1988) Sources, sinks, and population regulation. Am Nat 132: 652–661.

43. Angermeier PL, Karr JR (1994) Biological integrity versus biological diversity as policy directives – protecting biotic resources. BioSci 44: 690–697.

44. Caro TM, O'Doherty G (1999) On the use of surrogate species in conservation biology. Conserv Biol 13: 805–814.

45. Mesquita PCMD, Passos DC, Cechin SZ (2013) Efficiency of snake sampling methods in the Brazilian semiarid region. An Acad Bras Ciênc 2013: 1–13.

46. Pawar S (2003) Taxonomic chauvinism and the methodologically challenged. BioSci 53: 861–864.

47. Todd BD, Andrews KM (2008) Response of a reptile guild to forest harvesting. Conserv Biol 22: 753–761.

48. Patrick DA, Gibbs JP (2009) Snake occurences in grassland associated with roads versus forest edges. J Herpetol 43: 716–720.

49. Heino J. (2010). Are indicator groups and cross-taxon congruence useful for predicting biodiversity in aquatic ecosystems? Ecol Indicators 10(2010): 112–117.

50. Barlow J, Gardner TA, Araujo IS, Ávila-Pires TCS (2007) Quantifying the biodiversity value of tropical primary, secondary and plantation forests. Proc Natl Acad Sci USA 104(47): 18555–18560.

51. IUCN (2013) The IUCN Red List of Threatened Species. Version 2013.2. Available: http://www.iucnredlist.org. Accessed June 11 2014.

52. Bibby CJ (1999) Making the most of birds as environmental indicators. Ostrich 70: 81–88.

53. Spector S (2006) Scarabaeinae dung beetles (Coleoptera: Scarabaeidae: Scarabaeinae): an invertebrate focal taxon for biodiversity research and conservation. Coleoptera Bull 60: 71–83.

4

Temperate Pine Barrens and Tropical Rain Forests Are Both Rich in Undescribed Fungi

Jing Luo[1], Emily Walsh[1], Abhishek Naik[1], Wenying Zhuang[2], Keqin Zhang[3], Lei Cai[2], Ning Zhang[1,4]*

1 Department of Plant Biology & Pathology, Rutgers University, New Brunswick, New Jersey, United States of America, **2** State Key Laboratory of Mycology, Institute of Microbiology, Chinese Academy of Sciences, Beijing, China, **3** Laboratory for Conservation and Utilization of Bio-Resources and Key Laboratory for Microbial Resources of the Ministry of Education, Yunnan University, Kunming, China, **4** Department of Biochemistry & Microbiology, Rutgers University, New Brunswick, New Jersey, United States of America

Abstract

Most of fungal biodiversity on Earth remains unknown especially in the unexplored habitats. In this study, we compared fungi associated with grass (Poaceae) roots from two ecosystems: the temperate pine barrens in New Jersey, USA and tropical rain forests in Yunnan, China, using the same sampling, isolation and species identification methods. A total of 426 fungal isolates were obtained from 1600 root segments from 80 grass samples. Based on the internal transcribed spacer (ITS) sequences and morphological characteristics, a total of 85 fungal species (OTUs) belonging in 45 genera, 23 families, 16 orders, and 6 classes were identified, among which the pine barrens had 38 and Yunnan had 56 species, with only 9 species in common. The finding that grass roots in the tropical forests harbor higher fungal species diversity supports that tropical forests are fungal biodiversity hotspots. Sordariomycetes was dominant in both places but more Leotiomycetes were found in the pine barrens than Yunnan, which may play a role in the acidic and oligotrophic pine barrens ecosystem. Equal number of undescribed fungal species were discovered from the two sampled ecosystems, although the tropical Yunnan had more known fungal species. Pine barrens is a unique, unexplored ecosystem. Our finding suggests that sampling plants in such unexplored habitats will uncover novel fungi and that grass roots in pine barrens are one of the major reservoirs of novel fungi with about 47% being undescribed species.

Editor: Gabriele Berg, Graz University of Technology (TU Graz), Austria

Funding: The research was supported by the National Science Foundation (DEB 1145174, http://www.nsf.gov/) to NZ. The funders had no role in study design, data collection and analysis, decision to publish, or preparation of the manuscript.

Competing Interests: The authors have declared that no competing interests exist.

* Email: zhang@aesop.rutgers.edu

Introduction

Fungi comprise the second largest kingdom of eukaryotic life and include a diverse group of organisms that have vital functions as decomposers, pathogens, and as components of other symbioses in biomes such as endophytes and mycorrhizae [1,2]. It is hypothesized that there are 1.5 million to several million fungal species on Earth but after two centuries of active study, only about 100,000 (less than 10%) of these prognosticated fungal taxa have been discovered and described by scientists [3–5].

So where are all the undescribed fungi? Hawksworth and Rossman [6] speculated that the major reservoir of novel fungi is in association with plants. As a matter of fact, the widely used 1.5 million fungal species working hypothesis was calculated based on the average number of unique fungi per host plant species – there are approximately 250,000 plants on Earth and an estimated six fungal species for every native plant species (250,000×6) [4,7]. Plant-fungus symbiotic associations are very common but many plant-associated fungal communities have not been sampled, especially those in the roots.

Roots were an early development in plant life evolving on land during the Devonian Period (416 to 360 million years ago) [8]. The fossil record and molecular phylogenetic analysis suggest that

from the outset, mycorrhizal fungi played a crucial role in facilitating plant invasion of land, which was dry and poor in nutrients at the time of colonization [9,10]. Such drought and low nutrient stress continue to challenge plants living in many extant habitats, such as our selected study area, pine barrens.

Pine barrens is a general name for a unique type of ecosystem that is dry, acidic and nutrient-poor. Pines and oaks are the dominant trees in pine barrens, whereas the understory is composed of grasses (Poaceae), sedges (Cyperaceae), blueberries and other heath family members (Ericaceae). The largest and most uniform area of pine barrens in the United States is the 1.4 million acre (57,000 km^2) pine barrens of New Jersey (NJ) located in southern New Jersey. The podzolic soil in this region is highly acidic (pH~4.0 with very low cation exchange capacity), sandy, dry (low moisture holding capacity), nutrient poor (low in P, K, etc) and containing elevated levels of soluble aluminum [11–14]. During the 1600's and 1700's when settlers first came to this area they discovered most of the region's soils would not support the growing of vegetable and grain crops from traditional European agriculture. Therefore they named the region "Barrens" [12]. Scarce attention has been received on studies of fungi in the pine barrens, and much remains unknown about fungal diversity and function in this ecosystem [12,15]. The NJ pine barrens represents one of a series of barrens ecosystems along the eastern seaboard of the USA and one of a series of similar ecosystems around the world. For example the Hogue Veluwe of the central Netherlands is an oligotrophic sandy podzol soil supporting Scots pine (*Pinus sylvestris*) forest with grass and ericaceous understory.

Poaceae, the true grasses, are a large plant family with more than 10,000 wild and domesticated species. Grasslands compose approximately 20% of the vegetation of the Earth and play important roles in the ecosystem functioning. Poaceae plants also have had significance in human society, providing food (e.g. maize, wheat, rice), forage (e.g. tall fescue, annual ryegrass), ornamentals (turf grasses), as well as bioenergy (switchgrass, miscanthus). It is well known that the upper-ground tissues of cool season C3 grasses harbor vertically transmitted endophytic fungi, which belong to the *Epichloë* (*Neotyphodium*) group in Clavicipitaceae (Sordariomycetes, Ascomycota). These endophytes often increase drought and heat tolerance of the host grasses, but also produce alkaloids toxic to pest insects and livestock [16]. There is an increasing interest in fungal endophytes but our knowledge is biased toward their above-ground parts. The warm season C4 grasses such as switchgrass (*Panicum virgatum*) do not have this kind of endophytic symbiosis but rather contain root endophytes (e.g. dark septate endophytes), which are often found in the roots of various grasses from different climate and locations [17]. Despite the recognized ubiquity of plant root-associated fungi, their taxonomy, diversity and ecological functions in nature are understudied and enigmatic [17–21].

Albeit some exceptional cases [22,23], it is generally believed that tropical forests have higher fungal diversity [6,24]. Therefore, the tropical forests can be used as a reference to compare with other ecosystems in order to identify new biodiversity hotspots and set research priorities. In this paper, we compared grass root associated fungal communities between the NJ pine barrens and tropical rain forests in Yunnan, China, a traditional biodiversity hotspot [25], using the same sampling strategy and analysis methods, in order to address the following questions: 1) Do fungal communities in the two ecosystems have the same composition and structure? 2) Do grass root associated fungi in the tropical rain forests have higher diversity than the temperate pine barrens? and 3) Are there more undescribed fungi in the tropical rain forests than the temperate pine barrens?

Materials and Methods

Sample collection

A total of 80 samples were collected from four locations in July and August 2012 (Table 1). Daweishan Nature Reserve (DWS, N22°58′, E103°41′) and Western Hills (XS, N24°57′, E102°38′) are located in mountain forests of Yunnan province in southwest China, where the climate is tropical to subtropical. DWS has mean monthly temperature from 15.2 to 27.7°C and annual precipitation of 180 cm, while XS has mean monthly temperatures from 7.7 to 19.8°C and annual precipitation of 109.4 cm [26,27]. Grasses collected from the Yunnan locations include *Setaria* spp., *Alopecurus* spp. and *Digitaria* spp. Yunnan soil samples were not available for this study but according to previous research, typical Yunnan rain forests soil had pH of 5.5–6.0, organic matter 4.5–5.5%, P 7–14 ppm, and K 145–146 ppm [28]. Colliers Mills (CM, N40°04′, W74°26′) and Assunpink Lake (AL, N40°12′, W74°30′) are in the NJ pine barrens, which has a cool temperate climate, with mean monthly temperatures from 0.3 to 24.3°C and average annual precipitation of 116.5 cm (1981–2010, NJ State Climatologist). The pine barrens soil property measurements were: pH 4.6, organic matter 0.4%, P 2.5 ppm, and K 19 ppm. Grasses

collected from the NJ pine barrens include *Panicum virgatum*, *Eragrostis* spp., *Digitaria* spp., and *Schizachyrium scoparium*. The field studies did not involve endangered or protected species and no specific permissions were required for these locations. At each location 20 apparently healthy Poaceae grass root samples (5–10 individual plants per sample) were collected randomly, with a distance of at least 20 meters between each pair of sampled plants.

Isolation of fungi

Within 24 h after collection, the root samples were rinsed in tap water to remove soil particles on the surface and cut into about 5 mm long segments. The root segments were surface sterilized with 75% ethanol for 5 min, followed by 5 min in 0.6% sodium hypochlorite and two rinses in sterile distilled water. For each sample, 20 disinfected root segments were air dried and placed on malt extract agar (MEA, BD) with 0.07% lactic acid, and incubated at room temperature for two months. In the first two weeks, cultures will be observed daily and twice every week afterwards. Fungal cultures were isolated and purified by subculturing from emergent hyphal tips. Imprints of root fragments were made on MEA plates to confirm the effectiveness of the surface disinfection protocols. Spore morphology, if present, was examined for each fungal isolate. Colony characteristics including color, elevation, texture, mycelium type, margin shape, density also were examined. Based on these phenotypic features, all fungal isolates were classified into morphotypes. The representative isolates of each morphotype were selected for molecular identification [29,30].

DNA extraction, amplification and sequence analysis

Protocols described in Zhang et al [31] and Luo and Zhang [32] were used for DNA extraction, PCR amplification and sequencing of the internal transcribed spacer (ITS) of the fungal ribosomal RNA genes. The UltraClean Soil DNA Isolation Kit (MoBio, California) was used for DNA isolation following the manufacturer's protocol. Primers ITS1 and ITS4 were used for PCR and sequencing [33]. The ITS sequences were designated to phylotypes by using 97% similarity criterion in Usearch 7 [34]. The ITS sequences of representative phylotypes were aligned in Clustal X V.1.8 [35] and edited in BioEdit 7.0.5 [36]. Separate phylogenetic analyses were performed for three major groups identified from this collection: Sordariomycetes and Leotiomycetes; Dothideomycetes, Eurotiomycetes, and Pezizomycetes; and Basidiomycota and early diverging lineages. A Bayesian inference (BI) analysis was conducted with the Markov Chain Monte Carlo method in MrBayes 3.2.1 [37] under the nucleotide substitution model selected by using Hierarchical Likelihood Ratio Tests (hLRTs) and Akaike Information Criterion (AIC) in MrModeltest 2.3 [38]. The general time reversible with a proportion of invariable sites and gamma distributed rate variation among sites (GTR+I+G) was the selected model for Sordariomycetes and Leotiomycetes; and Dothideomycetes, Eurotiomycetes, and Pezizomycetes. The Hasegawa-Kishino-Yano model with a proportion of invariable sites (HKY+I) was used for Basidiomycota and the early diverging lineages. Trees were sampled every 100 generations from 10 000 000 generations resulting in 100 000 trees. The first 25 000 trees were discarded as the burn-in and the remaining 75 000 trees were chosen to calculate posterior probability values of clades in a consensus tree. Fungal phylotypes were identified by searching with Blastn in GenBank (http://www.ncbi.nlm.nih.gov/) and the AFTOL (http://aftol.org/) databases and morphology (Table S1). A number of phylogenetic analyses for each individual species/genus were performed to further confirm fungal identification. A species was categorized as undescribed when it had lower

Table 1. Comparison of fungal communities associated with grass roots from the four sampling locations.

Location	Number of samples	Number of fungal isolates	Number of fungal species	Species evenness (E)	Shannon's index (H')	Fisher's alpha	Dominant species
Daweishan Reserve (DWS)	20	81	33	0.93	3.26	20.75	*Nectria mauritiicola* (9.9%), *Fusarium oxysporum* (7.4%), *Pleosporales sp4.* (7.4%)
Western Hills (XS)	20	89	37	0.93	3.37	23.75	*Microdochium bolley* (9%), *Fusarium avenaceum* (6.7%), *Fusarium equiseti* (6.7%)
Colliers Mills (CM)	20	133	26	0.87	2.83	9.65	*Fusarium oxysporum* (15%), *Fusarium moniformis* (12.8%), *Verticillium leptobactrum* (9.8%)
Assunpink Lake (AL)	20	123	26	0.89	2.89	10.07	*Fusarium oxysporum* (13%), *Periconia macrospinosa* (12.2%), *Acephala sp2.* (8.1%), *Verticillium leptobactrum* (8.1%)

than 97% ITS sequence similarity with any known taxa in GenBank, and morphologically, there was no match with available fungal descriptions in the literature. Collected samples are preserved in both Zhang lab at Rutgers University, USA and Cai lab in China. Described new fungal taxa (Fig. 1) were also deposited in CBS-KNAW Fungal Biodiversity Centre (details and accession numbers in [39,40]). Fungal DNA sequences from this study were deposited in GenBank (Table S1).

Fungal diversity analysis

Species accumulation and rarefaction curves, and bootstrap estimates of total species richness were made using 50 randomizations of sample order in Estimate 9.1 [41]. Fungal diversity was measured using Shannon-Wiener index and Fisher's alpha. To compare the similarities among the fungal communities from different locations, Sørensen index and Jaccard coefficient index were calculated (Table 2).

Results

Fungal diversity and undescribed species

A total of 426 fungal isolates were obtained from 1600 root segments of the 80 grass samples (Table 1). They were identified as 85 fungal species (OTUs) belonging to 45 genera, 23 families, 16 orders and 6 classes (Figs. 1, 2, 3). The pine barrens had 38 fungal species, among which 18 (47.4%) were undescribed, whereas Yunnan had 56 species and 18 (32.1%) were undescribed (Table 1, Fig. 4). Shannon's index (H') for DWS and XS of Yunnan were 3.26 and 3.37, respectively, whereas H' for CM and AL of pine barrens were 2.83 and 2.89 (Table 1). Fisher's alpha showed even more difference: 20.75 and 23.75 for DWS and XS; while 9.65 and 10.07 for CM and AL of pine barrens. The protocol used for surface disinfection was efficient as no fungi were emerged from the root imprint plates after four weeks of incubation.

All rarefaction curves did not reach an asymptote, indicating that more fungal species will likely be uncovered with additional sampling in these locations (Fig. 5). Compared to CM and AL, steeper slopes were shown from DWS and XS. The species richness of bootstrap estimate did not exceed the upper bound of the 95% confidence intervals for estimated richness, which indicated that the sampling method sufficiently captured fungal species from the four sampling locations (Fig. 5).

Fungal community composition

The majority of the fungi isolated from the sampled grass roots were Ascomycota (99% of all isolates), with a few (0.5%) in Basidomycota and early diverging lineages (0.5%). Fungal assemblage in the two ecosystems is different, with only 9 species in common (Fig. 4). In Yunnan, 48.8%, 38.8% and 1.8% of the fungal isolates belong to Sordariomycetes, Dothideomycetes and Leotiomycetes, respectively. While in NJ pine barrens, 61.3%, 14.8% and 21.9% of the isolates belong to the three classes, respectively (Fig. 6).

The paired similarity of the four sampling locations ranged from 0.16 to 0.54 for the Sørensen index, and 0.09 to 0.37 for the Jaccard coefficient index. Among the 85 identified species, 20 species occurred in two or more locations. There were 16 species isolated only from DWS, 21 from CM, 12 from CM and 11 from AL. *Nectria mauritiicola* (9.9%), *Fusarium oxysporum* (7.4%) and *Pleosporales sp.4* (7.4%) were the most frequently isolated species in DWS. *Microdochium bolleyi* (9%), *Fusarium avenaceum* (6.7%) and *Fusarium equisetii* (6.7%) were the most frequent ones in XS. *Fusarium oxysporum* (15%), *Fusarium moniformis* (12.8%) and

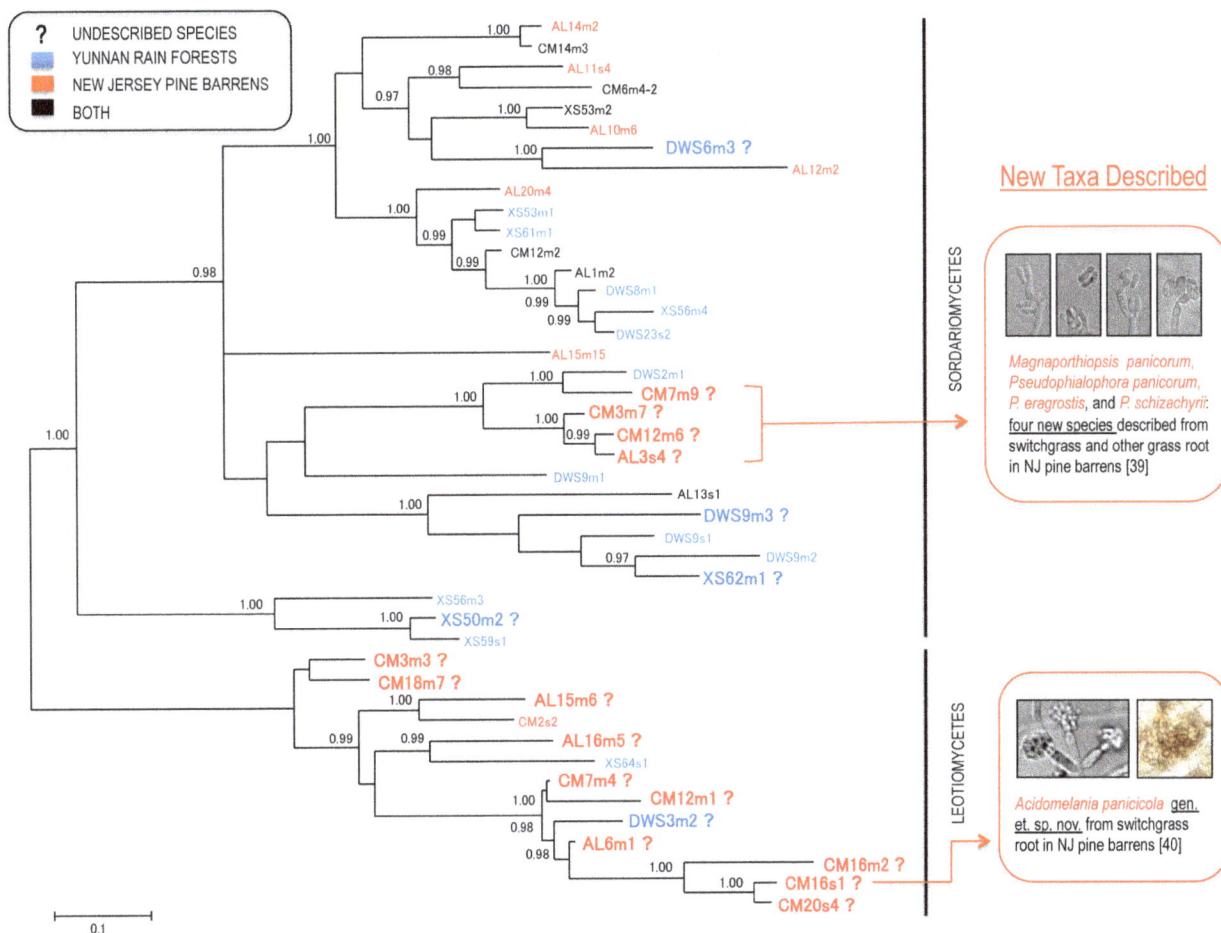

Figure 1. Consensus Bayesian tree based on ITS sequences for the 44 representative fungal OTUs of Sordariomycetes and Leotiomycetes obtained in this study. Bayesian posterior probability values ≥0.95 are shown on the internodes.

Verticillium leptobactrum (9.8%) were dominant in CM. *Fusarium oxysporum* (13%), *Periconia macrospinosa* (12.2%), *Acephala* sp.2 (8.1%) and *Verticillium leptobactrum* (8.1%) were dominant in AL. Among the four locations, CM and AL had the highest similarity indices and shared two dominant species. DWS and XS had the second highest similarity indices (Table 2).

Discussion

This study revealed that grass (Poaceae) roots in pine barrens are one of the reservoirs of novel fungi with about 47% being undescribed species. The results support Hawksworth and Rossman's speculation [6] that sampling plants in such unexplored habitats will uncover novel fungi. In our recent taxonomy papers based on these collections, we named and described a new genus in Leotiomycetes and four new species in Magnaporthaceae (Sordariomycetes) based on multi-locus phylogeny, morphological and ecological characters [39,40]. Magnaporthaceae is a family that includes the rice blast pathogen, take-all pathogen and other ecologically important fungi [39,40] (Fig. 1). Taxonomic work for other novel fungi discovered from this study is underway. The finding that grass roots in the tropical forests harbor higher total fungal species diversity corroborates previously published results on foliar endophytes and further demonstrated that tropical forests are biodiversity hotspots [24].

Table 2. Similarity of fungal communities among the sampling locations.

Location	Daweishan Reserve (DWS)	Western Hills (XS)	Colliers Mills (CM)	Assunpink Lake (AL)
Daweishan Reserve (DWS)		0.40	0.20	0.24
Western Hills (XS)	0.25		0.16	0.19
Colliers Mills (CM)	0.11	0.09		0.54
Assunpink Lake (AL)	0.14	0.11	0.37	

Sørenson's index is shown above the diagonal, and Jaccard's index is below the diagonal.

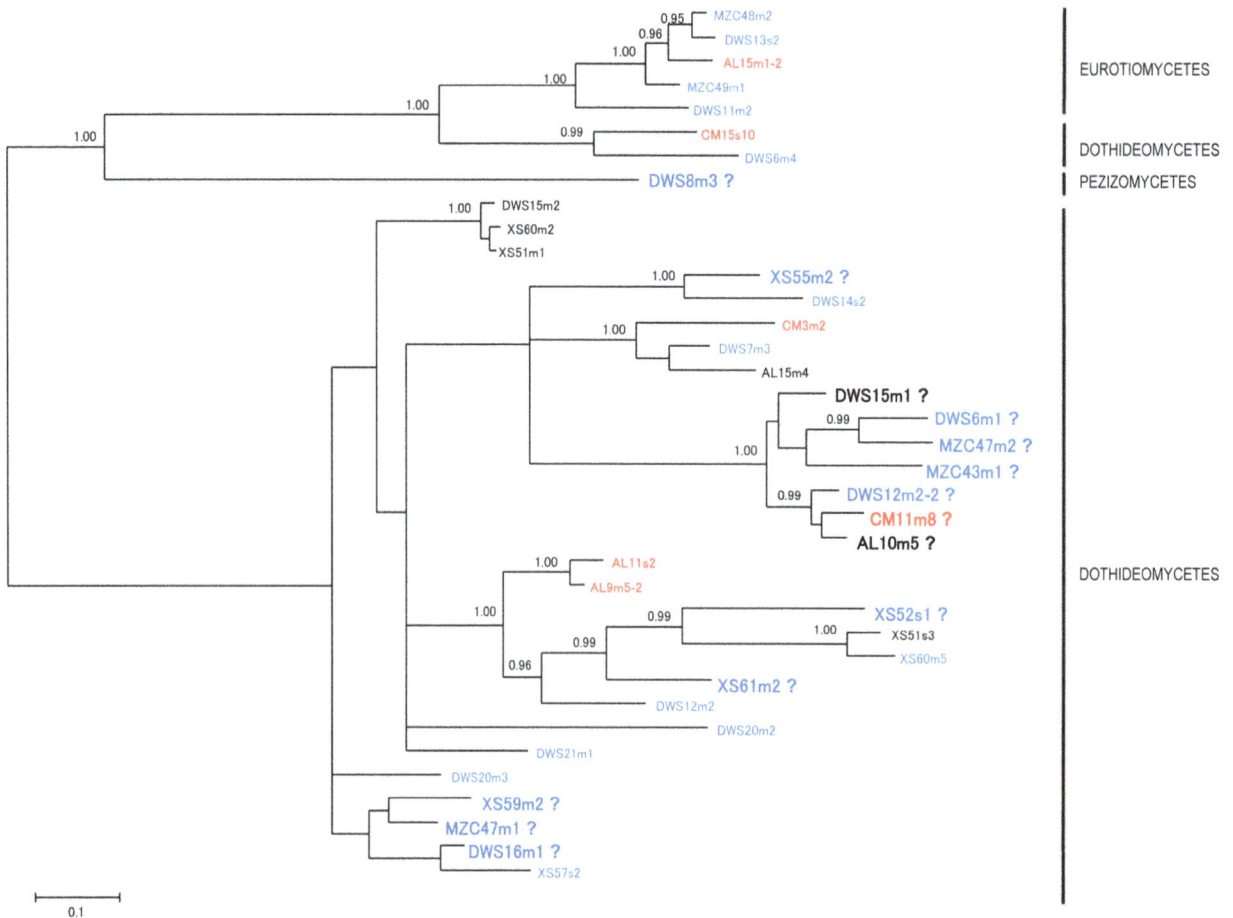

Figure 2. Consensus Bayesian tree tree based on ITS sequences for the 37 representative fungal OTUs of Dothideomycetes, Eurotiomycetes, and Pezizomycetes obtained in this study. Bayesian posterior probability values ≥0.95 are shown on the internodes.

The fungal assemblage is different between the NJ pine barrens and Yunnan tropical forests with only 9 fungal species found in both of the sampled ecosystems. In the pine barrens 21.9% of the fungal isolates belong to Leotiomycetes, a significantly higher report than the tropical Yunnan (1.8%) or other similar studies we have encountered. Usually Sordariomycetes and Dothideomycetes are the dominant classes in the fungal endophyte communities, regardless of the plant species, associated host tissue or geographically location [20,24,42,43]. Factors such as climate, geographical location and host species may have contributed to the difference in fungal community composition and structure [44]. The acidic, low nutrient and dry pine barrens soil may be the cause for the observed unique fungal assembly. The high occurrence of Leotiomycetes in pine barrens indicates that they may play a role

in stress tolerance of the host plants. Further studies are needed to test the ecological roles of these uncovered root-associated fungi.

We also found some common features between the two sampled fungal communities. Our results suggest that Sordariomycetes of Ascomycota were the most commonly isolated fungi from Poaceae grass roots in both tropical rain forests and temperate pine barrens, which was consistent with some previous studies on grasses (e.g., *Holcus lanatus*, *Dactylis glomerata*, *Ammophila arenaria* and *Elymus farctus*) [29,30,45]. Species of *Fusarium*, *Microdochium*, *Periconia* and *Verticillium*, the most frequently isolated fungi from this study were also found by other researchers from various grasses in different locations [29,30,45,46], indicating that they may serve as core group of fungal symbionts of grass roots. Basidiomycota predominance was reported in certain

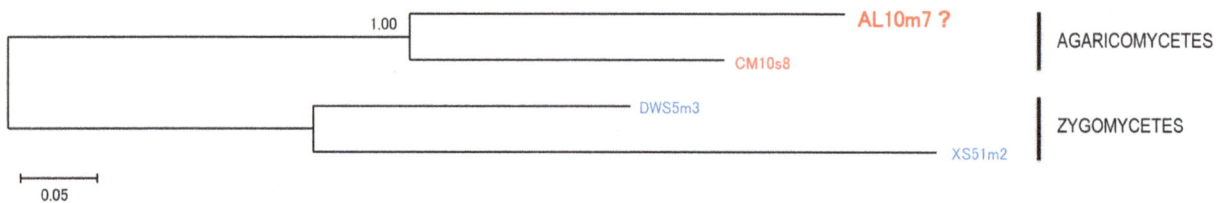

Figure 3. Consensus Bayesian tree tree based on ITS sequences for the 4 representative fungal OTUs of Agaricomycetes and Zygomycetes obtained in this study. Bayesian posterior probability values ≥0.95 are shown on the internodes.

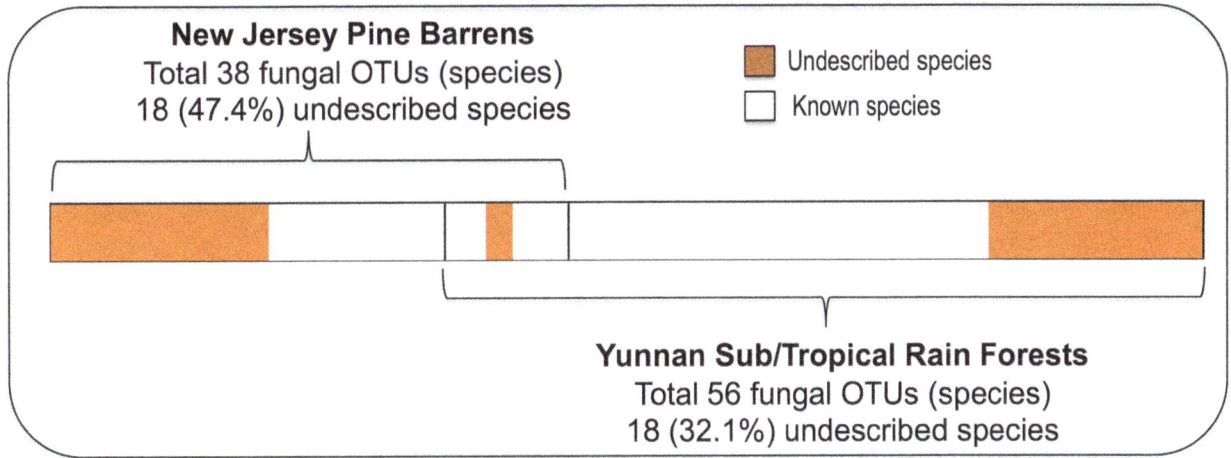

Figure 4. Number of undescribed and total fungal species from the NJ pine barrens and Yunnan rain forests.

grasses, such as *Bouteloua gracilis*, *Festuca paniculata* and *Sporobolus crytandrus* [47–49]. However the occurrence of Basidiomycota was very low in the samples of this study.

A number of potential fungal pathogens were among the isolated fungi from this study. For example, *Fusarium oxysporum* may cause Fusarium wilt of various plants, *Microdochium bolleyi* is the causal agent of root rot of wheat and flax, *Phoma tropica* can cause leaf spot, and *Gaeumannomyces graminis* is the causal agent of take-all disease of cereal crops. Under certain conditions, these fungi may change from endophytes to pathogens [50]; however, pathogenicity is beyond the scope of this paper.

ITS is the selected fungal DNA barcode, which is useful in species identification for most fungal lineages [2]. We used ITS in this study for species recognition but we are aware that ITS sequences are highly variable at family or higher taxonomic levels, and therefore are not appropriate to infer phylogenetic relations for all fungal taxa collected in the study. Figs. 1, 2, 3 are shown

only to demonstrate overall fungal diversity. Our species identification was based on a number of separate phylogenetic analyses for each individual OTU (not shown), rather than only based on BLAST. In our taxonomy papers, a six-locus (SSU, LSU, ITS, *EF-1alpha*, *MCM7* and *RPB1*) phylogenetic analysis was performed for the four proposed new Magnaporthaceae species [39], and a three-locus (ITS, LSU and *ACT*) phylogeny was performed for *Acidomelania* [40]. The species delimitation was based on the genealogical concordance method, the "gold standard" for species delimitation in fungi [51,52]. Previously, various ITS sequence similarity criteria (95–99%) have been used in delimiting fungal species [24,53,54] and our taxonomic work support that 97% is appropriate for species delimitation for this group of fungi [39,40]. Therefore, we adopted 97% in ITS sequence clustering analysis for species delimitation in this study. Re-evaluation of this criterion is needed when new taxonomic information is available.

In this study, a species was categorized as undescribed when it had lower than 97% ITS sequence similarity with any known taxa in GenBank, and morphologically, there was no match with available fungal descriptions in the literature. It is possible that some species may be described but have not been sequenced or

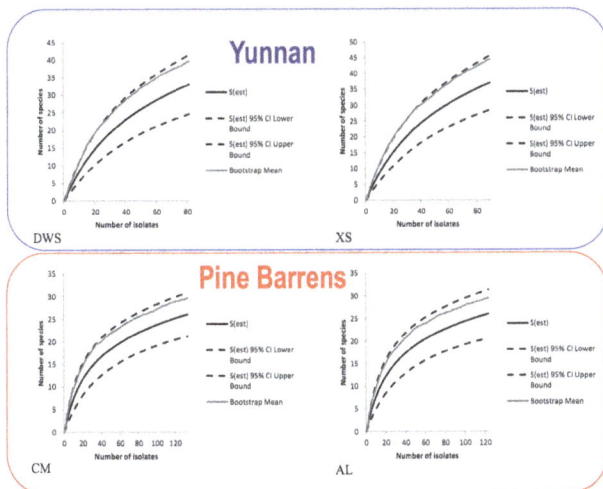

Figure 5. Species accumulation curves for grass root associated fungi from the NJ pine barrens: Collier's Mills (CM) and Assumpin Lake (AL) and Yunnan rain forests: Daweishan (DWS) and Western Hill (XS). Bootstrap estimates of total richness are also provided.

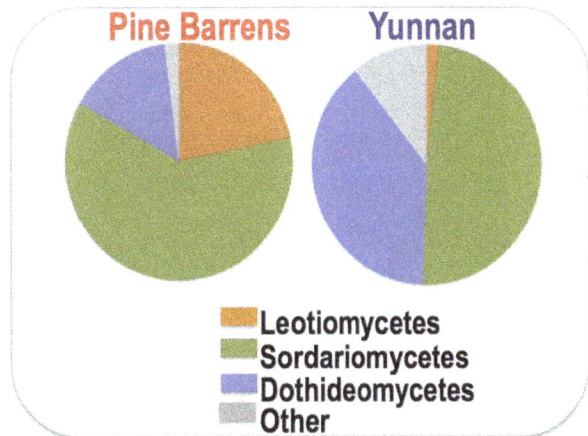

Figure 6. Taxonomic distribution of grass root associated fungi among classes in the NJ pine barrens and Yunnan rain forests.

deposit in GenBank. However, it was estimated that over 90% of fungal species have not been discovered and the unmatched sequences more likely are new taxa. Furthermore, we also checked morphology of the fungal cultures to confirm their novelty. Our results indicate that the current sampling effort has uncovered only a fraction of the fungal diversity in the pine barrens and tropical forests and future sampling likely will uncover more novel fungi. Solid taxonomic work, culture-independent metagenomic analysis and functional experiments are needed to have a holistic understanding of the fungal diversity in nature.

References

1. Alexopoulos CJ, Mims CW, Blackwell M (1996) Introductory Mycology: John Wiley & Sons, Inc.
2. Schoch CL, Seifert KA, Huhndorf S, Robert V, Spouge JL, et al. (2012) Nuclear ribosomal internal transcribed spacer (ITS) region as a universal DNA barcode marker for Fungi. Proceedings of the National Academy of Sciences of the United States of America 109: 6241–6246.
3. Blackwell M (2011) The Fungi: 1, 2, 3 ... 5.1 Million Species? American Journal of Botany 98: 426–438.
4. Hawksworth DL (1991) The fungal dimension of biodiversity - magnitude, significance, and conservation. Mycological Research 95: 641–655.
5. Mora C, Tittensor DP, Adl S, Simpson AGB, Worm B (2011) How many Species are there on earth and in the ocean? Plos Biology 9.
6. Hawksworth DL, Rossman AY (1997) Where are all the undescribed fungi? Phytopathology 87: 888–891.
7. Hawksworth DL (2001) The magnitude of fungal diversity: the 1.5 million species estimate revisited. Mycological Research 105: 1422–1432.
8. Gensel PG, Kotyk ME, Basinger JF (2001) Morphology of above- and below-ground structures in early Devonian (Pragian-Emsian) plants. In: PG . G, D . E, editors. Plants Invade the Land: Evolutionary and Environmental Perspectives. New York: Columbia University Press. pp. 83–102.
9. Blackwell M (2000) Evolution - Terrestrial life - Fungal from the start? Science 289: 1884–1885.
10. Redecker D, Kodner R, Graham LE (2000) Glomalean fungi from the Ordovician. Science 289: 1920–1921.
11. Joffe SE, Watson CW (1933) Soil profile studies: V. Mature podzols. Soil Science 35: 313–332.
12. Forman RT (1998) Pine Barrens: Ecosystem and Landscape: Academic Press.
13. Tedrow JCF (1952) Soil conditions in the Pine Barrens of New Jersey. Bartonia 26: 28–35.
14. Turner RS, Johnson AH, Wang D (1985) Biogeochemistry of aluminum in McDonalds Branch watershed, New Jersey pine barrens. J Env Qual 14: 314–323.
15. Tuininga AR, Dighton J (2004) Changes in ectomycorrhizal communities and nutrient availability following prescribed burns in two upland pine-oak forests in the New Jersey pine barrens. Canadian Journal of Forest Research-Revue Canadienne De Recherche Forestiere 34: 1755–1765.
16. Roberts CA, West CP, Spiers DE (2005) Neotyphodium in Cool-Season Grasses: Blackwell.
17. Sieber TN, Grunig CR (2013) Fungal root endophytes. In: Eshel A, Beeckman T, editors. Plant Roots - The Hidden Half. 4th ed. Boca Raton, FL, USA: CRC Press, Taylor & Francis Group.
18. Grünig CR, McDonald BA, Sieber TN, Rogers SO, Holdenrieder O (2004) Evidence for subdivision of the root-endophyte Phialocephala fortinii into cryptic species and recombination within species. Fungal Genetics and Biology 41: 676–687.
19. Mandyam K, Jumpponen A (2005) Seeking the elusive function of the root-colonising dark septate endophytic fungi. Studies in Mycology: 173–189.
20. Knapp DG, Pintye A, Kovacs GM (2012) The dark side is not fastidious - dark septate endophytic fungi of native and invasive plants of semiarid sandy areas. Plos One 7.
21. Grünig CR, Queloz V, Sieber TN, Holdenrieder O (2008) Dark septate endophytes (DSE) of the Phialocephala fortinii s.l. - Acephala applanata species complex in tree roots: classification, population biology, and ecology. Botany-Botanique 86: 1355–1369.
22. Berndt R (2012) Species richness, taxonomy and peculiarities of the neotropical rust fungi: are they more diverse in the Neotropics? Biodiversity and Conservation 21: 2299–2322.
23. Scholler M, Aime MC (2006) On some rust fungi (Uredinales) collected in an Acacia koa-Metrosideros polymorpha woodland, Mauna Loa Road, Big Island, Hawaii. Mycoscience 47: 159–165.

24. Arnold AE, Lutzoni F (2007) Diversity and host range of foliar fungal endophytes: Are tropical leaves biodiversity hotspots? Ecology 88: 541–549.
25. Lopez-Pujol J, Zhang FM, Ge S (2006) Plant biodiversity in China: richly varied, endangered, and in need of conservation. Biodiversity and Conservation 15: 3983–4026.
26. Li X, Song L, Chen J, Yuan C, Zhang L (2012) Seedling regeneration of the primary semi-humid evergreen broadleaved forest and its secondary succession communities in Xishan, Kunming. Guihaia 32: 475–482.
27. Wang J, Ma QY, Du F (2006) Flora diversity characteristics of seed plants of Dawei Mountain National Nature Reserve in Yunnan province, China. Scientia Silvae Sinicae 42: 7–15.
28. Zhu H, Xu ZF, Wang H, Li BG (2004) Tropical rain forest fragmentation and its ecological and species diversity changes in southern Yunnan. Biodiversity and Conservation 13: 1355–1372.
29. Marquez SS, Bills GF, Acuna LD, Zabalgogeazcoa I (2010) Endophytic mycobiota of leaves and roots of the grass Holcus lanatus. Fungal Diversity 41: 115–123.
30. Marquez SS, Bills GF, Zabalgogeazcoa I (2008) Diversity and structure of the fungal endophytic assemblages from two sympatric coastal grasses. Fungal Diversity 33: 87–100.
31. Zhang N, Zhao S, Shen QR (2011) A six-gene phylogeny reveals the evolution of mode of infection in the rice blast fungus and allied species. Mycologia 103: 1267–1276.
32. Luo J, Zhang N (2013) Magnaporthiopsis, a new genus in Magnaporthaceae (Ascomycota). Mycologia 105: 1019–1029.
33. White TJ, Bruns T, Lee S, Taylor J (1990) Amplification and direct sequencing of fungal ribosomal RNA genes for phylogenetics. In: Innis M, Gelfand D, Sninsky J, White T, editors. PCR Protocols: A Guide to Methods and Applications. New York: Academic Press.
34. Edgar RC (2010) Search and clustering orders of magnitude faster than BLAST. Bioinformatics 26: 2460–2461.
35. Thompson JD, Gibson TJ, Higgins DG (1997) The CLUSTAL-X windows interface: flexible strategies for multiple sequence alignment aided by quality analysis tools. Nucleic Acids Research 25: 4876–4882.
36. Hall TA (1999) Bioedit: a user-friendly biological sequences alignment editor analysis program for windows 95/98/NT. Nucleic Acids Symposium Series 41: 95–98.
37. Ronquist F, Teslenko M, van der Mark P, Ayres DL, Darling A, et al. (2012) MrBayes 3.2: efficient Bayesian phylogenetic inference and model choice across a large model space. Systematic Biology 61: 539–542.
38. Nylander JAA (2004) MrModeltest 2.2, Program distributed by the author.: Evolutionary Biology Centre, Uppsala University.
39. Luo J, Walsh E, Zhang N (2014) Four new species in Magnaporthaceae from grass roots in New Jersey Pine Barrens. Mycologia 106: 580–588.
40. Walsh E, Luo J, Zhang N (2014) Acidomelania panicicola gen. et. sp. nov. from switchgrass roots in acidic New Jersey Pine Barrens. Mycologia In press.
41. Colwell RK, Chao A, Gotelli NJ, Lin SY, Mao CX, et al. (2012) Models and estimators linking individual-based and sample-based rarefaction, extrapolation and comparison of assemblages. Journal of Plant Ecology 5: 3–21.
42. Yuan ZL, Lin FC, Zhang CL, Kubicek CP (2010) A new species of Harpophora (Magnaporthaceae) recovered from healthy wild rice (Oryza granulata) roots, representing a novel member of a beneficial dark septate endophyte. Fems Microbiology Letters 307: 94–101.
43. Ghimire SR, Charlton ND, Bell JD, Krishnamurthy YL, Craven KD (2011) Biodiversity of fungal endophyte communities inhabiting switchgrass (Panicum virgatum L.) growing in the native tallgrass prairie of northern Oklahoma. Fungal Diversity 47: 19–27.
44. Hoffman MT, Arnold AE (2008) Geographic locality and host identity shape fungal endophyte communities in cupressaceous trees. Mycological Research 112: 331–344.

Acknowledgments

The research was supported by the National Science Foundation (DEB 1145174) to N. Zhang. The authors thank Drs. J. Dighton, D. Bhattacharya and the anonymous reviewers for their helpful comments on the paper.

Author Contributions

Conceived and designed the experiments: NZ JL. Performed the experiments: JL EW AN. Analyzed the data: JL EW. Contributed reagents/materials/analysis tools: NZ WZ KZ LC. Contributed to the writing of the manuscript: NZ JL.

45. Marquez SS, Bills GF, Zabalgogeazcoa I (2007) The endophytic mycobiota of the grass Dactylis glomerata. Fungal Diversity 27: 171–195.

46. Macia-Vicente JG, Jansson HB, Abdullah SK, Descals E, Salinas J, et al. (2008) Fungal root endophytes from natural vegetation in Mediterranean environments with special reference to Fusarium spp. Fems Microbiology Ecology 64: 90–105.

47. Mouhamadou B, Molitor C, Baptist F, Sage L, Clement JC, et al. (2011) Differences in fungal communities associated to Festuca paniculata roots in subalpine grasslands. Fungal Diversity 47: 55–63.

48. Khidir HH, Eudy DM, Porras-Alfaro A, Herrera J, Natvig DO, et al. (2010) A general suite of fungal endophytes dominate the roots of two dominant grasses in a semiarid grassland. Journal of Arid Environments 74: 35–42.

49. Porras-Alfaro A, Herrera J, Sinsabaugh RL, Odenbach KJ, Lowrey T, et al. (2008) Novel root fungal consortium associated with a dominant desert grass. Applied and Environmental Microbiology 74: 2805–2813.

50. Eaton CJ, Cox MP, Scott B (2011) What triggers grass endophytes to switch from mutualism to pathogenism? Plant Science 180: 190–195.

51. Hibbett DS, Taylor JW (2013) Fungal systematics: is a new age of enlightenment at hand? Nature Reviews Microbiology 11: 129–133.

52. Taylor JW, Jacobson DJ, Kroken S, Kasuga T, Geiser DM, et al. (2000) Phylogenetic species recognition and species concepts in fungi. Fungal Genetics and Biology 31: 21–32.

53. O'Brien HE, Parrent JL, Jackson JA, Moncalvo JM, Vilgalys R (2005) Fungal community analysis by large-scale sequencing of environmental samples. Applied and Environmental Microbiology 71: 5544–5550.

54. Gazis R, Chaverri P (2010) Diversity of fungal endophytes in leaves and stems of wild rubber trees (Hevea brasiliensis) in Peru. Fungal Ecology 3: 240–254.

5

The Reduced Effectiveness of Protected Areas under Climate Change Threatens Atlantic Forest Tiger Moths

Viviane G. Ferro[1], Priscila Lemes[2], Adriano S. Melo[1], Rafael Loyola[1]*

1 Departamento de Ecologia, Universidade Federal de Goiás, Goiânia, Goiás, Brazil, **2** Programa de Pós-Graduação em Ecologia e Evolução, Universidade Federal de Goiás, Goiânia, Goiás, Brazil

Abstract

Climate change leads to species' range shifts, which may end up reducing the effectiveness of protected areas. These deleterious changes in biodiversity may become amplified if they include functionally important species, such as herbivores or pollinators. We evaluated how effective protected areas in the Brazilian Atlantic Forest are in maintaining the diversity of tiger moths (Arctiinae) under climate change. Specifically, we assessed whether protected areas will gain or lose species under climate change and mapped their locations in the Atlantic Forest, in order to assess potential spatial patterns of protected areas that will gain or lose species richness. Comparisons were completed using modeled species occurrence data based on the current and projected climate in 2080. We also built a null model for random allocation of protected areas to identify where reductions in species richness will be more severe than expected. We employed several modern techniques for modeling species' distributions and summarized results using ensembles of models. Our models indicate areas of high species richness in the central and southern regions of the Atlantic Forest both for now and the future. However, we estimate that in 2080 these regions should become climatically unsuitable, decreasing the species' distribution area. Around 4% of species were predicted to become extinct, some of them being endemic to the biome. Estimates of species turnover from current to future climate tended to be high, but these findings are dependent on modeling methods. Our most important results show that only a few protected areas in the southern region of the biome would gain species. Protected areas in semideciduous forests in the western region of the biome would lose more species than expected by the null model employed. Hence, current protected areas are worse off, than just randomly selected areas, at protecting species in the future.

Editor: Brock Fenton, University of Western Ontario, Canada

Funding: VGF work is funded by CNPq (grant #563332/2010-7). PL received a PhD studentship from CNPq. RDL and ASM received productivity fellowships granted by CNPq (grants #304703/2011-7 and #307479/2011-0, respectively). RDL work is funded by CAPES-FCT Program, by the Brazilian Research Network on Global Climate Change (Rede CLIMA), and Conservation International Brazil. This study benefited from resources provided by Site 13 (Parque Nacional das Emas) of the Brazilian Long Term Ecological Research Network (CNPq, grant #558187/2009-9). The funders had no role in study design, data collection and analysis, decision to publish, or preparation of the manuscript.

Competing Interests: The authors have declared that no competing interests exist.

* Email: rdiasloyola@gmail.com

Introduction

The implementation and maintenance of protected areas is still the cornerstone of conservation actions [1]. However, due to broad-scale environmental changes which will potentially shift the distribution of suitable habitats for many species across the geographic space [2,3], scientists have expressed concern that existing networks of protected areas might not be able to guarantee the long-term persistence of the species they are supposed to protect [4,5].

Climate change poses a new challenge to the traditionally static way conservation planning is usually done, by forcing planning to become more dynamic [6]. Several species have already shifted their ranges to cooler regions, both in temperate regions and in the tropics, as a response to a warming climate [7–9]. Most solutions offered by conservation scientists and practitioners to deal with spcies' range shifts focus on the establishment of new protected areas that should cope with the effects of climate change on species distribution [6,10–16]. However, the effectiveness of protected areas may decrease as they become climatically unsuitable for most species [5] and more suitable to invasive species [17,18].

Climate-driven modifications in species composition within protected areas may disrupt species interactions [19] by altering ecosystem functioning [20]. Therefore, it is necessary to take into account forecasted changes in species distributions to evaluate the future effectiveness of protected areas [5,6,21].

The effects of climate change on species distributions have been generally inferred through ecological niche models (ENM) [22], also referred to as "bioclimatic envelope models" (BEM) [23] or "species distribution models" (SDM) [24]. Araújo and Peterson [25], Peterson and Soberón [26], and Rangel and Loyola [27] have all provided recent clarifications on their conceptual differences. However, different modeling methods and climate models may produce very different outputs, increasing uncertainties in projected distributions and their applicability to conservation efforts [3,28]. In the last decade, new modeling techniques have been developed that take into consideration such impediments (e.g. model ensemble forecasting). Nonetheless, few studies have applied modern techniques to predict the distribution of invertebrates (but see Diniz-Filho et al., 2010a, b), despite an urgent need to evaluate the consequences of climate change to this hyperdiverse group in order to plan for its conservation [29].

Lepidoptera is the second richest order of insects, with 150,000 species recorded in the world [30]. Butterflies and moths are exclusive pollinators of many plant species [31,32]. The vast majority of Lepidoptera larvae are herbivores, consuming almost all orders of gymnosperms and angiosperms as well as mosses and ferns [33]. Herbivory can influence the fitness [34], distribution, composition and abundance of plant species [35], as well as the rate of litter decomposition [36]. Lepidopterans are also important food items for arthropods (in particular spiders and other insects) and vertebrates (especially birds and bats). Changes in Lepidoptera diversity, abundance, phenology, distribution and assemblage composition driven by climate changes may affect ecosystem functions and services, species interactions, as well as the structure of plant communities and economic losses due insect infestation and pestilence [37].

Many studies have found that climate change will alter patterns of phenology, horizontal or vertical range (distribution and area), and the abundance of Lepidoptera species (reviewed in [38]). For example, Conrad et al. [39] studied the population dynamics of the tiger moth *Arctia caja* (Arctiinae) between 1968 and 1999 in Great Britain and observed a decrease of about 30% of abundance and proportion of occupied sites after 1984. Arctiinae (Erebidae) (classification following [40]), in particular, comprises almost 11,000 species of moths, of which about 6,000 are found in the Neotropics [41]. In Brazil, there are records for 1,391 species [42]. The Atlantic Forest has the richest Arctiinae fauna among all Brazilian biomes (1193 species) and approximately 40% of these species are endemic to this biome [43]. Arctiinae larvae feed on angiosperms and gymnosperms, as well as algae, lichens, and mosses [44]. Although Arctiinae are among the most polyphagous lepidopterans [44], the proportion of generalist species decreases toward the tropics. Many Arctiinae larvae and adults have conspicuous coloration, are diurnal and many adults form mimetic rings with Hemiptera, Hymenoptera, Coleoptera and unpalatable butterflies [45]. As well as exhibiting warning coloration, most of these moths are also toxic or unpalatable. Several secondary compounds were found in all stages of Arctiinae (eggs, larvae, pupae and adults), including pyrrolizidine alkaloids, which are generally sequestered from their host plants during the larval stage [45]. We choose this group because of its important roles in ecosystems and also because of their extremely high diversity in the region, so that we can run models using a great amount of data and keeping the taxonomic group narrow.

Here, we evaluated the current and future climatic suitability of protected areas located in the Atlantic Forest Biodiversity Hotspot (in Brazil) based on species' ecological niche models and diversity patterns of tiger moths (Lepidoptera: Erebidae: Arctiinae). More specifically, we addressed the following questions: (1) how will climate change affect the geographical pattern of Arctiinae species richness in the region? And (2) how does the spatial location of a given protected area determine if it will gain or lose species under different climate change scenarios?

We selected tiger moths as our case-study because they comprise a species-rich subfamily, are well represented in Brazilian collections (especially in the Atlantic Forest, which is the geographical area covered in this study) [42] and because there are several active researchers currently working with Neotropical Arctiinae. These researchers helped with the identification of some species and allowed for a nomenclatural update, incorporating recent occurrence records in our dataset.

Methods

Study region

We focused our analyses on the Atlantic Forest Biodiversity Hotspot [46]. The Atlantic Forest originally covered around 150 million ha (Fig. 1) with heterogeneous environmental conditions. Its latitudinal range extends into tropical and subtropical regions, and its wide longitudinal range harbors differences in forest composition due to a diminishing gradient in rainfall from coast to interior [47]. Although the Atlantic Forest has high diversity and endemism (with more than 20,000 plant species, 261 mammal species, 688 bird species, 200 reptile species, 280 amphibian species, to name well-studied taxonomic groups), currently, only *ca.* 1% of the original forests are legally protected [47]. An effective reserve network, taking into account climate change to ensure the species persistence in the long-term, is therefore imperative to address conservation investment in appropriate sites.

Ecological niche models

We obtained occurrence records for 703 tiger moth species inhabiting the Atlantic Forest from field surveys and museum records. Tiger moth records included the period of 1920 to2008. We overlaid these point-locality records for each species into an equal-area grid (10 km×10 km of spatial resolution) that covered the full extent of the Atlantic Forest. Then, we built a species by grid cell matrix, considering presences of species inside grid cells. Species with less than five occurrences were excluded to avoid model bias, and therefore, a total of 507 species were studied.

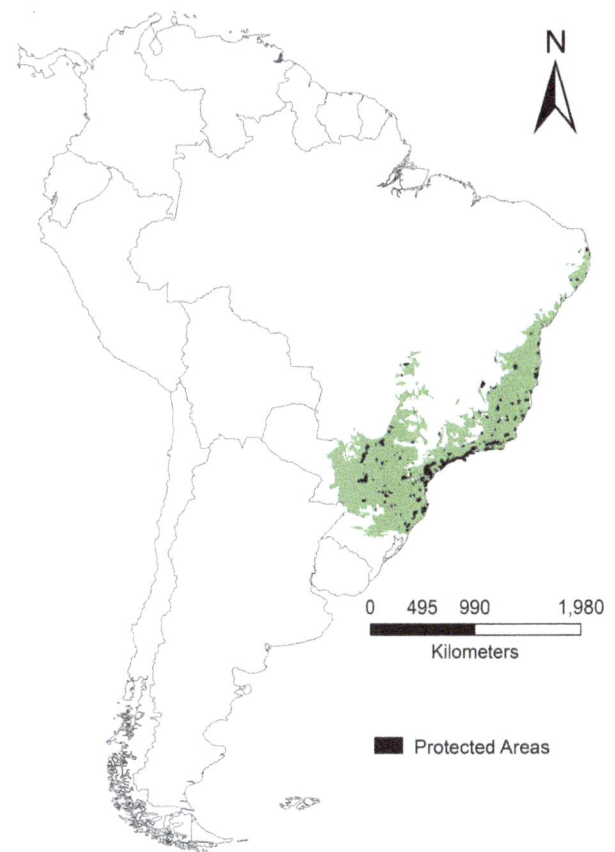

Figure 1. Original extent of the Atlantic Forest Biodiversity Hotspot in Brazil.

We obtained current climatic data from the WorldClim database (http://www.worldclim.org/current) and future climatic scenarios from CIAT (http://ccafs-climate.org). These future scenarios were developed by the Intergovernmental Panel on Climate Change (IPCC) Fourth Assessment Report (AR4). For each species, we modeled its distribution as a function of four climatic variables: annual mean temperature, temperature seasonality (standard deviation * 100), annual precipitation and precipitation seasonality (coefficient of variation). These variables represent interpolated climate data from 1950 to2000 [48]. For future climatic conditions we used the same climate variables for the year 2080, obtained from three Atmosphere-Ocean General Circulation Models (AOGCMs) of the A2 emission scenario (CCCMA-CGCM2, CSIRO-MK2.0 and UKMO-HadCM3), that were generated by application of delta downscaling method on the original data from IPCC's report. Data original resolution was 30 arc-seconds and both current and future climate variables were re-scaled to our grid resolution.

We used presence derived from species' occurrences and climatic variables to model species' ecological niche and project their distributions. As reliable absence data were not available, we fitted six presence-only modeling methods (which differ both conceptually and statistically [27]), grouped them into two separate sets (distance methods and machine-learning methods) and applied the ensemble forecasting approach within each set (see text below). Distance methods were Euclidian and Gower distances [49] and Ecological Niche Factor Analysis (ENFA) [50]. Machine learning methods were Maximum Entropy (MaxEnt) [51], Genetic Algorithm for Rule Set Production (GARP) [52] and Artificial Neural Networks (ANN) [53]. These presence-only methods for modeling species' ecological niches can be grouped into three types of presence-only methods [54]. Firstly, into methods based solely on presence records (e.g. Euclidian and Gower distances), the prediction is made without reference to other samples from the study area. Secondly, into methods using "background" climatic data for the whole study area (e.g. ENFA, MaxEnt), which evaluate how the climatic conditions where species are known to occur relates to the climate across the rest of the study area (the 'background'). Thirdly, into methods that generate (sample) "pseudo-absences" from the study area (e.g. GARP, ANN), assessing differences between occurrence sites and a set of sites chosen from the study area which are used instead of real absence data. In this case, the set of "pseudo-absences" were selected randomly [52].

For all models, we randomly partitioned presence and pseudo-absence data of each species in 75% for calibration (or training) and 25% for validation (or test); repeating this process 10 times (i.e. a cross-validation) and maintaining the observed prevalence of each species. We converted continuous predictions in presence and pseudo-absences finding the threshold with maximum sensitivity and specificity values in the receiver operating characteristic (ROC) curve and calculated the True Skill Statistics (TSS) to evaluate model performance [55]. The ROC Curve is created by plotting the fraction of true positives out of the positives vs. the fraction of false positives out of the negatives, at various threshold settings. The TSS range from -1 to $+1$, where values equal $+1$ is a perfect prediction and values equal to or less than zero is a prediction no better than random [55]. Although the area under the receiver operating characteristic curve (AUC) is the most common method to evaluate the accuracy of predictive distribution models, we decided to use TSS. There are several reasons why AUC should not be used for this purpose [56]. In particular, AUC weighs omission and commission errors equally

and the total geographic extent of the study highly influences its scores [56].

We did an ensemble of forecasts to produce consensual predictions of species distributions [3,10,23,28,57–59]. We projected distributions to current climatic conditions and obtained 30 projections per species within each set of methods (3 modeling methods ×10 randomly partitioned data). We also projected distributions to future climate, obtaining 90 projections per species (3 modeling methods ×3 climate models ×10 randomly partitioned data). This allowed us to generate a frequency of projections in the ensemble. We then generated the frequency of projections weighted by the TSS statistics for each species and timeframe within each set of methods, i.e. best models have more weight in our consensus projections. We considered the presence of a species only in cells with 50% or more of frequency of projections, but a continuous value was held when this occurred. Then, we defined species richness as the sum of the ranges which overlapped (predicted by ENMs) for each cell. So, for example, if twenty different species were projected in a given cell the species richness for that cell was considered as twenty. Finally, we calculated species turnover between contemporary and future species distributions in each cell (G+L/S)/S+G, where "G" was the number of species gained, "L" the number of species lost and "S" is the contemporary species richness found in the cell.

Evaluation of protected area effectiveness under climate change

The locations of the 187 protected areas (IUCN Categories I–IV) currently established in the Atlantic Forest were obtained from the United Nations Environmental Programme, World Conservation Monitoring Cemntre (UNEP-WCMC) [60]. These protected areas comprised 820 cells distributed in 11,461 Atlantic Forest cells. We overlaid protected area polygons onto our grid considering a grid cell as "protected" even if only a portion of it is protected, assuming that all species occurring in that cell could potentially benefit from the occurrence of a protected area in that cell.

Our aim was to evaluate whether current locations of protected areas are better than random allocations in protecting tiger moth diversity in the face of climate change. For this, we generated a null model that maintained size, form and orientation of protected areas but removed other intrinsic effects that likely will affect their suitability in the face of climate change (i.e. latitude, altitude). The null model allocated the protected areas randomly in the Atlantic Forest and obtained species richness in the present and the future based on the projections of species distribution models. Since there are many distinct possibilities to randomly allocate protected areas in the Atlantic Forest, this procedure was repeated 1,000 times and the average species richness obtained.

Results

For most species, TSS values were relatively high (mean TSS ± SD = 0.55±0.05 for distance methods; and 0.64±0.14, for machine-learning methods) indicating good model fit (Table S1). Different modeling methods projected similar patterns of tiger moth species richness both for current and future climates, except for ENFA and ANN (Fig. 2). For all modeling methods, ensemble projections indicated areas of high species richness in the central and southern regions of the Atlantic Forest both for now and for 2080 (Fig. 3). However, these regions should become climatically unsuitable in 2080 for many species, decreasing species' distribution areas (Fig. 3, Table S2 and S3). We also found high species temporal turnover (up to 100% for all methods) across the

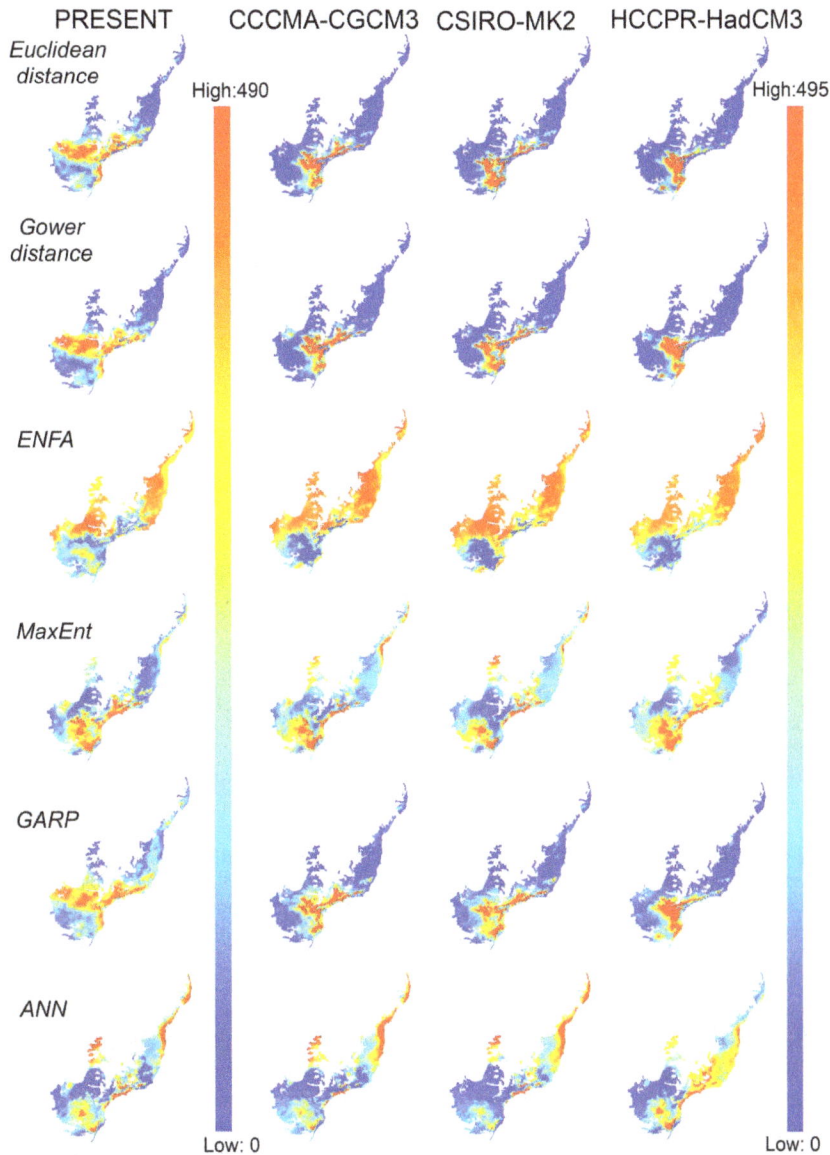

Figure 2. Species richness patterns for tiger moths in the Atlantic Forest Biodiversity Hotspot, Brazil. Tiger moth species richness patterns in the Atlantic Forest, Brazil (present and future, 2080, climate models CCCMA-CGCM3, CSIRO-MK2, and HCCPR-HadCM3) forecasted by ecological niche models generated by different distance modeling methods (Euclidian and Gower distances, Ecological Niche Factor Analysis, ENFA) and machine learning methods (Maximum Entropy, MaxEnt; Genetic algorithm for Rule set Production, GARP; Artificial Neural Networks, ANN).

projections between current and future climates (Fig. 4). This means that future scenarios showed dramatic changes not only in species richness but also in turnover of species composition (Table S4).

Central and northern sections of the Atlantic Forest should face higher temperatures, with lower seasonality, whereas the south should receive more rainfall, yet also have lower seasonality (Fig. 5). Most species had a significant range contraction (Table S3). Consensus of machine-learning methods projected up to 4.3% of species that had their ranges reduced by 100%, whereas consensus of distance methods projected 0.4% of species in the same situation (Table S3).

Future species richness should be lower than contemporary richness for most of the protected areas (Fig. 6). Differences between future and contemporary species richness were more varied for Distance than Machine Learning methods (Fig. 6A1

and 6B1). Distance models show a more optimistic result, with some protected areas holding more species (green dots in Fig. 6A1 and protected areas in the Fig. 6A2) in the future than in current time. These protected areas predicted to gain species richness are mostly located in the mountainous southeastern portion of the biome.

Machine-learning methods predicted that almost all protected areas will lose species in the future (Figs. 6B1 and 6B2). The main expectation of the null model of random location of protected areas is the loss of species in the future, both for Distance (Fig. 6A1) and Machine Learning (Fig. 6B1) methods. Among the protected areas predicted to lose species (orange and red dots and protected areas in Fig. 6), around half of them should experience losses worse than those predicted by the null model (red dots and protected areas in Fig. 6). These protected areas are concentrated in the southwestern and northern sections of the biome.

Present
Distance methods

Future

Machine Learning methods

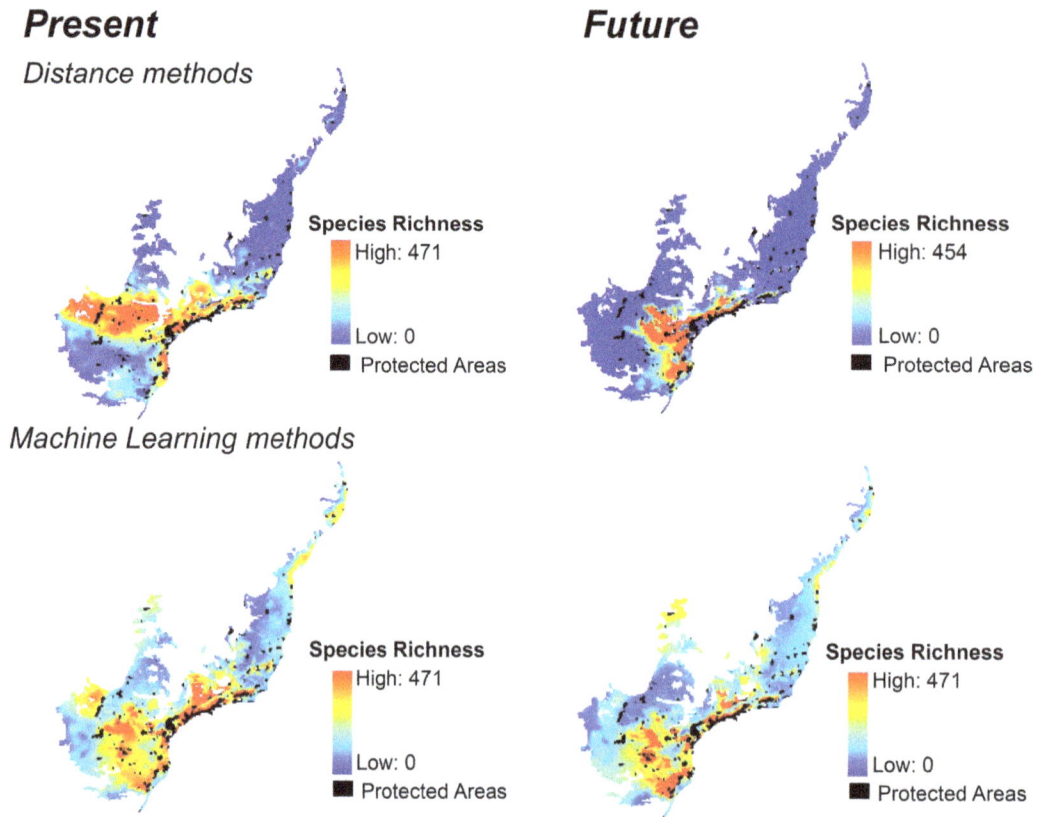

Figure 3. Consensus maps for tiger moth species richness in the Atlantic Forest Biodiversity Hotspot, Brazil. Maps of modeled tiger moth species richness based on consensus projections of 507 species predicted to occur in the Atlantic Forest Biodiversity Hotspot, Brazil, for current time (1950–2000) and 2080 (2051–2080) according to two different types of modeling methods and climate models. Models from distance and machine-learning methods were combined through an ensemble of forecasts to generate these consensus maps.

In addition to the general trend of species loss, the baseline comparison provided by the null model of random allocation of protected areas (blue dots) indicated great variation between how many species protected areas would lose (and also gain in case of distance methods). For machine-learning methods, species richness for current and future climates for the real locations of protected areas was generally lower (red dots and protected areas in Fig. 6) than those obtained from the random locations of protected areas (blue dots in Fig. 6). Nearly half of currently established protected areas should become climatically unsuitable at rates higher than expected by a random distribution of protected areas (red dots and protected areas in Fig. 6).

Protected areas predicted to gain species (particularly for distance methods) are mostly located in the cooler southern region of the Atlantic Forest (Fig. 6). Protected areas along the coast or in adjacent mountainous areas should lose species, although those in the north would do so at rates higher than those expected by our null model (protected areas indicated in red). The western region of the biome, which includes semideciduous forests, will experience a severe reduction in species richness due to climate change (Fig. 6).

Discussion

We forecasted species' range shifts and range contractions for most tiger moth species inhabiting the Atlantic Forest to evaluate the effectiveness of existing protected areas [61,62]. Our findings indicate that most protected areas should become climatically unsuitable for sustaining their current number of species under climate change. As Atlantic Forest protected areas will become less effective in safeguarding moths, it is important to anticipate how climatic changes will lead to a decreasing species representation across the entire network of protected areas.

Our results agree with Pearson and colleagues [63] which detected similar accuracy among species distribution models, although the spatial pattern in the predictions was different. Distance methods are simple methods that do not consider complex relationships between species occurrence and predictors, use presence-only data and tend to underestimate the distribution in novel conditions like those expected to occur under climate change [24,63]. Alternatively, machine-learning methods are very complex, assume different relationships and can underestimate or overestimate distributions in novel conditions [24,63]. These features help explain the differences in our results obtained by distance methods when compared with machine-learning methods, especially regarding the pattern of species richness and turnover rates.

The choice of a perfect set of modeling methods, however, is not an easy task. When predicting climate change effects on species distributions, commission errors lead to the overestimation of range expansions whereas omission errors produce overestimates of range contractions. If one is employing such models to predict regions of climatic stability to guide conservation actions, both omission and commission errors are of particular interest, as these errors are likely to produce huge bias in the results of gap-analysis as well. Distance models tend to inflate commission errors and

Distance methods

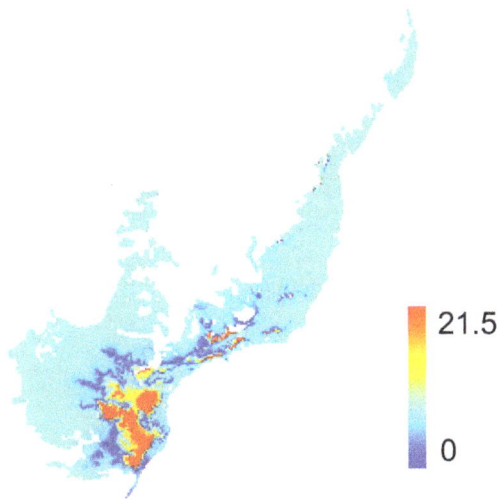

Machine Learning methods

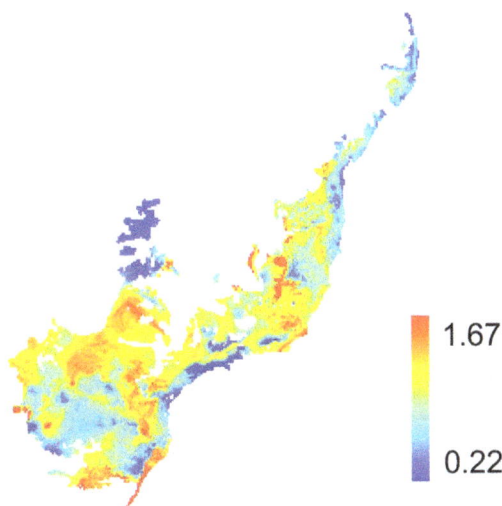

Figure 4. Species turnover pattern for tiger moths in the Atlantic Forest Biodiversity Hotspot, Brazil. Maps of modeled tiger moth species turnover based on consensus projections of 507 species predicted to occur in the Atlantic Forest Biodiversity Hotspot, Brazil, for current time (1950–2000) and 2080 (2051–2080) according to two different types of modeling methods and climate models. Models from distance and machine-learning methods were combined through an ensemble of forecasts to generate these consensus maps.

individual species distribution, which are usually larger than predicted by other models. Machine-learning models usually result in smaller distributions when compared to other models. Thus, if conservation practitioners want to reduce commission errors, machine-learning models are perhaps the best option. Although, in this case, one will probably face the inconvenience of a lack of clarity and in turn, a difficulty of explaining results to stakeholders and decision makers involved in the conservation process [27].

Our study predicts that up to 4.3% of tiger moth species should face 100% of range contraction in the Atlantic Forest's future. Among these species that would disappear from the region, three of them are endemic (*Clemensia marmorata*, *Cosmosoma cingla*,

and Arctiini NI92 morphospecies). The majority (n = 18) of species predicted to have a 100% of range contraction belong to the Arctiini tribe (in general robust, medium-sized moths and polyphagous larvae) and the Lithosiini tribe (in general slender, small-sized moths, and lichen-feeding larvae). Unfortunately, there islittle, scattered information on the natural history of and the appropriate climatic conditions for these species available (as for the vast majority of Neotropical species). It is therefore hard to make recomendations for conservation strategies for these species.

In general, the consequences of these extinctions can spread through networks of interaction causing extinction cascades and consequent disrupted ecological functions in ecosystems [19,64]. In addition to causing local and regional extinctions, climate change can alter the size and location of the range and patterns of abundance and phenology of species [38]. Many lepidopterans, including Arctiinae species [65], are agricultural pests and changes in range, abundance and phenology of these species may increase the rate of invasion and increase the damage intensity they can cause [66,67]. Furthermore, climate change can modify patterns of synchronization between larvae and host plants, among adults and flowers, and among the two life stages and their natural enemies [68]. Some tiger moth species are important pollinators of Atlantic Forest plants, including orchids [69], and the lack of synchronization between flowering and pollinating may cause more extinctions.

We also found an increase in species richness in the future in a few southern areas of the biome, due to the expansion of the range toward higher latitudes. This pattern has been observed in studies on Lepidoptera [9,70] and other groups [8]. It is known, however, that species with wide latitudinal ranges could be preadapted to cope with climate change because they already find considerable temperature variation across their habitats. However, low dispersal capability and difficulties to cross open habiats may prevent or at least slow down range shifts of Arctiinae [38]. For instance, several studies have found that Arctiinae species disperse over short distances in natural conditions [71,72]. The average distance traveled by *Dysauxes ancilla* L. moths in two consecutive days (at mark-released-recapture study), for example, was only 43 m [72]. The same author observed that moths rarely leave their breeding area and did not cross very open areas or dense forest. In fact, two other studies in Brazilian Biodiversity Hotspots, the Atlantic Forest [73] and the Cerrado [74] found that Arctiinae fauna of open vegetation (grassland) differed from the denser vegetation (forest) and a very small number of species co-occurred in these vegetation types. More importantly, these differences were evident even when the sample sites were close to each other (about 100 m). Accordingly, the low dispersal capability of the species of Arctiinae compounded with the low ability of the species to cross open areas in fragmented landscapes of the Atlantic Forest should further complicate the future of these species. This is particularly concerning in the Atlantic Forest, given that human activity has degraded more than 85% of the biome [47].

One strategy to mitigate the loss of species in the Atlantic Forest protected areas in the future would be the creation of protected areas in southern regions and at higher altitudes of the Atlantic Forest. Klorvuttimontara and colleagues [75], for example, despite having recorded a reduction of approximately 30% of butterfly richness in the protected areas of Thailand in the future (A2 scenario), observed that the effectiveness of protected areas remained almost the same in the future. The explanation for this was that most protected areas in Thailand are located at high altitudes, allowing the butterflies to migrate from lowland areas to higher altitudes as the climate becomes unsuitable in the future. Of course, critical questions should be addressed to actually imple-

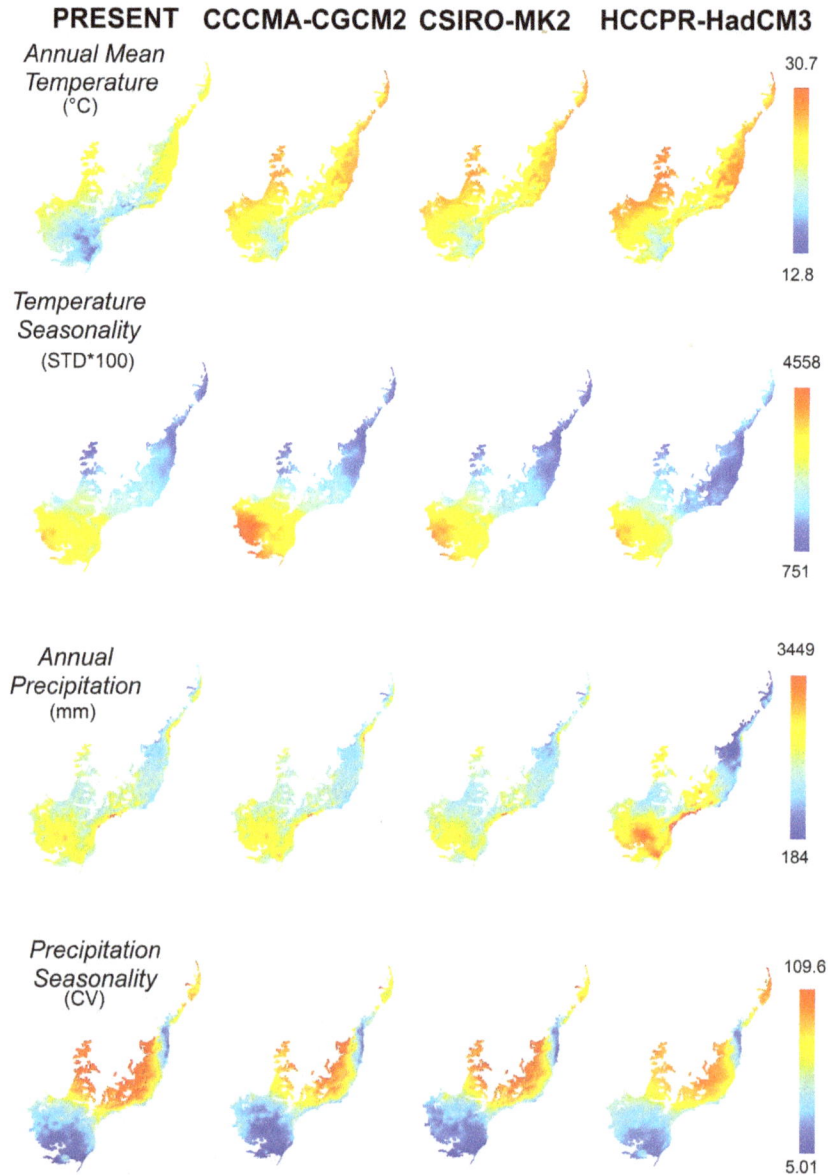

Figure 5. Expected changes for the four climatic variables used to model species' ecological niches in the Atlantic Forest Biodiversity Hotspot, Brazil. Maps show present conditions and values for future climate models CCCMA-CGCM3, CSIRO-MK2, and HCCPR-HadCM3, in 2080).

ment a decision like this, such as infostering the implementation of protected areas in high-altitude regions. Some questions would be, for example: what are the indications that these regions would protect other taxa than moths? Weel, it seems defensible to predict that they likely will be, because many species and communities are likely to migrate south and to higher elevations in response to climate change [7,8,10,17]. Should protected areas be placed to span altitudinal ranges within the protected area itself, so that species can start low and move steadily higher as the climate changes? These questions are still difficult to answer, but we hope our study is a step towards scientifically driven decisions in future protected area establishment and management.

The establishment of protected areas is still one of our best conservation actions to protect biodiversity. The Atlantic Forest has been indicated as aconservation priority by several world studies using different criteria [76,77]. Our approach is one of the

first to incorporate climate change threats on the long-term assessment of protected areas (see [5,17,21,77,78]). In fact, climatic changes are increasingly driving species out of protected areas due to species range shifts [5,79,80].

It is important to highlight some of the caveats of our study. Firstly, our database has a large number of records, which were gathered from field studies and museum records. Although the data is dense and trustworthy, there could be some bias towards easily assessed sampling regions or toward large, colorful species, as in all kinds of samplings. Further, some records were from as early as 1920 and climate has likely shifted already since then. Secondly, our models assume tiger moths are in equilibrium with the current climate and have unlimited dispersal to tackle suitable climates as they move in the geographic space. These are simple assumptions allowing us to model all species distribution at a time. Thirdly, we predicted future species distribution assuming that the

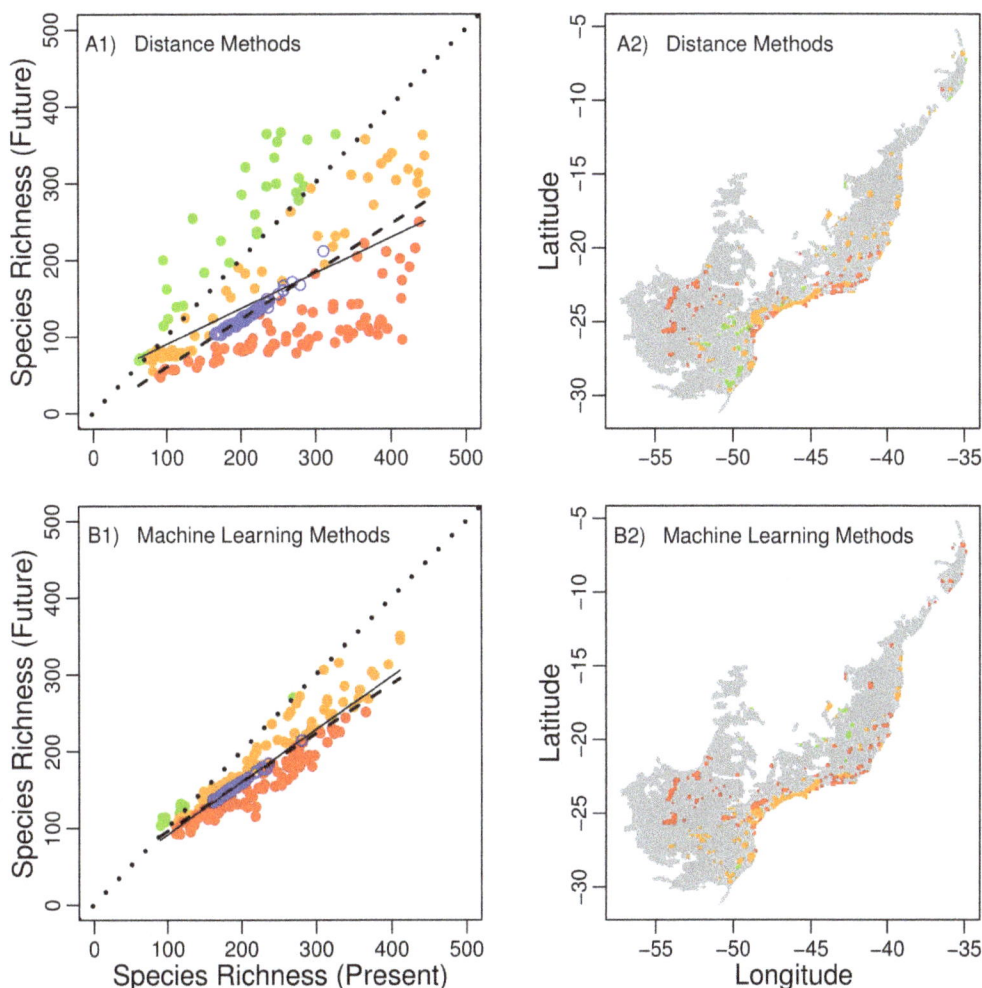

Figure 6. Relationship between present and future tiger moth species richness in the Atlantic Forest Biodiversity Hotspot, Brazil. Modeled present and future species richness of protected areas (filled circles) in the Atlantic Forest biome (A1 and B1). Open blue circles indicate expected species richness according to a null model of random location of protected areas in the biome. Dashed lines are the extrapolated regressions of the expected species richness according to the null model (A1: $y \sim -1.983 + 0.623*x$; B1: $y \sim 32.178 + 0.640*x$). Filled red circles indicate protected areas predicted to have severe species richness losses, defined as those in which future species richness will be lower than the predicted by a null model of random location of protected areas (below dashed regression line). Orange filled circles indicate protected areas predicted to have mild species richness losses, defined as those in which future species richness will be higher than the predicted by a null model of random location of protected areas. Green filled circles indicate protected areas predicted to gain species richness. Solid lines indicates the regression of modeled species richness in the future against modeled species richness in the present (A1: $y \sim 43.817 + 0.464*x$, $R^2 = 0.362$, $F_{1,185} = 105.2$, $P < 0.001$; B1: $y \sim 22.736 + 0.688*x$, $R^2 = 0.850$, $F_{1,185} = 1051.0$, $P < 0.001$). Maps of protected areas predicted to gain (green) or lose (orange, red) species in future changing climate (A2 and B2).

vegetation types in the Atlantic Forest will remain in the same regions of the current distribution. This last assumption can affect our species distribution model predictions. Species' range shifts outside of the current limits of the biome or their preferential habitat cannot be measured by our methods. Fourthly, the effectiveness of protected areas tends to be overestimated given that species were considered to be protected if any part of their range overlapped with protected areas polygons [81]. However, simple presence within a protected area is insufficient to ensure the long-term persistence of many species. Most existing protected areas in the Atlantic Forest were created without following any ecological criterion and species representation in this network is highly variable. Clearly, there is a need to update the current protected areas to improve species representation and climate change is an important factor to be considered.

As a final message, it is important to remember that species will generally alter their distributions independently. How do we prioritize which species to try to track through these changes? In this paper we recognize this conundrum and selected Tiger moths as our model group because of available data and diversity of species. It is now cear that we simply can't model or predict the response of all species on Earth to climate change. However, there are some generalities, such as the general tendency of species to mode southward or uphill in the southern hemisphere that could be helpful, if combined with models like the one we presented here, to guiding decisions about how to better allocate finite resources for biodiversity conservation.

Supporting Information

Table S1 Tiger moth tribes, species, True Skill Statistics (TSS) for each ecological niche modeling method and the ensembles among distance methods (Euclidian and Gower distance and ENFA) and machine-learning methods (MaxEnt, GARP, and Artificial Neural Networks).

Table S2 Tiger moth tribes, species, and number of grid cells predicted to be occupied during baseline climate (1950–2000) and future climate (2051–2080) under different climate models and ecological niche modeling methods.

Table S3 Tiger moth tribes, species, and percent of range contraction predicted to during baseline climate (1950–2000) and future climate (2051–2080) under different climate models and ecological niche modeling methods.

Table S4 Turnover measures for each of 11,464 grid cells overlaying the Atlantic Forest, Brazil, in this study. Turnover is shown for each modeling method (Euclidian and Gower distance, ENFA, MaxEnt, GARP, and Artificial Neural Networks), and climate model (CCCMA-CGCM, CSIRO-MK2, and HCCPR-HadCM3). Turnover found across the ensemble of distance and machine learning methods are also shown.

Author Contributions

Conceived and designed the experiments: VF RL. Analyzed the data: PL ASM. Wrote the paper: RL VF ASM PL.

References

1. Loucks C, Ricketts TH, Naidoo R, Lamoreux J, Hoekstra J (2008) Explaining the global pattern of protected area coverage: relative importance of vertebrate biodiversity, human activities and agricultural suitability. J Biogeogr 35: 1337–1348. doi:10.1111/j.1365-2699.2008.01899.x.
2. Parmesan C (2006) Ecological and Evolutionary Responses to Recent Climate Change. Ann Rev Ecol Evol Sys 37: 637–669. doi:10.1146/annurev.ecolsys.37.091305.110100.
3. Diniz-Filho JAF, Bini LM, Fernando Rangel T, Loyola RD, Hof C, et al. (2009) Partitioning and mapping uncertainties in ensembles of forecasts of species turnover under climate change. Ecography 32: 897–906. doi:10.1111/j.1600-0587.2009.06196.x.
4. Scott JM, Davis FW, McGhie RG, Wright RG, Groves C, et al. (2001) Nature reserves: do they capture the full range of america's biological diversity? Ecol Appl 11: 999–1007.
5. Araujo MB, Cabeza M, Thuiller W, Hannah L, Williams PH (2004) Would climate change drive species out of reserves? An assessment of existing reserve-selection methods. Global Change Biol 10: 1618–1626. doi:10.1111/j.1365-2486.2004.00828.x.
6. Hannah L (2010) A global conservation system for climate-change adaptation. Conserv Biol 24: 70–77. doi:10.1111/j.1523-1739.2009.01405.x.
7. Colwell RK, Brehm G, Cardelús CL, Gilman AC, Longino JT (2008) Global warming, elevational range shifts, and lowland biotic attrition in the wet tropics. Science 322: 258–261.
8. Chen I, Hill JK, Ohlemüller R, Roy DB, Thomas CD (2011) Rapid range shifts of species associated with high levels of climate warming. Science 333: 1024–1026. doi:10.1126/science.1206432.
9. Parmesan C, Ryrholm N, Stefanescu C, Hill JK, Thomas CD, et al. (1999) Poleward shifts in geographical ranges of butterfly species associated with regional warming. Nature 399: 579–583. doi:10.1038/21181.
10. Lemes P, Loyola RD (2013) Accommodating Species Climate-Forced Dispersal and Uncertainties in Spatial Conservation Planning. PLoS ONE 8: e54323. doi:10.1371/journal.pone.0054323.
11. Faleiro F V., Machado RB, Loyola RD (2013) Defining spatial conservation priorities in the face of land-use and climate change. Biol Conserv 158: 248–257. doi:10.1016/j.biocon.2012.09.020.
12. Loyola RD, Lemes P, Nabout JC, Trindade-Filho J, Sagnori MD, et al. (2013) A straightforward conceptual approach for evaluating spatial conservation priorities under climate change. Biodivers Conserv 22: 483–485. doi:10.1007/s10531-012-0424-x.
13. Dawson TP, Jackson ST, House JI, Prentice IC, Mace GM (2011) Beyond predictions: biodiversity conservation in a changing climate. Science 332: 53–58. doi:10.1126/science.1200303.
14. Williams P, Hannah L, Andelman S, Midgley G, Araújo M, et al. (2005) Planning for Climate Change: Identifying Minimum-Dispersal Corridors for the Cape Proteaceae. Conserv Biol 19: 1063–1074. doi:10.1111/j.1523-1739.2005.00080.x.
15. Mawdsley JR, O'Malley R, Ojima DS (2009) A review of climate-change adaptation strategies for wildlife management and biodiversity conservation. Conserv Biol 23: 1080–1089. doi:10.1111/j.1523-1739.2009.01264.x.
16. Mawdsley J (2011) Design of conservation strategies for climate adaptation. Wiley Interdisciplinary Reviews: Climate Change 2: 498–515. doi:10.1002/wcc.127.
17. Nori J, Urbina-Cardona JN, Loyola RD, Lescano JN, Leynaud GC (2011) Climate change and American Bullfrog invasion: what could we expect in South America? PloS ONE 6: e25718. doi:10.1371/journal.pone.0025718.
18. Loyola RD, Lemes P, Faleiro FV, Trindade-Filho J, Machado RB (2012) Severe loss of suitable climatic conditions for marsupial species in Brazil: challenges and opportunities for conservation. PLoS ONE 7: e46257. doi:10.1371/journal.pone.0046257.
19. Tylianakis JM, Didham RK, Bascompte J, Wardle DA (2008) Global change and species interactions in terrestrial ecosystems. Ecol Lett 11: 1351–1363. doi:10.1111/j.1461-0248.2008.01250.x.
20. Terborgh J, Lopez L, Nuñez P, Rao M, Shahabuddin G, et al. (2001) Ecological meltdown in predator-free forest fragments. Science 294: 1923–1926. doi:10.1126/science.1064397.
21. Hannah L, Midgley G, Andelman S, Araújo M, Hughes G, et al. (2007) Protected area needs in a changing climate. Frontiers Ecol Env 5: 131–138.
22. Peterson AT, Soberón J, Pearson RG, Anderson RP, Martínez-Meyer E, et al. (2011) Ecological Niches and Geographic Distributions. Levin SA, Horn HS, editors Princeton University Press. p.
23. Garcia RA, Burgess ND, Cabeza M, Rahbek C, Araújo MB (2012) Exploring consensus in 21st century projections of climatically suitable areas for African vertebrates. Global Change Biol 18: 1253–1269. doi:10.1111/j.1365-2486.2011.02605.x.
24. Franklin J (2009) Mapping species distributions: spatial inference and prediction. Cambridge: Cambridge University Press.
25. Araújo MB, Peterson AT (2012) Uses and misuses of bioclimatic envelope modeling. Ecology 93: 1527–1539.
26. Peterson AT, Soberón J (2012) Species Distribution Modeling and Ecological Niche Modeling: Getting the Concepts Right. Nat Conservação 10: 102–107. doi:10.4322/natcon.2012.019.
27. Rangel TF, Loyola RD (2012) Labeling Ecological Niche Models. Nat Conservação 10: 119–126. doi:10.4322/natcon.2012.030.
28. Araújo MB, New M (2007) Ensemble forecasting of species distributions. Trends Ecol Evol 22: 42–47.
29. Moilanen A, Possingham HP, Wilson KA (2009) Spatial conservation prioritization-past, present and future. In: Moilanen A, Wilson KA, Possingham HP, editors. Spatial Conservation Prioritization: Quantitative Methods and Computational Tools. Oxford, UK: Oxford University Press. pp. 260–268.
30. Grimaldi D, Engel MS (2005) Evolution of the Insects. Grimaldi D, Engel MS, editors Cambridge University Press.
31. Wasserthal LT (1997) The pollinators of the Malagasy star orchids Angraecum sesquipedale, A. sororium and A. compactum and the evolution of extremely long spurs by pollinator shift. Acta Botanica 110: 343–359.
32. Pellmyr O (2003) Yuccas, yucca moths, and coevolution: A review. Ann Mo Bot Gard 90: 35–55. doi:10.2307/3298524.
33. Duarte M, Marconato G, Specht A, Casagrande M (2012) Lepidoptera. In: Rafael J, Melo G, Carvalho C, Casari S, Constantino R, editors. Insetos do Brasil Diversidade e Taxonomia. Ribeirão Preto: Holos Editora. pp. 625–682.
34. Marquis RJ (1984) Leaf herbivores decrease fitness of a tropical plant. Science 226: 537–539.
35. Bouchard M, Kneeshaw D, Bergeron Y (2006) Forest dynamics after successive spruce budworm outbreaks in mixedwood forests. Ecology 87: 2319–2329.
36. Chapman SK, Hart SC, Cobb NS, Whitham TG, Koch GW (2003) Insect Herbivory Increases Litter Quality and Decomposition: an Extension of the Acceleration Hypothesis. Ecology 84: 2867–2876. doi:10.1890/02-0046.
37. Speight M, Hunter M, Watt A (2008) Ecology of Insects: Concepts and Applications. Oxford, UK: Wiley Blackwell.
38. Kocsis M, Hufnagel L (2011) Impacts of climate change on lepidoptera species and communities. Applied Ecology and Environmental Reserach 9: 43–72.
39. Conrad KF, Woiwod IP, Perry JN (2002) Long-term decline in abundance and distribution of the garden tiger moth (Arctia caja) in Great Britain. Biol Conserv 106: 329–337. doi:10.1016/S0006-3207(01)00258-0.
40. Zahiri R, Holloway JD, Kitching IJ, Lafontaine JD, Mutanen M, et al. (2012) Molecular phylogenetics of Erebidae (Lepidoptera, Noctuoidea). Syst Entomol 37: 102–124. doi:10.1111/j.1365-3113.2011.00607.x.
41. Heppner JB (1991) Faunal regions and the diversity of Lepidoptera. Tropical lepidoptera 2.

42. Ferro V, Diniz I (2010) Riqueza e composição das mariposas Arctiidae (Lepidoptera) no Cerrado. In: Diniz I, Marinho-Filho J, Machado R, Cavalcanti R, editors. Cerrado: conhecimento científico quantitativo como subsídio para ações de conservação. Brasília: Theraurus Editora. pp. 255–313.

43. Ferro VG, Melo AS (2010) Diversity of tiger moths in a Neotropical hotspot: determinants of species composition and identification of biogeographic units. J Insect Conserv 15: 643–651. doi:10.1007/s10841-010-9363-6.

44. Wagner DL, Conner WE (2008) The immature stages: structure, function, behavior, and ecology. In: Conner WE, editor. Tiger moths and wolly bears: Behavior, Ecology, and Evolution of the Arctiidae. Oxford: Oxford University Press. pp. 31–55.

45. Weller SJ, Jacobson NL, Conner WE (1999) The evolution of chemical defences and mating systems in tiger moths (Lepidoptera: Arctiidae). Biol J Linn Soc 68: 557–578. doi:10.1111/j.1095-8312.1999.tb01188.x.

46. Myers N, Mittermeier Ra, Mittermeier CG, da Fonseca GA, Kent J (2000) Biodiversity hotspots for conservation priorities. Nature 403: 853–858. doi:10.1038/35002501.

47. Ribeiro MC, Metzger JP, Martensen AC, Ponzoni FJ, Hirota MM (2009) The Brazilian Atlantic Forest: How much is left, and how is the remaining forest distributed? Implications for conservation. Biol Conserv 142: 1141–1153. doi:10.1016/j.biocon.2009.02.021.

48. Hijmans RJ, Cameron SE, Parra JL, Jones PG, Jarvis A (2005) Very high resolution interpolated climate surfaces for global land areas. International Journal of Climatology 25: 1965–1978. doi:10.1002/joc.1276.

49. Carpenter G, Gillison AN, Winter J (1993) DOMAIN: a flexible modelling procedure for mapping potential distributions of plants and animals. Biodivers Conserv 2: 667–680. doi:10.1007/BF00051966.

50. Hirzel AH, Hausser J, Chessel D, Perrin N (2002) Ecological-niche factor analysis: hw to compute habitat-suitability maps without absence data? Ecology 83: 2027–2036.

51. Phillips S, Anderson R, Schapire R (2006) Maximum entropy modeling of species geographic distributions. Ecol Model 190: 231–259. doi:10.1016/j.ecolmodel.2005.03.026.

52. Stockwell DRB, Noble IR (1992) Induction of sets of rules from animal distribution data: a robust and informative method of data analysis. Mathematics and Computers in Simulation 33: 385–390. doi:10.1016/0378-4754(92)90126-2.

53. Manel S, Dias JM, Buckton ST, Ormerod SJ (1999) Alternative methods for predicting species distribution: an illustration with Himalayan river birds. J Appl Ecol 36: 734–747. doi:10.1046/j.1365-2664.1999.00440.x.

54. Pearson RG (2007) Species' Distribution Modeling for Conservation Educators and Practitioners: 1–50.

55. Allouche O, Tsoar A, Kadmon R (2006) Assessing the accuracy of species distribution models: prevalence, kappa and the true skill statistic (TSS). J Appl Ecol 43: 1223–1232. doi:10.1111/j.1365-2664.2006.01214.x.

56. Lobo JM, Jiménez-Valverde A, Real R (2008) AUC: a misleading measure of the performance of predictive distribution models. Glob Ecol Biogeogr 17: 145–151. doi:10.1111/j.1466-8238.2007.00358.x.

57. Marmion M, Parviainen M, Luoto M, Heikkinen RK, Thuiller W (2009) Evaluation of consensus methods in predictive species distribution modelling. Divers Distrib 15: 59–69. doi:10.1111/j.1472-4642.2008.00491.x.

58. Diniz-Filho JAF, Nabout JC, Bini LM, Loyola RD, Rangel TF, et al. (2010) Ensemble forecasting shifts in climatically suitable areas for *Tropidacris cristata* (Orthoptera: Acridoidea: Romaleidae). Insect Conserv Div 3: 213–221. doi:10.1111/j.1752-4598.2010.00090.x.

59. Faleiro FV., Machado RB, Loyola RD (2013) Defining spatial conservation priorities in the face of land-use and climate change. Biol Conserv 158: 248–257. doi:10.1016/j.biocon.2012.09.020.

60. IUCN, UNEP (2010) World Database on Protected Areas (WDPA). Annual Release: 1–9.

61. Griffith B, Scott JM, Adamcik R, Ashe D, Czech B, et al. (2009) Climate change adaptation for the US National Wildlife Refuge System. Environ Manag 44: 1043–1052.

62. Araújo MB, Alagador D, Cabeza M, Nogués-Bravo D, Thuiller W (2011) Climate change threatens European conservation areas. Ecol Lett 14: 484–492.

63. Pearson RG, Thuiller W, Araújo MB, Martinez-Meyer E, Brotons L, et al. (2006) Model-based uncertainty in species range prediction. J Biogeogr 33: 1704–1711. doi:10.1111/j.1365-2699.2006.01460.x.

64. Cardinale BJ, Srivastava DS, Duffy JE, Wright JP, Downing AL, et al. (2006) Effects of biodiversity on the functioning of trophic groups and ecosystems. Nature 443: 989–992. doi:10.1038/nature05202.

65. Zanuncio JC, Mezzomo JA, Guedes RNC, Oliveira AC (1998) Influence of strips of native vegetation on Lepidoptera associated with Eucalyptus cloeziana in Brazil. For Ecol Manag 108: 85–90. doi:10.1016/S0378-1127(98)00215-1.

66. Hódar J, Zamora R (2004) Herbivory and climatic warming: a Mediterranean outbreaking caterpillar attacks a relict, boreal pine species. Biodivers Conserv 13: 493–500.

67. Gutierrez AP, D'Oultremont T, Ellis CK, Ponti L (2006) Climatic limits of pink bollworm in Arizona and California: effects of climate warming. Acta Oecol 30: 353–364. doi:10.1016/j.actao.2006.06.003.

68. Logan JA, Régnière J, Powell JA (2003) Assessing the impacts of global warming on forest pest dynamics. Frontiers Ecol Env 1: 130–137.

69. Singer RB, Breier TB, Flach A, Farias-Singer R (2007) The Pollination Mechanism of *Habenaria pleiophylla* Hoehne & Schlechter (Orchidaceae: Orchidinae). Functional Ecosystems and Communities 1: 10–14.

70. Diniz Filho JAF, Ferro VG, Santos T, Nabout JC, Dobrovolski R, et al. (2010) The three phases of the ensemble forecasting of niche models: geographic range and shifts in climatically suitable areas of Utetheisa ornatrix (Lepidoptera, Arctiidae). Rev Bras Entomol 54: 339–349. doi:10.1590/S0085-56262010000300001.

71. Yamanaka T, Tatsuki S., Shimada M (2001) Flight characteristics and dispersal patterns of fall webworm (Lepidoptera: Arctiidae) males. Env Entomol 30: 1150–1157.

72. Betzholtz PE (2002) Population structure and movement patterns within an isolated and endangered population of the moth *Dysauxes ancilla* L. (Lepidoptera, Ctenuchidae): implications for conservation. J Insect Conserv 6: 57–66.

73. Ferro VG, Romanowski HP (2012) Diversity and composition of tiger moths (Lepidoptera: Arctiidae) in an area of Atlantic Forest in southern Brazil: is the fauna more diverse in the grassland or in the forest? ZOOLOGIA 29: 7–18. DOI: 10.1590/S1984-46702012000100002

74. Ferro VG, Diniz IR (2007) Composição de espécies de Arctiidae (Insecta, Lepidoptera) em áreas de Cerrado. Revta Bras Zool 24: 635–646.

75. Klorvuttimontara S, McClean CJ, Hill JK (2011) Evaluating the effectiveness of Protected Areas for conserving tropical forest butterflies of Thailand. Biol Conserv 144: 2534–2540. doi:10.1016/j.biocon.2011.07.012.

76. Brooks T, Mittermeier R, Fonseca Gda (2006) Global Biodiversity Conservation Priorities. Science 313: 58–61.

77. Loyola RD, Kubota U, da Fonseca GAB, Lewinsohn TM (2009) Key Neotropical ecoregions for conservation of terrestrial vertebrates. Biodivers Conserv 18: 2017–2031. doi:10.1007/s10531-008-9570-6.

78. Loyola RD, Nabout JC, Trindade-filho J, Lemes P, Urbina-Cardona JN, et al. (2012) Climate change might drive species into reserves: a case study of the American bullfrog in the Atlantic Forest Biodiversity Hotspot. Alytes 29: 61–74.

79. Monzón J, Moyer-Horner L, Palamar MB (2011) Climate Change and Species Range Dynamics in Protected Areas. BioScience 61: 752–761. doi:10.1525/bio.2011.61.10.5.

80. Wiens JA, Seavy NE, Jongsomjit D (2011) Protected areas in climate space: What will the future bring? Biol Conserv 144: 2119–2125. doi:10.1016/j.biocon.2011.05.002.

81. Rodrigues A, Andelman S, Bakarr M, Boitani L (2004) Effectiveness of the global protected area network in representing species diversity. Nature 428. doi:10.1038/nature02459.1.

6

Mean Annual Precipitation Explains Spatiotemporal Patterns of Cenozoic Mammal Beta Diversity and Latitudinal Diversity Gradients in North America

Danielle Fraser[1,2]*, **Christopher Hassall**[1,3], **Root Gorelick**[1,4,5], **Natalia Rybczynski**[1,2]

1 Department of Biology, Carleton University, Ottawa, Ontario, Canada, 2 Palaeobiology, Canadian Museum of Nature, Ottawa, Ontario, Canada, 3 School of Biology, University of Leeds, Leeds, United Kingdom, 4 Department of Mathematics and Statistics, Carleton University, Ottawa, Ontario, Canada, 5 Institute of Interdisciplinary Studies, Carleton University, Ottawa, Ontario Canada

Abstract

Spatial diversity patterns are thought to be driven by climate-mediated processes. However, temporal patterns of community composition remain poorly studied. We provide two complementary analyses of North American mammal diversity, using (i) a paleontological dataset (2077 localities with 2493 taxon occurrences) spanning 21 discrete subdivisions of the Cenozoic based on North American Land Mammal Ages (36 Ma – present), and (ii) climate space model predictions for 744 extant mammals under eight scenarios of future climate change. Spatial variation in fossil mammal community structure (β diversity) is highest at intermediate values of continental mean annual precipitation (MAP) estimated from paleosols (~450 mm/year) and declines under both wetter and drier conditions, reflecting diversity patterns of modern mammals. Latitudinal gradients in community change (latitudinal turnover gradients, aka LTGs) increase in strength through the Cenozoic, but also show a cyclical pattern that is significantly explained by MAP. In general, LTGs are weakest when continental MAP is highest, similar to modern tropical ecosystems in which latitudinal diversity gradients are weak or undetectable. Projections under modeled climate change show no substantial change in β diversity or LTG strength for North American mammals. Our results suggest that similar climate-mediated mechanisms might drive spatial and temporal patterns of community composition in both fossil and extant mammals. We also provide empirical evidence that the ecological processes on which climate space models are based are insufficient for accurately forecasting long-term mammalian response to anthropogenic climate change and inclusion of historical parameters may be essential.

Editor: Alistair Robert Evans, Monash University, Australia

Funding: D. Fraser was supported by a Natural Science and Engineering Research Council of Canada (NSERC) postgraduate scholarship, a Fulbright Traditional Student Award, a Mary Dawson Pre-Doctoral Fellowship grant, an Ontario Graduate Scholarship (OGS), and a Koningstein Scholarship for Excellence in Science and Engineering. C. Hassall was supported by an Ontario Ministry of Research and Innovation Postdoctoral Fellowship. R. Gorelick was supported by an NSERC Discovery Grant (#341399). N. Rybczynski was supported by an NSERC Discovery Grant (#312193). The funders had no role in study design, data collection and analysis, decision to publish, or preparation of the manuscript.

Competing Interests: The authors have declared that no competing interests exist.

* Email: danielle_fraser@carleton.ca

Introduction

Terrestrial species from all major taxonomic groups show dramatic changes in richness and diversity across the landscape [1]. One of the fundamental goals in ecology is therefore to ascertain why there are more species in some places than in others. A satisfactory answer would identify and disentangle the drivers of biodiversity at all spatial scales, from the microhabitat to the globe, as well as explain changes through time. Attempts to provide such an answer have produced many studies of species richness patterns and community composition in extant organisms [1–8]. Prime examples are the numerous studies of latitudinal richness gradients (LRGs), which have been observed in many terrestrial groups including angiosperms, birds, mammals, insects and other invertebrates. The best supported hypotheses show that richness declines toward the poles in correlation with reductions in precipitation, temperature, and net primary productivity [9]. Correlation of global climate with animal richness over the past 65 Ma, specifically a decline in richness as climates cooled, similarly supports a link between diversity and climate [10–12]. However, of the spatial and temporal dimensions of diversity, spatial patterns of community differences ("β diversity") are infrequently studied despite considerable variation on both local and regional scales [2,13,14] and their influential role in the structuring of continental-scale richness patterns including LRGs [3,4].

β diversity has been defined most broadly as the differentiation in community composition (i.e. the species that make up the community) among regions or along environmental gradients [15]. Similar to LRGs, β diversity generally declines from the tropics to the poles in correlation with climate [2]. However, temporal changes in β diversity remain poorly studied despite their potential power for illuminating the drivers of past and present richness patterns and importance in modern conservation [16–18]. This study therefore tests the hypothesis that climatic influences on mammalian β diversity apply equally to temporal patterns, i.e. that the underlying ecological processes are "ergodic" (dynamic processes that are the same in both time and space).

The mid to late Cenozoic (36 Ma to present) has been a time of dramatic mammalian diversity change, shaped in part by the transition from the productive ice-free ecosystems of the early to mid Cenozoic to the more temperate glaciated ecosystems of the late Cenozoic. Under these changing climatic conditions, mammalian communities show dramatic reductions in richness, changes in community composition, and morphology [10,19–24]. The most dramatic changes occurred at high latitudes, where ecosystems transitioned from *Metasequoia* forests during the early to mid Cenozoic [25,26] to boreal-type forests during the later Cenozoic and to modern tundra [27]. Associated with Cenozoic climate change, were changes in latitudinal climate gradients; overall, the intensity of latitudinal climate gradients increased toward the present, reflecting disproportionate polar cooling due to the formation of permanent Arctic glaciation [28,29]. We therefore predict that latitudinal diversity gradients increased in strength under cooler, less productive environmental conditions just as modern LRGs are steeper in temperate than in tropical regions. Further, we predict that β diversity declined under cooler, less productive environmental conditions just as modern β diversity declines toward the poles [2,7].

Quaternary (2.6 Ma to present) climates have been cool relative to the majority of the late Cenozoic. Recently, however, high latitudes have experienced disproportionate increases in annual temperature (up to 2°C to date), increases in plant primary productivity, and loss of large areas of perennial ice under anthropogenic global warming [30]. Flora and fauna have responded through shifts in phenology [31], *in situ* evolution [32], and, in some cases, extinction [33]. However, perhaps the most often recorded response is the climatically-correlated pattern of extirpations and colonization that manifest as shifts in the location of a species' geographic range. Distributional studies over ecological timescales (<100 yrs) have recorded dramatic poleward range shifts and expansions for a wide range of terrestrial taxa in response to northern warming [34,35]. Projections (i.e. Special Report on Emissions Scenarios) for the next 100 years predict levels of global warming similar to the middle Miocene (+ 6°C) − a time of reduced or absent perennial Arctic glaciation [36,37] − or warmer (+11°C for the most extreme case; Table S1). We therefore expect continued range expansion, extinction, evolution, and community level changes among North American animals and plants.

A common approach to predicting the long-term outcomes of climate change for terrestrial organisms is climate space modeling (CSM). CSMs use distributional information and climate data to project species ranges into the future, usually under the assumption of no evolution and without adjustment for dispersal differences among species [38–40]. Rapid evolutionary changes on very short timescales and high degrees of variation in dispersal ability under climate change have been observed across a wide range of organisms [34,39,41], therefore CSMs are unlikely to generate accurate forecasts of climate change response. The fossil record, which encompasses many disparate environments and climates, might serve as record of a natural experiment by which ecological hypotheses can be tested in the temporal dimension. Fossil collections are a rich historical record of response to various climatic events that can be incorporated into predictive models, and mammals, in particular, are an excellent group for testing the generality of ecological hypotheses because they have an extensive Cenozoic fossil record. However, studies of extinct organisms have focused largely on richness [12,22,23,42,43] or morphology [44],

with limited focus on community composition [20,22]. Because changes in biological communities are not always associated with changes in richness, spatiotemporal patterns of community composition may be better indicators of climate change response [13,18].

We propose that integrating the study of fossil, modern, and projected spatiotemporal patterns of community composition i) allows for the testing of ecological principles in the temporal dimension, ii) provides the most complete picture of diversity responses to climate change, and iii) enables evaluation of the performance of commonly employed CSMs. Our approach of combining the study of fossil, modern, and projected diversity patterns provides novel insights into the ecological and evolutionary processes that drive continental patterns of biodiversity in space and time.

Methods

Data collection and preparation

We downloaded occurrences for modern North American mammals from NatureServe Canada. The extant mammal dataset included 744 species after the exclusion of a small number of unreadable or corrupted files [45]. We restricted our study of fossil mammals to the late Eocene through Pleistocene, thus avoiding the confounding effects of the early Paleogene mammal radiation. We partitioned the fossil mammal occurrence data by North American Land Mammal Age (NALMA) subdivisions because they delineate relatively temporally stable community assemblages and allowed us to obtain a nearly continuous sequence of mammal community change without large intervening gaps. Using NALMA subdivisions leads to time averaging of mammal communities and to differences in sampling (i.e. intensity, geographic coverage etc.) among time periods. However, we use a statistical approach to reduce these biases, described below. We based the dates for all NALMA subdivisions on Woodburne (2004). Further, we combined data for the entire Clarendonian and excluded for the Whitneyan, late Late Hemphillian, and early Chadronian due to poor sampling (Table 1).

We downloaded fossil mammal occurrence data for the Eocene, Oligocene, Pliocene, and Pleistocene from the the Paleobiology Database using the Fossilworks Gateway (fossilworks.org) in July and August, 2012, using the group name 'mammalia' and the following parameters: time intervals = Cenozoic, region = North America, paleoenvironment = terrestrial (primary contributor: John Alroy; literature sources summarized in Appendix S1). We downloaded Miocene mammal occurrence data from the Miocene Mammal Mapping Project in March 2011 [46] using the NALMA subdivision as our search criterion. For all analyses, with the exception of the Miocene, we used paleolatitudes and paleolongitudes. We chose to use MIOMAP for the Miocene data because it is the most complete Miocene dataset. However, MIOMAP does not provide paleo-coordinates. Fortunately, there are only small differences between modern and Miocene latitudes for the downloaded localities. We removed all taxa with equivocal species identifications (e.g. *Equus* sp.) unless they were the only occurrence for a genus. We assumed all occurrences of open nomenclature (e.g. *Equus* cf. *simplicidens*) were correct identifications.

We did not use latitudinal grids for fossil or extant mammals as in previous studies of latitudinal richness gradients [1,47] because our study is focused on community composition. We therefore do not need to clump localities by spatial proximity to employ rarefaction methods. In addition, the uneven spatial distribution of fossil localities makes the use of a grid method impractical. Instead,

Table 1. Summary of sampled North American Land Mammal Age (NALMA) subdivisions.

Epoch	NALMA subdivision	Age Range (Ma)	Midpoint Age (M)	Number of species	Number of fossil localities	Area (km^2)
Pleistocene	Rancholabrean	0.25–0.011	0.1305	222	180	176615.9
Pliocene	Irvingtonian II	0.85–0.25	0.55	189	94	144745.5
Pliocene	Irvingtonian I	1.72–0.85	1.285	102	37	60361.4
Pliocene	Blancan V	2.5–1.72	2.11	165	130	125042.6
Pliocene	Blancan III	4.1–2.5	3.3	183	163	122839.5
Pliocene	Blancan I	4.9–4.1	4.5	85	66	140433.4
Miocene	Early late Hemphillian	6.7–5.9	6.3	68	46	20108.2
Miocene	Late early Hemphillian	7.5–6.7	7.1	63	55	29446.7
Miocene	Early early Hemphillian	9–7.5	8.25	65	47	31455.8
Miocene	Clarendonian	12.5–9	10.75	104	90	36139.8
Miocene	Late Barstovian	14.8–12.5	13.6	195	194	33789.1
Miocene	Early Barstovian	15.9–14.8	15.5	150	168	51753.3
Miocene	Late Hemingfordian	17.5–15.9	16.7	100	83	25478.4
Miocene	Early Hemingfordian	18.8–17.5	18.15	107	105	45531.3
Miocene	Late late Arikareean	19.5–18.8	19.15	108	123	38307.2
Oligocene/Miocene	Early late Arikareean	23.8–19.5	21.65	71	67	37892.2
Oligocene	Late early Arikareean	27.9–23.8	25.85	95	65	20927.8
Oligocene	Early early Arikareean	30–27.9	28.95	116	124	15382.3
Oligocene	Late Orellan	33.1–32	32.55	38	36	17725.7
Oligocene	Early Orellan	33.7–33.1	33.4	88	130	5579.8
Eocene	Middle Chadronian	35.7–34.7	35.3	88	37	10349.7

we created taxon-by-locality occurrence matrices for extant and fossil mammals at the species taxonomic level excluding *Homo sapiens* [20,22]. In all cases, taxa and localities with fewer than two occurrences were removed from the dataset. Final numbers of localities and species are summarized in Table 1.

To make direct comparisons with modern mammals, we created occurrence matrices for extant mammals by pseudo fossil localities, which were generated using an iterative procedure in R with the maptools, sp, gpclib, ggplot2, rgeos, and MASS packages [48–54] (contact corresponding author for R code). To generate pseudo fossil localities and to ensure that we created pseudo fossil localities with the same spatial distributions as the fossil localities, we fit frequency distributions (normal, gamma, or β) to fossil localities for each NALMA subdivision (Fig. S1). We then generated point samples based on the frequency distributions and the number of fossil localities from which we created occurrence matrices (taxon-by-pseudo locality), repeating the procedure 100 times for each NALMA sub-age for a total of 2100 occurrence matrices. Fossil localities do not record the entire community and so show reduced richness compared to the actual communities (however, note that time averaging also increases richness at fossil localities). Further, most fossil localities, unless intensively screen washed, are biased against small species. Therefore, we also intentionally tested for the effects of sampling bias by removing 25%, 50%, and 75% of species from the extant mammal occurrence matrices for a total of 6300 occurrence matrices. Further, we tested for the effects of body mass bias by 25%, 50%, and 75% of species smaller than 5 kg for a total of 6300 occurrence matrices.

Climate space models

To create climate space models, we sampled the ranges of extant North and South American mammals at a series of 5066 points corresponding to a 1° grid (which we only used to project mammal occurrences under climate change models, but not to calculate biodiversity). Due to the focus on North America, we omitted any species with southern hemisphere ranges that did not cross the equator (n = 602; Table S2). We also excluded rare species (present in <20 cells) for which accurate species distribution models could not be generated (n = 361), leaving 706 species for the climate change projections. We extracted mean annual and winter (December, January, February) temperature and mean annual precipitation data from Climate Wizard (www.climatewizard.org) for the period of 1951–2006 and the following SRES scenarios and time periods: B1 2050s, A1b 2050s, A1b 2080s, A2 2050s, and A2 2080s [55] (Table S1). Each of these projections is based on an ensemble of 16 global circulation models [56]. However, to ensure that we sampled a range of potential warming, we also extracted the ensemble lowest B1 2050s projection (hereafter "B1 2050s low") and the ensemble highest A2 2080s projection (hereafter "A2 2080s high"). This gave a range of warming in North America from 1.49°C (B1 2050s low) to 6.78°C (A2 2080s high, see Table S1 for the full range).

We modeled species' ranges with the BIOMOD package in R using generalized linear models, generalized boosted models, classification tree analysis, artificial neural networks, surface range envelopes, flexible discriminant analysis, multiple adaptive regression splines, and random forests [57] (contact corresponding author for R code). We then used these models to make consensus forecasts for each of the projections described above, as well as current climate to evaluate the performance of the models. We tested model performance using area under the receiver operating curve (AUC), true skill statistic (TSS), and proportion correct classification (PCC, Fig. S2). Species and generic presences were determined across the 1° latitude-longitude grid to give presence or absence in each location at each time and SRES scenario.

Using the projections described above, we created pseudo localities, as before. From this, we created occurrence matrices as described above. We repeated this process 100 times for each projection for a total of 16,800 occurrence matrices.

Latitudinal turnover gradients (LTGs) and β diversity

We calculated β diversity as the change in mammalian communities across the North American landscape using multivariate dispersion and the Jaccard index for each NALMA subage, for modern mammals, and for the climate projections [58]. We calculated Euclidean distances from the centroid for localities using the R package vegan [59]. Larger distances from the centroid indicate greater spatial community turnover and thus higher β diversity. We did not regress the Jaccard index values against distance, as has been used for modern species [2] because we have found such an approach to be highly influenced by species-area relationships.

To estimate ancient, modern, and projected LTG strength for North American mammals, we calculated the amount of community change with latitude using detrended correspondence analysis (DCA; an ordination technique) in the vegan R package [59]. We used explained variance (R^2; how much of the variation in community change is explained by latitude) as a measure of LTG strength [13]. High values of explained variance indicate strong LTGs [60]. We did not compute latitudinal richness gradients because sampling bias (e.g. loss of taxa, body mass bias) is too great (Fraser, D. unpub.).

Sampling bias control

Although we have chosen methods that minimize the effects of sampling bias, we still used multiple methods to control for the non-independence of β diversity from the number of localities, the geographic area sampled, and the number of sampled taxa. We used three approaches. Firstly, we used a re-sampling approach wherein we sub-sampled (without replacement) each NALMA 100× using a standardized number of localities (thirty) and limited to localities occurring between 30° and 50° North latitude. We also re-sampled the extant mammal ranges under various conditions of bias (taxonomic bias through the removal of 25%, 50%, 75% of taxa and body mass bias where we removed 25%, 50%, and 75% of species with a body mass lower than 5 kg) as above to test for direct causality of sampling bias. We also used a method of detrending whereby we regressed LTG strength and β diversity against statistically significant sampling bias metrics and further analyzed the residuals from the model. Finally, we used multivariate linear models to simultaneously account for the model variance explained by sampling and biological phenomena. The last multivariate method is similar to [61] and [62] (also addressed in [63]) who combine the predictive properties of models of biodiversity change and taphonomic bias.

Correlation with climate

We tested for correlations of β diversity and LTG strength with stable oxygen isotopes from benthic foraminifera ($\delta^{18}O$ ‰) [64,65], mean annual precipitation estimated from paleosols [66], number of localities, sampling area (km^2), number of species, latitudinal range (degrees), and length of the sampled interval (Ma) of the fossil localities using generalized least squares and using an autocorrelation structure of order one (corAR1) to account for

temporal autocorrelation in R [67,68]. Best fit models were selected using automated model selection in the MuMIn R package [69] and the Akaike Information Criterion (ΔAIC).

Results

Fossil mammal β diversity showed considerable variation with the warmest intervals (late Eocene, mid-late Oligocene, mid Miocene, and mid Pliocene), but showing generally higher β diversity than with cooler intervals (early Oligocene, late Miocene) (Fig. 1C). The best fit model includes mean annual precipitation (MAP squared), length of the NALMA subdivision, and number of taxa, which together accounts for 67% of model variance (Table 2). β diversity is statistically significant for all three predictors (p<0.05). Residual β diversity is significantly explained by MAP only (Table 2; Fig. 2B). Re-sampling did not alleviate the effects of sampling bias; re-sampled β diversity is significantly explained by MAP-squared, number of taxa, and NALMA subdivision length (Table 2). The remainder of the manuscript will discuss the results from the analyses of raw and residual β diversity only.

Mammalian latitudinal turnover gradients (LTGs) are weak prior to the late Miocene (Fig. 1D). Raw LTG strength (i.e. not detrended) peaks during late Miocene (Hemphillian) and late Pleistocene (Rancholabrean) (Fig. 1D). The best fit model includes mean annual precipitation (MAP) [66], number of taxa, area (km^2) and an the interaction of area and the number of taxa, which explains 47% of the model variance (Table 2; Fig. 2C). LTG strength of late Cenozoic mammal species is statistically significantly explained by all four metrics (p<0.001; Table 2). Residual LTG strength is significantly explained only by MAP (p<0.05; Table 2; Fig. 2D). As above, re-sampling did not alleviate the effects of sampling bias on LTG strength (Table 2). In other words, even accounting for variables that describe potential sources of bias, a climatic variable (MAP) still explains a significant proportion of the variance.

β diversity is much lower for extant mammals than for extinct mammals (Fig. 3A). LTG strength for extant mammals is also greater than for early to mid Cenozoic fossil mammals, but similar to the values for the late Miocene and Pleistocene (Fig. 3B). Extant mammal β diversity shows a slight decrease under incomplete sampling and a slight increase under body-mass–bias sampling (Fig. 3A), but the change is much smaller than observed for fossil mammals. LTG strength does not appear to be significantly affected by the sample size reduction.

Our forecast models (which showed a strong fit to modern mammalian distributions, see Fig. S2A–C) show a slight increase in β diversity for extant mammals (Fig. 3C), but no substantial change in LTG strength compared to the present (Fig. 3D).

Discussion

Spatiotemporal patterns of β diversity remain poorly studied despite being potentially very useful in conservation biology [17,18,70] and linkage to well-studied biogeographic phenomena such as latitudinal richness gradients [4]. Using an extensive analysis of past and present mammalian communities, we demonstrate that, over the past 36 Ma, spatiotemporal patterns of mammal community composition have varied by orders of magnitude in North America. Specifically, Cenozoic spatial turnover of mammal communities is explained by continental mean annual precipitation (MAP) (Fig. 2A–B), broadly supporting predictions drawn from published studies of modern terrestrial organisms [2,70,71] and our predictions outlined above.

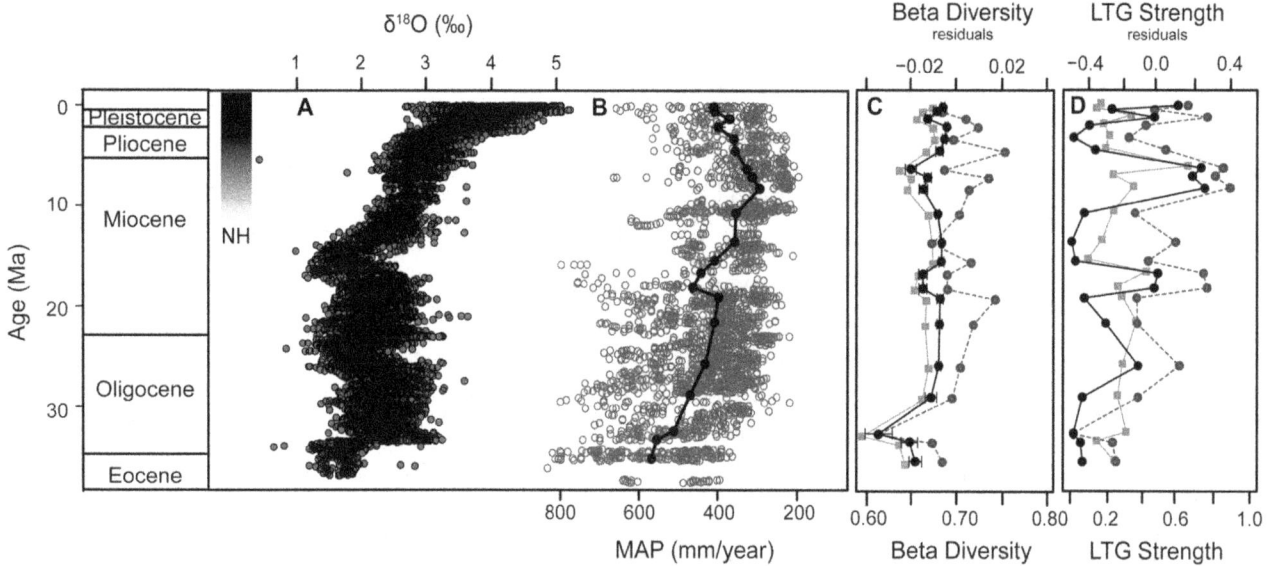

Figure 1. Mid to late Cenozoic trends of (A) $\delta^{18}O$ (‰) from benthic foraminifera (Zachos et al. 2008), (B) mean annual precipitation estimated from paleosols (Retallack, 2007), (C) β diversity of North American mammal species measured using multivariate dispersion (average distance from the centroid), and (D) strength of latitudinal turnover gradients (LTGs) measured as gradient strength for North American fossil mammals. Black lines are raw values, gray lines are residuals from significant sampling bias predictors, and gray dashed lines are re-sampled. Standard errors for re-sampled data are too small to display.

Contemporary ecological theory predicts that mammal diversity either declines monotonically with productivity or shows a unimodal pattern, declining with both low and high productivity [1,2,70,72]. Further, stronger latitudinal diversity gradients are associated with cooler, less productive environments [71] and steeper latitudinal climate gradients [1,70]. Both sets of predictions assume that changes in climate, productivity, and seasonality influence rates of origination and extinction [72,73], niche breadths [74], as well as the carrying capacity of the ecosystem [75], all factors that change the spatial turnover of terrestrial faunas [70]. Specifically, terrestrial organisms in low latitude, high productivity environments show low rates of speciation and extinction [73], high β diversity [2,76], and weak or absent latitudinal diversity gradients [71]. In contrast, high latitude organisms show high rates of speciation and extinction [73], low β diversity [2,76], and strong latitudinal diversity gradients [71]. Evolutionary history also plays a role in determining rates of spatial community turnover. Modern tropical organisms show

Table 2. Results of best fit generalized least squares models relating β diversity and latitudinal turnover gradient (LTG) strength to mean annual precipitation from paleosols (Retallack, 2007), $\delta^{18}O$ (‰) from benthic forams (mm/year; Zachos et al. 2001; 2008), length of North American Land Mammal Age subdivision, number of taxa sampled, sampling area (km^2), and number of fossil localities.

Dependent Variable	Parameters of Best Fit Model	Variance explained by model (%)	t value	p
Beta Diversity	Mean annual precipitation (quadratic)	66.51	−3.25	0.005
	Length of NALMA subdivision		2.43	0.027
	Number of taxa		5.30	<0.001
Beta Diversity Residuals	Mean annual precipitation (quadratic)	26.48	−3.50	0.002
Beta Diversity Re-sampled	Mean annual precipitation (quadratic)	66.04	−2.39	0.029
	Length of NALMA subdivision		2.51	0.023
	Number of taxa		5.47	<0.001
Latitudinal Turnover Gradient Strength (LTGs)	Mean annual precipitation (quadratic)	46.76	−5.65	<0.001
	Area		−4.62	<0.001
	Number of taxa		−4.36	<0.001
	Area : Number of taxa		4.85	<0.001
LTG Residuals	Mean annual precipitation (linear)	37.48	−3.79	0.001
LTG Re-sampled	Number of taxa	28.59	−2.55	0.020

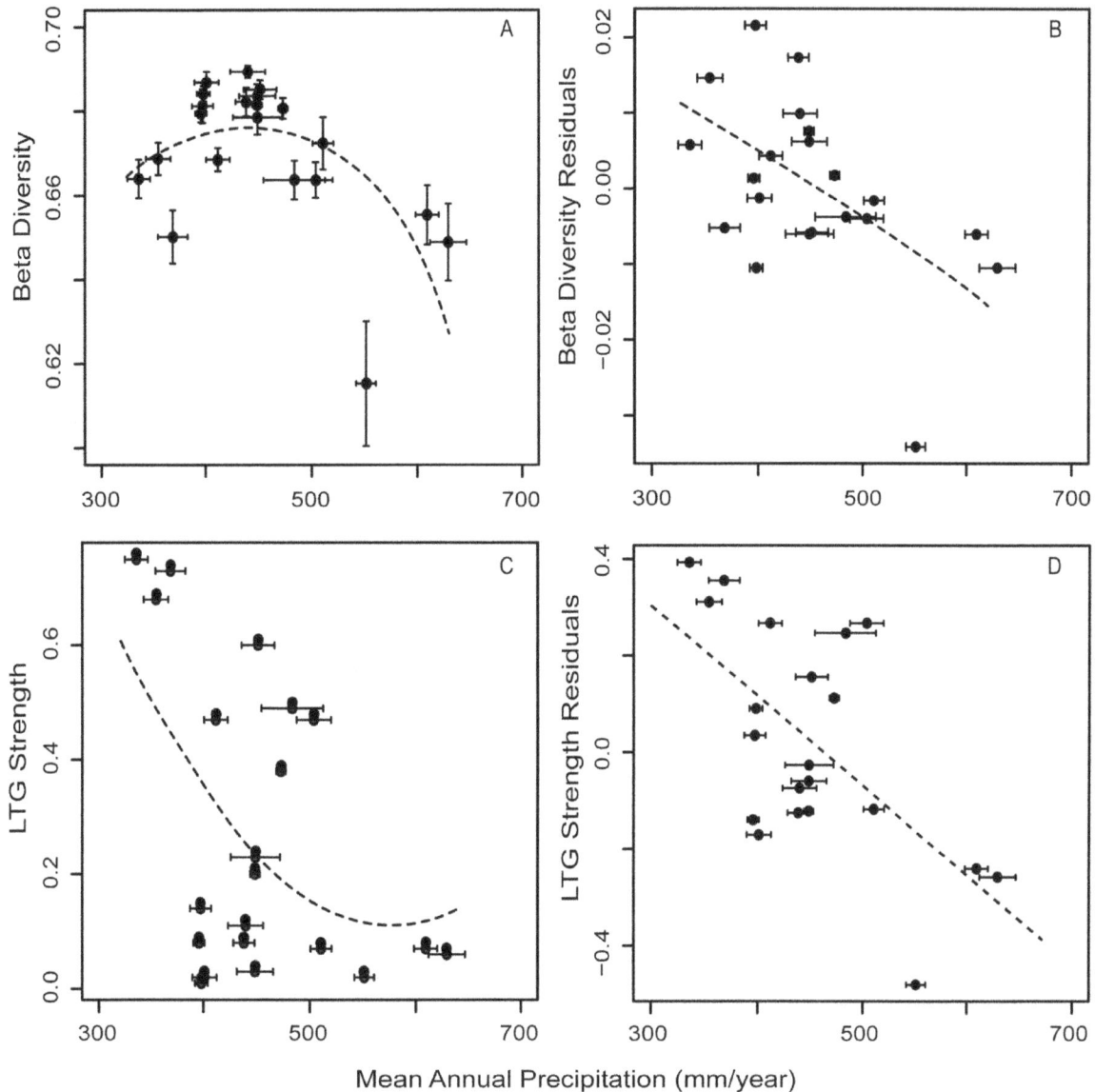

Figure 2. Relationship of mean annual precipitation estimated from paleosols (Retallack, 2007) with North American fossil mammal (A) raw β diversity ($R^2 = 0.43$), (B) residual beta diversity ($R^2 = 0.26$) and (C) raw latitudinal turnover gradient (LTG) strength ($R^2 = 0.25$), and (D) residual LTG strength ($R^2 = 0.37$).

faster turnover than their temperate counterparts regardless of the rate of environmental change [70]. Spatial and, by extension, temporal patterns of β diversity are the result of a mosaic of ecological and evolutionary processes.

Cenozoic fossil mammal β diversity peaked at intermediate values of mean annual precipitation and declined under both drier and wetter conditions (MAP; ~450 mm per year; Fig. 2B), showing a similar shape to latitudinal diversity curves for modern mammals [71]. Mammal β diversity was similarly lowest during periods of relative cooling, including the early Oligocene and late Miocene, coincident with declining atmospheric CO_2 [77–80] and, in the latter case, the expansion of ice sheets in the Northern Hemisphere [27,36], strengthening of thermohaline circulation [27,37,81–84], and transition from C_3 to C_4 dominated ecosystems at middle latitudes [66,85,86]. Declining β diversity during the late Miocene is also coincident with increased maximum body

mass [87], an ecologically relevant characteristic linked to lower ecosystem energy [88,89]. Water is a key component in photosynthesis and therefore net primary productivity (NPP) and MAP are correlated at a global scale, showing an asymptotic relationship [90]. Our results therefore suggest that putatively lower energy ecosystems (e.g. early Oligocene, late Miocene) supported more spatially homogenous mammal faunas than putatively higher energy ecosystems (e.g. late Eocene, mid Miocene, mid Pliocene). Temporal changes in fossil mammal β diversity (this study) are therefore conceptually similar to spatial patterns observed in extant mammals.

Early Oligocene mammals had lower β diversity than expected based on MAP (Fig. 1C; Fig. 2A). The early Oligocene is associated with rapid global cooling [64] and expansion of open grassy ecosystems [91], which may have resulted in lower ecosystem energy. However, our taxonomic sample is the poorest

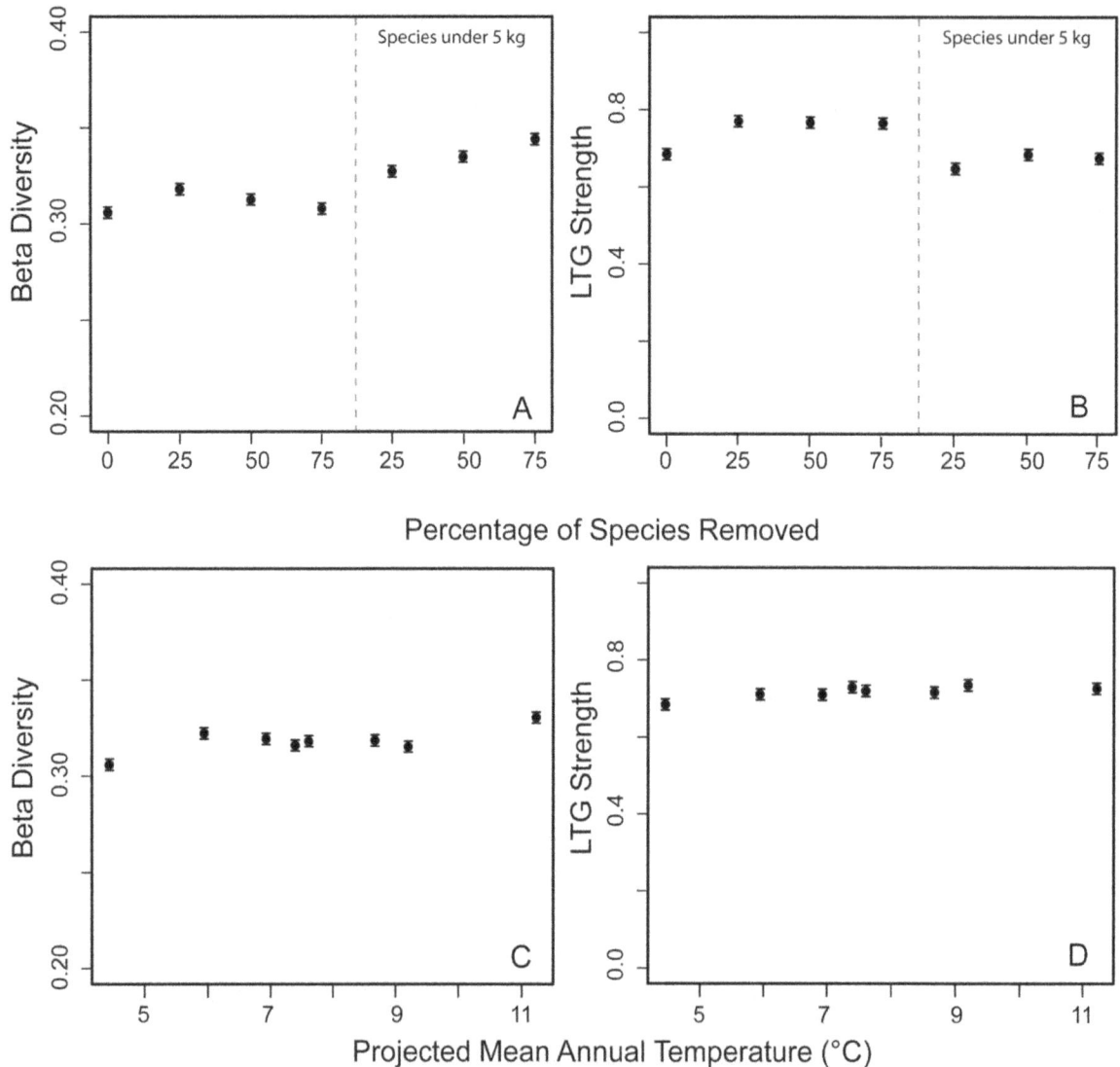

Figure 3. (A) β diversity (distance from centroid) and (B) latitudinal turnover gradients (LTG) strength of extant North American mammals under incomplete taxonomic sampling (removal of 25, 50, and 75% of species in sample) and body mass bias (removal of 25, 50, 75% of species smaller than 5 kg) and (C) β diversity (distance from centroid) and (D) latitudinal turnover gradients (LTG) strength of extant North American mammals under several International Panel on Climate Change scenarios (Special Reports on Emissions Scenarios).

for the early Oligocene; number of taxa is a significant predictor of fossil mammal β diversity (Table 2), suggesting some variation in preservation of species among NALMA subdivisions. Rarefied diversity also shows little change from the late Eocene to the early Oligocene [10]. However, our incomplete sampling trials show that removing even 75% of species reduces β diversity by a negligible amount (Fig. 3A), suggesting that at least some (but not all) of the observed decline in early Oligocene β diversity may have been climatically driven.

The magnitude of the latitudinal turnover gradient (LTG) for fossil mammals shows a temporally cyclic pattern that increases in amplitude during the late Cenozoic as well as a general trend toward stronger LTGs (Fig. 1D), coincident with the formation of ice on Svalbard at ~15 Ma and perennial Arctic sea ice at ~14 Ma, declining atmospheric CO_2 [37], and declining terrestrial MAP (Fig. 2B). Specifically, LTGs are strongest when precipitation is lowest (putatively lower productivity environments)

and weakest at when precipitation is highest (putatively high productivity environments; Fig. 2B), similar to modern mammals that show weak or absent latitudinal diversity gradients in the tropics and strong diversity gradients at mid to high latitudes [71]. Climate gradients are steeper at mid to high latitudes in North America due to the albedo of high latitude glaciation. Northern glaciation is an important means by which solar radiation is reflected from high latitudes, resulting in cool, low productivity Arctic environments [92,93]. Mammal communities are sorted along a latitudinal axis according to their climatic tolerances and the process of abiotic filtering, whereby taxa meet the limits of their environmental tolerances and are excluded from communities farther north [94]. Although late Miocene sea and land ice thickness and extent were reduced compared to the modern, increasing northern albedo and strengthening of thermohaline circulation are coincident with that strengthening of mammal

LTGs during the late Miocene (25–60% stronger than for any preceding NALMA; Fig. 1D) [27,81–84].

At first glance, the Pliocene appears to be anomalous because the magnitude of the mammalian LTG declines dramatically (60–70% reduction in the magnitude of the LTG; Fig. 1D). However, evidence from fossil deposits on Ellesmere Island show that approximately 3.5 Ma the Pliocene Arctic was ~14–22°C warmer than present [83,95,96] with an associated reduced volume of Arctic sea ice [27,82]. Pliocene Arctic warming is similarly coincident with reduced richness gradients of marine zooplankton [81]. The Pliocene might therefore be the "exception" that proves the rule.

Under modern global warming, Arctic winter temperatures have increased at a greater rate than at southern latitudes [97]. Long-term projections suggest boosts in high latitude net primary productivity due to increasing nitrogen fertilization and increases in mean annual precipitation of 100–150 mm per year or 5–20% at middle to high latitudes [98]. From our analyses of fossil North American mammals and published studies of beta diversity [18], we therefore expect weakened climate gradients and thus weakened LTGs due to northward range shifting, and, in the long-term, declining β diversity under the influence of modern anthropogenic climate change. β diversity decline may be facilitated by the homogenization of communities due to any of the following (note the lack of mutual exclusivity): i) extinction of species with small geographic ranges and replacement with wide-ranging species, ii) evolution toward larger range sizes within species, and, iii) invasion by wide-ranging species even without the extinction of residents [18]. However, our climate space models that are based on SRES scenarios corresponding to absolute mean annual temperatures of 4.4–11.2°C (averaged across North and South America) did not show changes in mammal LTGs or β diversity (Fig. 3C–D). We suggest that climate space models (CSMs) are unlikely to accurately forecast the outcomes of anthropogenic climate change for modern mammals because current CSM algorithms do not incorporate microevolutionary, macroevolutionary, or ecological processes, such as niche shifts, niche creation, and differences in dispersal abilities that are inherent in the response of animals to climate change. However, even on modern ecological timescales, rapid evolutionary changes and niche shifts have been observed in native and invasive populations [41], and this local adaptation complicates the prediction of range shifts. On longer timescales, taxa adapt to new climates and the processes of speciation and extinction help form new terrestrial communities. Without the explicit inclusion of evolutionary parameters and historical data for the taxa of interest, we are unlikely to accurately predict long-term changes in terrestrial biodiversity patterns.

We have shown here that macroecological patterns of North American mammal community composition varied considerably over the past 35 million years in response to changes in global climate change and Arctic glaciation (Fig. 1C–D). Furthermore, our comparison of fossil evidence with climate-space forecast models (CSMs) suggests that CSMs (in which species are modeled to simply track climate variables) may distort the degree of community composition change we should expect in the future. A unifying ecological theory relating diversity to climate must address both the spatial and temporal dimensions of diversity, as well as both richness and community composition. However, studies of organismal richness are far more common than studies of community composition (β diversity), despite the importance of the latter in conservation and their vast potential for contributing to our understanding of the processes underlying modern biodiversity. Studying the community composition of fossil animals represents a new frontier in paleontological research with potential to truly inform modern conservation.

Supporting Information

Figure S1   Maps of North America showing the distribution of fossil localities for all sampled North American Land Mammal Age subdivisions.

Figure S2   Model fit statistics for climate space models of extant North American mammals. Model performance was tested using area under the operating curve (A; AUC), the true skill statistics (B; TSS), and the proportion of correct classification (C).

Table S1   Summary of Special Emissions Report Scenarios (SERs) to which we fit climate models for extant mammalian species.

Table S2   List of mammalian taxa included and excluded from the species distribution models.

Appendix S1   Sources for the majority of mammal occurrence data downloaded from the Fossilworks database.

Acknowledgments

We thank John P. Hunter for a thorough review of this paper. Further, we thank John Alroy for his substantial contributions to the fossil data used in this analysis, accessed via his Fossilworks website, and his detailed review of the paper. We would also like to thank two anonymous reviewers, D. Currie, M. Clementz, M. Churchill, R. Haupt, J. Hoffmann, and E. Lightner for reviewing earlier versions of this manuscript, as well as L. Fahrig and S. Kim for constructive comments on this project.

Author Contributions

Conceived and designed the experiments: DF CH NR. Performed the experiments: DF. Analyzed the data: DF. Contributed reagents/materials/analysis tools: CH RG. Contributed to the writing of the manuscript: DF. Manuscript copyediting: CH RG NR.

References

1. Hawkins BA, Field R, Cornell HV, Currie DJ, Guegan JF, et al. (2003) Energy, water, and broad-scale geographic patterns of species richness. Ecology 84: 3105–3117.

2. Qian H, Badgley C, Fox DL (2009) The latitudinal gradient of beta diversity in relation to climate and topography for mammals in North America. Global Ecology and Biogeography 18: 111–122.

3. Condit R, Pitman N, Leigh EG Jr, Chave J, Terborgh J, et al. (2002) Beta-diversity in tropical forest trees. Science 295: 666–669.

4. Baselga A, Lobo JM, Svenning JC, Aragón P, Araújo MB (2012) Dispersal ability modulates the strength of the latitudinal richness gradient in European beetles. Global Ecology and Biogeography 21: 1106–1113.

5. Engle VD, Summers JK (1999) Latitudinal gradients in benthic community composition in Western Atlantic estuaries. Journal of Biogeography 26: 1007–1023.

6. Condamine FL, Sperling FAH, Wahlberg N, Rasplus JY, Kergoat GJ (2012) What causes latitudinal gradients in species diversity? Evolutionary processes and ecological constraints on swallowtail biodiversity. Ecology Letters 15: 267–277.

7. Currie DJ, Fritz JT (1993) Global patterns of animal abundance and species energy use. Oikos 67: 56–68.

8. Currie DJ, Francis AP, Kerr JT (1999) Some general propositions about the study of spatial patterns of species richness. Ecoscience 6: 392–399.

9. Mittelbach GG, Schemske DW, Cornell HV, Allen AP, Brown JM, et al. (2007) Evolution and the latitudinal diversity gradient: speciation, extinction and biogeography. Ecology Letters 10: 315–331.

10. Figueirido B, Janis CM, Pérez-Claros JA, Renzi MD, Palmqvist P (2012) Cenozoic climate change influences mammalian evolutionary dynamics. Proceedings of the National Academy of Sciences USA 109: 722–727.

11. Sepkoski JJ (1998) Rates of speciation in the fossil record. Philosophical Transactions of the Royal Society of London B 353: 315–326.

12. Mayhew PJ, Bell MA, Benton TG, McGowan AJ (2012) Biodiversity tracks temperature over time. Proceedings of the National Academy of Sciences USA 109: 15141–15145.

13. Kent R, Bar-Massada A, Carmel Y (2011) Multiscale analyses of mammal species composition-environment relationship in the contiguous USA. PLoS One 6: e25440.

14. Legendre P, Borcard D, Peres-Neto PR (2005) Analyzing beta diversity: partitioning the spatial variation of community composition data. Ecological Monographs 75: 435–450.

15. Whittaker RJ, Willis KJ, Field R (2001) Scale and species richness: towards a general, hierarchical theory of species diversity. Journal of Biogeography 28: 453–470.

16. Soininen J (2010) Species turnover along abiotic and biotic gradients: patterns in space equal patterns in time? BioScience 60: 433–439.

17. Hassall C, Hollinshead J, Hull A (2012) Temporal dynamics of aquatic communities and implications for pond conservation. Biodiversity and Conservation 21: 829–852.

18. Dornelas M, Gotelli NJ, McGill B, Shimadzu H, Moyes F, et al. (2014) Assemblage time series reveal biodiversity change but not systematic loss. Science 344: 296–299.

19. Janis CM, Damuth J, Theodor JM (2000) Miocene ungulates and terrestrial primary productivity: where have all the browsers gone? Proceedings of the National Academy of Sciences USA 97: 7899–7904.

20. Atwater AL, Davis EB (2011) Topographic and climate change differentially drive Pliocene and Pleistocene mammalian beta diversity of the Great Basin and Great Plains provinces of North America. Evolutionary Ecology Research 13: 833–850.

21. Finarelli JA, Badgley C (2010) Diversity dynamics of Miocene mammals in relation to the history of tectonism and climate. Proceedings of the Royal Society of London, Series B 277: 2721–2726.

22. Davis EB (2005) Mammalian beta diversity in the Great Basin, western USA: palaeontological data suggest deep origin of modern macroecological structure. Global Ecology and Biogeography 14: 479–490.

23. Barnosky AD, Hadly EA, Bell CJ (2003) Mammalian response to global warming on varied temporal scales. Journal of Mammalogy 84: 354–368.

24. Barnosky AD (2005) Effects of Quanternary climatic change on speciation in mammals. Journal of Mammalian Evolution 12: 247–264.

25. Eberle J, Fricke H, Humphrey J (2009) Lower-latitude mammals as year-round residents in Eocene Arctic forests. Geology 37: 499–502.

26. Eberle JJ, Fricke HC, Humphrey JD, Hackett L, Newbrey MG, et al. (2010) Seasonal variability in Arctic temperatures during early Eocene time. Earth and Planetary Science Letters 296: 481–486.

27. Polyak L, Alley RB, Andrews JT, Brigham-Grette J, Cronin TM, et al. (2010) History of sea ice in the Arctic. Quaternary Science Reviews 29: 1757–1778.

28. Clementz MT, Sewall JO (2011) Latitudinal gradients in greenhouse seawater $\delta18O$: evidence from Eocene sirenian tooth enamel. Science 332: 455–458.

29. Micheels A, Bruch A, Mosbrugger V (2009) Miocene climate modelling sensitivity experiments for different CO_2 concentrations. Palaeontologia Electronica 12: 5A.

30. Post E, Forchhammer MC, Bret-Harte MS, Callaghan TV, Christensen TR, et al. (2009) Ecological dynamics across the Arctic associated with recent climate change. Science 325: 1355–1358.

31. Primack RB, Ibáñez I, Higuchi H, Lee SD, Miller-Rushing AJ, et al. (2009) Spatial and interspecific variability in phenological responses to warming temperatures. Biological Conservation 142: 2569–2577.

32. Bradshaw WE, Holzapfel CM (2006) Evolutionary response to rapid climate change. Science 312: 1477–1478.

33. Parmesan C (2006) Ecological and evolutionary responses to recent climate change. Annual Review of Ecology and Systematics 37: 637–639.

34. Chen IC, Hill JK, Ohlemüller R, Roy DB, Thomas CD (2011) Rapid range shifts of species associated with high levels of climate warming. Science 333: 1024–1026.

35. Parmesan C, Yohe G (2003) A globally coherent fingerprint of climate change impacts across natural systems. Nature 421: 37–42.

36. Foster GL, Lunt DJ, Parrish RR (2009) Mountain uplift and the threshold for sustained Northern Hemisphere glaciation. Climate of the past discussions 5: 2439–2464.

37. Foster GL, Lear CH, Rae JWB (2012) The evolution of pCO₂, ice volume and climate during the middle Miocene. Earth and Planetary Science Letters 341–344: 243–254.

38. Lawler JJ, White D, Neilson RP, Blaustein AR (2006) Predicting climate-induced range shifts: model differences and model reliability. Global Change Biology 12: 1568–1584.

39. Hoffmann AA, Sgró CM (2011) Climate change and evolutionary adaptation. Nature 470: 479–485.

40. Thuiller W, Münkemüller T, Lavergne S, Mouillot D, Mouquet N, et al. (2013) A road map for integrating eco-evolutionary processes into biodiversity models. Ecology Letters 16: 94–105.

41. Lavergne S, Mouquet N, Thuiller W, Ronce O (2010) Biodiversity and climate change: integrating evolutionary and ecological responses of species and communities. Annual Review of Ecology, Evolution, and Systematics 41: 321–350.

42. Sepkoski JJ (1997) Biodiversity: past, present, and future. Journal of Paleontology 71: 533–539.

43. Rose PJ, Fox DL, Marcot J, Badgley C (2011) Flat latitudinal gradient in Paleocene mammal richness suggests decoupling of climate and biodiversity. Geology 39: 163–166.

44. Secord R, Bloch JI, Chester SGB, Boyer DM, Wood AR, et al. (2012) Evolution of the earliest horses driven by climate change in the Paleocene-Eocene thermal maximum. Science 335: 959–962.

45. Patterson BD, Ceballos G, Sechrest W, Tognelli MF, Brooks T, et al. (2007) Digital distribution maps of the mammals of the Western Hemisphere, version 3.0. NatureServe, Arlington, Virginia, USA.

46. Carrasco MA, Kraatz BP, Davis EB, Barnosky AD (2005) Miocene mammal mapping project (MIOMAP). University of California Museum of Paleontology.

47. McCoy ED, Connor EF (1980) Latitudinal gradients in the species diversity of North American mammals. Evolution 34: 193–203.

48. Lewin-Koh NJ, Bivand R (2008) maptools package version 0.8–16.

49. Pebesma EJ, Bivand RS (2005) Classes and methods for spatial data in R. R News 5.

50. Bivand RS, Pebesma EJ, Gomez-Rubio V (2008) Applied spatial data analysis with R. New York: Springer.

51. Peng RD (2007) The gpclib package version 1.5–5.

52. Wickham H (2009) ggplot2: elegant graphics for data analysis. New York: Springer.

53. Bivand R, Rundel C (2012) rgeos: interface to geometry engine version 0.2–16.

54. Venables WN, Ripley BD (2002) Modern and applied statistics with S. New York: Springer.

55. Nakicenovic N, Swart R (2000) Emissions scenarios: a special report of Working Group III of the Intergovernmental Panel on Climate Change. Cambridge: Cambridge University Press.

56. Girvetz EH, Zganjar C, Raber GT, Maurer EP, Kareiva P, et al. (2009) Applied climate-change analysis: the climate wizard tool. PLoS One 4: e8320.

57. Thuiller W, Georges D, Engler R (2012) BIOMOD: Ensemble platform for species distribution modeling. Ecography 32: 369–373.

58. Anderson MJ, Ellingsen KE, McArdle BH (2006) Multivariate dispersion as a measure of beta diversity. Ecology Letters 9: 683–693.

59. Oksanen J, Blanchet FG, Roeland Kindt PL, Minchin PR, O'Hara RB, et al. (2012) Package vegan version 2.0–7.

60. Tuomisto H, Ruokolainen K (2006) Analyzing and explaining beta diversity? understanding the targets of different methods of analysis. Ecology 87: 2697–2708.

61. Benson RBJ, Mannion PD (2012) Multi-variate models are essential for understanding vertebrate diversification in deep time. Biology Letters 8: 127–130.

62. Mannion PD, Upchurch P, Carrano MT, Barrett PM (2011) Testing the effect of the rock record on diversity: a multidisciplinary approach to elucidating the generic richness of sauropodomorph dinosaurs through time. Biological Reviews 86: 157–181.

63. Benton MJ, Dunhill AM, Lolyd GT, Marx FG (2011) Assessing the quality of the fossil record: insights from vertebrates. In: A. J McGowan and A. B Smith, editors. Comparing the geological and fossil records: implications for biodiversity studies. London: Geological Society of London. 63–94.

64. Zachos JC, Dickens GR, Zeebe RE (2008) An early Cenozoic perspective on greenhouse warming and carbon-cycle dynamics. Nature 451: 279–283.

65. Zachos J, Pagani M, Sloan L, Thomas E, Billups K (2001) Trends, rhythms, and aberrations in global climate 65 Ma to present. Science 292: 686–693.

66. Retallack GJ (2007) Cenozoic paleoclimate on land in North America. Journal of Geology 115: 271–294.

67. Development core team R (2012) R: A language and environment for statistical computing. Vienna, Austria: Foundation for Statistical Computing.

68. Dornelas M, Magurran AE, Buckland ST, Chao A, Chazdon RL, et al. (2013) Quantifying temporal change in biodiversity: challenges and opportunities. Proceedings of the Royal Society B 280: 1–10.

69. Bartoń K (2013) Multi-model inference package 'MuMIn' version 1.10.0 (http://cran.r-project.org/web/packages/MuMIn/MuMIn.pdf).

70. Buckley LB, Jetz W (2008) Linking global turnover of species and environments. Proceedings of the National Academy of Sciences USA 105: 17836–17841.

71. Currie DJ (1991) Energy and large-scale patterns of animal- and plant-species richness. American Naturalist 137: 27–49.

72. VanderMeulen MA, Hudson AJ, Scheiner SM (2001) Three evolutionary hypotheses for the hump-shaped productivity–diversity curve. Evolutionary Ecology Research 3: 379–392.

73. Weir JT, Schluter D (2007) The latitudinal gradient in recent speciation and extinction rates of birds and mammals. Science 315: 1574–1576.

74. Vázquez DP, Stevens RD (2004) The latitudinal gradient in niche breadth: concepts and evidence. American Naturalist 164: E1–E19.

75. Buckley LB, Davies J, Ackerly DD, Kraft NJB, Harrison SP, et al. (2010) Phylogeny, niche conservatism and the latitudinal diversity gradient in mammals. Proceedings of the Royal Society, Series B 277: 2121–2138.

76. Qian H, Xiao M (2012) Global patterns of the beta diversity energy relationship in terrestrial vertebrates. Acta Oecologica 39: 67–71.

77. Franks PJ, Beerling DJ (2009) Maximum leaf conductance driven by CO$_2$ effects on stomatal size and density over geologic time. Proceedings of the National Academy of Sciences USA 106: 10343–10347.

78. DeConto RM, Pollard D, Wilson PA, Pälike H, Lear CH, et al. (2008) Thresholds for Cenozoic bipolar glaciation. Nature 455: 652–657.

79. Tripati AK, Roberts CD, Eagle RA (2009) Coupling of CO$_2$ and ice sheet stability over major climate transitions of the last 20 million years. Science 326: 1394–1397.

80. Zhang YG, Pagani M, Liu Z, Bohaty SM, DeConto R (2013) A 40-million-year history of atmospheric CO$_2$. Philosophical Transactions of the Royal Society, Series A 371: 1–20.

81. Yasuhara M, Hunt G, Dowsett HJ, Robinson MM, Stoll DK (2012) Latitudinal species diversity gradient of marine zooplankton for the last three million years. Ecology Letters 15: 1174–1179.

82. Haywood AM, Valdes PJ, Sellwood BW, Kaplan JO, Dowsett HJ (2001) Modelling middle Pliocene warm climates of the USA. Palaeontologia Electronica 4: 1–21.

83. Ballantyne AP, Greenwood DR, Damsté JSS, Csank AZ, Eberle JJ, et al. (2010) Significantly warmer Arctic surface temperatures during the Pliocene indicated by multiple independent proxies. Geology 38: 603–606.

84. Ballantyne AP, Rybczynski N, Baker PA, Harington CR, White D (2006) Pliocene Arctic temperature constraints from the growth rings and isotopic composition of fossil larch. Palaeogeography, Palaeoclimatology, Palaeoecology 242: 188–200.

85. Fox DL, Honey JG, Martin RA, Peláez-Campomanes P (2012) Pedogenic carbonate stable isotope record of environmental change during the Neogene in the southern Great Plains, southwest Kansas, USA: Oxygen isotopes and paleoclimate during the evolution of C$_4$-dominated grasslands. Geological Society of America Bulletin 124: 431–443.

86. Strömberg CAE, McInerney FA (2011) The Neogene transition from C$_3$ to C$_4$ grasslands in North America: assemblage analysis of fossil phytoliths. Paleobiology 37: 50–71.

87. Smith FA, Boyer AG, Brown JH, Costa DP, Dayan T, et al. (2010) The evolution of maximum body size of terrestrial mammals. Science 330: 1216–1219.

88. Freckleton RP, Harvey PH, Pagel M (2003) Bergmann's rule and body size in mammals. American Naturalist 161: 821–825.

89. Blackburn TM, Gaston KJ, Loder N (1999) Geographic gradients in body size: a clarification of Bergmann's rule. Diversity & Distributions 5: 165–174.

90. Del Grosso S, Parton W, Stohlgren T, Zheng D, Bachelet D, et al. (2008) Global potential net primary production predicted from vegetation class, precipitation, and temperature. Ecology 89: 2117–2126.

91. Jacobs BF, Kingston JD, Jacobs LL (1999) The origin of grass-dominated ecosystems. Annals of Missouri Botanical Garden 86: 590–643.

92. Alexeev VA, Langen PL, Bates JR (2005) Polar amplification of surface warming on an aquaplanet in "ghost forcing" experiments without sea ice feedbacks. Climate Dynamics 24: 655–665.

93. Holland MM, Bitz CM (2003) Polar amplification of climate change in coupled models. Climate Dynamics 21: 221–232.

94. Soininen J, McDonald R, Hillebrand H (2007) The distance decay of similarity in ecological communities. Ecography 30: 3–12.

95. Csank AZ, Tripati AK, Patterson WP, Eagle RA, Rybczynski N, et al. (2011) Estimates of Arctic land surface temperatures during the early Pliocene from two novel proxies. Earth and Planetary Science Letters 304: 291–299.

96. Rybczynski N, Gosse JC, Harington CR, Wogelius RA, Hidy AJ, et al. (2013) Mid-Pliocene warm-period deposits in the High Arctic yield insight into camel evolution. Nature Communications 4: 1–9.

97. Kaplan JO, Bigelow NH, Prentice IC, Harrison SP, Bartlein PJ, et al. (2003) Climate change and Arctic ecosystems: 2. Modeling, paleodata-model comparisons, and future projections. Journal of Geophysical Research 108: 1–17.

98. Oechel WC, Vourlitis GL (1994) The effects of climate change on land-atmosphere feedbacks in arctic tundra regions. Trends in Ecology and Evolution 9: 324–329.

7

Soil Bacterial Community Response to Differences in Agricultural Management along with Seasonal Changes in a Mediterranean Region

Annamaria Bevivino[1]*, Patrizia Paganin[1], Giovanni Bacci[2,3], Alessandro Florio[2],

Maite Sampedro Pellicer[1], Maria Cristiana Papaleo[3], Alessio Mengoni[3], Luigi Ledda[4], Renato Fani[3],

Anna Benedetti[2], Claudia Dalmastri[1]

1 ENEA (Italian National Agency for New Technologies, Energy and Sustainable Economic Development) Casaccia Research Center, Technical Unit for Sustainable Development and Innovation of Agro-Industrial System, Rome, Italy, 2 Consiglio per la Ricerca e la Sperimentazione in Agricoltura - Research Centre for the Soil-Plant System, Rome, Italy, 3 Laboratory of Microbial and Molecular Evolution, Department of Biology, University of Florence, Florence, Italy, 4 Dipartimento di Agraria, University of Sassari, Sassari, Italy

Abstract

Land-use change is considered likely to be one of main drivers of biodiversity changes in grassland ecosystems. To gain insight into the impact of land use on the underlying soil bacterial communities, we aimed at determining the effects of agricultural management, along with seasonal variations, on soil bacterial community in a Mediterranean ecosystem where different land-use and plant cover types led to the creation of a soil and vegetation gradient. A set of soils subjected to different anthropogenic impact in a typical Mediterranean landscape, dominated by *Quercus suber* L., was examined in spring and autumn: a natural cork-oak forest, a pasture, a managed meadow, and two vineyards (ploughed and grass covered). Land uses affected the chemical and structural composition of the most stabilised fractions of soil organic matter and reduced soil C stocks and labile organic matter at both sampling season. A significant effect of land uses on bacterial community structure as well as an interaction effect between land uses and season was revealed by the EP index. Cluster analysis of culture-dependent DGGE patterns showed a different seasonal distribution of soil bacterial populations with subgroups associated to different land uses, in agreement with culture-independent T-RFLP results. Soils subjected to low human inputs (cork-oak forest and pasture) showed a more stable bacterial community than those with high human input (vineyards and managed meadow). Phylogenetic analysis revealed the predominance of *Proteobacteria*, *Actinobacteria*, *Bacteroidetes*, and *Firmicutes* phyla with differences in class composition across the site, suggesting that the microbial composition changes in response to land uses. Taken altogether, our data suggest that soil bacterial communities were seasonally distinct and exhibited compositional shifts that tracked with changes in land use and soil management. These findings may contribute to future searches for bacterial bio-indicators of soil health and sustainable productivity.

Editor: Jack Anthony Gilbert, Argonne National Laboratory, United States of America

Funding: This research was funded by MIUR (Integrated Special Fund for Research - FISR) in the frame of the Italian National Project SOILSINK "Climate change and agro-forestry systems, impacts on soil carbon sink and microbial diversity", and partially supported by MIUR (Research Department of Italian Government) in the framework of the Agreement Program ENEA-CNR (Articolo 2, comma 44, Legge 23.12.2009 n. 191 - Legge Finanziaria 2010). The funders had no role in study design, data collection and analysis, decision to publish, or preparation of the manuscript.

Competing Interests: The authors have declared that no competing interests exist.

* Email: annamaria.bevivino@enea.it

Introduction

Soil microorganisms play an important role as regulators of major biogeochemical cycles and can significantly affect the ecosystem functioning [1], being involved in organic matter dynamics, nutrient cycling and decomposition processes [2]. The anthropogenic activities affect the diversity of natural habitats modifying the number of species occurring in the environment at the landscape scale. Soil management strongly influences soil biodiversity in agricultural ecosystems. Different practices can alter the below-ground ecosystem, often leading to depletion of soil carbon and loss of biodiversity, and thus affecting the structure of

the resident microbial communities [3]. Therefore, characterizing genetic and functional diversity of soil bacterial communities in response to agricultural practices and/or climate is fundamental to better understand and manage the ecosystem processes.

The Mediterranean area is one of the most important biodiversity hotspots in the world and is increasingly threatened by intensive land use [4]. The high environmental diversity that characterizes the Mediterranean region is related to the integration of natural ecosystems and traditional human activities such as the agroforestry practices [5]. The collapse of the traditional agro-silvo-pastoral system that occurred during the past century has led to major changes in the extension of woodlands dominated by typical Mediterranean species, i.e. cork oak (*Quercus suber*) and/ or holm oak (*Quercus ilex*) woodlands [6], [7]. Research on the influence of management practices on the biodiversity of these agro-silvo-pastoral systems is increasing but it has focused mostly on plants [8] and vertebrates [5], [9]. In the frame of the Italian Project SOILSINK (Climatic changes and agricultural and forest systems: impact on C reservoirs and on soil microbial diversity), a hilly basin in Gallura (Berchidda site, Sardinia, Italy) was selected as a reference Mediterranean site for studying the influence of land-use changes on diversity, function and seasonal variations of soil microbial communities [10], [11]. The site is within an area of about 1,450 ha and is characterised by extensive agro-silvo-pastoral systems, typical of north-eastern Sardinia (Italy) and similar areas of the Mediterranean basin [12]. The chosen site represents a sustainable balance between human activities and natural resources that have created a landscape of high heterogeneity and cultural value, whose importance has been recognized at the European level [13], [14]. Indeed, it is considered climatically (Mediterranean zone) and pedologically homogeneous with vegetation patterns similar to those called *dehesas* or *montados* of the south-western Iberian Peninsula [15], [16], [17]. In the past, this area was covered by cork-oak forests, which gradually were subjected to increasing under-storey grazing and usage for the extraction of cork. Today, there are different land-use and plant cover types that lead to a soil and vegetation gradient with an ecological progression: from a cork-oak forest undergoing minimum disturbance to managed vineyards with an intensive agricultural practice (grass covered and ploughed), passing through areas with temporary grassland, and pasture.

The different land uses altered soil potential, making possible to discriminate the role of human management on soil functioning. When forests are converted to grasslands, and grasslands turned into agricultural lands, a sharp switch from one type of soil microbial community to another one occurs. Since the ability of an ecosystem to withstand serious disturbances may partly depend on its microbial component(s), characterizing bacterial community composition and/or structure might help to better understand and manipulate ecosystem processes. The aim of the present study was to investigate the effects of soil characteristics and different agricultural managements on soil bacterial community in two seasons. Sampling was carried out in spring and autumn 2007,

when the plant cover-growing season usually starts and ends. A combination of culture-based and molecular techniques along with statistical analysis of data obtained was applied to interrogate the diversity, function, and ecology of soil bacterial communities. The results obtained in the present study, along with the other ones obtained within the SOILSINK Project [11], [18], [19], [20], provide useful data on the impact of soil type, cover vegetation, and human activities on the distribution of the bacterial genetic resources in soil communities for this Mediterranean region. Our results confirm that the environments with low inputs (cork-oak forest and pasture) show a more stable soil microbial community than those subjected to increasing human input (vineyards and managed meadow) and suggest that soil bacterial communities are seasonally distinct with compositional shifts that track with changes in land use and soil management.

Materials and Methods

Ethics Statement

We carried out the study on the hilly basin in Gallura (Olbia-Tempio municipalities, Sardinia, Italy). Five soils uses were identified in private farms within Berchidda site (40°49′ 15″N, 9°17′ 32″ E) and were obtained. The soil sampling was carried out in the frame of a national research project (SOILSINK Project) and soils used in this study were collected under consent of the landowners. The responsible of the study site was Prof. P.P. Roggero (University of Sassari, Italy). We confirm that our study did not harm the environment and did not involve endangered or protected species. Specific geographic coordinates (referred to World Geodetic System, 1984) of our study area are reported in Table S1.

Sampling site

The study area (Olbia-Tempio) is representative of the climate, vegetation type and management of some of the most common agro-forestry systems in the Mediterranean basin [16].

The Berchidda site is made up of hydromorphic and granitic soil with a loamy sand texture. The altitude ranges from 275 m to 300 msl. This area is referred to as a meso-thermo Mediterranean, subhumid phytoclimatic belt with a mean annual rainfall of 862 mm and the mean annual temperature of 13.8°C [10]. In the

SPRING

AUTUMN

Figure 1. Long-term effects of different land-use with increasing level of intensification in spring and autumn. Both pasture and managed meadow included spotted cork oak trees, which are key components of the Dehesatype landscape typical of this area of Sardinia. The cork-oak formation, pasture, and managed meadow have been converted to the current use and maintained unchanged for more than 30 years, whereas the non-tilled cover cropped vineyard and the tilled one were planted in 1985 and 1994, respectively. From the left to right: cork-oak forest (CO), hayland pasture rotation (PA), managed meadow (MM), grass covered vineyard (CV), tilled vineyard (TV).

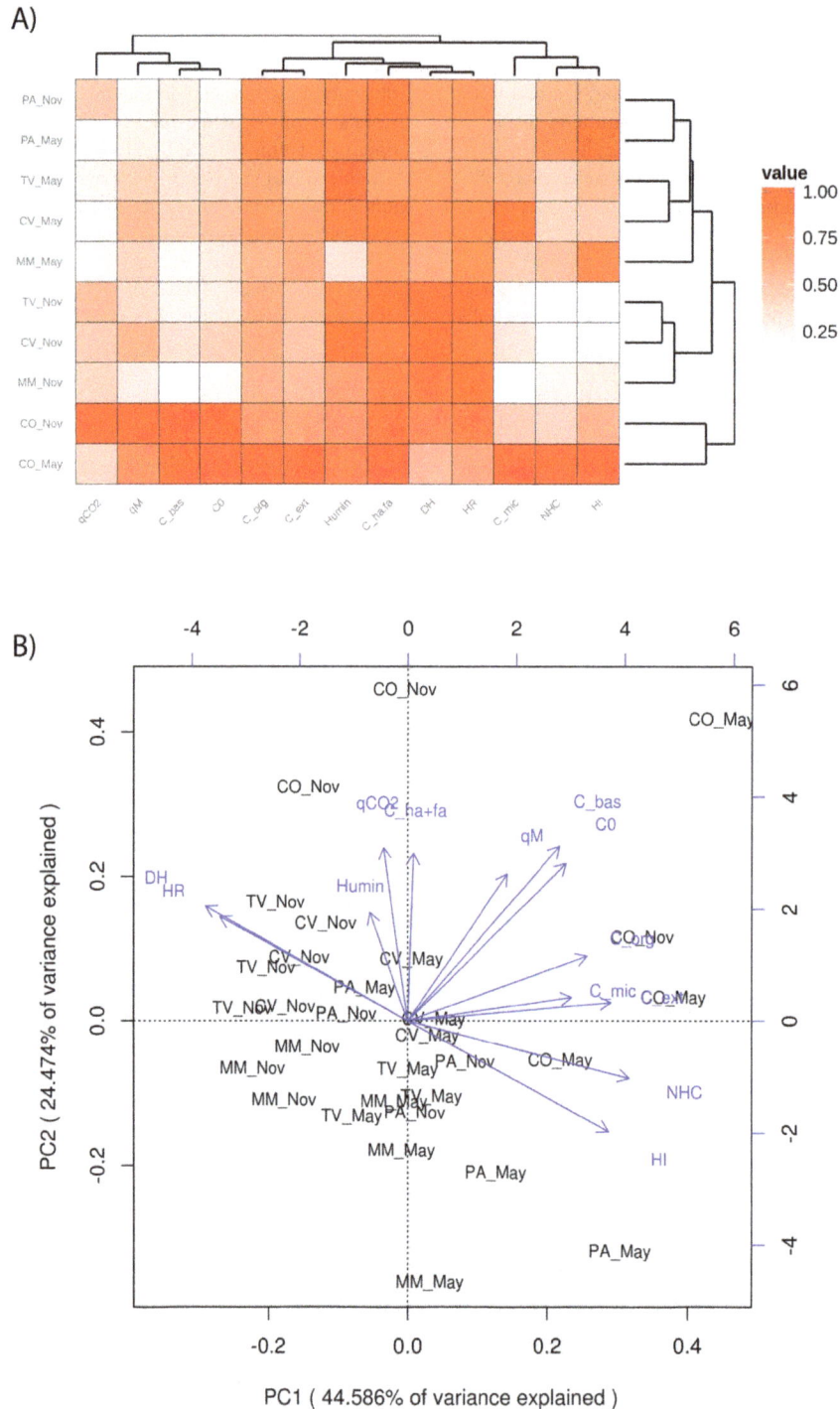

Figure 2. Effect of land-use and season on soil physical-chemical and biological parameters. A) Heat map with hierarchal clustering of physical-chemical and biological parameters across the five Sardinia soils with different land uses at the two different sampling time points (May and November). The heat map was constructed using a maximum-minimum normalization of the data in order to represent each value in a range between 0 and 1. Higher values are represented by darker colors whereas lower ones are represented by lighter colors. CO = cork-oak forest; PA = hayland-pasture rotation; MM = managed meadow; TV = tilled vineyard; CV = grass covered vineyard. B) PCA ordination of data (axes 1 and 2) generated from physical-chemical and biological properties of the different types of land use in May and November.

past, the Berchidda area was covered by cork-oak forests (dominated by *Quercus Suber* L.), which were subjected to intense usage for the extraction of cork. Today, there are five main different land-use units close together: cork-oak forest (CO),

hayland-pasture rotation (PA), managed meadow (MM), and tilled (TV) and grass covered vineyard (CV). The cork-oak formation, pasture, and managed meadow have been converted to the current use and maintained unchanged for more than 30 years,

A)

B)

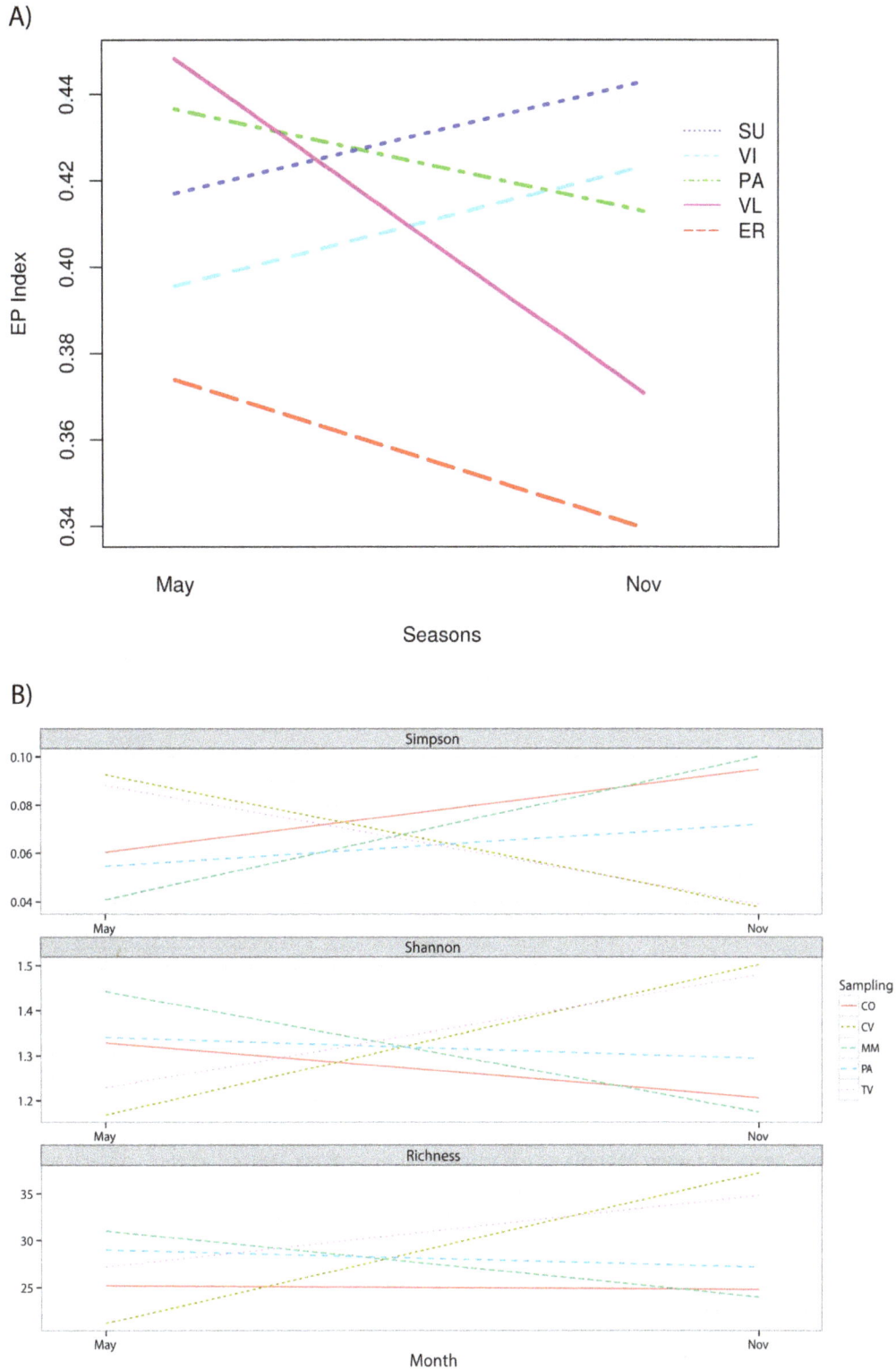

Figure 3. Effect of land-use and season on Eco-Physiological (EP) index of culturable bacteria (A) and diversity indices from CD-DGGE profiles (B).

whereas the non-tilled cover cropped vineyard and the tilled one were planted in 1985 and 1994, respectively. The five soils are located inside an area of 161.5 km². Detailed characteristic and management of the five soils have been previously described [12], [20], [21], [22], [23].

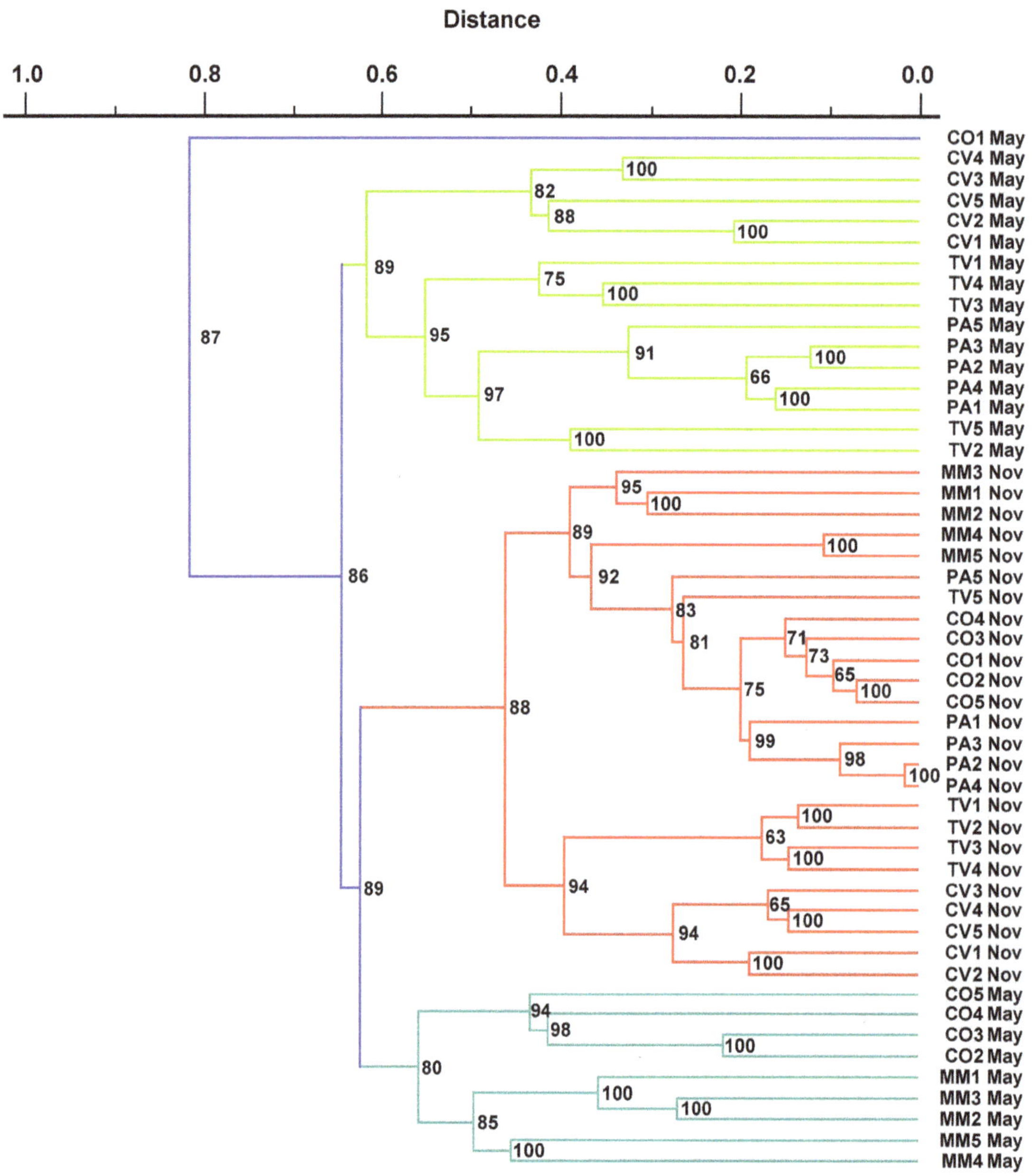

Figure 4. UPGMA dendrogram of DGGE profiles of amplified 16S rDNA of bacterial communities recovered in CO, PA, MM, CV and TV, in spring and autumn, generated using Phoretix ID advanced analysis package. Designation of samples is the same as in Fig. 1. The scale bar represents dissimilarity among samples. Consistency of each cluster was measured by the Cophenetic correlation coefficient shown at each node.

Pedological characterization of the study area

The pedogenic substrate of the study area consists of medium-grained granite, affected by localized presence of veins of quartz and porphyry. The morphology of the Berchidda area varies from flat to undulating. The processes of soil erosion by water channeled are evident only in the short-term forage crops made on soil with more than 15% of slope. All profiles have a horizons sequence of the type A-Bw-C or A-Bw-BC-C or more rarely, A-Bw-C-R (Table S1). The power of the profiles, limited to horizons A and Bw, varies from a minimum of 38 cm to a maximum of 100 cm. In three soil profiles in TV, the sequence between the horizon A and Bw and the substrate is gradually altered by the presence of a horizon BC, characterised by coarse texture, the

power of which varies from about 30 cm to 90 cm. Direct contact with the unaltered rock, R horizon, was observed only in one profile in CV. The prevailing textural classes are sandy and sandy loam. The content of organic substance in the horizons A is never very high, with average value around to about 3%. The maximum values of 11.8% and 8.6% were observed on the A horizons in two soil profiles in CO. The exchange complex, in agreement with the reduced content in clays, is never high. The degree of base saturation is predominantly less than 60% (Dystric conditions of the USDA Soil Taxonomy). The profiles with the exchange complex with a degree of base saturation below 60% in all horizons between 25 and 75 cm (TV, CV, MM, and PA) were

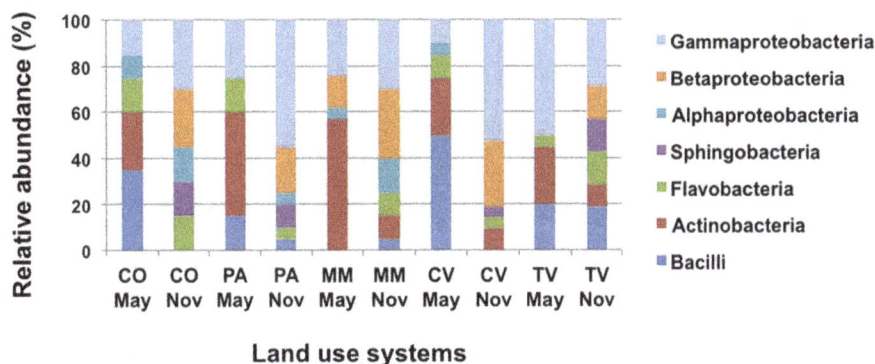

Figure 5. Relative abundances of major taxonomic groups across land use systems in spring and autumn. Detailed data of each class are listed in Additional file 1: Table S6. CO = cork-oak forest; PA = hayland-pasture rotation; MM = managed meadow; TV = tilled vineyard; CV = grass covered vineyard. Values presented are the mean percent.

classified as Typic Dystroxerepts [24]. The other one (CO) was classified as Lithic Xerorthents.

Soil sampling

Five soil replicates were collected from bulk soil of the five different managements (CO, PA, MM, CV, and TV) (Fig. 1). Soils were collected in May and November 2007. After removal of litter layer, soil core samples (50 to 100 g; diameter, 5 cm) were taken from each of the five locations, using a 5-on-dice sampling pattern with ca. 70 m distance between each sampling point. Sampling was performed at 20 cm depth, where most microbial activity is known to occur [25], [26]. In the vineyard soils, samples from along the rows and between the rows were pooled together to form a field replicate. The other soil samples (CO, PA, MM) were collected out of trees influence. At each season and for each soil type, five randomly field replicates were collected for a total of 25 soil samples (5 replicates × 5 land uses), each one being a composite sample of five soil cores. A total of 50 composite samples were taken for the two seasons. Soil samples were immediately sieved (<2 mm) to remove fine roots and large organic debris, air dried, and transported to the labs for microbiological analysis. The moisture content was adjusted to 60% of their water holding capacity (WHC) and soil samples were then left to equilibrate at room temperature in the dark for one day prior to analyses, in order to restore, within limits [27], the microbial activity of air- dried soils to that of soils in the field.

Chemical and biochemical analyses of soil samples

The chemical and biochemical analyses were performed on three replicates for each land use for a total of 15 soil samples (3 replicates × 5 land uses), each one being a composite sample of five soil cores. Total organic carbon (C_{org}) was estimated after oxidation with $K_2Cr_2O_7$ and subsequent titration of unreduced $Cr_2O_7^{2-}$ with $Fe(NH_4)_2(SO_4)_2$, as reported by Springer and Klee [28]. Soil Organic Matter (SOM) was determined by C_{org} multiplied by 1.724 van Belem coefficient. The C_{org} fractionation was set up as reported by Ciavatta and co-workers [29]. In particular, solid samples were extracted at 65°C for 24 h using 0.1 mol 1^{-1} NaOH plus 0.1 mol 1^{-1} $Na_4P_2O_7$ solution (1:50, solid:liquid ratio). The samples were then centrifuged at $5000 \times g$ and the supernatants were filtered through a 0.20-μm Millipore filter (Millipore, Billerica, MA) (total extractable C, C_{ext}). The humic-like acid (HA) fraction was separated from the fulvic-like acid (FA) and the non-humified carbon fractions (NHC) by

precipitation after acidification of the alkaline solution (supernatant) to pH<2. Chromatography on a column of polyvinylpyrrolidone (PVP, Aldrich, Germany) was used to separate the NHC from the FA. The FA was then combined with the HA to obtain total humified fraction (HA+FA). Total extractable C (C_{ext} %), humic and fluvic acid C (C_{HA+FA} %) were determined by the dichromate oxidation method. The non-humified carbon (C_{NH}) was determined as the difference between C_{ext} and C_{HA+FA}. Humification indexes HI, DH, and HR were determined according to previous works [29], [30].

Microbial biomass C (C_{mic}) was determined by the fumigation-extraction method of Vance and co-workers [31] with some slight modifications. The measurements were performed on air-dried soils, pre-conditioned by a 10-d incubation in open glass jars, at − 33 kPa water tension, and 30°C. The incubation was employed for restoring, within limits [27], the microbial activity of air-dried soils to that of soils in the field. Four replicates of each soil sample were used. Average values are given in mg C kg^{-1} of soil. For measuring microbial respiration, 20 g (oven-dry basis) of moist sample were placed in 1 L stoppered glass jars. The CO_2 evolved was trapped, after 1, 2, 4, 7, 10, 14, 21, 28 days of incubation, in 2 ml 1 M NaOH and determined by titration of the excess NaOH with 0.1 M HCl [32].

Non-linear least square regression analysis was used to calculate parameters affecting C mineralization from daily CO_2 evolution data (Stat Win 4.0 for Windows). The best fit was obtained with the exponential model of CO_2-C accumulation according to the negative exponential decay model:

$$C_m = C_0(1 - e^{-kt}),$$

where C_m is the cumulative value of mineralized C during t days, k is the rate constant, and C_o is the potentially mineralizable C [33]. The CO_2 emitted in 28 days of incubation was used as cumulative respiration (C_{cum}). The CO_2 evolved during the 28th day of incubation was used as the basal respiration value (C_{bas}). Microbial indices were calculated as follows [34], [35], [36]:

$$qCO_2 = [(\mu g\ CO_2 - C_{bas} \times \mu g^{-1} C_{mic})h^{-1}]10^3,$$
$$qM_{cum} = (\mu g\ CO_2 - C_0 \times \mu g^{-1} C_{org}),$$
$$qM_{bas} = (\mu g\ CO_2 - C_{bas} \times \mu g^{-1} C_{org})$$

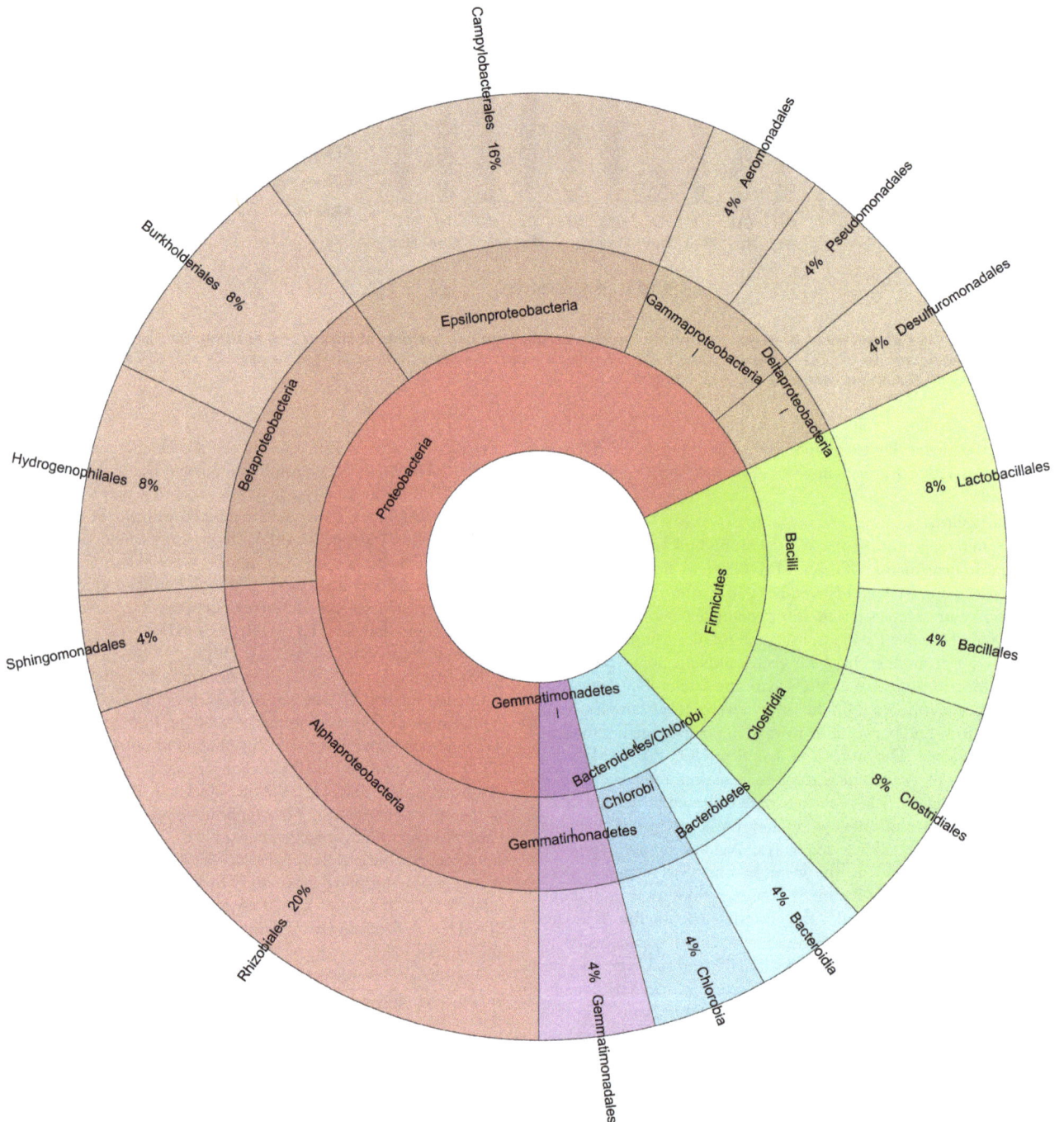

Figure 6. Plot of the taxonomic composition of total bacterial community as inferred from taxonomic interpretation of T-RFLP profiles.

Recovery of cultured bacterial cells from soil samples

Soil bacterial community analysis was performed on five soil replicates for each land use, as described above. Bacterial cell extraction was performed according to the recommendations of Smalla et al. [37] with minor modifications. Briefly, 1 g of soil was placed in a sterile 15 ml plastic tube containing 10 ml of phosphate buffered saline (PBS, pH 7.0). This mixture was homogenized for 30″ at low speed by using the Ultra-Turrax Thyristor Regle 50 (Janke & Kunkel IKA-Labortechnik). After homogenization, suspension was placed into a Erlenmeyer flask

(100 ml) containing 10 g of glass beads (0,2 mm) and shaken for 1 h at 180 r.p.m. and 28°C to disperse bacteria. The flasks and glass beads were autoclaved for 20 min at 121°C before use. The soil suspension was transferred into a sterile 15 ml plastic tube and serially diluted with sterile saline solution (9 g l^{-1} NaCl) from 10^{-1} up to 10^{-7}. The, 100 μl aliquots of serially diluted soil suspensions were plated in triplicate on 0.1 Triptic Soy Broth (TSB; Difco) amended with 15 g l^{-1} agar (0.1 TSA) and 100 g $μl^{-1}$ of cycloheximide (Sigma) to inhibit fungal growth. Plates were incubated at 28°C for 6 days.

Growth strategy and total bacterial populations

To determine the changes in the structure of culturable fraction of soil bacteria, the r/K-strategy concept proposed by De Leij and co-workers [38] was used. Bacterial colonies appearing within 48 h were designated as r-strategists, and the remaining as K-strategists. Colonies were enumerated at 1, 2 and 6 days of growth on 0.1 TSA; in this way, three counts (or classes) were generated per sample. Plates containing between 30 and 300 colonies were then selected for enumeration. Total bacterial counts obtained were expressed as colony forming units (CFU) per gram of soil. Distribution of bacteria in each class as a percentage of the total counts gave insight into the distribution of r- and K-strategists in each sample.

To evaluate the changes in the biodiversity of bacterial populations in soils, the eco-physiological (EP) index [38] was used. The EP index of each soils tested was calculated using the equation:

$$H' = - \sum (P_i)(log_e P_i),$$

where P_i represents the CFU on each day (1, 2 and 6 days of incubation) as a proportion of the total CFU in that sample after 6 days incubation, i.e., the proportion of colonies appearing on counting day i ($i = 1, 2, 6$) with $EP_{min} = 0$. Higher values of EP index imply a more even distribution of proportions of bacteria developing on different days (i.e., different classes of bacteria).

Terminal-Restriction Fragment Length Polymorphism (T-RFLP)

DNA was extracted from soil samples by using the FastDNA SPIN Kit for Soil (QBiogene). Terminal-Restriction Fragment Length Polymorphism (T-RFLP) was performed on 16SrRNA genes amplified from extracted DNA with primer pairs P0 and P6 as previously reported [39], [40]. Purified amplification products were digested separately with restriction enzymes RsaI and MspI and digestions were resolved by capillary electrophoresis on an ABI310 Genetic Analyzer (Applied Biosystems, Foster City, CA, USA) using LIZ 500 (Applied Biosystems) as size standard. T-RFLP analysis was performed as previously reported [41]. Diversity indices were calculated with PAST software [42] as previously reported [19], taking into account peak intensities of T-RFLP fragments. Taxonomic interpretation of T-RFLP profiles was performed by querying the Ribosomal Database Project Database by using MiCA3 web tool (http://mica.ibest.uidaho.edu/), as previously described [40].

Culture-dependent DGGE (CD DGGE) analysis

Culture-dependent DGGE (CD DGGE) fingerprinting of 16S rRNA gene was used to characterize mixed bacterial communities recovered on agar plates.

Collection of cultured bacterial communities for DGGE analysis. Cultured bacterial communities were collected following procedure proposed by Duineveld and co-workers [43] with minor modifications. Briefly, after one week of incubation at 28°C, colonies were removed from plates containing between 100 and 1000 colonies by adding 3.0 ml sterile physiological solution (0.9% NaCl) on each plate and scraping off all grown colonies with a sterile Drigalski spatula. The cell suspensions thus obtained were aliquoted into 1.5 ml Eppendorf tubes and centrifuged at 8,000 r.p.m. for 10 minutes. The pellets were stored at −80°C for subsequent DNA extraction and PCR-DGGE analysis.

DNA extraction and PCR amplification of 16S rRNA genes. Genomic DNA was extracted with sodium dodecyl sulfate-proteinase K lysis buffer, followed by a treatment with cetyl-trimethylammonium bromide (CTAB) as described in *Current Protocols for Molecular Biology* [44]. Briefly, the pellet was resuspended in 567 μl of TE [10 mM Tris-HCl - 1 mM EDTA (pH 8.0)] buffer, and glass beads 0.3 mm in diameter (250 mg) were added, followed by bead beating for 20 s. Then, 30 μl of 10% sodium dodecyl sulfate and 3 μl proteinase K 20 mg/ml (Sigma) were added and samples were incubated at 37°C for 1 h. Glass beads were removed by centrifugation 2 min at 2,800 rpm and samples were then incubated at 65°C for 10' with100 μl of 5 M NaCl prepared with sterile water and 80 μl of CTAB/NaCl (10% CTAB in 0.7 M NaCl). Following incubation, extracts were purified by using phenol/phenol-chloroform/isoamyl alcohol (49.5:49.5:1) extraction and DNA was recovered by isopropanol precipitation at 4°C o/n. Pelleted DNA was washed twice with cold 70% ethanol, allowed to air dry, and re-suspended in 50 μl of sterile water. Quantity and purity of DNA were checked by NanoDrop (NanoDrop Technologies, USA) and gel electrophoresis. The DNA samples were stored at –20°C until required for use.

The 16S rRNA gene was amplified using 20 ng of lysate suspension and the universal bacterial primers P0 and P6 [45]. Dilutions 1:100 (2 μl) of the1450 bp PCR products were then used as template for the second PCR amplification with the forward primer 63F (5′- AGGCCTAACACATGCAAGTC -3′), with a GC clamp (5′-CGCCCGCCGCGCGCGGCGGGCGGGGCG-GGGGCACGGGGGG -3′) incorporated at the 5′ end, and the reverse primer 518R (5′-ATTACCGCGGCTGCTGG-3′), to produce 495 bp fragments suitable for DGGE analysis [46]. Both PCR reactions were performed in Qiagen Taq buffer (10X) containing 1.5 mM MgCl₂, with 150 ng of each primer, 250 μM (each) deoxynucleoside triphosphates, and 0.5 U of *Taq* DNA polymerase (Qiagen, Hilden, Germany) in a 25 μl reaction volume. Cycle parameters for PCR with the primer pairs P0–P6 and 63F-GC and 518R were previously described by Di Cello and co-workers [45] and El Fantroussi and co-workers [46], respectively.

Denaturing Gradient Gel Electrophoresis. 16S rRNA gene amplicons were separated by double gradient denaturing gradient gel electrophoresis (DG-DGGE) as described by Cremonesi et al. [47], in a DCode universal mutation detection system (Bio-Rad, CA, USA). Separation of purified PCR products (700 ng) was achieved in6%–12% polyacrylamide (acrylamide: N,N-methylenebisacrylamide, 37.5:1) gels containing an increasing linear gradient of denaturants ranging from 30% to 60% (100% denaturant corresponds to 7 M urea and40% deionized formamide). Each gel also included marker lanes represented by DGGE profiles containing a large number of discrete bands spanning the entire gradient, suitable for within- and between-gel alignment. Electrophoresis were carried out for 16 h at 75 mV in 1X TAE buffer at 60°C, stained with 50 μg/ml ethidium bromide for 30 min, destained in water and photographed with the UVIpro Platinum Gel Documentation System (GAS7500/7510; Eppendorf, Cambridge,UK).

Cluster analysis and diversity indices

Quantity One software package (Bio-Rad) and Phoretix 1D PRO software (Phoretix International, Newcastle upon Tyne, United Kingdom) were used for CD-DGGE profile analysis. The cluster analysis and dendrogram generation were carried out by using the Phoretix 1D Pro software according to the manufacturer's instructions (Phoretix International, Newcastle upon Tyne,

United Kingdom). Bands of CD-DGGE patterns were aligned and normalized using reference lanes. Background noise was subtracted by rolling ball algorithm with a radius of 50 pixels; the automatic band detection was performed with a minimum slope of 200 and a noise reduction of 10, and peaks smaller than 2% of the maximum peak were discarded. Bands were manually corrected and matched to create an absent/present binary matrix. The similarity between the band patterns was calculated using the Dice coefficient and the clustering analysis was performed with the unweighted pair group method with arithmetic averages (UPGMA) to generate a dendrogram by using mathematic averages algorithm programs integral to the Phoretix 1D Pro software. Coefficient of cophenetic correlation was used to measure the consistency of clusters.

DGGE banding data were used to estimate three diversity indices by treating each band as an individual operational taxonomic unit (OTU). The number of DGGE bands present in each sample was used to measure the Richness index (R). The Shannon-Weaver index of general diversity (H') [48] and the Simpson index of dominance (D) [49] were calculated from the number of bands present and the relative intensities of each band (P_i) in each lane. Relative signal intensities of detected bands, in each gel track, were determined by using the Quantity One software package (Bio-Rad) and calculated from the peak area of the densitometric curves.

The Shannon-Weaver diversity (H') was calculated using the following equation:

$$H' = -\sum (P_i)(log_e P_i),$$

where P_i (the proportion of abundances of the ith band) is measured as:

$$P_i = n_i/N,$$

where n_i is the peak height of a band, and N is the sum of all peak heights in the densitometry profiles.

The Simpson index (D) was calculated with the formula:

$$D = \sum (P_i x P_i);$$

it measures the strength of dominance because it weights towards the abundance of the OTUs and varies inversely with species diversity [50].

Bacterial isolation and identification

A total of 100 bacterial colonies were randomly picked up for each soil sample from 0.1 TSA plates (containing approximately 50 to 500 colonies), previously used for the determination of the CFU counts and EPI-index, and repeatedly streaked onto 0.1 TSA fresh plates to obtain pure cultures. Isolated colonies were then grown overnight (o/n) in TSB medium at 28°C and 200 r.p.m., and stored at −80°C in 30% glycerol until further analysis. From all five soil samples, 500 colonies were isolated in each season, for a total of 1000 bacterial colonies. A total of 203 colonies with different morphologies (about 20 colonies per each sample) were taken up to investigate their taxonomic affiliation.

Genomic DNA and PCR amplification of the 16S rRNA gene were performed as described above. Sequencing reactions were prepared from PCR products using an Applied Biosystem Big Dye Terminator sequencing kit version 3.1, according to the manufacturer's instructions and analysed using a 3730 DNA Analyzer

Applied Biosystem apparatus. The sequences were compared with those in the GenBank databases by using the BLAST program and Seqmatch tool of the RDP (http://www.ncbi.nlm.nih.gov/BLAST/and http://rdp.cme.msu.edu/, respectively) and aligned with the closest relatives with the Clustal W function of the BioEdit package [51]. Bacterial identification by 16S rRNA gene sequences assignment was performed using the RDP Classification Algorithm (http://rdp.cme.msu.edu/classifier/classifier.jsp).

Statistical analysis

Bacterial population data (CFU/g of soil) were log transformed and subsequently analysed by one-way ANOVA (STATISTICA, Release 3.0b, Copyright StatSoft Inc., CA, USA). Percentage data of EP index value were *logit*-transformed, as follows:

$$Logit\,(p) = log[p/(1-p)]$$

for the proportion p, and compared using one-way ANOVA (STATISTICA, Release3.0b, Copyright StatSoft Inc., CA, USA).

Analyses of variance (ANOVA) on biodiversity indexes (Shannon-Weaver, Richness and Simpson), principal component analysis (PCA) and clustering analysis on biochemical data were performed using R packages "stats" and "vegan" (http://cran.r-project.org/and http://cran.r-project.org/web/packages/vegan/index.html). All data clustering were performed using the "UPGMA" algorithm implemented in the "hclust" function of the R "stats" package. Distances among samples were calculated using "Bray-Curtis" distance implemented in "vegan" package as:

$$i = sample\,``i"; \, j = sample\,``j", BC_{ij} = 2C_{ij}/(S_i + S_j),$$

where C_{ij} = sum of the smaller value for species in common between samples i and j, and S_i and S_j = total number of species in samples i and j, respectively [52]. Variation of biodiversity indexes of cultured bacteria was inspected using ANOVA analysis. Biodiversity indexes were first divided into groups depending on sampling season and different managements of soils and then the analysis was performed. Clustering analysis (UPGMA) on biochemical parameters was performed. Each parameter was first divided in groups, in the same way of previous ANOVA analysis, and averaged. Then, each result obtained was normalized using the maximum-minimum normalization technique in order to make the data comparable. PCA analysis using each biochemical data was performed.

Results and Discussion

Effect of land use on soil chemical and biochemical properties

Soil organic matter (SOM) represents a dynamic system influenced by several factors, including climate, clay content, mineralogy and soil management, which all affect the processes of organic matter transformation and evolution in soil [53], [54]. Both soil fertility and stability are related to the organic matter content of soil. Many functions of SOM are due to its more stabilised fractions, humified materials and balance between the labile and the stabilised fractions [55]. Changes in SOM content are related to changes in microbial biomass turnover, because they reflect the balance between rates of microbial organic matter accumulation and rates of organic matter degradation. The extent of organic matter's organization not only impacts the amount of carbon mineralized but also the type of carbon that is consumed by the microorganisms.

In this study two categories of soil quality indicators were used: organic matter quality indicators and microbial biomass activity indicators (Tables S2 and S3). Our results (Fig. S1) revealed that both land use and sampling season affect the chemical and structural composition of the most stabilised fractions of SOM. A higher content of C_{org} occurred in CO soil in May, and C_{ext} and C_{HA+FA} showed the same pattern, whereas the average values of these parameters were slightly higher in PA soil in November. These results were reflected in the humification parameters, where the humification rate (HR%) can provide quantitative information about the humic substances content normalised with respect to total SOM, while the degree of humification (DH%) provides the amount of the humified carbon in the extracted organic fraction and the humification index (HI) can be considered as an index of soil humification activity as well as of availability of non humified labile fractions [29]. Overall, land use change reduced soil C stocks and labile organic matter at both sampling times except for PA in November. Pasture has a great potential soil organic C stock and, in the long term, grass management systems have nearly equivalent potential to store soil organic C as forest [56]. Potentially mineralizable C (C_0), which indicates the amount of C in the labile fraction of soil organic matter, decreased over sampling time in all soils, and C_{mic} similarly declined. Cultivated soils are characterised by low microbial activity, mainly due to the disappearance of easily decomposable organic compounds through tillage and soil disturbance. As previously found for chemical properties, the metabolic activity responds to the different land uses; in both May and November samplings, qCO_2 was higher in CO soil when compared to the others, resulting in increases stress. Hence, unfavourable conditions result in a decrease in the size of the microbial biomass and the efficiency of C substrates degradation, conducting to an increase in respiration rate per unit of microbial biomass [57].

As beneficial and negative effects of soil management practices are strongly linked to microbial activities and regulate soil quality and functioning [58] the relationship existing between soil management practices and variation in chemical and biochemical parameters was evaluated, by performing a clustering analysis (UPGMA) as reported in Tables S2 and S3. The dendrogram obtained was linked to a heat-map representing all the biochemical parameters variation (Fig. 2A). Data analysis revealed that chemical and biochemical parameters clustered in two different groups each of which corresponding to one of the two parameters analyzed (chemical and biochemical). In both sampling times, TV and CV clustered together, suggesting that seasonal change rather than management regime was a major driving force contributing to vineyard soil fertility. Furthermore, according to the UPGMA clustering, CO formed a separate cluster from that of the other soil uses (Fig. 2A). Interestingly, the chemical and biochemical parameters clustered together in both CO and PA, regardless of the season, in contrast to MM, CV and TV, where they clustered separately in relation to the sampling season. This finding suggests that soils subjected to low human inputs (pasture and cork-oak forest) showed a more stable chemical and biochemical soil composition than those with high human input (managed meadow and vineyard).

The different positions of the variables in the plane of the first two principal components, as revealed by PCA analysis (Fig. 2B), indicated that chemical and biochemical parameters were differentially affected by the various land use types. The first and the second principal components (PC1 and PC2) accounted for 44.59% and 24.47% of the total variance in the data, respectively (Fig. 2B). Both chemical and biochemical parameters were positively affected by PC1 except for HR and DH, whereas PC2 was able to discriminate between the two variable sets, with the most of chemical and biochemical parameters being positively and negatively affected, respectively. The above analyses revealed an important effect of land uses on both chemical and biochemical; on the contrary, only biochemical parameters were affected by seasons.

Influence of land use on soil bacterial community

Community structure analysis by EP-index and r-K strategy. Environmental conditions select organisms that either grow rapidly in uncrowded, nutrient-rich conditions (r-strategists), or can efficiently exploit resources in crowded conditions (K-strategists). In this work, we applied the method developed by De Leij et al. [38], who used the r/K-strategy concept for the characterization of soil bacterial communities. First, we compared the microbial community structures found in the different land uses during each of the two seasons. Since sampling was performed at 20 cm depth, where most microbial activity is known to occur [26], the main changes are expected through conversion from one soil management system to another one [59]. Microbial community structure found in the different land uses during each of the two seasons was investigated by means of EP index that is a measure of both richness (i.e. total number of species in the community) and evenness (i.e. how evenly individuals in the community are distributed over the different species) of groups of microorganisms with similar developmental characteristics.

The cultured bacteria belonging to fast-growing organisms, especially the r-strategists detected on day 2, dominated in all land uses and in both seasons (Table S4). The MM soil exhibited a lower EP index and a lower percentage of r-strategists in both seasons, possibly due to amensalism from a dominant bacterial group in the community. Variation of EP index of cultured bacteria was inspected using an ANOVA analysis of variance (Fig. 3A). Interestingly, statistical analysis revealed a significant effect of land uses on bacterial community structure ($P<0.001$) as well as an interaction effect between land uses and season change ($P<0.001$) on EP index. Variation of EP index due to seasonal changes was significant in the soil with higher human impact (TV) when compared with the other soils. Total bacterial concentrations varied significantly in respect of season ($P<0.001$) being higher in spring than in autumn (Table S2) in all but CO soils, with significant differences between CO and PA in spring, and between CO and MM, and MM and PA in autumn.

Overall results suggest that land uses affected cultured bacterial communities. Soils with low human impact (cork-oak forest) have a more stable bacterial density and show a less variation of bacterial community structure across seasons than soils subjected to high human impact such as tilled vineyard.

Bacterial community profiling by CD-DGGE and T-RFLP analyses. Culture dependent DGGE (CD-DGGE) fingerprinting of the 16S rRNA was used to characterize mixed bacterial communities recovered on agar plates. CD-DGGE represents a useful technique to follow the dynamics of distinct culturable fractions of the soil bacterial community in relation to physical, chemical and biological changes in the soil environment [60]. Since culture-dependent and culture-independent methods likely profile distinct fractions of the soil bacterial community with unique ecological roles [61], culturable bacteria may provide an ecologically relevant complement to culture-independent community characterization [62].

By pooling the bacterial cells growing on individual agar plates, we obtained a culture-dependent bacterial community. The analysis of DGGE profiles revealed clear banding patterns for each land-use and plant-cover types of sufficient complexity to investigate differences in soil microbial communities and identified stable communities with highly reproducible profiles (Fig. S2). A different community composition among land-use types was found as evidenced by the presence of different dominant signals; this finding indicated a compositional shift among soils examined in spring and autumn and subjected to different anthropogenic impact. Bacterial diversity was investigated through richness (R), Shannon-Weaver (H'), and Simpson (D) indices. Analysis of variance (ANOVA) confirmed the differences in the distribution of bacterial species due to land uses and sampling seasons ($P<0.001$). Both R and H' indices decreased from May to November in MM samples and increased in TV and CV samples, and the complementary opposite trend was observed for D index (Fig. 3B). Otherwise, cultured bacterial community in CO and PA soils did not vary significantly over seasons. Most likely, these results reflect the impact of both land-use and vegetation type and coverage on soil microbial communities. In particular, the low shift of biodiversity indices observed in CO and PA seems to be correlated to a higher stability of bacterial populations in natural habitats with low human impact, whereas populations inhabiting more anthropogenic areas tend to be more variable. Land-use type and, in particular, differences in vegetation dynamics may have a large role in modulating the temporal variability in soil bacterial communities. As observed by Lauber and co-workers [3], soils from the different land-use types did not exhibit identical temporal dynamics even though all the soils were located in close proximity and exposed to the same climatic conditions. Diversity indices obtained from T-RFLP analysis performed on total bacterial DNA (Fig. S3) partially confirmed the trends of diversity shown by CD-DGGE, though observed differences were not statistically significant (Table S5).

The unweighted pair-group method using arithmetic averages (UPGMA) of bacterial community profiling by CD-DGGE revealed a high diversity of the bacterial communities in each land-use and plant-cover soils. The similarity between the DGGE patterns of the soil bacterial communities revealed three distinct clusters (Fig. 4). Samples collected in autumn grouped into a separate cluster with about 52% similarity, whereas samples collected in spring grouped into two separate clusters, each of which composed by samples sharing about 40% similarity. The highest similarity values were shown by clusters based on soil land uses, suggesting a low bacterial variation within each soil type. This finding is consistent with previous reports showing that land-use type was the most important factor in determining the composition of soil microbial community [3].

The clustering results based on CD-DGGE were in agreement with culture-independent T-RFLP analysis that confirmed a different seasonal distribution of soil bacterial populations with subgroups associated to different land uses. Most of samples retrieved from the same soil (CO, PA, and MM) clustered together, whereas samples retrieved from TV and CV were often intermixed in the UPGMA dendrogram (Fig. S4). As culture-dependent and culture-independent profiles can separately resolve unique, diverse, and equally complex fractions of the soil bacterial community [60], our combined results from both CD-DGGE and T-RFLP methods, applied on culturable and total fractions, respectively, revealed an interaction effect between land uses and season change in affecting soil bacterial communities.

Taxa responses to land use determined by phylogenetic affiliation of soil bacterial isolates. Taxonomic affiliation was investigated on a total of 203 bacterial isolates. In detail, 20–21 colonies recovered from each soil sample at each season and showing different morphologies (for a total of 101 bacterial isolates in spring and 102 in autumn) were subjected to DNA extraction and PCR amplification of 16S rRNA gene. An amplicon of the expected size (about 1500 bp) was obtained from each isolate, and its nucleotide sequence was determined and submitted to GenBank (Table S6). Although it is generally accepted that not all bacteria, including types of soil bacteria, are culturable, the isolation of bacteria by agar plate cultivation and subsequent phylogenetic analysis permit to isolate and identify previously uncultured representatives or even new members of certain bacterial species for further analysis of their metabolic function. Therefore, even if the metagenome sequencing is becoming the most powerful tool to investigate microbial communities, the ability to isolate indigenous strains actually remains the unique way to further characterize and select soil bacteria showing interesting properties. Additionally, standard cultivation techniques have been shown to be able to capture members of the soil rare biosphere which could not be detected by metagenome sequencing [63].

The 203 sequences obtained were compared with those present in the GenBank databases by using the BLAST [64] program. Results showed that most of 16S rRNA gene sequences matched NCBI database sequences at 99–100% of similarity at the genus level, with *Arthrobacter*, *Bacillus*, *Stenotrophomonas*, *Pseudomonas* and *Burkholderia* as the most representative genera. So, identification at the genus level was achieved in 103 isolates and at the species level in 72 isolates, while 12 isolates were only affiliated to taxa level higher than genus and 16 remained unidentified (Table S6). Further comparison with GenBank databases by using the Seqmatch tool of the RDP indicated that 16S rRNA gene sequences were affiliated to four phyla: *Proteobacteria* (classes α, β and γ), *Bacteroidetes* (classes *Flavobacteria* and *Sphingobacteria*), *Actinobacteria* and *Firmicutes*. Even though the RDP analysis does not permit to affiliate a bacterial isolate to a given species, it allowed to assess the genus of almost all isolates, including those not identified through the BLAST search. In fact, a boostrap value of at least 80% is satisfactory for RDP requirement. Almost all our sequences gave rise to 100% boostrap, 12 ranged from 89% to 99%, and only two of them only resulted in low values (24 and 54% respectively, referred to two spring isolates).

Data obtained from both approaches allowed to assess that Berchidda soil is colonized by bacteria included in the classes of γ-*Proteobacteria* (with a prevalence of *Pseudomonas* and *Stenotrophomonas* genera), *Actinobacteria* (in particular, the genus *Arthrobacter*), β-*Proteobacteria* (with a prevalence of *Burkholderia* genus), *Bacilli* (with the genus *Bacillus* as the dominant one), Flavobacteria and Sphingobacteria. The relative abundances of the different classes identified by RDP across the different samples related to soil uses and season are represented in Figure 5. A putative taxonomic description of total bacterial community, derived from the interpretation of T-RFLP data (Fig. S3) is reported in Figure 6, in which the largest fraction of the community is represented by members of *Proteobacteria*, with α-*Proteobacteria* as the most abundant class.

Differences in class composition across the site were observed suggesting that the microbial composition changes in response to land uses. In fact, all the seven classes (α, β and γ *Proteobacteria*, *Sphingobacteria* *Flavobacteria*, *Actinobacteria* and *Bacilli*) were present in CV, PA, and CO soils, while all classes but α-*Proteobacteria* in TV and all classes but *Sphingobacteria* in MM were found. The observed large diffusion of *Proteobacteria* in all soil-uses is in agreement with previously reported data concerning soil bacterial communities in different land-use systems [65]. Within *Proteobacteria*, the majority of the isolates fell into the gamma subgroup, which showed higher relative abundance compared to that of any other taxa at the class level while a few were alfa proteobacteria, especially in vineyards and pasture. In all soil-uses, were also found *Actinobacteria*, in agreement with previous work [65], and β-*Proteobacteria*, like *Burkholderia* sp., previously detected by Pastorelli and co-workers [19] who analysed the denitrifying bacterial communities present in the same Berchidda soil samples. Among Bacteroidetes, it has to be noted the relative abundance of the genera *Chryseobacterium* sp. (in all soil uses) and *Flavobacterium* sp. (in all soil uses, but MM), which include isolates already observed in Korea soils [66], and bacteria with plant growth promoting properties recovered in Iran soil [67], respectively. As already pointed out by Fierer and co-workers [68], the β-*Proteobacteria* and *Bacteroidetes* follow copiotrophic lifestyles and their relative abundance were highest in soils with high C availability. In general, copiotrophic bacteria should have higher growth rates and traditional culturing methods are likely to select for microorganisms that can grow rapidly in high resource environments.

Strong differences in class composition were observed in each soil when sampled in the two seasons: for instance, *Bacilli* and *Actinobacteria* dominate in spring (particularly, *Bacilli* in CV and CO, while *Actinobacteria* in MM and PA), whilst β-*Proteobacteria* tend to dominate in autumn. Six out of the seven identified classes (β and γ-*Proteobacteria*, *Flavobacteria*, *Actinobacteria* and *Bacilli*) were recovered in both seasons whilst *Sphingobacteria* were recovered only in autumn. In all soils, *Actinobacteria* were prevalent in spring, β *Proteobacteria* predominated in autumn, while β and γ-*Proteobacteria* in both spring and autumn. When genera composition was used to infer the diversity indices of cultured bacteria (Table S7) an increase of diversity (estimated as Shannon H and Evenness) from May to November was present for all soils, but CO, where Shannon H was slightly higher in May than in November. In particular, MM soil showed the highest increases for most indices, especially for alpha diversity (*i.e.* the species richness and evenness within a sample) that has often been correlated with ecosystem stability and functionality [69]. It must be noted that cultivated bacterial populations did not fluctuate with seasonal changes in soils with low human input (cork-oak forest). The bacterial communities of the cork-oak forest soil were similar in richness and composition, furthering the point that in a community with moderate disturbance, new individuals and groups could be introduced in a manner that promotes competition and diversity of the community, thus establishing a more stable community [65].

These data suggest that shifts from forest to managed meadow and vineyard result in changes of bacterial communities composition. Previous studies have also shown that shifts from forest to grassland soil [70], from cultivated system to pasture [71] and from grazed pine forest to cultivated crop and grazed pasture [65] resulted in significant changes in bacterial community composition. Our results suggest that the use of culture-dependent 16S rRNA gene sequencing along with traditional analysis of soil physiochemical properties may provide insight into the ecological relevance of soil bacterial taxa.

Conclusion

Overall, data obtained in this work revealed an important effect of land uses on both chemical and biochemical soil parameters. Soil bacterial communities were seasonally distinct and exhibited compositional shifts that tracked with changes in land use and soil management. This study, combining the pedological and biochemical data with microbiological and molecular analysis, furnishes a good methodological approach to describe the influence of different soil managements on soil microbial community structure. In fact, the results demonstrate that, in the same pedological conditions, long-term soil management influence the community structure; i.e., soils subjected to low human inputs (cork-oak forest and pasture) showed a more stable chemical and biochemical soil composition as well as bacterial community than those with high human input (vineyards and managed meadow). Further research is required to determine whether the observed shifts in bacterial community composition produce parallel changes in the functional attributes of these communities across soil types under different long-term management regimes. The use of culture-independent approaches, like metabarcoding and metagenome sequencing, will make it possible to identify the specific drivers of land-use dynamics exhibited by soil bacterial communities and to give a complete picture of the bacterial communities in a typical Mediterranean agro-silvo-pastoral system.

Supporting Information

Figure S1 Box-plot analysis showing the frequency distribution of physical-chemical and biological properties of the five Sardinia soils.

Figure S2 Examples of CD-DGGE profiles of the soil bacterial communities associated to the different land uses in May and November. A) From the left to right: hayland pasture rotation (PA), tilled vineyard (TV), grass covered vineyard (CV) in May; B) managed meadow (MM), cork-oak forest (CO), hayland pasture rotation (PA) in May; C) grass covered vineyard (CV) in May, grass covered vineyard (CV) in November, tilled vineyard (TV) in May; D) cork-oak forest (CO), hayland pasture rotation (PA), managed meadow (MM) in November.

Figure S3 Examples of T-RFLP profiles obtained after digestion with *Msp*I (A) and *Rsa*I (B) restriction enzymes of amplified of 16S rRNA gene sequences.

Figure S4 Cluster analysis of T-RFLP patterns generated by *Msp*I and *Rsa*I digestion of 16S rRNA gene sequences. The UPGMA cluster analysis based on Jaccard similarity matrix was calculated for each set of samples using the "hclust" function of the R "stats" package. The scale bar represents the percent of dissimilarity.

Table S1 Pedological profiles and classification of the soils investigated.

Table S2 Determination of total organic carbon soil (C_{org}), extractable carbon (C_{ext}), humified carbon (C_{HA+FA}), non humified carbon (C_{NH}) and humification parameters of the five Sardinian soils.

Table S3 Biochemical parameters measured in the five Sardinian soils.

Table S4 r/k bacterial strategists, total culturable bacteria and EPI index measured in soils under different long-term management practices.

Table S5 Diversity indices of total bacterial communities as inferred from T-RFLP profiles.

Table S6 Phylogenetic affiliations of 203 randomly selected soil bacterial isolates based on comparative analysis of their 16S rRNA gene sequences.

Table S7 The ratio of diversity indices related to the abundance of the different genera detected in cultured isolates between November and May.

Acknowledgments

We are grateful to P.P. Roggero (University of Sassari) and his team (G. Seddaiu, G. Urracci and L. Doro) for their contribution to collect the Sardinian soils and for providing data on experimental area. The experimental site was chosen on the basis of vegetation and soil surveys made in collaboration with S. Madrau, S. Bagella, R. Filigheddu, M.C. Caria, and E. Farris (University of Sassari). We thank R. Pastorelli for her help in soil sampling, S. Tabacchioni and L. Chiarini for bacterial isolation, C. Cantale and M. Sperandei for figure preparation, M.T. Rubino for chemical-biochemical analysis, and S. Cesarini and L. Pirone for their valuable suggestions. The research was carried out in the context of the FISR SOILSINK research project coordinated by R. Francaviglia (CRA-RPS, Rome) (?http://soilsink.entecra.it).

Author Contributions

Conceived and designed the experiments: A. Bevivino RF CD A. Benedetti. Performed the experiments: A. Bevivino PP MSP MCP CD LL. Analyzed the data: A. Bevivino PP GB AF AM CD LL. Contributed reagents/materials/analysis tools: A. Bevivino PP GB AF MSP AM LL MCP RF CD A. Benedetti. Contributed to the writing of the manuscript: A. Bevivino RF GB AF CD LL A. Benedetti.

References

1. Tiedje JM, Asuming-Brempong S, Nusslein K, Marsh TL, Flynn SJ (1999) Opening the black box of soil microbial diversity. Appl Soil Ecol 13: 109–122. doi:10.1016/S0929-1393(99)00026-8.

2. Nannipieri P, Ascher J, Ceccherini MT, Landi L, Pietramellara G, et al. (2003) Microbial diversity and soil functions. Eur J Soil Sci 54: 655–670. doi:10.1046/j.1365-2389.2003.00556.x.

3. Lauber CL, Ramirez KS, Aanderud Z, Lennon J, Fierer N (2013) Temporal variability in soil microbial communities across land-use types. ISME J 7: 1641–1650. doi:10.1038/ismej.2013.50.

4. Myers N, Mittermeier RA, Fonseca GAB, Fonseca GAB, Kent J (2000) Biodiversity hotspots for conservation priorities. Nature 403: 853–858. doi:10.1038/35002501.

5. Puddu G, Falcucci A, Maiorano L (2011) Forest changes over a century in Sardinia: implications for conservation in a Mediterranean hotspot. Agrofor Syst 85: 319–330. doi:10.1007/s10457-011-9443-y.

6. Médail F, Quézel P (1999) Biodiversity Hotspots in the Mediterranean Basin: Setting Global Conservation Priorities. Conserv Biol 13: 1510–1513. doi:10.1046/j.1523-1739.1999.98467.x.

7. Blondel J, Aronson J (1999) Biology and wildlife of the Mediterranean region. Oxford University Press, Oxford.

8. Salis L, Marrosu M, Bagella S, Sitzia M RP (2010) Grassland management, forage production and plant biodiversity in a Mediterranean grazing system. In: Porqueddu C, Ríos S, editors. The contributions of grasslands to the conservation of Mediterranean biodiversity. Zaragoza: CIHEAM/CIBIO/FAO/SEEP, Vol. 185. 181–185.

9. Gonçalves P, Alcobia S, Simões L, Santos-Reis M (2012) Effects of management options on mammal richness in a Mediterranean agro-silvo-pastoral system. Agrofor Syst 85: 383–395. doi:10.1007/s10457-011-9439-7.

10. Bacchetta G, Bagella S, Biondi E, Farris E, Filigheddu R, et al. (2004) A contribution to the knowledge of the order Quercetalia ilicis Br. -Bl. ex Molinier 1934 of Sardinia. Fitosociologia 41: 29–51.

11. Orgiazzi A, Lumini E, Nilsson RH, Girlanda M, Vizzini A, et al. (2012) Unravelling soil fungal communities from different Mediterranean land-use backgrounds. PLoS One 7: e34847. doi:10.1371/journal.pone.0034847.

12. Francaviglia R, Benedetti A, Doro L, Madrau S, Ledda L (2014) Influence of land use on soil quality and stratification ratios under agro-silvo-pastoral Mediterranean management systems. Agric Ecosyst Environ 183: 86–92.

13. Council of the European Communities (1992) Council Directive 92/43/EEC of 21 May 1992 on the conservation of natural habitats and of wild fauna and flora. Off J Eur Communities 35: 7–50.

14. Council of the European Communities (2001) Commission Regulation (EC) No 1808/2001 of 30 August 2001 laying down detailed rules concerning the implementation of Council Regulation (EC) No 338/97 on the protection of species of wild fauna and flora by regulating trade therein. Off J Eur Communities L250: 1–43.

15. Aru A, Baldaccini P, Delogu G, Dessena M, Madrau S, et al. (1990) Carta dei Suoli della Sardegna, scala 1/250.000. Assessorato alla programmazione e all'assestamento del territorio, Centro Regionale Programmazione, Dip. Sc. della Terra, Università di Cagliari, Italy.

16. Bacchetta G, Bagella S, Biondi E, Farris E, Filigheddu R, et al. (2009) Forest vegetation and serial vegetation of Sardinia (with map at the scale 1:350,000). Fitosociologia 46: 3–82.

17. Bagella S, Caria MC (2011) Vegetation series: a tool for the assessment of grassland ecosystem services in Mediterranean large-scale grazing systems. Fitosociologia 48: 47–54.

18. Lumini E, Orgiazzi A, Borriello R, Bonfante P, Bianciotto V (2010) Disclosing arbuscular mycorrhizal fungal biodiversity in soil through a land-use gradient using a pyrosequencing approach. Environ Microbiol 12: 2165–2179. doi:10.1111/j.1462-2920.2009.02099.x.

19. Pastorelli R, Landi S, Trabelsi D, Piccolo R, Mengoni A, et al. (2011) Effects of soil management on structure and activity of denitrifying bacterial communities. Appl Soil Ecol 49: 46–58. doi:10.1016/j.apsoil.2011.07.002.

20. Lagomarsino A, Benedetti A, Marinari S, Pompili L, Moscatelli MC, et al. (2011) Soil organic C variability and microbial functions in a Mediterranean agro-forest ecosystem. Biol Fertil Soils 47: 283–291. doi:10.1007/s00374-010-0530-4.

21. Francaviglia R, Coleman K, Whitmore AP, Doro L, Urracci G, et al. (2012) Changes in soil organic carbon and climate change - Application of the RothC model in agro-silvo-pastoral Mediterranean systems. Agric Syst 112: 48–54.

22. Seddaiu G, Porcu G, Ledda L, Roggero PP, Agnelli A, et al. (2013) Soil organic matter content and composition as influenced by soil management in a semi-arid Mediterranean agro-silvo-pastoral system. Agric Ecosyst Environ 167: 1–11. doi:http://dx.doi.org/10.1016/j.agee.2013.01.002.

23. Lai R, Lagomarsino A, Ledda L, Roggero PP (2014) Variation in soil C and microbial functions across tree canopy projection and open grassland microenvironments. Turkish J Agric For 38: 62–69. doi:10.3906/tar-1303-82.

24. Soil Survey Staff (2006) Keys to soil taxonomy. In: United States Department of Agriculture NRCS, editor. Natural Resoiurces Conceration Service.

25. Doran JW, Elliott ET, Paustian K (1998) Soil microbial activity, nitrogen cycling, and long-term changes in organic carbon pools as related to fallow tillage management. Soil Tillage Res 49: 3–18.

26. O'Brien HE, Parrent JL, Jackson JA, Moncalvo J-M, Vilgalys R (2005) Fungal community analysis by large-scale sequencing of environmental samples. Appl Environ Microbiol 71: 5544–5550.

27. Stotzky G, Goos RD, Timonin MI (1962) Microbial changes occurring in soil as a result of storage. Plant Soil 16: 1–18. doi:10.1007/BF01378154.

28. Springer U KJ (n.d.) Prüfung der Leistungsfähigkeit von einigen wichtigeren Verfahren zur Bestimmung des Kohlenstoffs mittels Chromschwefelsäure sowie Vorschlag einer neuen Schnellmethode. J Plant Nutr Soil Sci 64: 1–26 (in German).

29. Ciavatta C, Govi M, Vittori Antisari L, Sequi P (1990) Characterization of humified compounds by extraction and fractionation on solid polyvinylpyrrolidone. J Chromatogr 509: 141–146.

30. Sequi P, De Nobili M, Leita L, Cercignani G (1986) A new index of humification. Agrochimica 30: 175–179.

31. Vance ED, Brookes PC, Jenkinson DS (1987) An extraction method for measuring soil microbial biomass C. Soil Biol Biochem 19: 703–707. doi:10.1016/0038-0717(87)90052-6.

32. Badalucco L, Grego S, Dell'Orco S, Nannipieri P (1992) Effect of liming on some chemical, biochemical, and microbiological properties of acid soils under spruce (Picea abies L.). Biol Fertil Soils 14: 76–83. doi:10.1007/BF00336254.

33. Riffaldi R, Saviozzi A, Levi-Minzi R (1996) Carbon mineralization kinetics as influenced by soil properties. Biol Fertil Soils 22: 293–298. doi:10.1007/BF00334572.

34. Dilly O, Munch J-C (1998) Ratios between estimates of microbial biomass content and microbial activity in soils. Biol Fertil Soils 27: 374–379. doi:10.1007/s003740050446.

35. Anderson T-H, Domsch KH (1989) Ratios of microbial biomass carbon to total organic carbon in arable soils. Soil Biol Biochem 21: 471–479. doi:10.1016/0038-0717(89)90117-X.

36. Pinzari F, Trinchera A, Benedetti A, Sequi P (1999) Use of biochemical indices in the mediterranean environment: Comparison among soils under different forest vegetation. J Microbiol Methods 36: 21–28.

37. Smalla K, Wieland G, Buchner A, Zock A, Parzy J, et al. (2001) Bulk and Rhizosphere Soil Bacterial Communities Studied by Denaturing Gradient Gel Electrophoresis: Plant-Dependent Enrichment and Seasonal Shifts Revealed. 67: 4742–4751. doi:10.1128/AEM.67.10.4742.

38. De Leij FAAM, Whipps JM, Lynch JM (1994) The use of colony development for the characterization of bacterial communities in soil and on roots. Microb Ecol 27: 81–97.

39. Grifoni A, Bazzicalupo M, Di Serio C, Fancelli S, Fani R (1995) Identification of Azospirillum strains by restriction fragment length polymorphism of the 16S rDNA and of the histidine operon. FEMS Microbiol Lett 127: 85–91. doi:10.1016/0378-1097(95)00042-4.

40. Trabelsi D, Mengoni A, Elarbi Aouani M, Mhamdi R, Bazzicalupo M (2009) Genetic diversity and salt tolerance of bacterial communities from two Tunisian soils. Ann Microbiol 59: 25–32. doi:10.1007/BF03175594.

41. Mengoni A, Tatti E, Decorosi F, Viti C, Bazzicalupo M, et al. (2005) Comparison of 16S rRNA and 16S rDNA T-RFLP approaches to study bacterial communities in soil microcosms treated with chromate as perturbing agent. Microb Ecol 50: 375–384. doi:10.1007/s00248-004-0222-4.

42. Hammer Ø, Harper DAT, Ryan PD (2001) Past: Paleontological Statistics Software Package for Education and Data Analysi. Palaeontol Electron 4: 1–9. doi:10.1016/j.bcp.2008.05.025.

43. Duineveld B, Kowalchuk G, Keijzer A, van Elsas J, van Veen J (2001) Analysis of bacterial communities in the rhizosphere of chrysanthemum via denaturing gradient gel electrophoresis of PCR-amplified 16S rRNA as well as DNA fragments coding for 16S rRNA. Appl Environ Microbiol 67: 172–178.

44. Ausubel FM, Brent R, Kingston RE, Moore DD, Seidman JG, et al. (1987) Unit 2.4 Preparation of genomic DNA from bacteria. Curr Protoc Mol Biol: 2.4.1–2.4.2.

45. Di Cello F, Pepi M, Baldi F, Fani R (1997) Molecular characterization of an n-alkane-degrading bacterial community and identification of a new species, Acinetobacter venetianus. Res Microbiol 148: 237–249.

46. El Fantroussi S, Verschuere L, Verstraete W, Top EM (1999) Effect of phenylurea herbicides on soil microbial communities estimated by analysis of 16S rRNA gene fingerprints and community-level physiological profiles. Appl Environ Microbiol 65: 982–988.

47. Cremonesi L, Firpo S, Ferrari M, Righetti PG, Gelfi C (1997) Double-gradient DGGE for optimized detection of DNA point mutations. Biotechniques 22: 326–330.

48. Shannon C, Weaver W (1963) The mathematical theory of communication. University of Illinois Press, Urbana. doi:10.1145/584091.584093.

49. Simpson E (1949) Measurement of diversity. Nature 163: 688–688. doi:10.1038/163688a0.

50. Whittaker RH (1972) Evolution and measurement of species diversity. Taxon 21: 213. doi:10.2307/1218190.

51. Hall TA (1999) BioEdit: a user-friendly biological sequence alignment editor and analysis program for Windows 95/98/NT. Nucleic Acids Symp 41: 95–98. doi:citeulike-article-id:691774.

52. Bray J, Curtis J (1957) An ordination of the upland forest communities of southern Wisconsin. Ecol Monogr 27: 325–349.

53. Haider K (1992) Problems related to the humification processes in soils of temperate climates. In: Stotzky G, Bollag J-M, editors. Soil Biochemistry. Marcel Dekker, New York. 55–94.

54. Oades JM (1995) An overview of processes affecting the cycling of organic carbon in soils. In: Zepp R, Sonntag C, editors. Role of non living organic matter in the earth's carbon cycle. John Wiley, New York. 293–303.

55. Stevenson FJ, Cole MA (1999) Cycles of soil: carbon, nitrogen, phosphorus, sulfur, micronutrients. 2nd ed. John Wiley, New York.

56. Franzluebbers AJ, Stuedemann JA, Schomberg HH, Wilkinson SR (2000) Soil organic C and N pools under long-term pasture management in the Southern Piedmont USA. Soil Biol Biochem 32: 469–478. doi:10.1016/S0038-0717(99)00176-5.

57. Anderson T, Domsch KH (1993) The metabolic quotient for CO2 (qCO2) as a specific activity parameter to assess the effects of environmental conditions, such as pH, on the microbial biomass of forest soils. Soil Biol Biochem 25: 393–395.

58. Benedetti A, Brookes P, Lynch J (2005) Conclusive remarks. Part I: Approaches to defining, monitoring, evaluating and managing soil quality. In: Bloem J, Hopkins D, Benedetti A, editors. Soil Quality. CABI; Wallingford. 63–70.

59. Conant RT, Paustian K, Elliott ET (2001) Grassland management and conversion into grassland: effects on soil carbon. Ecol Appl 11: 343–355. doi:10.1890/1051-0761(2001)011[0343:GMACIG]2.0.CO;2.

60. Edenborn SL, Sexstone AJ (2007) DGGE fingerprinting of culturable soil bacterial communities complements culture-independent analyses. Soil Biol Biochem 39: 1570–1579.

61. Garland JL (1999) Potential and limitations of BIOLOG for microbial community analysis. In: Bell C, Brylinsky M, Johnson-Green P, editors. Microbial Biosystems: new frontiers. Proceedings of the 8th international symposium on microbial ecology. Atlantic Canada Society for Microbial Ecology. Halifax, Canada. 1–7.

62. Ellis RJ, Morgan P, Weightman AJ, Fry JC (2003) Cultivation-dependent and - independent approaches for determining bacterial diversity in heavy-metal-contaminated soil. Appl Environ Microbiol 69: 3223–3230. doi:10.1128/AEM.69.6.3223.

63. Shade A, Hogan CS, Klimowicz AK, Linske M, McManus PS, et al. (2012) Culturing captures members of the soil rare biosphere. Environ Microbiol 14: 2247–2252. doi:10.1111/j.1462-2920.2012.02817.x.

64. Altschul SF, Madden TL, Schäffer AA, Zhang J, Zhang Z, et al. (1997) Gapped BLAST and PSI-BLAST: a new generation of protein database search programs. Nucleic Acids Res 25: 3389–3402.

65. Shange RS, Ankumah RO, Ibekwe AM, Zabawa R, Dowd SE (2012) Distinct soil bacterial communities revealed under a diversely managed agroecosystem. PLoS One 7: e40338. doi:10.1371/journal.pone.0040338.

66. Weon H-Y, Kim B-Y, Yoo S-H, Kwon S-W, Stackebrandt E, et al. (2008) Chryseobacterium soli sp. nov. and Chryseobacterium jejuense sp. nov., isolated from soil samples from Jeju, Korea. Int J Syst Evol Microbiol 58: 470–473. doi:10.1099/ijs.0.65295-0.

67. Soltani A, Khavazi K, Asadi-Rahmani H, Omidvari M, Abaszadeh Dahaji P (2010) Plant growth promoting characteristics in some Flavobacterium spp. isolated from soils of Iran. J Agri Sci 2: 106–115. Available: http://www.ccsenet.org/journal/index.php/jas/article/view/8404/6221.

68. Fierer N, Bradford MA, Jackson RB (2007) Toward an ecological classification of soil bacteria. Ecology 88: 1354–1364. doi:10.1890/05-1839.

69. Girvan MS, Campbell CD, Killham K, Prosser JI, Glover LA (2005) Bacterial diversity promotes community stability and functional resilience after perturbation. Environ Microbiol 7: 301–313. doi:10.1111/j.1462-2920.2005.00695.x.

70. Nacke H, Thürmer A, Wollherr A, Will C, Hodac L, et al. (2011) Pyrosequencing-based assessment of bacterial community structure along different management types in German forest and grassland soils. PLoS One 6: e17000. doi:10.1371/journal.pone.0017000.

71. Acosta-Martínez V, Dowd S, Sun Y, Allen V (2008) Tag-encoded pyrosequencing analysis of bacterial diversity in a single soil type as affected by management and land use. Soil Biol Biochem 40: 2762–2770. doi:10.1016/j.soilbio.2008.07.022.

8

The Public and Professionals Reason Similarly about the Management of Non-Native Invasive Species: A Quantitative Investigation of the Relationship between Beliefs and Attitudes

Anke Fischer[1]*, Sebastian Selge[1,2¤], René van der Wal[2], Brendon M. H. Larson[3]

1 Social, Economic and Geographical Sciences Group, James Hutton Institute, Aberdeen, Scotland, United Kingdom, 2 Aberdeen Centre for Environmental Sustainability (ACES) and School of Biological Sciences, University of Aberdeen, Aberdeen, Scotland, United Kingdom, 3 Department of Environment and Resource Studies, University of Waterloo, Waterloo, Ontario, Canada

Abstract

Despite continued critique of the idea of clear boundaries between scientific and lay knowledge, the 'deficit-model' of public understanding of ecological issues still seems prevalent in discourses of biodiversity management. Prominent invasion biologists, for example, still argue that citizens need to be educated so that they accept scientists' views on the management of non-native invasive species. We conducted a questionnaire-based survey with members of the public and professionals in invasive species management (n = 732) in Canada and the UK to investigate commonalities and differences in their perceptions of species and, more importantly, how these perceptions were connected to attitudes towards species management. Both native and non-native mammal and tree species were included. Professionals tended to have more extreme views than the public, especially in relation to nativeness and abundance of a species. In both groups, species that were perceived to be more abundant, non-native, unattractive or harmful to nature and the economy were more likely to be regarded as in need of management. While perceptions of species and attitudes towards management thus often differed between public and professionals, these perceptions were linked to attitudes in very similar ways across the two groups. This suggests that ways of reasoning about invasive species employed by professionals and the public might be more compatible with each other than commonly thought. We recommend that managers and local people engage in open discussion about each other's beliefs and attitudes *prior* to an invasive species control programme. This could ultimately reduce conflict over invasive species control.

Editor: Zoe G. Davies, University of Kent, United Kingdom

Funding: This study was funded by the University of Aberdeen, Scotland's RERAD (Rural and Environment Research and Analysis Directorate) Programme 3 and a Standard Research Grant from the Social Sciences and Humanities Research Council of Canada (SSHRC). The funders had no role in study design, data collection and analysis, decision to publish, or preparation of the manuscript.

Competing Interests: The authors have declared that no competing interests exist.

* Email: anke.fischer@hutton.ac.uk

¤ Current address: Department of Psychology, Otto-von-Guericke-Universität, Magdeburg, Germany

Introduction

There is a longstanding debate about the differences between lay and expert knowledge in the ecological realm [1], and about the challenges for biodiversity governance where expert and lay perspectives are incongruent with each other [2]. Several studies have addressed differences between lay and expert knowledges and perspectives in relation to nature conservation and the management of landscapes and biodiversity [3,4]. For example, previous research has concluded that the general public perceives biodiversity inaccurately relative to the 'actual' biodiversity levels assessed by experts, and thus diagnosed a public lack of ecological knowledge [5]. According to a recent review of scientists' views of the public [6], such conclusions may indicate a 'deficit model' of public understanding that is still prevalent among many scientists, who tend to see 'the public' as a homogenous, amorphous body that needs to be educated.

However, there is an increasingly strong critique of the deficit model. This critique originates in the sociology of science and argues that the boundary between laypeople and experts is diffuse and permeable [7–9]: "Expert knowledge is open to reappropriation by anyone with the necessary time and resources to become trained, and [...] there is a continuous filtering back of expert theories, concepts and findings to the lay population" [10]. Recognition is growing, also in the ecological realm, that the dichotomy between experts and laypeople is less clear than it might have seemed twenty years ago [1], and several recent studies in conservation science provide evidence for this alternative view by highlighting the links between concepts used in both expert and lay domains [4,11,12]. Most empirical research in this context focuses on comparisons of these understandings, and reports that experts and laypeople differ in some aspects, for example, in their risk perceptions, preferences for landscapes and management scenarios, or knowledge of plant and animal species [3,5,13,14], whereas they might concur in others, for example, in their appreciation of woodland features [15]. While such findings might be interesting, insights into structures of expert and lay thought [4] and the mechanisms behind such similarities and differences remain scarce.

In this study, we investigate the beliefs and attitudes of laypeople and experts regarding the perceived non-nativeness and invasiveness of animal and plant species. This is relevant to the debate over the deficit model because, even in recent publications, some invasion biologists still maintain a clear dichotomy between themselves and the public: "Although most invasion scientists endorse a normative commitment towards biodiversity, their proper role as scientists, in terms of public discourse, is to educate citizens in a way that informs debate within society about how to *think about* and manage invasions" [16] (emphasis added; see [17] for other examples). Such statements do not seem to acknowledge the permeability described above [10], the potential diversity of public views of invasive species, or the range of perspectives even among invasion biologists [18,19]. One consequence attributed to the divide between laypeople and experts is that invasive species control has sometimes been delayed by public opposition, which has led to explicit calls for more insights into the obstacles to "public buy-in" to species management [20–23].

To date, a number of studies have investigated either public [24–29] or expert [18,19,30,31] views of non-native invasive species. In some instances, such studies have focused on one species, for example, buffel grass (*Cenchrus ciliaris*) [23], and it remains unclear whether their results can be generalised. Only very few studies have examined views of experts and the public simultaneously. While the terminology is sometimes used ambiguously, the term 'experts' can include practitioners, academics and hobby experts in invasive species management, whereas 'scientists' are a subset of experts active in academic research on this topic. 'Professionals' can be defined as a subset of experts who have specific skills or knowledge related to invasive species and whose work engages directly with them [10]. A study in the Doñana region of Spain included both professionals and non-professionals, and found that some groups of the public, such as nature tourists, held similar knowledge and attitudes as conservation professionals [32]. Another study, in Scotland, identified the arguments that members of the public and scientists used to support their views on a range of management options for invasive species [33]. These arguments consisted of perceived species characteristics, such as the harmfulness of a species in relation to nature or the economy.

The present study builds on this literature and quantitatively contrasts how members of the public and professionals, from a large and cross-cultural sample, view the management of multiple species. Most critically, and unlike most previous studies (see [4] for an exception), we do not solely concentrate on comparisons of single constructs, for example, knowledge about species or attitudes towards management options, because such an approach necessarily remains at a somewhat descriptive level. Instead, we investigate the arguments for and against certain management options by explicitly investigating the statistical *relationships* between beliefs about species and attitudes towards their management. In the terminology used in science education, we thus distinguish between content knowledge and ways of reasoning [34,35]. This allows us to compare the *ways* in which beliefs inform attitudes, rather than just either beliefs or attitudes on their own.

According to social psychological frameworks, such as the Theory of Planned Behaviour [36], beliefs are conceptualised as factors influencing attitudes, and ultimately, behaviour. We define beliefs here as the subjective probability that an object (here: an animal or plant species in a given context) has a certain attribute [36]; for example, that a person considers red deer (*Cervus elaphus*) to be rare in Scotland. Beliefs, whether held by laypeople or experts, can thus be considered as a subjective form of knowledge. Attitudes are evaluations of objects or behaviours with some degree of favour or disfavour [37]; for example, that red deer should be managed or not. A strong relationship between relevant beliefs and attitudes, for example, a link between the belief that "red deer cause economic damage in Scotland" and the attitude that "red deer should be controlled", suggests that individuals' views are well embedded in their cognitive contexts and thus not volatile and unstable [38].

Previous research on attitudes towards biodiversity management has identified several beliefs that are likely to be relevant to the management of invasive species, and that form a useful basis to study differences between professionals and the public [33,39,40]. Here, we examine the relationships between beliefs and attitudes in a quantitative manner, and hypothesise that species believed to be non-native will be seen in greater need of management than those perceived to be native. Moreover, we predict that species regarded as invasive – here operationalised as three separate concepts, namely (a) abundance, (b) harm caused to nature and (c) harm caused to the economy – will be seen to require greater management efforts [25,33,39,41]. We thus explicitly distinguish between four different aspects (nativeness, abundance, harm caused to nature, and harm caused to the economy) of the complex of notions associated to non-nativeness and invasiveness. Finally, we hypothesise that there will be strong support for management of species that are regarded as unattractive [29,40,41] or easy to control [13].

In addition, these relationships between beliefs and attitudes towards management might vary between professionals and laypeople. For example, aesthetic attractiveness has been described as a factor that is considered especially by laypeople, rather than experts [42], whereas we would expect factors related to conservation science, such as nativeness and detrimental effects on nature, to constitute a stronger influence on professionals' attitudes than public ones [33]. Here, we explicitly investigate the degree to which the strength and the nature of these relationships differ between professionals and members of the public.

Methods

Ethics statement

The project underwent an ethics review and received full clearance, specific to the study, from the Office of Research Ethics of the University of Waterloo (#16311). In Scotland, no additional ethics approval was obtained, as neither the James Hutton Institute nor the School of Biological Sciences at Aberdeen University had an ethics committee for social scientific research at the time (2010). However, although the ethics clearance from the University of Waterloo referred only to the Canadian participants, the survey instrument used in Scotland was the same as the one employed in Ontario. Ethical issues were thus minimised. The return of completed questionnaires was considered as inferred informed consent, given that anonymity and confidentiality were explicitly granted and questionnaires did not include any information that could be used to identify individual respondents. All data was thus anonymised prior to analysis.

Survey administration and sampling

We conducted a questionnaire-based survey in two geographic areas, Scotland (UK) and Ontario (Canada), to increase the generality of our results (and obtain a sufficiently large sample size of professionals for quantitative analysis, see below). Both areas are English-speaking (a factor that facilitates cross-country application of questionnaires), and both have an industrialised Western cultural context, a landscape composed of both agricultural and semi-natural habitat, and broadly similar climates and thus functional types of biota (Köppen-Geiger classification for Scotland: Cfb; Ontario: Dfb, i.e., similar rainfall patterns (f) and warm summers (b), while temperature ranges largely overlap (C/D) [43]).

Rather than investigating differences in views between the two study areas, we aimed to compare beliefs, attitudes and, importantly, their relationships across the two sub-samples, i.e., the public and professionals (hereafter called 'groups'). We focus here on professionals, who we define as a sub-group of experts, not necessarily with scientific backgrounds, whose work engages directly with invasive species [10]. This includes members of governmental, non-governmental and research organisations as well as company representatives. Note that our public sample, as it is randomly drawn, may include not only laypeople, but also the occasional professional.

Samples of the general public (Table 1) were designed to be representative of their target population, that is, the adult residents of Scotland and Ontario, respectively. In Scotland, we used a commercial dataset that built on the electoral roll and had been complemented by other data sources that also included non-voters. Addresses were randomly selected from the dataset. A printed copy of the questionnaire (Questionnaire S1) was then sent to 1,500 addresses in June 2010, together with a prepaid return-envelope and a cover letter that referred to species management in general and did not mention biological invasions. A reminder postcard was sent two weeks after the initial questionnaire. About 20% of the addressees returned completed questionnaires.

In Ontario, we used an online version of the questionnaire that was distributed to a survey panel administered by a market research company, GMI (Global Market Insite, Inc., http://www.gmi-mr.com). The respondents were sampled from southwestern Ontario telephone area codes across rural and urban areas, but excluding Toronto (a disproportionally large metropolitan centre compared to those in Scotland). The panel completed the survey in November and December 2010.

Professionals involved with invasive species were directly contacted through pertinent e-mail lists, namely the list of the Non-Native Species Secretariat for Great Britain (https://secure.fera.defra.gov.uk/nonnativespecies/home/index.cfm) and the Ontario Invasive Plant Council (http://www.ontarioinvasiveplants.ca). A cover e-mail and a link to an on-line version of the questionnaire were sent to all members of these mailing lists. For consistency with the public sample in Scotland, the professionals (who were distributed over the whole of Britain) were asked to specifically consider the species in question in a Scottish context. As the surveys were also advertised in two relevant newsletters, response rates are difficult to estimate but were a minimum of 59% and 16% for Scotland and Ontario, respectively. The overall sample included n = 732 respondents (n = 186 professionals; n = 546 public, Table 1).

Questionnaire design

Our survey was structured by two sets of species (one for each country) that included a range of non-native, native, invasive and non-invasive species in order to obtain variation in the respective beliefs. Each set contained five species representing the following types: native mammal (two species), non-native mammal, native tree, and non-native tree/tall shrub. 'Non-native' refers here to species introduced from one geographical region into another one and is synonymous with 'alien' [44]. The term 'invasion', by contrast, denotes a species' spread [45,46], often with implicit or explicit connotations of harm caused by this spread [44]. The main criterion for species selection was familiarity: species had to be well known to a broad lay audience as widespread lack of familiarity would have resulted in meaningless data. We also aimed to identify species that were as similar as possible across the two study areas. From a bigger set of species explored in several rounds of pre-testing, we chose five species per study area for inclusion in the final version of the questionnaire (Table 2). Although the opossum, a rabbit-sized marsupial, has been found in eastern Canada for about 150 years, it has recently been spreading northward in response to climate change [47] and can thus be considered as non-native. We also included the beaver (*Castor fiber/canadensis*), a mammal that is found in both study areas and which is culturally significant; the beaver is native to Ontario, and reintroduced to Scotland after historical extirpation from the UK. All species were shown in a small photograph.

Draft versions of the questionnaire were jointly developed by the authors and extensively pilot-tested, both qualitatively and quantitatively. Belief and attitude statements had been identified in earlier qualitative work [33]. The final version included six items that captured *beliefs* about species, phrased as semantic differentials [48] that included the opposite ends of a spectrum, namely: (i) ugly – beautiful; (ii) beneficial – detrimental to the economy; (iii) beneficial – detrimental to nature; (iv) non-native – native; (v) uncontrollable – controllable; and (vi) rare – overabundant. Three additional pairs of attributes elicited *attitudes* towards species management: (i) not a severe problem – a severe problem; (ii) no need to reduce species numbers – need to reduce species numbers;

Table 1. Overview of sample sizes and demographic characteristics (gender, age) per sub-sample.

	Public		Professionals	
	Scotland	Ontario	Scotland*	Ontario
	(n = 276)	(n = 270)	(n = 93)	(n = 93)
Gender (female)	58.3%	55.2%	32.3%	40.9%
Age				
18–30 years	12.8%	20.1%	15.2%	22.6%
31–60 years	53.1%	60.4%	75%	70.9%
>60 years	34.1%	19.5%	9.8%	6.5%

*Professionals in the 'Scotland' sample were based across the whole of Great Britain.

Table 2. Species used in the questionnaire, per study area.

Species type	Scotland	Ontario
Native mammal	Red deer (*Cervus elaphus*)	White-tailed deer (*Odocoileus virginianus*)
Non-native mammal	Grey squirrel (*Sciurus carolinensis*)	Virginia opossum* (*Didelphis virginiana*)
Native tree	Scots pine (*Pinus sylvestris*)	Eastern white pine (*Pinus strobus*)
Non-native tree/tall shrub	Rhododendron (*Rhododendron ponticum*)	Scots pine (*Pinus sylvestris*)
Significant mammal	Eurasian beaver* (*Castor fiber*)	American beaver (*Castor canadensis*)

*See text for discussion.

and (iii) killing this species is not ok – killing this species is ok. For each belief and attitude item, following common practice in questionnaire surveys, five response options were offered (-2, -1, 0, 1, 2). The belief items also allowed 'don't know/not applicable' answers.

Data analysis

To obtain a robust measure of attitudes towards species management, we combined the three five-level attitude items (see above) into a single attitude index by a simple computation of the average score. Inter-item reliability as measured by Cronbach's α across the whole sample and all five species was $\alpha = 0.801$. Cronbach's α captures the degree to which different items reflect the same construct and is thus a measure of internal consistency of a group of items that are intended to address the same idea. Although α tends to vary with the number of items (the more items, the higher α), values above 0.6 are commonly considered acceptable for studies on attitudinal concepts [49].

We ran linear mixed models in SPSS Version 21 including the data for all five species pairs, both study areas (Ontario and Scotland) and both groups (public and professionals), with the attitude index as a response variable. As fixed effects we entered the categorical variables 'study area' and 'group', and the interval-scaled variables 'beauty', 'impact on economy', 'impact on nature', 'nativeness', 'controllability' and 'abundance' (i.e., the six beliefs). To explicitly test whether professionals employ beliefs differently from members of the public we fitted all two-way interactions between the individual beliefs and 'group' (e.g., beauty×group).

In an exploratory step of the analysis, we also included all interactions between these beliefs and 'study area'. However, only the term 'study area×abundance' was significant (F = 60.04, estimate = 0.19, p<0.001), which was not surprising given the large overall effect of abundance on the response variable (F = 834.3, estimate = 0.38, p<0.001). The direction of the effect suggested that the relationship between attitudes and perceptions of abundance were stronger in the Scottish sample than in the Canadian one. As no further significant impacts of study area could be detected and the investigation of country differences was not the focus of this study, we omitted these interaction terms from the analysis to concentrate on a comparison between the views of professionals and the public.

Because of our specific interest in the significance levels of the interactions between the beliefs and 'group', we used Type III calculation of the sums of squares. The order of the explanatory variables entering the model was thus based on the strength of the partial correlation with the dependent variable, starting with the strongest. The procedure gives – unlike Type I – an equal chance for main effects and interactions to be entered first in the model and is thus more likely to detect interactions present in the data. The factors 'species' and 'individual respondent' were entered as random effects to reflect the study design and thus structure of the data (i.e., each respondent expressed their beliefs and attitude towards the management of five different species). Parameters were computed using maximum likelihood estimation. Visual inspection of residual plots did not reveal any obvious deviations from assumptions of normality and homoscedasticity. Models that combined both the public and professional samples (with interactions fitted) yielded the same results as a model with a split sample; we therefore report only the combined model.

Results

Beliefs about species and attitudes towards management

While the public's and the professionals' belief scores for the five species were usually on the same side of the spectrum (Fig. 1), differences were often statistically significant. This was, in many cases, due to the tendency of professionals' beliefs to be more extreme, especially in relation to nativeness and abundance of a species (see e.g., the native mammal, where differences between all other belief scores were n. s.). The divergence in beliefs was particularly pronounced in the case of the non-native plants. On average, professionals perceived rhododendron (Scotland) and Scots pine (Ontario) to be less beautiful, more abundant and detrimental, and less controllable than the public perceived them to be. Not surprisingly, the professionals rated these two species clearly as non-native, whereas the public considered them, on average, as neither native nor non-native (all differences p<0.01, t-test).

There was also a clear divergence in attitudes: for all species except the native plant (Scots pine in Scotland, white pine in Ontario), professionals were significantly (t-test, p<0.001) more in favour of intervention than the public.

How beliefs inform attitudes: A comparison between professionals and the public

All beliefs except controllability were closely related to the attitude scores for both professionals and the public (Fig. 2): when respondents perceived a species to be more abundant (or less beautiful, or less native), they were more likely to strongly support its management. This relationship was also visible for controllability in the public sample – the more people perceived a species to be controllable, the less they felt that management was required – but this was less clear for the professionals. For all other beliefs, these relationships were strikingly similar for the two groups. Overall, professionals tended to have more favourable attitudes towards species management, as indicated by a consistently higher average score in the attitude index.

A linear mixed model (Fig. 3, Table 3) confirmed the role of beliefs as explanatory factors: perceived beauty, impact on economy and on nature, nativeness and abundance were all

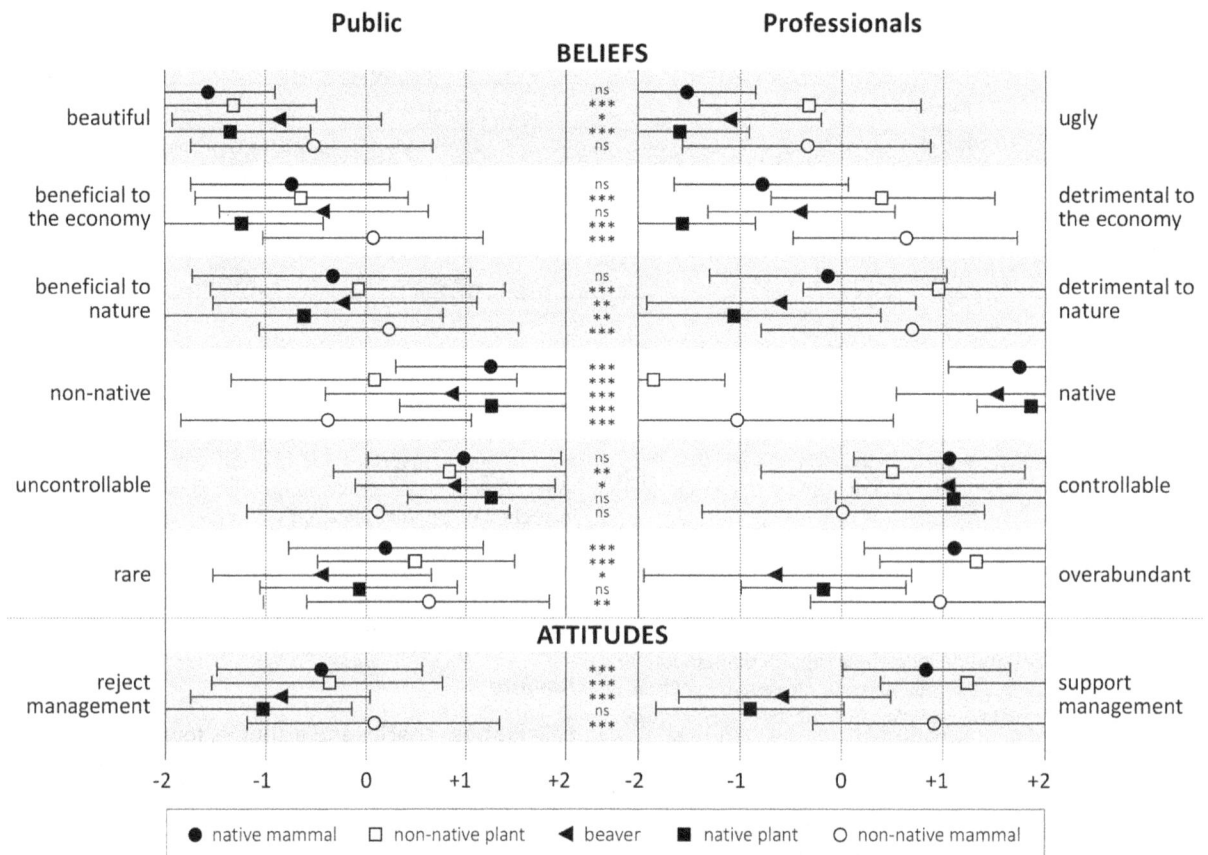

Figure 1. Mean semantic differential scores for six beliefs and for attitude towards management of five species among both public and professionals in Scotland and Ontario. Error bars show standard deviations. Asterisks show significance of difference between public and professionals (t-test, two-tailed): ***: $p<0.001$, **: $p<0.01$, *: $p<0.05$. Public sample n = 564, professionals n = 186.

found to contribute significantly ($p<0.001$) to variance of the dependent variable, i.e., attitudes towards species management. Abundance was by far the most influential factor (F = 1101.71), and even more strongly pronounced in the professionals (parameter estimate = 0.51, i.e., a 0.51 unit increase in the predicted attitude score per one unit increase in the abundance score) than the public (parameter estimate = 0.35). The effects of perceived nativeness, beauty, and impact on the economy and nature had F-values ranging between 50 and 82. Although of similar size to each other, they were thus substantially less powerful explanatory variables than abundance. The only belief that did not add explanatory value was controllability in the professional sample (as already suggested by Fig. 2d). In terms of their direction, significance and relative size, the effects of most beliefs included in the models, except controllability, were thus strikingly similar for the public and professional samples, indicating that people from both groups used beliefs in a similar fashion in their reasoning about species management.

The significance of interaction terms (Table 3) adds further detail to the comparison of how beliefs and attitudes relate to one another in the public and professional samples. Interactions of group with 'controllability' and 'beauty' were significant, with the public more strongly relating these beliefs with their attitudes towards species management, though the effects were relatively small. Because abundance as a main effect was extremely strong, the interaction between abundance and group was also significant, but effectively a rather small 'correction' on abundance as a main

effect. Importantly, F-values for the interaction between group and the three other beliefs were very small and non-significant. There was therefore no evidence that professionals utilised the beliefs 'impact on economy', 'impact on nature' and 'nativeness' differently than the public when considering the need for management of the focal species.

The variable 'study area' took account of variation in attitudes towards management that could not be associated to (beliefs about) species and suggested that on the whole, respondents from Scotland tended to support management action more strongly than those from Ontario. This held true for both professionals and the public as was confirmed by the small and non-significant F-value for 'study area×group'.

Discussion

While beliefs and attitudes towards species often differed between the public and professionals (Fig. 1), the way in which these beliefs informed attitudes were very similar across the two groups (Figs. 2, 3; Table 3). In other words, while there were differences in average content knowledge between the two samples, ways of reasoning (as captured in this study) largely concurred. For example, for many species, public and professional views on abundance or nativeness diverged. Yet, relationships between these beliefs and attitudes towards the management of these species were essentially the same among both professionals and the public, except in the case of perceived controllability, as both groups made the same connections between species attributes

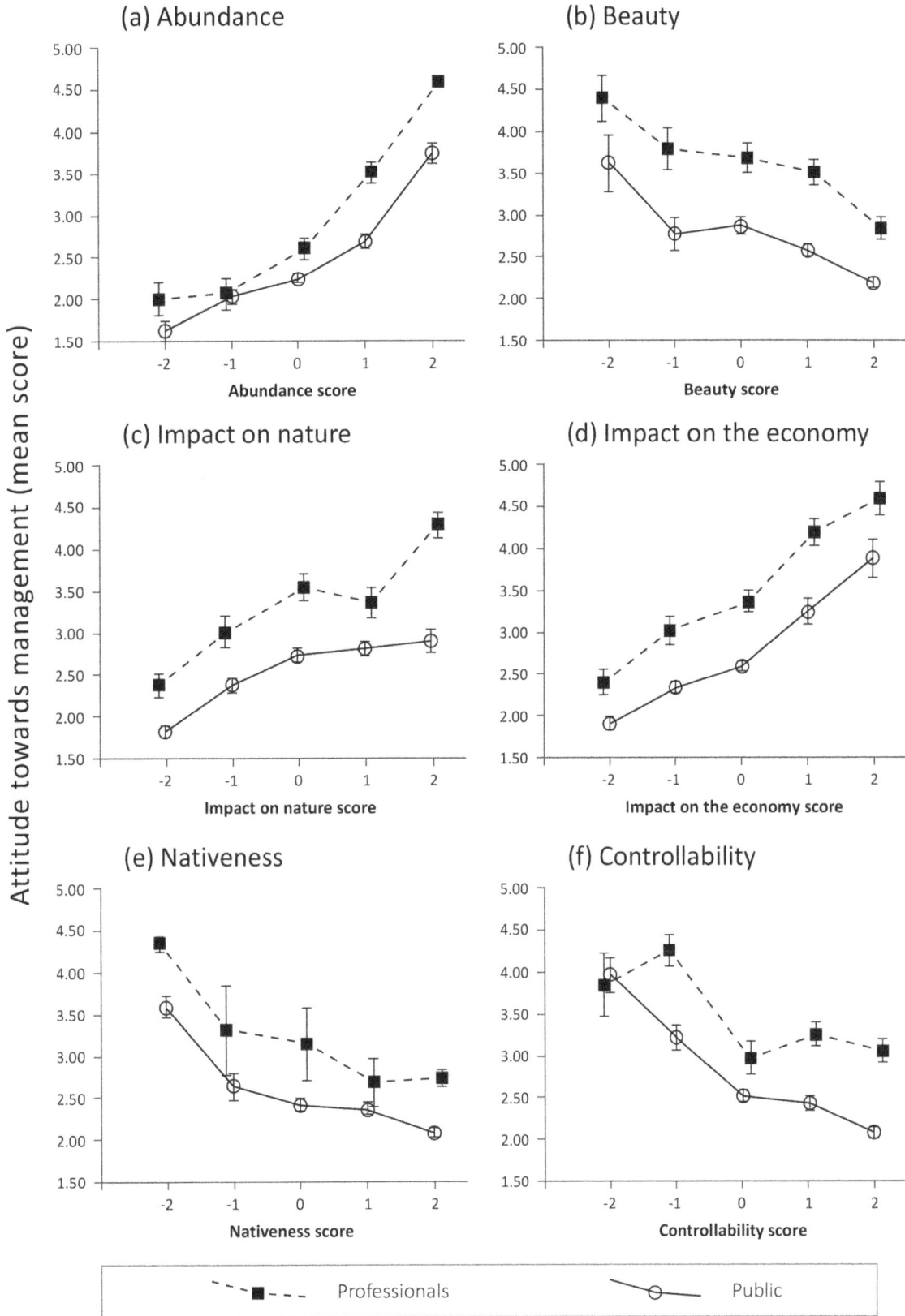

Figure 2. Mean attitude index scores (higher scores = stronger support for management) among both public and professionals in Scotland and Ontario (pooled) towards management of all five species types, in relation to beliefs: (a) abundance, (b) beauty, (c) impact on nature, (d) impact on the economy, (e) nativeness and (f) controllability. Belief variables were coded such that label indicates direction: the higher the score, the more abundant, beautiful, detrimental, native, or controllable. Error bars show 95% confidence interval. Note that lines between points do not denote interpolation, but are added to improve legibility.

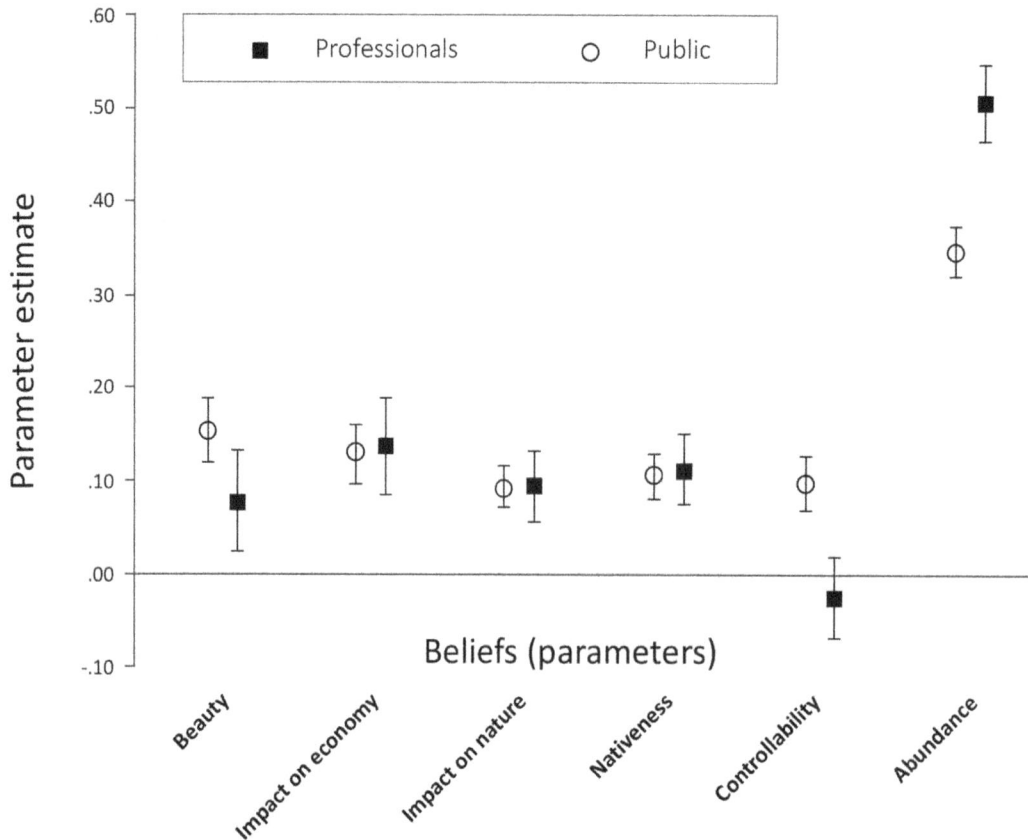

Figure 3. Parameter estimates of fixed effects in a linear mixed model. Type III sums of squares. The dependent variable is attitude towards species management, with higher scores indicating stronger support for management. Belief variables recoded such that parameter estimates are all positive: the higher the score, the more ugly, detrimental, non-native, uncontrollable, abundant the species. The parameter 'study area' (Scotland/Ontario) was included in the analysis as a fixed effect (not shown in diagram for clarity). Error bars show 95% confidence interval.

Table 3. Fixed effects statistics for linear mixed model.

Parameter	F-value	Significance (p)
Intercept	156.11	<0.001
Beauty	50.95	<0.001
Impact on economy	72.52	<0.001
Impact on nature	71.86	<0.001
Nativeness	81.45	<0.001
Controllability	8.35	0.004
Abundance	1101.71	<0.001
Study area (Scotland vs. Canada)	51.76	<0.001
Group	9.17	0.002
Group×Beauty	5.45	0.018
Group×Impact on economy	0.05	0.819
Group×Impact on nature	0.01	0.940
Group×Nativeness	0.09	0.770
Group×Controllability	21.04	<0.001
Group×Abundance	41.82	<0.001
Study area×Group	0.38	0.540

Type III sums of squares. Dependent variable: Attitude towards species management.

and the need for management. This means that ways of reasoning about animal and plant species employed by professionals and the public might be much more compatible with each other than commonly thought by invasion biologists, a finding that further expands on previous results that, while the *content* of their thoughts might diverge, ecological professionals and the lay public share the same *structure* of thought about the natural environment in general [4]. Our study thus offers support for a new critical perspective on the 'deficit model' of public understanding (see Introduction) that focuses on structure rather than merely on content.

Such commonalities in reasoning could form the starting point for discussions on invasive species management in which both professionals and the public participate on more equal terms. An increase in experts' trust in non-experts' ability to form well-founded attitudes towards invasive species management could facilitate the joint development of more widely accepted management options, and thus reduce the current level of public objection to invasive species control [22]. For example, we found that members of the public, in line with widespread thinking in invasion biology, clearly linked the harm caused by a species with the need to manage it. However, our participants' assessments were made on a case-by-case basis, which might be at odds with a heuristic in invasion biology that contends that all non-native species are potentially harmful and thus require management per se [16] (see [50] for a critical debate): Although there is a widespread understanding that only a fraction of non-native species turn into pests, the uncertainties associated with non-native management have led some invasion biologists to stress the need for general prevention [20] or to argue that all alien species should be considered "guilty until proven innocent" [51]. Our data suggests that the public will appreciate communication on species management that also addresses factors other than non-nativeness and invasiveness – notably a species' abundance and its effects on nature and the economy (Fig. 3). However, often 'invasion' appears to be used as a shorthand that implies a range of other (normative) factors such as harmfulness, which should be made explicit. Some of the apparent differences between public and scientific views, which have been ascribed to "divergent ethical frameworks" [16], could instead be due to such shortcuts in the arguments of invasion biologists that neglect species characteristics which for many people, including professionals in the field, are of importance when assessing the need for species management.

This is not to say that diverging beliefs about species are irrelevant and should be ignored. For example, where perceptions of a species' abundance differ, these should be addressed, and the causes for conflicting perceptions investigated in more detail. Such conflicts can occur between professionals and the public, as well as between professionals or between members of the public, such as when assessments of the economic value of a species diverge. Yet, a discussion that starts on common ground – namely the recognition that patterns of reasoning are shared, even if content knowledge may differ – has higher chances of success than a polarised debate based on the assumption that the public simply has to be told how to think by professionals in a uni-directional manner (see also [52]).

Our study shows that such patterns of reasoning can also be found implicitly, in the relationships between beliefs and attitudes as expressed in a questionnaire, and not only in the explicit connections made in discussions, as documented in qualitative studies with usually small samples [33,34]. While our study was built on a simple framework that included only two types of constructs, namely beliefs and attitudes, we have elsewhere investigated more complex mental structures related to biodiver-

sity management, their discursive contexts and a range of decision-making processes [33,38,53]. Both beliefs and attitudes can, for example, be conceptualised as part of social representations that develop and are negotiated in individuals' social context [54]. The degree to which they are actively linked in a concrete situation, such as responding to a questionnaire, can depend on a range of factors, for example, a person's emotional involvement (i.e., their affective engagement) with the topic [53]. Our previous findings are compatible with the reduced framework adopted in this study, and we thus argue that, where the context is kept in mind, simple frameworks can produce valid insights into more complex relationships.

Our hypotheses proposed that professionals do not consider attractiveness (or beauty) of a species in their thinking about management (see Introduction). However, our results clearly show that they did, albeit to a lesser degree than the public (Figs. 2 and 3). Conversely, we hypothesised that a species' nativeness and the harm it caused to nature and the economy would play a very strong role in informing professionals' attitudes towards management, whereas this role would be less pronounced for the general public. Yet, differences between the two groups were not significant (Table 2), again suggesting that conservation biological thinking might be more mainstream among the public than often thought.

Our results also add further evidence to previous studies regarding individual beliefs about invasive species management. Overall, we found that beliefs about nativeness appeared more important for the formation of attitudes than in previous studies [41]. This might be due to the fact that in our analysis, nativeness and abundance were not explicitly separated from the concept of human responsibility in a species' spread – a notion that earlier qualitative research had identified as important in participants' talk about invasive species [33]. Future analyses could address the relationship between these closely related and often confounded concepts [52,55] in more depth. Our results also add evidence to previous studies with members of the general public that identified perceived abundance as the most important factor in informing attitudes towards a species [41]: For professionals, this effect was even stronger than among members of the public, although we can assume that the statistical significance of the difference between the two groups was largely due to the exceptionally high F-value for abundance compared to the F-values for other beliefs (Table 3).

Only in the case of controllability did the relationship between beliefs and attitudes really diverge between the two groups. While there was no significant relation between controllability and support for management among the professionals, members of the public tended to perceive more controllable species as requiring less management than less controllable ones – a result that seems to contradict ideas of risk research in other environmental domains [13]. Nevertheless, this finding is plausible as it suggests that respondents saw a greater cause for concern and need for intervention in those species they considered difficult to control. The absence of a measurable effect in the professional sample might be due to the simultaneous presence of these competing conceptualisations of controllability.

Our strict focus on a particular set of species and two psychological constructs allowed us to identify patterns among human populations through a quantitative approach, but also constrained the scope of the findings. The target species had to be widely familiar because large proportions of 'don't know' responses among both the public and often highly specialised professionals would have rendered the data meaningless. The inclusion of other study areas with possibly more controversial species, such as large carnivores with expanding populations [56], less well-known and less charismatic, or conversely, culturally or economically significant non-native plant species [23], could have led to even larger differences between professionals and the public in beliefs or attitudes. However, there is no reason to suspect that the *relationships* between beliefs and attitudes would be affected by specific species, because similar relationships have also been found for other species and other European countries [41]. While our set of species would ideally be larger for the conclusions to be more readily transferable (but for pragmatic reasons, the length of a questionnaire is obviously limited), our study goes substantially beyond previous research that often addressed only one or two species [23,25]. This notwithstanding, qualitative research will always be required to provide more in-depth insights into cultural or symbolic meanings of species.

In summary, we conclude that in countries such as Great Britain and Canada, differences between public and professional views on invasive species management are unlikely to result from fundamental differences in reasoning (as captured in this study). Instead, such divergences may either be caused by diverging beliefs about species, or by procedural aspects related to communication, decision-making or species management. In both cases, we suggest that disputes about invasive species management may be reduced by increased transparency and a more differentiated debate that makes use of shared understandings of relationships between species characteristics and the need for intervention.

Acknowledgments

The authors would like to thank all survey participants for their contributions, Nicole Spiegelaar for assistance with survey administration in Ontario, Mark Brewer for statistical advice, and Robin Pakeman and several anonymous reviewers for comments on earlier versions of this manuscript. Ian Williamson improved the graphical quality of the figures.

Author Contributions

Conceived and designed the experiments: AF SS RvdW. Performed the experiments: AF SS RvdW BL. Analyzed the data: AF SS RvdW. Contributed reagents/materials/analysis tools: AF SS RvdW BL. Wrote the paper: AF SS RvdW BL.

References

1. Wynne BE (1996) May the sheep safely graze? A reflexive view of the expert-lay knowledge divide. In: Lash S, Szerszynksi B, Wynne B, editors. Risk, environment and modernity: towards a new ecology. London: Sage. 44–82.
2. Rauschmayer F, van den Hove S, Koetz T (2008) Participation in EU biodiversity governance: How far beyond rhetoric? Environment and Planning C 27: 42–58.
3. Hunziker M, Felber P, Gehring K, Buchecker M, Bauer N, et al. (2008) Evaluation of landscape change by different social groups. Mountain Research and Development 28: 140–147.
4. Buijs AE, Elands BHM (2013) Does expertise matter? An in-depth understanding of people's structure of thoughts on nature and its management implications. Biological Conservation 168: 184–191.
5. Dallimer M, Irvine KN, Skinner AMJ, Davies ZG, Rouquette JR, et al. (2012) Biodiversity and the feel-good factor: Understanding associations between self-reported human well-being and species richness. BioScience 62: 47–55.
6. Besley JC, Nisbet M (2011) How scientists view the public, the media and the political process. Public Understanding of Science 20: 1–16.
7. Gieryn TF (1983) Boundary-work and the demarcation of science from non-science: strains and interests in professional ideologies of scientists. American Sociological Review 48: 781–795.
8. Trumbull DJ, Bonney R, Bascom D, Cabral A (2000) Thinking scientifically during participation in a citizen-science project. Science Education 84: 265–275.
9. Bell M, Sheail J (2005) Experts, publics and the environment in the UK: twentieth-century translations. Journal of Historical Geography 31: 496–512.
10. Giddens A (1994) Living in a post-traditional society. In: Beck U, Giddens A, Lash S, editors. Reflexive modernization: politics, tradition and aesthetics in the modern social order. Cambridge: Polity Press. 56–109.
11. Roux DJ, Rogers KH, Biggs HC, Ashton PJ, Sergeant A (2006) Bridging the science-management divide: Moving from unidirectional knowledge transfer to knowledge interfacing and sharing. Ecology and Society 11: 4.
12. Larson B (2011) Metaphors for Environmental Sustainability: Redefining our Relationship with Nature. New Haven: Yale University Press.
13. McDaniels TL, Axelrod LJ, Cavanagh NS, Slovic P (1997) Perception of ecological risk to water environments. Risk Analysis 17: 341–352.
14. Tveit MS (2009) Indicators of visual scale as predictors of landscape preference: a comparison between groups. Journal of Environmental Management 90: 2882–2888.
15. Dandy N, Van der Wal R (2011) Shared appreciation of woodland landscapes by land management professionals and lay people: An exploration through field-based interactive photo-elicitation. Landscape and Urban Planning 102: 43–53.
16. Simberloff D, Martin JL, Genovesi P, Maris V, Wardle DA, et al. (2013) Impacts of biological invasions: what's what and the way forward. Trends in Ecology and Evolution 28: 58–66.
17. Larson BMH (2007) An alien approach to invasive species: Objectivity and society in invasion biology. Biological Invasions 9: 947–956.
18. Young AM, Larson BMH (2011) Clarifying debates in invasion biology: A survey of invasion biologists. Environmental Research 111: 893–898.
19. Humair F, Edwards PJ, Siegrist M, Kueffer C (2014) Understanding misunderstandings in invasion science: why experts don't agree on common concepts and risk assessments. NeoBiota 20: 1–30.
20. Hulme PE (2006) Beyond control: Wider implications for the management of biological invasions. Journal of Applied Ecology 43: 835–847.
21. DEFRA (Department for Environment, Food and Rural Affairs) (2008) The Invasive Non-Native Species Framework Strategy for Great Britain. DEFRA, London.
22. Perry D, Perry G (2008) Improving interactions between animal rights groups and conservation biologists. Conservation Biology 22: 27–35.
23. Marshall NA, Friedel M, Van Klinken RD, Grice AC (2011) Considering the social dimension of invasive species: the case of buffel grass. Environmental Science and Policy 14: 327–338.
24. Bremner A, Park K (2007) Public attitudes to the management of invasive non-native species in Scotland. Biological Conservation 139: 306–314.
25. Fischer A, Van der Wal R (2007) Invasive plant suppresses charismatic seabird – the construction of attitudes toward biodiversity management options. Biological Conservation 135: 256–267.
26. Somaweera R, Somaweera N, Shine R (2010) Frogs under friendly fire: How accurate can the general public recognize invasive species? Biological Conservation 143: 1477–1484.
27. Sharp RL, Larson LR, Green GT (2011) Factors influencing public preferences for invasive alien species management. Biological Conservation 144: 2097–2104.
28. Schüttler E, Rozzi R, Jax K (2011) Towards a societal discourse on invasive species management: A case study of public perceptions of mink and beavers in Cape Horn. Journal for Nature Conservation 19: 175–184.
29. Verbrugge LNH, Van den Born RJG, Lenders HJR (2013) Exploring public perceptions of non-native species from a visions of nature perspective. Environmental Management 52: 1562–1573.
30. Andreu J, Vilá M, Hulme P (2009) An assessment of stakeholder perceptions and management of noxious alien plants in Spain. Environmental Management 43: 1244–1255.
31. Bardsley DK, Edwards-Jones G (2007) Invasive species policy and climate change: Social perceptions of environmental change in the Mediterranean. Environmental Science and Policy 10: 230–242.
32. García-Llorente M, Martín-López B, González JA, Alcorlo P, Montes C (2008) Social perceptions of the impacts and benefits of invasive alien species: Implications for management. Biological Conservation 141: 2969–2983.
33. Selge S, Fischer A, Van der Wal R (2011) Public and professional views on invasive non-native species - a qualitative social scientific investigation. Biological Conservation 144: 3089–3097.
34. Sadler TD, Zeidler DL (2005) The significance of content knowledge for informal reasoning regarding socioscientific issues: Applying genetics knowledge to genetic engineering issues. Science Education 89: 71–93.
35. Jordan RC, Gray SA, Howe DV, Brooks WR, Ehrenfeld JG (2011) Knowledge gain and behavioural change in citizen-science programs. Conservation Biology 25: 1148–1154.
36. Ajzen I (1988) Attitudes, Personality, and Behavior. Open University Press, Milton Keynes.
37. Milfont TL, Duckitt J (2010) The environmental attitudes inventory: A valid and reliable measure to assess the structure of environmental attitudes. Journal of Environmental Psychology 30: 80–94.

38. Fischer A, Langers F, Bednar-Friedl B, Geamana N, Skogen K (2011) Mental representations of animal and plant species in their social contexts: Results from a survey across Europe. Journal of Environmental Psychology 31: 118–128.

39. Montgomery CA (2002) Ranking the benefits of biodiversity: An exploration of relative values. Journal of Environmental Management 65: 313–326.

40. Gobster PH (2011) Factors affecting people's responses to invasive species management. In: Rotherham ID, Lambert RA, editors. Invasive and introduced plants and animals – human perceptions, attitudes and approaches to management. London: Earthscan. 249–263.

41. Fischer A, Bednar-Friedl B, Langers F, Geamana N, Skogen K, et al. (2011) Universal criteria for species conservation priorities? Findings from a survey of public views across Europe. Biological Conservation 144: 998–1007.

42. Buijs AE, Arts BJM, Elands BHM, Langkeek J (2011) Beyond environmental frames: The social representation and cultural resonance of nature in conflicts over a Dutch woodland. Geoforum 42: 329–341.

43. Peel MC, Finlayson BL, McMahon TA (2007) Updated world map of the Köppen-Geiger climate classification. Hydrology and Earth System Sciences 11: 1633–1644.

44. Colautti RI, MacIsaac HJ (2004) A neutral terminology to define 'invasive' species. Diversity and Distribution 10: 135–141.

45. Jeschke JM, Strayer DL (2005) Invasion success of vertebrates in Europe and North America. Proceedings of the National Academy of Sciences 102: 7198–7202.

46. Lodge DM, Williams S, MacIsaac H, Hayes KR, Leung B, et al. (2006) Biological invasions: Recommendations for U.S. policy and management. Ecological Applications 16: 2035–2054.

47. Naughton D (2012) The Natural History of Canadian Mammals. Toronto: Canadian Museum of Nature and University of Toronto Press.

48. Osgood CE (1952) The nature and the measurement of meaning. Psychological Bulletin 49: 197–237.

49. Fulton DC, Manfredo MJ, Lipscomb J (1996) Wildlife Value Orientations: A Conceptual and Measurement Approach. Human Dimensions of Wildlife 1: 24–47.

50. Davis MA, Chew MK, Hobbs RJ, Lugo AE, Ewel JJ, et al. (2011) Don't judge species on their origins. Nature 474: 153–154.

51. Simberloff D (2007) Given the stakes, our *modus operandi* in dealing with invasive species should be "guilty until proven innocent". Conservation Magazine 8: 18–19.

52. Davis MA (2009) Invasion biology. Oxford: Oxford University Press.

53. Fischer A, Glenk K (2011) One model fits all? On the moderating role of emotional engagement and confusion in the elicitation of preferences for climate change adaptation policies. Ecological Economics 70: 1178–1188.

54. Buijs AE, Hovardas T, Castro P, Devine-Wright P, Figari H, et al. (2012) Understanding people's ideas on natural resource management: research on social representations of nature and the environment. Society and Natural Resources 25: 1167–1181.

55. Warren CR (2007) Perspectives on the 'alien' versus 'native' species debate: A critique of concepts, language and practice. Progress in Human Geography 31: 427–446.

56. Skogen K, Thrane C (2008) Wolves in context. Using survey data to situate attitudes within a wider cultural framework. Society and Natural Resources 21: 17–33.

9

Uncertainties in Ecosystem Service Maps: A Comparison on the European Scale

Catharina J. E. Schulp[1]*, **Benjamin Burkhard**[2,4], **Joachim Maes**[3], **Jasper Van Vliet**[1], **Peter H. Verburg**[1]

1 Faculty of Earth and Life Sciences, VU University Amsterdam, Amsterdam, the Netherlands, 2 Institute for Natural Resource Conservation, Kiel University, Kiel, Germany, 3 European Commission - Joint Research Centre, Institute for Environment and Sustainability, Ispra, Varese, Italy, 4 Leibniz Centre for Agricultural Landscape Research ZALF, Müncheberg, Germany

Abstract

Safeguarding the benefits that ecosystems provide to society is increasingly included as a target in international policies. To support such policies, ecosystem service maps are made. However, there is little attention for the accuracy of these maps. We made a systematic review and quantitative comparison of ecosystem service maps on the European scale to generate insights in the uncertainty of ecosystem service maps and discuss the possibilities for quantitative validation. Maps of climate regulation and recreation were reasonably similar while large uncertainties among maps of erosion protection and flood regulation were observed. Pollination maps had a moderate similarity. Differences among the maps were caused by differences in indicator definition, level of process understanding, mapping aim, data sources and methodology. Absence of suitable observed data on ecosystem services provisioning hampers independent validation of the maps. Consequently, there are, so far, no accurate measures for ecosystem service map quality. Policy makers and other users need to be cautious when applying ecosystem service maps for decision-making. The results illustrate the need for better process understanding and data acquisition to advance ecosystem service mapping, modelling and validation.

Editor: Gen Hua Yue, Temasek Life Sciences Laboratory, Singapore

Funding: The authors acknowledge financial support of ERA-net BiodivERsA, with the national funder NWO, part of the 2011 BiodivERsA call within the project CONNECT (http://www.biodiversa.org/; http://www.nwo.nl/) (CJES, PV), the European Commission-funded FP7 project OPERAs (PV) and the German BMBF-funded project LEGATO (FKZ 01LL0917F) (http://www.bmbf.de/en/)(BB). The funders had no role in study design, data collection and analysis, decision to publish, or preparation of the manuscript.

Competing Interests: The authors have declared that no competing interests exist.

* Email: nynke.schulp@vu.nl

Introduction

The benefits that ecosystems provide to society are increasingly acknowledged. Safeguarding these benefits and maintaining, restoring and enhancing ecosystem services (ES) in the future is included as a target in several international policies, such as the 2020 targets of the Convention on Biological Diversity [1,2]. The European Union (EU) elaborates this target in the European Biodiversity Strategy that aims at maintaining and enhancing ecosystems and their services [3].

Decisions or policies on ES are made based on available information on the status, trends, and spatial distribution of ecosystem service provision. To support such policies, there is, consequently, an increasing demand for accurate maps of the supply and demand of ecosystem services [4,5]. The European Commission therefore aims to "map and assess the state of ecosystems and their services (...) by 2014" [3] and several attempts to map the supply and demand of ecosystem services have been presented in the literature [5,6,7,8].

To support policy design on ES, indicators are needed to quantify specific targets on maintaining ES and to monitor progress towards these targets. These indicators should pass basic quality criteria. The Impact Assessment Guidelines of the European Commission summarize such criteria in the RACER framework [9]: indicators should be *Relevant* to the objectives to be reached by the target, *Accepted* by staff and stakeholders, *Credible*, unambiguous and easy to interpret, *Easy* to monitor and *Robust* against manipulation. Although not explicitly stated, the "Accepted, Credible and Robust" criteria acknowledge that indicators should provide accurate data.

Due to the lack of direct monitoring data on ES, they are commonly mapped using model-based proxies. Although it is frequently recognized that such maps are crude estimates, there is little discussion on the magnitude of the errors associated with them [10,11]. Eigenbrod et al. [12] were the first to quantify the magnitude of errors in ES maps for part of the UK and raised concerns about the accuracy of ES maps and about inconsistencies among mapping approaches [12]. Later reviews of ES mapping studies indicated that only a small fraction of the ES mapping studies did address uncertainty in a quantitative way [11,13]. For example, uncertainties in ES models were addressed by Kozak et al [14] and Lautenbach et al [15] for small case studies. Schulp et al [16] do address uncertainties in a map for pollination services at a EU scale. Nevertheless, there is little knowledge about the influence of the mapping method and input data on the representation of spatial patterns of ecosystem service supply [17]. Most of the mapping studies pay little attention to error propagation [18].

Studies on ecosystem service map validation are lacking [11]. This is attributed to the fact that many ecosystem services cannot be measured directly, resulting in a lack of direct data on the provision of the service. Several ES mapping studies provide a qualitative comparison with independent proxies, e.g. [7,19], but these are rather indications for the credibility than full validations.

In this paper, we identify uncertainties in continental-scale ecosystem service maps. Based on a systematic comparison of maps for the EU territory for five ecosystem services (climate regulation, flood regulation, erosion protection, pollination, and recreation), we map spatial patterns of agreement and disagreement for the provision of these five services. Secondly, we evaluate the magnitude and sources of uncertainty in current ecosystem service maps. Based on the quantitative evaluation of ES map consistency and validity, we evaluate to what extent the current ES indicators suffice for target setting and evaluation and other forms of policy support and recommend best practices for ES mapping.

Methods

Considered studies

To provide insight in the uncertainty in ecosystem service maps, we made an inventory of ES maps that cover the EU extent (Figure 1). For five ES (climate regulation, flood regulation, erosion protection, pollination, and recreation), we identified a wide range of existing maps representing a variety of sources and approaches. For other ES insufficient different maps were available to be included in the analysis.

We analysed uncertainties in ES maps building on four consistent and published sets of ecosystem service maps (Table 1). First, Burkhard et al. [6] map the capacity to provide ES at the European scale using an expert-based classification of land cover data (hereafter referred to as: LC approach). In this approach, an expert evaluation of the level of ES provision for each land cover type into five classes is used to map ES provision. Second, Kienast et al. [7] provide an expert-based map of landscape capacities to provide ES. These maps are based on a similar expert evaluation as the maps of Burkhard et al [6] but include a wide range of environmental variables like relief and landscape types (EV approach). A third set of ES maps originates from a hybrid approach. A hybrid approach uses both process-based models as well as empirical models. This set of ES maps aims to take optimal advantage of available data on land use and environmental indicators [5] (JRC approach). The fourth set of ES maps consists of maps of intermediate complexity. In an intermediate complexity approach, process-based model results are upscaled to a wider extent using empirical relationships with other spatial data [16,19,20,21,22] (IVM approach).

In addition to these four sets that all contain the same five ES, other studies are available that map one single ES (Table 1). Together, the maps represent the range of complexity of approaches to create ES maps for larger areas.

Map preparation

Available ES maps strongly differ in representation of the services, units, range of output values and spatial resolution (Table 1). To enable comparison, the maps were made consistent by aggregation to a common spatial resolution and normalizing the ecosystem service indicator values to a common range and unit. Firstly, the categorical LC maps were converted into numerical maps. Following Burkhard et al [33], who indicate that the ES provision categories can be translated into numerical variables using an equal interval classification, we reclassified the ES provision categories of the LC maps into: no ES provision = 0, very low ES provision = 0.2, low = 0.4, moderate = 0.6, high = 0.8, very high = 1. Secondly, all maps were aggregated to NUTS2 regions by calculating the mean value for the ES for each region. The NUTS2 level was chosen as it represents the resolution of the least detailed map. NUTS (the EU Nomenclature of Territorial Units for Statistics) is the standard regional subdivision of the European territory for statistical purposes. NUTS2 regions are the basic regions for the application of regional policies [34]. Thirdly, all maps were normalised using a min-max

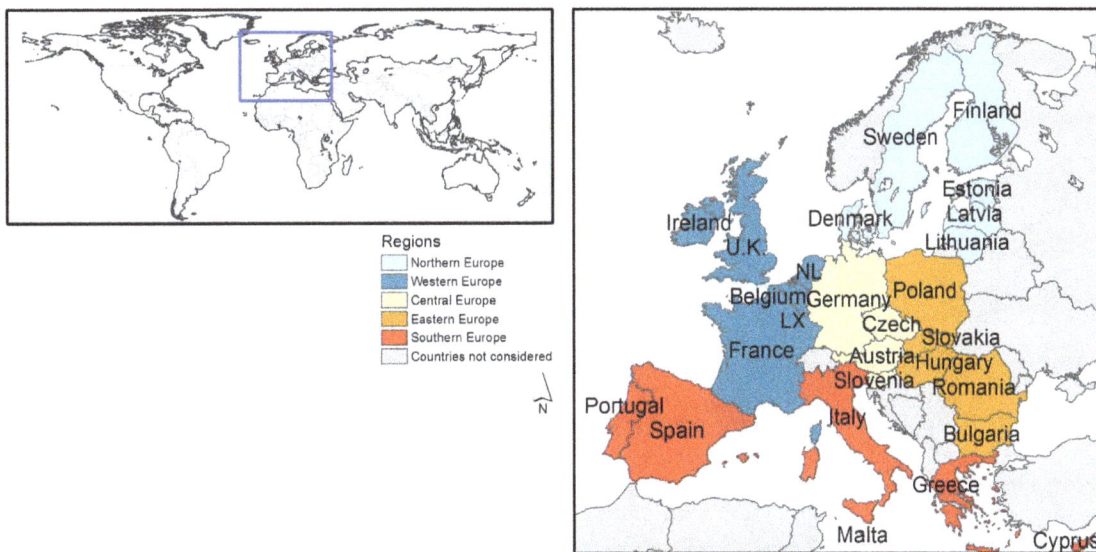

Figure 1. Study area location (left) and regional subdivision and country names as referred to in the results (right). U.K. = United Kingdom, NL = Netherlands, LC = Luxembourg.

Table 1. Overview of the ecosystem service datasets analysed in this study.

Dataset	Climate regulation	Flood regulation	Pollination	Erosion protection	Recreation
Datasets included in full analysis					
LC approach [6]	Capacity of the landscape to provide the service. Based on categorical links between land cover and the service, using CORINE land cover data [23] Categorical, 6 classes ranging from no capacity to very high capacity. 100 m resolution.				
EV approach [7]	Capacity of the landscape to provide the service, expressed as an index based on a set of binary links between environmental variables (including CORINE land cover [23]) and the ecosystem service. Continuous (Dimensionless). NUTS2 resolution.				
JRC approach	Carbon flow, expressed as Net Ecosystem Productivity (NEP). Based on a model based on RS image interpretation [24].	Water quantity regulation: Annually aggregated soil infiltration, derived from a pollutant pathway model. 1 km² resolution [24].	Visitation probability, based on distance decay function from pollinator habitat, multiplied with dependency level of pollinator dependent crops. Based on a crop type map and CORINE land cover [23]. 1 km² resolution [25].	Area based indicator to express the protective function of forests and semi-natural areas based on CORINE land cover [23] in areas with high erosion risk. 1 km² resolution [24]	Capacity of the landscape to provide recreational services. Dimensionless index based on the degree of naturalness, presence of protected areas, distance to coasts, lakes and rivers and bathing water quality. 1 km² resolution [26].
IVM approach	Carbon sequestration, expressed as NEP. Bookkeeping model where detailed flux measurements and simulations are aggregate to country-specific, land use type (based on aggregated CORINE land cover [23]) specific emission factors. 1 km² resolution [20].	Index of flood regulation provision. Based on upscaling of catchment-scale simulations with a process-based hydrological model, to EU scale, using catchment characteristics like land use, topography and soil characteristics. 1 km² resolution [21].	Visitation probability, based on distance decay function from pollinator habitat. Based on CORINE land cover [23] and a map of green linear elements [16].	Protection against erosion by vegetation, based on the Universal Soil Loss Equation and an aggregated version of CORINE land cover [23]. 1 km² resolution [22].	Capacity of the landscape to provide recreational services. Dimensionless index, based on the degree of naturalness; presence of protected areas, presence of coasts, lakes and rivers, presence of High Nature Value farmlands [19].
Additional maps					
	Carbon storage: Coupling of global-scale carbon stocks to European-scale land use maps (CORINE land cover [23]) 250 m resolution [24].	Natural hazard reduction: Influence of ecosystem structure on dampening environmental disturbances. Capacity of the landscape to provide the service, following EV approach [7].	Habitat percentage: Area percentage of pollinator habitat. Based on CORINE land cover [23] and a map of green linear elements. 1 km² resolution [16].		
	Net Ecosystem Productivity (NEP) as calculated with the process-based LPJ model for the global carbon cycle. 0.5° resolution [27].		Habitat percentage: Pollinator habitat within a 2 km range of croplands. 1 km² resolution [28].		
Independent proxy data used for validation					
Dataset	Global-scale map of NPP, 0.25° resolution [29]. No data for Ireland.	Global-scale map of flood frequency, 1985–2012 [30].	Density of occurrence of wild Apis and Bombus species in northwest Europe [31].	Global-scale map of NPP change over 1980–2003 [32]. No data for Ireland.	Density of inland camping sites [19]
Relation assumed to represent good fit with independent proxy	High NPP coincides with high values of the ecosystem service map	Low frequency coincides with high flood regulation ecosystem service	High pollinator density coincides with high pollination provision	Low NPP loss coincides with high erosion protection	High density coincides with high recreation potential

LC approach: set of ecosystem service maps based on land cover; EV approach: set of ecosystem service maps based on environmental variables. JRC approach: set of data driven ecosystem service maps. IVM approach: set of ecosystem service maps of intermediate complexity.

normalisation to cover the range [0,1] with 0 indicating the lowest value for an ES. The EV maps were computed using a linear combination of explaining variables, and therefore we assumed linearity for the normalization [7].

To summarise maps of individual services, we calculated an ES bundle map for each of the four sets of maps (LC, EV, JRC, IVM). An ES bundle represents an overall level of the provisioning of the five services considered and was calculated as the sum of the five normalised ecosystem service maps. High values thus indicate locations with a relatively high supply or multiple services, while low values indicate the opposite. These bundle maps were included in the comparison because policies aim at protecting the overall level of ecosystem service provision rather than or additional to the provision of individual services [3,35].

Map comparison and analysis

Maps for individual ES as well as the bundles were analysed both pair-wise and for all maps together. For the pair-wise comparisons, we use one index that summarizes the relative differences between the maps, the Map Comparison Statistic (MCS) (Equation 1):

$$MCS = \frac{\sum_{n=1}^{N} (|a-b|/\max(a,b))}{N} \qquad \text{(Equation 1)}$$

Where MCS is the Map Comparison Statistic, a and b are the normalized values of an ecosystem service in a particular NUTS2 region, N is the number of NUTS2 regions considered. MCS values were computed for all available ES maps. This comparison statistic was chosen because it is symmetric (yielding the same result independent of which of the maps is map a or b) and has a defined range (zero for two equal maps; one for two completely contrasting maps) [36]. The MCS thus indicates the average difference between any pair of ES values, expressed as a fraction of the highest value. Two random maps would have an MCS of 0.5 while two opposing maps would have an MCS of 1 and two identical maps would have an MCS of zero.

To analyse the agreement in spatial patterns of ES in the four sets of maps (LC, EV, JRC, IVM), we calculated hotspot and coldspot maps. Hotspots and coldspots are areas providing, respectively, high and low amounts of a particular ES [37,38] and are defined as areas where the ES supply values fall within the upper or lower quartile of its value distribution. Agreement between the hotspot and coldspot maps of the four mapping approaches was calculated by counting the number of maps that indicated a hotspot or coldspot at a certain NUTS2 region. In addition, we calculated the mean value per NUTS2 region over the four included maps, as well as the coefficient of variation (CV), which is the standard deviation divided by mean. The mean and CV maps were calculated to give a general impression of the spatial patterns of ES supply and their related uncertainty. The mean value over the four included maps gives an indication of the ES supply in each NUTS2 region, while the CV provides information on the relative range of values provided by the different maps, and is therefore an indicator for the patterns of uncertainty in the ES maps.

To support analysis of the sources of uncertainty, ES maps were compared with spatial patterns of land cover. We calculated Pearson's correlations between the percentage of a specific land cover per NUTS2 region and the mean and CV of the mapped supply of each ES. This analysis aimed to explore relations between land cover and ES supply values and the variation in reported values of the ES indicators per land cover type. Although relations between land cover and ES are frequently used for mapping ES, they are largely untested [13] and the exploration therefore was expected to provide useful insights into the variation between ES models for different land cover types.

Validation

The map comparison methods described in the section "Map comparison and analysis" provide insight in the agreement among the maps and give an indication of the overall level of uncertainty in ES mapping at EU scale. However, it does not provide insight in the deviations from the actual provision of the ES. Therefore, additional to the systematic comparison of ecosystem services maps, the ES maps were compared against independent data that provide a proxy for the ES. As most ES are difficult to measure directly, the independent proxies do not fully match the definition of the ES and not all independent proxies cover the full European Union. However, in all cases association between the spatial patterns of the ES map and the independent proxy can be expected and we interpret such a coincidence association as an indication of the model quality. Table 1 provides an overview of the independent data used for comparison and the assumed relations between the values of the independent map and the ES maps. All independent maps were aggregated to NUTS2 resolution and transformed to ensure the highest values in the independent maps matched the highest values expected from the ES maps. Consistent with Kienast et al. [7], we subdivided the ES map data and the independent data into four quartiles and made a cross-tabulation to calculate the overlap between the independent

data and the ES maps. We counted the regions where there was agreement and subdivided this by number of regions where agreement could be expected by chance.

All analyses were performed in R using the packages rgdal [39] and raster [40].

Results and Interpretation

Pair-wise map comparisons

Differences among the climate regulation maps are small with MCS values of 0.27 and lower (Table 2). However, when the four maps included in this study are compared to a process-based estimate derived from the LPJ carbon cycle model [27], larger differences are found, with MCS values up to 0.46 for the comparison with the LC map. A map of carbon stocks [41] (Table 1) compares reasonably with all other climate regulation maps; MCS values range between 0.13 (EV map) and 0.24 (LC map). The recreation maps also show relatively small differences among the maps. MCS values for pollination are higher and range up to 0.49 for the comparison between the JRC map and the IVM map. The maps were also compared to two other maps that are an indicator of the available potential pollinator habitat. The map by Serna-Chavez et al. [28] is close to the LC map (MCS: 0.19) but deviates from the JRC map (MCS: 0.44). The habitat map by Schulp et al. [16] is most similar to the JRC map (MCS: 0.19) and deviates most from the IVM map (MCS: 0.38). Flood regulation and erosion protection show high MCS values, indicating that the maps are more different from each other than maps for the other ecosystem services.

For the ecosystem service bundles, the MCS values are lower than for the individual service maps (Table 2). While for the individual services the JRC maps and IVM maps are most deviating, the bundle maps of JRC and IVM are the most similar because differences amongst ecosystem services average out.

Spatial patterns

For *climate regulation*, there is agreement on the location of a coldspot in the north-western EU while there is reasonable agreement on hotspots in the central southern region (Figure 2; see Figure 1 for regional subdivision). These coldspots and hotspots can also be seen in the average climate regulation map (Figure 3). High climate regulation capacities are found in Sweden and Finland because of the high percentage forest cover, but here the maps strongly disagree. The provision of climate regulation is strictly defined [42], and is normally quantified based on the rate of carbon sequestration (e.g. in the Common International Classification of Ecosystem Services, CICES (http://cices.eu)). This service is to a large extent provided by natural vegetation. Consequently, the climate regulation maps depend largely on land cover data. All the analysed maps use the same land cover map [23]. The process of carbon sequestration is well-researched [11] and there is consensus on the direction and magnitude of drivers for climate regulation. All maps assume that arable land and urban areas do not provide the service in a relevant amount, and assume that forests and areas that are more natural do. Although the parameterisation of the land cover types differs among the studies, the consistency in input data, well-established process knowledge and strict indicator definition result in the highest level of agreement among the ES assessed here.

Pollination maps agree on hotspots in southern Europe and coldspots in western and eastern Europe, while disagreement is seen in central and northern Europe (Figure 2). The areas where the maps disagree have a high average level of pollination provisioning (Figure 3). Indicators for pollination are all based on

Table 2. Map comparison statistics of individual ecosystem services and bundles.

Map comparison	Service					
	Climate	Flood regulation	Pollination	Erosion protection	Recreation	Bundle
LC-EV	*0.27*	0.28	0.23	**0.26**	0.28	0.18
LC-JRC	0.18	0.44	0.30	0.26	0.25	0.14
LC-IVM	0.27	**0.17**	0.29	0.45	**0.14**	0.15
EV-JRC	0.20	0.22	0.44	0.40	0.16	0.17
EV-IVM	**0.15**	0.37	**0.20**	0.27	*0.28*	*0.20*
JRC-IVM	0.19	*0.53*	*0.49*	*0.64*	0.26	**0.11**
Average	0.21	0.34	0.32	0.38	0.23	0.16

For each service, the *highest* (least similar) and **lowest** (most similar) map comparison statistic are indicated.

land cover only, but the maps differ strongly. First, there are differences in definition of the service and in its parameterisation. The JRC approach uses a joint indicator for demand and supply while the indicators used in the LC, EV and IVM approaches focus on supply. The IVM indicator quantifies the probability that a pollinator visits a location while the EV and LC indicators are based on pollinator habitats as proxies. Secondly, pollination depends on landscape configuration and small landscape elements, and with that on the resolution of the input data. The analysed maps are based on different input land cover data and differ in parameterisation of small patches of nature, forest edges, linear landscape elements, and the role of pastures, olive groves and other permanent crops. The EV and LC approach only use the EU-scale CORINE land cover map [23]. The JRC approach supplements this with a map of the density of small patches of nature while the IVM map includes the density of linear landscape elements as an input. As a consequence of the different indicators, parameterisations and input data, high variation among the maps is seen as well as high CV values (Table 3) and a positive correlation of the CV with pasture areas (Table 4).

The *erosion protection* maps show no agreement in regions identified as a hotspot (Figure 2). A few regions show agreement between the coldspots for erosion protection, especially within strongly urbanised regions. This disagreement between the maps for this service is also reflected in the high minimum coefficient of variation (0.31, Table 3). On average, high erosion protection is expected in Sweden and Finland, and in central Europe, due to the high amount of natural vegetation. Low values are found in Hungary, the UK and parts of Spain. In most of the areas with a high average level of erosion protection, the variation among the estimates is large. A variety of indicator definitions is available for this service, that quantify the reduction of soil loss or provide a general indication for the protective effect of natural vegetation [7]. The service depends on many variables, including precipitation, water flow, soil, relief, vegetation and management. This leads to a large variation in input data, as well as model definition and parameterisation, and consequently to a large disagreement between the maps.

The *flood regulation* maps show large differences in their spatial pattern but agree on hotspots in Sweden and Finland and coldspots in Hungary. High mean values are also found in large parts of central Europe while low values dominate in the UK. In considerably large areas with low flood regulation, the maps are in agreement. A variety of indicator definitions is available for this service. Flood regulation can be quantified as the water storage capacity, the reduction of flood danger or damage [43] or as the

role of land cover in regulating runoff, discharge, or retention of water [7,44]. Flood regulation depends on many variables, including precipitation, water flow, soil, relief, vegetation and management. This leads to a large variation in input data, as well as model definition and parameterisation. Finally, flood regulation is a directional service (the service flows from a point of production to a point of use in a specific direction [45,46,47]. This is accounted for in the IVM and JRC flood regulation maps but not in the LC and EV maps.

The *recreation* service maps only show small areas of disagreement scattered across Europe. High values are seen along the southern margin of the EU, in the north of the UK, northern Spain and in Sweden and Finland. There is reasonable agreement on the values between the maps (Figure 3) with low coefficients of variation (Table 3). Recreation ES supply is strongly dependent on land cover and the four mapping approaches use the same variables to quantify supply, such as the presence of coasts, protected areas and relief. Also the input datasets are similar in the different approaches, resulting in similar maps.

High overall ES provision is expected in Sweden and Finland and parts of southern Europe while a low provision of the selected services is seen in large urban areas, and in Hungary and the southeast of the UK (Figure 3). The maps do agree on the areas with low values for the ecosystem service bundle. Agreement on areas with high provision of the total bundle is lower (Figure 2).

Figure 4 summarizes the area percentage in which the analysed ES maps agree. The erosion protection maps disagree in half of the area considered, meaning that in half of the EU territory some maps expect a hotspot for erosion protection while other maps expect a coldspot at the same location. The recreation maps (partly) agree in >80% of the EU territory. In about 5% of the area, all four analysed maps expect a coldspot for recreation. For all ES, there tends to be more agreement on the locations of coldspots than on the locations of hotspots (Figures 2 and 4). Coldspots for all ES coincide with urban or arable areas. This is supported by more detailed studies that have focussed on the provision of services in arable and urban areas: carbon sequestration [48], pollinator habitat [49], protection against erosion and floods [21] and landscape features for recreational activities [19] are often observed at lower levels in urban or arable areas.

Table 4 summarizes correlations between the mean ES provision values and CVs, and the percentage per region covered by particular land cover types. The provision of all five ES is negatively correlated with regions with a high coverage of urban and arable land, and positively correlated with forests and natural areas. For all individual ES, except recreation, the CVs are

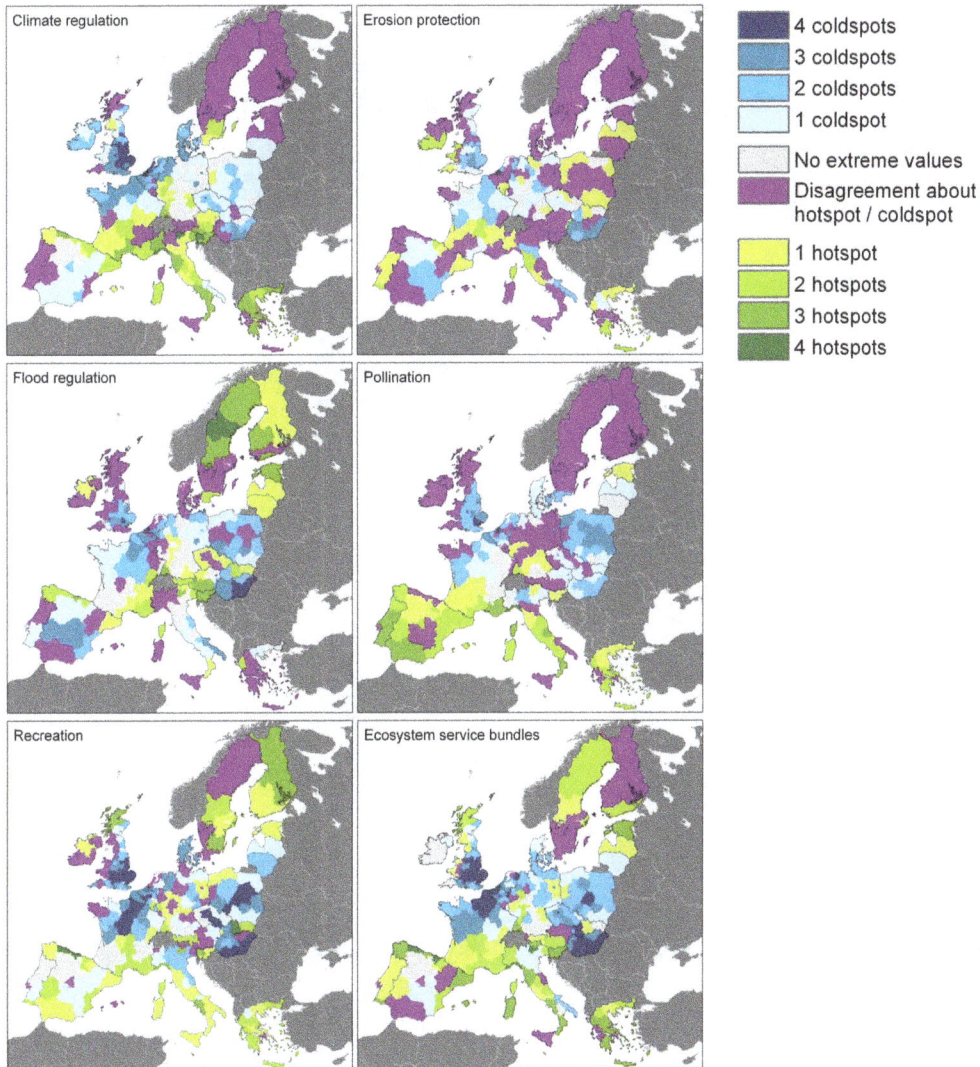

Figure 2. Agreement between maps for each ecosystem service. The maps indicate the number of maps that have a hotspot or coldspot per NUTS2 region. Dark grey areas were not considered.

positively correlated with the area covered by built-up land, indicating that the maps disagree on ES provision in urban areas. The maps agree on a high level of ES provision in forest areas, indicated by the negative correlations between CVs and forest areas. The positive correlations between CVs and arable land area indicate that the maps disagree on the ES provision in arable land, while for pasture the CV differs per service. In areas with more pasture, the maps disagree more on the provision of pollination and recreational services, while for the other services no relations were found. However, the individual estimates for these areas still show a large variation due to a lack of process understanding and different parameterisations in the models used. The higher agreement of the ES bundle maps is due to averaging out differences between the services.

Validation

Table 5 summarizes the agreement of the ES maps with independent datasets. For climate regulation, all ES maps compare better with the independent proxy than could be expected by chance. Agreement is mainly seen in parts of northern and southern Europe. For pollination and erosion prevention only one of the ES maps corresponds better with the independent proxy than would be expected by chance. Correspondence of the pollination maps with the independent proxy is seen in mainland western Europe while the erosion maps only show agreement in southern Europe. The flood regulation maps show some agreement with the independent proxy mainly in northern Europe while the recreation maps mainly show some agreement in western Europe.

Discussion

Sources of uncertainty

The considerable disagreement among spatial patterns of ecosystem service provision across Europe is an indication of the uncertainties in large-scale ecosystem service assessments. Five sources can contribute to these uncertainties (classified after [18]).

First, the definition of the ecosystem service indicators is not consistent. Different categorisations of ecosystem services are available, the Millennium Ecosystem Assessment [50], the TEEB

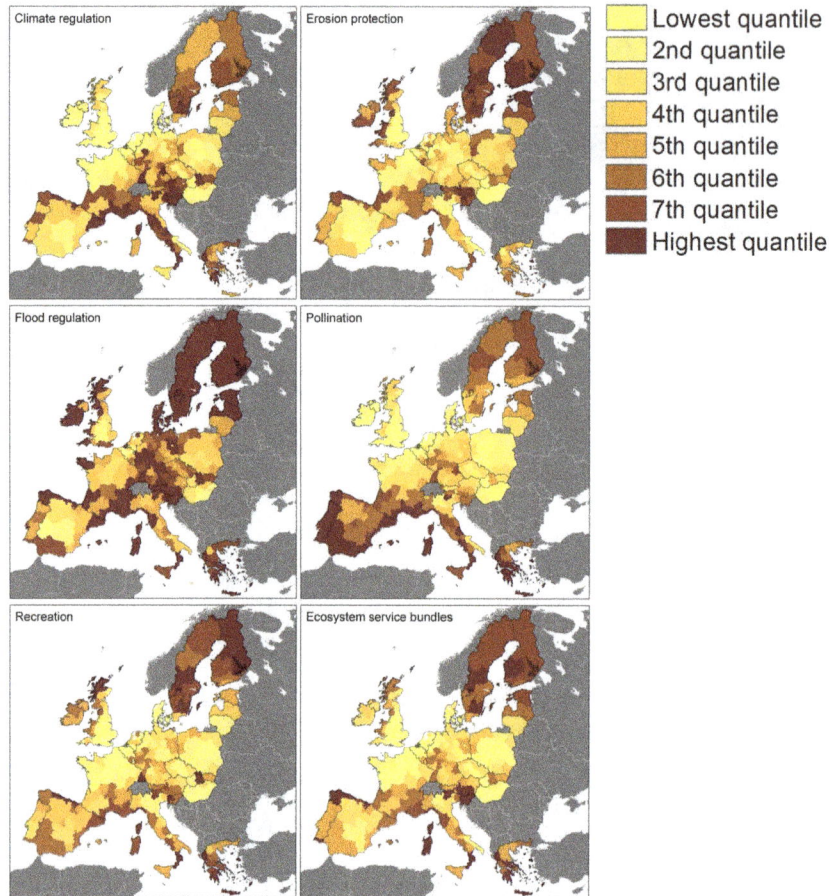

Figure 3. Mean ecosystem service provision per NUTS2 region. Dark grey areas were not considered.

classification [42] and the CICES classification [51] being the most common. Due to differences in the definition of services in these classification systems, the same service does not necessarily address the same aspects [52,53].

Secondly, the level of process understanding can cause uncertainty in quantification and mapping. Ecosystem services are supplied by ecosystems to humans through a variety of biophysical and socio-economic processes. Not all these processes are completely understood or quantified [44]. Different levels of understanding and the inherent uncertainty in understanding lead

to different quantification methods and different choices regarding the inclusion of determinants.

Third, the aim of mapping influences the selection of the most relevant indicators, the data that are used, and the parameterization of the models.

Fourth, the data sources themselves influence the uncertainty of ES maps. An important data source for all ES maps are land cover data, but also several other biophysical or socio-economic data sources are used. Different ES maps are often based on different data sources to quantify the same input variable. These differences

Table 3. Minimum and maximum coefficients of variation for NUTS2 regions between service estimates; low values indicate agreement between the different ES estimates, high values indicate large variation between reported values.

Service	CV			
	Minimum	Location of low values	Maximum	Location of high values
Carbon	0.164	Germany	1.786	Southeast UK, Sweden, Finland
Pollination	0.136	Greece, Spain, Portugal	1.516	Northwest Europe
Erosion protection	0.306	Central Europe and France-Spain border region	1.318	Netherlands, Germany, UK
Flood regulation	0.090	Northern UK, Ireland, Sweden, Finland, Portugal	1.373	Spain, Poland, Hungary
Recreation	0.039	Southern fringes, Germany, Estonia	1.000	Poland, Hungary, UK

Table 4. Correlations between area percentages of land cover classes* per NUTS2 region and mean and CV of ecosystem service provision.

		Urban	Pasture	Nature	Forest	Arable
Carbon	Mean	−0.499	−0.120	0.311	0.777	−0.398
	CV	0.370	0.059	−0.123	−0.439	0.144
Pollination	Mean	−0.525	−0.077	0.438	0.455	−0.307
	CV	0.329	0.340	−0.379	−0.336	0.152
Erosion prevention	Mean	−0.570	0.254	0.304	0.583	−0.428
	CV	0.347	−0.093	−0.466	−0.424	0.548
Flood protection	Mean	−0.533	0.055	0.283	0.609	−0.321
	CV	0.256	−0.084	−0.248	−0.314	0.334
Recreation	Mean	−0.476	0.137	0.481	0.572	−0.570
	CV	−0.013	0.192	−0.292	−0.229	0.363
Bundle	Mean	−0.504	0.082	0.402	0.550	−0.420
	CV	−0.028	0.271	−0.078	0.234	−0.177

*: Urban: all artificial surfaces (CORINE classes 111–142).
Pasture: CORINE class 231. Nature: scrublands, herbaceous vegetation and open spaces (CORINE classes 321–335). Forest: All coniferous/deciduous/mixed forests (CORINE classes 311–313). Arable: All rainfed and irrigated arable land (CORINE classes 211–213).

in input data propagate into differences in the resulting ecosystem service maps.

Finally, the methodology for mapping ES is a source of uncertainty. Mapping methods have different levels of complexity, ranging from process-based simulation to expert based value-transfer methods. Different methods result in different ES maps.

The systematic comparison of ES maps indicates that the agreement among ES maps increases when the sources of uncertainty described above are lower. Climate regulation is a clearly defined and well-understood ES which indicators are mostly based on land cover only, resulting in a high agreement. On the other hand, for erosion protection the indicators used for

mapping diverge strongly and use a wide variety of input data. As a consequence, the agreement among the maps is lower.

Validation

By the intercomparison of different ES maps, we provide insight in the spatial patterns of uncertainties and the level of agreement between the maps across Europe. Such intercomparisons of various model outcomes are a common methodology used for different types of global environmental change models, especially for those where independent validation data are scarce [54,55]. However, model intercomparison does not provide insight in the validity of the models and deviations from the actual ecosystem

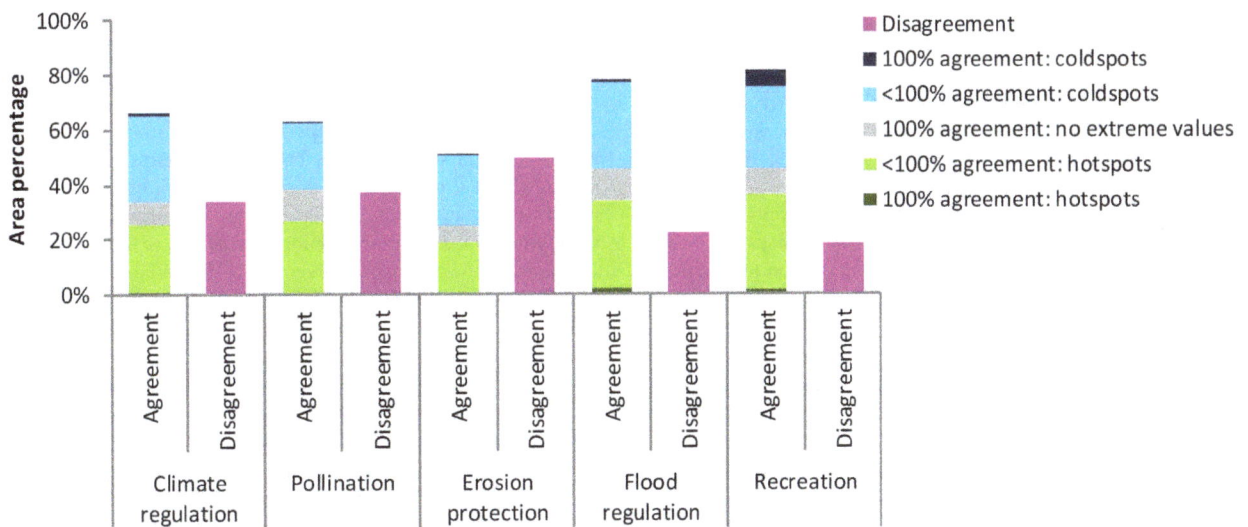

Figure 4. Agreement between the ecosystem service maps. 100% agreement indicates the area where all maps indicate a hotspot, a coldspot or no extreme values, <100% agreement indicates regions where one to three of the maps have a hotspot or coldspot and the other maps do not demonstrate extreme values. Disagreement indicates the regions where at least one map indicates a hotpot and at least one other map indicates a hotspot.

Table 5. Agreement between ecosystem service maps and independent maps.

Map	Climate regulation	Flood regulation	Erosion protection	Pollination	Recreation
LC	1.22	1.33	0.87	0.57	1.18
EV	1.20	1.48	0.92	0.54	1.33
JRC	1.89	0.97	1.14	1.25	0.66
IVM	1.08	1.13	0.95	0.69	0.86

The table shows the ratio between the regions that agree, and the number of regions that would agree by chance.

service provision. Only a comparison of the maps with observed data can help to determine absolute validity levels of the ecosystem service maps.

For the five ecosystem services considered in this paper, available data that can serve as independent proxies to validate the results are collected and compared with the ecosystem service maps. Generally, the studied ES maps are poorly or moderately similar to the independent proxy maps. This does not necessarily provide evidence of the quality of the ES maps assessed here and results have to be interpreted with care. First, the independent proxies are highly different from the ES maps. For none of the ES maps, independent proxies could be identified that exactly match the definition of the service and the indicators used for mapping them. As many ES cannot be measured directly [56], independent proxies that better fit the definitions of the ES indicators were not available. Additionally, there are differences in scale between the maps and independent data. It is not possible to disentangle the relative importance of these two causes from the importance of the map quality upon comparison with independent proxy maps. For a proper validation, independent data covering the variety of conditions throughout the EU would be needed that match the ES definition as described in Table 1. Such data are lacking.

We compared ES maps at the level of administrative units using normalised values of ES provision. Comparing normalized values is potentially less accurate than comparing absolute values. For example, comparison of normalized values would not yield any difference in the case were values for ES provisioning are overestimated by a certain percentage in all locations. However, because the ESs compared in this study were expressed using different units for the same ES, a comparison based on absolute values was not possible. The harmonization of the data to allow comparison has a few other disadvantages as well. Aggregating maps to the resolution of administrative units can lead to a reduction of the spatial variation in ES provision and is likely to result in higher levels of agreement between the maps, especially for ecosystem services showing high spatial variation (e.g. erosion risk, pollination), this can have a large impact.

Importance for policy making

Robust, reliable and comparable data on ES are important for the advancement of biodiversity objectives and to inform the development and implementation of related policies, on water, climate, agriculture, forest, and regional planning. As elaborated by Maes et al. [51], ES maps at EU scale can support decision-making and implementation in multiple ways. First and foremost, mapping ES is an essential part of the EU Biodiversity Strategy to 2020 and ES maps are used to identify priorities for ecosystem restoration and enable the development of an initiative on no net loss of biodiversity and ES. Importantly, the biodiversity strategy aims to mainstream ES into other policies [41], which also entails

the use of ES maps for implementation and targeting of those policies. In the EU, these are notably the common agriculture policy (CAP) and the cohesion policy. The CAP has a profound influence on ecosystems and biodiversity [57] that is recognised in a reform of the policy in 2014, by introducing ecological focus areas aiming at protecting farmland biodiversity and small landscape elements. ES maps will prove to be crucial elements in the spatial identification of those areas where regulating ES support (e.g. pollination, control of pest species, erosion control) and enhance sustainable agricultural production. The cohesion policy, which is essentially responsible for the main share of the EU's investments in the regional economy, is aligning its objectives with goals on sustainable growth. In particular, the conservation and enhancement of natural assets through the development of a green infrastructure network at multiple scales can give rise to socio-economic benefits, which is a priority of cohesion policy. Hence again the need for high quality and consistent spatial information on the levels of services provided by ecosystems which are essential to decision making on future investments using regional EU funds.

Conclusions

This study showed that a different definition of an ecosystem service or a different mapping approach could lead to strongly different spatial patterns of ecosystem service provision. The systematic intercomparison of four EU-scale maps of different ecosystem services demonstrated that there is an overall agreement among the climate regulation maps and the recreation potential maps. The erosion protection and flood regulation maps differed strongly, the pollination maps showed intermediate variation among the maps. Differences between the maps are caused by differences in the mapping aim, indicator definitions, input data and mapping approaches. The sources of uncertainty differ in their importance for the mapping of different ecosystem services. For services with larger differences in definition and mapping approaches, larger differences between individual maps emerge.

Due to the lack of independent data on ecosystem service provision, ecosystem service maps cannot be properly validated and there are, so far, no appropriate measures for map quality.

The choice of a specific ecosystem service map to support policy will influence the specification of policy targets and implementation priorities. Together with the lack of insight in ecosystem service map quality, varying map compilation and interpretation skills, this indicates that mapmakers and end-users should be cautious when applying ecosystem service maps for decision-making. Mapmakers need to clearly underpin the indicators used, the methods, and related uncertainties. Finally, there is an urgent need for better process understanding and data acquisition for ecosystem service mapping, modelling and validation.

Acknowledgments

Felix Kienast and Veiko Lehsten are thanked for making their maps available.

Author Contributions

Conceived and designed the experiments: CJES PV JV. Performed the experiments: CJES. Analyzed the data: CJES PV JV. Contributed reagents/materials/analysis tools: CJES BB JM. Wrote the paper: CJES PV JV BB JM.

References

1. Larigauderie A, Prieur-Richard AH, Mace GM, Lonsdale M, Mooney HA, et al. (2012) Biodiversity and ecosystem services science for a sustainable planet: the DIVERSITAS vision for 2012–20. Curr Opin Environ Sustain 4: 101–105.
2. Mace GM, Cramer W, Díaz S, Faith DP, Larigauderie A, et al. (2010) Biodiversity targets after 2010. Curr Opin Environ Sustain 2: 3–8.
3. European Commission (2011) Communication from the Commission to the European Parliament, the Council, the Economic and Social Committee and the Committee of the Regions: Our life insurance, our natural capital: an EU biodiversity strategy to 2020. Brussel: European Commission.
4. Crossman ND, Burkhard B, Nedkov S, Willemen L, Petz K, et al. (2013) A blueprint for mapping and modelling ecosystem services. Ecosystem Services 4: 4–14.
5. Maes J, Egoh BN, Willemen L, Liquete C, Vihervaara P, et al. (2012) Mapping ecosystem services for policy support and decision making in the European Union. Ecosystem Services 1: 31–39.
6. Burkhard B, Kroll F, Nedkov S, Müller F (2012) Mapping ecosystem service supply, demand and budgets. Ecol Indic 21: 17–29.
7. Kienast F, Bolliger J, Potschin M, De Groot RS, Verburg PH, et al, (2009) Assessing Landscape Functions with Broad-Scale Environmental Data: Insights Gained from a Prototype Development for Europe. Environ Manage 44: 1099–1120.
8. Schulp CJE, Alkemade JRM, Klein Goldewijk K, Petz K (2012) Mapping ecosystem functions and services in Eastern Europe using global-scale data sets. Int J Biodiv Sci, Ecosyst Serv & Manage 8: 156–168.
9. European Commission (2009) Impact Assessment Guidelines. Brussels: European Commission, SEC(2009) 92
10. Egoh BN, Reyers B, Rouget M, Richardson DM, Le Maitre DC, et al. (2008) Mapping ecosystem services for planning and management. Agric Ecosyst Environ 127: 135–140.
11. Seppelt R, Dormann CF, Eppink FV, Lautenbach S, Schmidt S (2011) A quantitative review of ecosystem service studies: approaches, shortcomings and the road ahead. J Appl Ecol 48: 630–636.
12. Eigenbrod F, Armsworth PR, Anderson BJ, Heinemeyer A, Gillings S, et al. (2010) The impact of proxy-based methods on mapping the distribution of ecosystem services. J Appl Ecol 47: 377–385.
13. Pagella TF, Sinclair FL (2014) Development and use of a typology of mapping tools to assess their fitness for supporting management of ecosystem service provision. Landsc Ecol 29: 383–399.
14. Kozak J, Lant C, Shaikh S, Wang G (2011) The geography of ecosystem service value: The case of the Des Plaines and Cache River wetlands, Illinois. Appl Geogr 31: 303–311.
15. Lautenbach S, Kugel C, Lausch A, Seppelt R (2011) Analysis of historic changes in regional ecosystem service provisioning using land cover data. Ecol Indic 11:676–687.
16. Schulp CJE, Lautenbach S, Verburg PH (2014) Quantifying and mapping ecosystem services: Demand and supply of pollination in the European Union. Ecol Indic 36: 131–141.
17. Kandziora M, Burkhard B, Müller F (2013) Mapping provisioning ecosystem services at the local scale using data of varying spatial and temporal resolution. Ecosystem Services 4: 47–59.
18. Hou Y, Burkhard B, Müller F (2013) Uncertainties in landscape analysis and ecosystem service assessment. J Environ Manage in press.
19. Van Berkel DB, Verburg PH (2011) Sensitising rural policy: Assessing spatial variation in rural development options for Europe. Land Use Policy 28: 447–459.
20. Schulp CJE, Nabuurs GJ, Verburg PH (2008) Future carbon sequestration in Europe - Effects of land use change. Agric Ecosyst Environ 127: 251–264.
21. Sturck J, Poortinga A, Verburg PH (2014) Mapping ecosystem services: The supply and demand of flood regulation services in Europe. Ecol Indic 38: 198–211.
22. Tucker G, Allen B, Conway M, Dickie I, Hart K, et al. (2013) Policy Options for an EU No Net Loss Initiative. London: Institute for European Environmental Policy,.
23. EEA (2000) CORINE Land Cover 2000. Available: http://www.eea.europa.eu/data-and-maps/data/corine-land-cover-2000-raster-2. Accessed 2012 Apr 2.
24. Maes J, Paracchini ML, Zulian G (2011) A European assessment of the provision of ecosystem services. Towards an atlas of ecosystem services. Ispra: JRC.
25. Zulian G, Maes J, Paracchini ML (2013) Linking Land Cover Data and Crop Yields for Mapping and Assessment of Pollination Services in Europe. Land 2: 472–492.
26. Paracchini ML, Zulian G, Kopperoinen L, Maes J, Schägner JP, et al (2014) Mapping cultural ecosystem services: A framework to assess the potential for outdoor recreation across the EU. Ecol Indic 45:371–385.
27. Lehsten V, Sykes MT, Kallimanis A, Mazaris A, Verburg PH, et al (2014) Disentangling the effects of expected land use, climate and CO_2 on European habitat types, carbon stocks and primary productivity. Global Ecology and Biogeography: Submitted.
28. Serna-Chavez HM, Schulp CJE, Van Bodegom PM, Bouten W, Verburg PH, et al. (2014) A quantitative framework for assessing spatial flows of ecosystem services. Ecol Indic 39: 24–33.
29. Imhoff ML, Bounoua L, Ricketts TH, Loucks C, Harriss R, et al. (2004) HANPP Collection: Global Patterns in Net Primary Productivity (NPP). Palisades, NY: NASA Socioeconomic Data and Applications Center (SEDAC). http://dx.doi.org/10.7927/H40Z715X. Accessed 15 February 2012.
30. Brakenridge GR (2010) Global Active Archive of Large Flood Events. Available: http://floodobservatory.colorado.edu/. Accessed 2013 Jul 27.
31. GBIF (2007) Global Biodiversity Information Facility data portal. Available: http://gbif.org/datasets/resource/14567. Accessed 2013 Jul 27.
32. Bai ZG, Dent DL, Olsson L, Schaepman ME (2008) Global Assessment of Land Degradation and Improvement 1. Identification by remote sensing (No. - GLADA Report 5). Wageningen: ISRIC.
33. Burkhard B, Kandziora M, Hou Y, Müller F (2014) Ecosystem service potentials, flows and demands – concepts for spatial localization, indication and quantification. Landscape Online 34:1–32.
34. EUROSTAT (2011) Regions in the European Union. Nomenclature of territorial units for statistics NUTS2010/EU-27. EUROSTAT methodologies and working papers. DOI: 10.2785/15544.
35. BIO by Deloitte (2014) Study supporting potential land and soil targets under the 2015 Land Communication, Report prepared for the European Commission, DG Environment in collaboration with AMEC, IVM and WU.
36. Hagen-Zanker A (2006) Comparing Continuous Valued Raster Data. Maastricht: RIKS BV.
37. García-Nieto AP, García-Llorente M, Iniesta-Arandia I, Martín-Lopez B (2013) Mapping forest ecosystem services: From providing units to beneficiaries. Ecosystem Services 4: 126–138.
38. Gimona A, Horst D (2007) Mapping hotspots of multiple landscape functions: a case study on farmland afforestation in Scotland. Landsc Ecol 22: 1255–1264.
39. Bivand R, Keitt T, Rowlingson B, Pebesma M, Summer M, et al. (2014) Rgdal: Bindings for the Geospatial Data Abstraction Library. Available at http://raster.r-forge.r-project.org/. Accessed: 2013 Apr 8.
40. Hijmans RJ, Van Etten J (2012) Raster: Geographic analysis and modelling with raster data. Available at http://raster.r-forge.r-project.org/. Accessed: 2013 Apr 8.
41. Maes J, Hauck J, Paracchini ML, Ratamäki O, Hutchins M, et al (2013) Mainstreaming ecosystem services into EU policy. Curr Opin Environ Sustain In Press.
42. De Groot RS, Fisher B, Christie M, Aronson J, Braat L, et al. (2010) Integrating the ecological and economic dimensions in biodiversity and ecosystem service valuation. In: Kumar P, editor. The Economics of Ecosystems and Biodiversity: Ecological and Economic Foundations. London and Washington, Earthscan, pp. 9–40.
43. De Groot RS, Alkemade JRM, Braat L, Hein L, Willemen L (2010) Challenges in integrating the concept of ecosystem services and values in landscape planning, management and decision making. Ecological Complexity 7: 260–272.
44. Nedkov S, Burkhard B (2012) Flood regulating ecosystem services—Mapping supply and demand, in the Etropole municipality, Bulgaria. Ecol Indic 21: 67–79.
45. Bagstad KJ, Johnson GW, Voigt B, Villa F (2013) Spatial dynamics of ecosystem service flows-A comprehensive approach to quantifying actual services. Ecosystem Services 4: 117–125.
46. Costanza R (2008) Ecosystem services: Multiple classification systems are needed. Biol Conserv 141: 350–352.
47. Syrbe R, Walz U (2012) Spatial indicators for the assessment of ecosystem services: Providing, benefiting and connecting areas and landscape metrics. Ecol Indic 21: 80–88.
48. Janssen IA, Freibauer A, Schlamadinger B, Ceulemans R, Ciais P, et al. (2004) The carbon budget of terrestrial ecosystems at country-scale–a European case study. Biogeosci Discuss 1: 167–193.
49. Carré G (2009) Landscape context and habitat type as drivers of bee diversity in European annual crops. Agric Ecosyst Environ 133: 40–47.
50. Millennium Assessment (2005) Ecosystems and human well-being.
51. Maes J, Teller A, Erhard M (2013) Mapping and assessment of ecosystems and their services: an analytical framework for ecosystem assessments under action 5 of the EU Biodiversity Strategy to 2020. Luxembourg: Publications office of the European Union.
52. Hernández-Morcillo M, Plieninger T, Bieling C (2013) An empirical review of cultural ecosystem service indicators. Ecol Indic 29: 434–444.

53. Villamagna AM, Angermeier PL, Bennett EM (2013) Capacity, pressure, demand, and flow: A conceptual framework for analyzing ecosystem service provision and delivery. Ecological Complexity 15: 114–121.

54. Cheaib A, Badeau V, Boe J, Chuine I, Delire C, et al. (2012) Climate change impacts on tree ranges: model intercomparison facilitates understanding and quantification of uncertainty. Ecol Lett 15: 533–544.

55. Rosenzweig C, Elliott J, Deryng D, Ruane AC, Müller C, et al. (2014) Assessing agricultural risks of climate change in the 21st century in a global gridded crop model intercomparison. Proc Natl Acad Sci U S A 111: 3268–3273.

56. Bennett EM, Peterson GD, Gordon LJ (2009) Understanding relationships among multiple ecosystem services. Ecol Lett 12: 1394–1404.

57. Pe'er G, Dicks LV, Visconti P, Arlettaz R, Báldi A, et al (2014) EU agricultural reform fails on biodiversity. Science 344:1090–1092.

10

DNA Barcode Authentication of Wood Samples of Threatened and Commercial Timber Trees within the Tropical Dry Evergreen Forest of India

Stalin Nithaniyal[1,2], Steven G. Newmaster[3]*, Subramanyam Ragupathy[3], Devanathan Krishnamoorthy[1], Sophie Lorraine Vassou[1], Madasamy Parani[1]*

1 Department of Genetic Engineering, Center for DNA Barcoding, SRM University, Chennai, India, **2** Interdisciplinary School of Indian System of Medicine, SRM University, Chennai, India, **3** Centre for Biodiversity Genomics, University of Guelph, Guelph, Ontario, Canada

Abstract

Background: India is rich with biodiversity, which includes a large number of endemic, rare and threatened plant species. Previous studies have used DNA barcoding to inventory species for applications in biodiversity monitoring, conservation impact assessment, monitoring of illegal trading, authentication of traded medicinal plants etc. This is the first tropical dry evergreen forest (TDEF) barcode study in the World and the first attempt to assemble a reference barcode library for the trees of India as part of a larger project initiated by this research group.

Methodology/Principal Findings: We sampled 429 trees representing 143 tropical dry evergreen forest (TDEF) species, which included 16 threatened species. DNA barcoding was completed using *rbcL* and *matK* markers. The tiered approach (1st tier *rbcL*; 2nd tier *matK*) correctly identified 136 out of 143 species (95%). This high level of species resolution was largely due to the fact that the tree species were taxonomically diverse in the TDEF. Ability to resolve taxonomically diverse tree species of TDEF was comparable among the best match method, the phylogenetic method, and the characteristic attribute organization system method.

Conclusions: We demonstrated the utility of the TDEF reference barcode library to authenticate wood samples from timber operations in the TDEF. This pilot research study will enable more comprehensive surveys of the illegal timber trade of threatened species in the TDEF. This TDEF reference barcode library also contains trees that have medicinal properties, which could be used to monitor unsustainable and indiscriminate collection of plants from the wild for their medicinal value.

Editor: Arndt von Haeseler, Max F. Perutz Laboratories, Austria

Funding: Support for this study was provided by the following institutions: Funding Institute: Department of Science and Technology, Government of India. Grant No:VI-D&P/372/10-11/TDT URLs: (http://www.dst.gov.in) Author: SN. Funding Institute: Department of Biotechnology, Cutting-edge Research Enhancement and Scientific Training Fellowship. Government of India. Grant No: BT/IN/DBT-CREST Awards/09/PM/2011-12. URLs: (http://dbtindia.nic.in/index.asp). Author: PM. Funding Institute: Biodiversity Institute of Ontario, University of Guelph, Canada. Grant No: nil. URL: (http://www.biodiversity.uoguelph.ca) Author: MP, SN, SR 2. Funding Institute: SRM University, India. Grant No: nil. URL: (www.srmuniv.ac.in) Author: MP. The funders had no role in study design, data collection and analysis, decision to publish, or preparation of the manuscript.

Competing Interests: The authors have declared that no competing interests exist.

* Email: snewmast@uoguelph.ca (SGN); parani.m@ktr.srmuniv.ac.in (MP)

Introduction

India is the custodian for considerable biodiversity as it intersects four global biodiversity hotspots, and is the eighth largest country among the 17 mega biodiversity countries [1]. According to the India's fourth report to the convention on biological diversity, it harbors nearly 11% of the world's floral diversity, which includes ca. 6000 endemic species and over 246 globally threatened species [2]. India's biogeography is diverse with ten different bio-geographic zones, of which 23.4% of the land area is forested [3,4]. The forests in India are classified into 6 major types and 16 minor types on the basis of structure, physiognomy and floristic diversity [5]. The tropical dry evergreen forest (TDEF) is one of the minor forest types classified within the major forest type, tropical dry forest (TDF). The TDEF of India is part of the costal bio-geographic zone that is narrowly confined to the East coast, which is under considerable development pressures. Tropical forest ecosystems are known as critical habitats for the conservation of biodiversity, and these ecosystems are threatened by urbanization and climatic change resulting in species extinction at the rate of 0.8% to 2% per year [6]. The TDEF in India is particularly vulnerable because of its very narrow geographic boundaries. The forest cover under TDEF is rapidly declining due to overexploitation for timber, fuel wood, and construction of infrastructure such as buildings, dams, and roads. This has recently resulted in substantial media calling for conservation measures within the TDEF of India. Notable scientists have reported that the impending threat to the rich native biodiversity in the TDEF of India is partly due to its inherent abundance in natural resources [7,8]. The TDEF needs to be given high priority for natural resource planning strategies that conserve biodiversity as envisioned in National Environment Policy [9].

Quick and reliable species identification is needed in order to facilitate the large-scale biodiversity inventories required for conservation strategies [10]. Taxonomic identification of tropical trees can be challenging; individual trees of a species may vary morphologically according to their age and growing conditions, and at the same time, closely related species can look morphologically similar [11]. Traditional taxonomic methods based on morphological identifications are costly and require a considerable amount of time in order to provide accurately identified plants [12–15]. There are only a few taxonomists in India with botanical field experience who can reliably identify all the tree species in TDEF. Moreover, it is extremely difficult to identify the species when the specimen is incomplete, damaged or derived from plant parts such as leaves, roots, bark, wood and seeds. It is desirable to utilize an alternate method for species identification that can use specimens in different forms (e.g., wood) and life stages. Recent advances in DNA sequencing and molecular diagnostic tools for plants [16,17] have the capacity to improve upon traditional methods of species identification [18].

DNA barcoding is emerging as a valuable tool for quick assessments of biodiversity that provides high quality data for developing conservation strategies [19,20]. A recent study reported assessment data from the same site wherein DNA barcoding survey provided more accurate estimates (42% more species) than traditional morphological taxonomic survey, which was 37% more expensive than barcoding [21]. DNA barcoding uses a short standardized DNA sequence for species identification that is divergent between species but conserved within species [22]. While cytochrome c oxidase I (COI) gene is widely regarded as a universal DNA barcode to identify most groups of animals, a different approach has been taken for plants. This is due to the fact that there is little COI variation in plants and there has been difficulty in identifying a single universal barcode marker for plants; plants have inherently low nucleotide variation in recently evolved species, and undergo complex evolutionary processes such as hybridization and polyploidy [23,24]. Although many researchers have searched for a single region for barcoding plants, it is generally agreed that a multi-locus barcode combination would be required to discriminate plant species [25–29]. Newmaster *et al.* [17] and Purushothaman *et al.* [30] described this as the multigene tiered approach wherein barcodes are constructed from two 'tiered' gene regions; an easily amplified and aligned region is used for the first tier (*rbcL*) that acts as a scaffold on which data from a more variable second-tier region are interpreted for species identification. The chloroplast *rbcL* was proposed as the first tier marker because of its universality and demonstrated success for differentiating congeneric plant species [17,31]. The second tier variable marker may be chloroplast *trnH-psbA* (non-coding) and *matK* (coding) or nuclear ITS2.

DNA barcoding has been used in many botanical studies ranging from detailed study on single genus to ecosystem level surveys in tropical, subtropical and temperate forests. DNA barcoding of all the 1073 trees in two hectares of a tropical forest in French Guiana showed that it could increase the quality and the speed of biodiversity surveys [13]. It was found to be useful for detecting errors in morphological identifications and increased the identification rate of juveniles from 72% to 96%. DNA barcoding of 200 accessions from two 0.1 hectare tropical forest plots in

Northeast Queensland also showed that it could rapidly estimate species richness in forest communities [12]. Tripathi *et al.* [32] have studied 300 specimens from tropical trees of North India, and suggested that DNA barcoding will be useful in large-scale biodiversity inventories. Vegetation surveys in four equally sized temperate forest plots in the Italian pre-alpine region of Lombardy, Valcuvia by morphological identification and DNA barcoding revealed that the later could save time and resources [18]. Parmentier *et al.* [11] have assessed the accuracy of DNA barcoding in assigning a specimen to a species or genus by studying 920 trees from five lowland evergreen forest plots in Korup and Gabon, Africa. DNA Barcoding was found to be useful in assigning unidentified trees to a genus, but assignment to a species was less reliable, especially in species-rich clades. In a large study that included 2,644 individuals representing 490 vascular plant species, mostly from the Canadian Arctic zone, again showed that DNA barcoding differentiated the taxa more at the genus level than at the species level [33].

In another interesting study of tropical forest, DNA barcoding was applied on 1,035 samples representing all the 296 species of a Forest Dynamics Plot on Barro Colorado Island in Panama [34]. Barcode data from *rbcL*, *matK* and *trnH-psbA* were found to be sufficient to reconstruct evolutionary relationships among the plant taxa that were congruent with the broadly accepted phylogeny of flowering plants. The same research group studied another Forest Dynamics Plot in the Luquillo Mountains of Northeast Puerto Rico that encompassed a mix of old growth and secondary forest that has been largely free from human disturbance since the 1940 s. This study again reinforced the congruence of the barcode phylogeny with the phylogeny of flowering plants as per APG III classification [35]. DNA barcoding was also used to construct community phylogeny in order to understand the patterns of species occurrence in forest habitats [36]. Community phylogeny which was constructed for the Dinghushan Forest Dynamics Plot in China by sequencing *rbcL*, *matK*, and *trnH-psbA* loci from 183 species showed that closely related species tend to prefer similar habitats. The patterns of co-occurrence within habitats are typically non-random with respect to phylogeny. While phylogenetic clustering was observed in valley and low-slope, phylogenetic over-dispersion was characteristic of high-slope, ridge-top and high-gully habitats.

Our study reports DNA barcoding of tree species from the TDEF in India. The specific objectives of this project are to 1) Develop a TDEF reference barcode library for 143 tropical tree species, 2) Utilize the TDEF reference barcode library for species identification of lumber from logged timber sites, 3) To monitor the endemic and threatened species in timber trade, and 4) To prevent indiscriminate collection of non-timber forest products. This research seeks to provide a DNA reference barcode library for floristic assessments of tropical dry evergreen forests in biodiversity rich countries like India, which can be utilized for the conservation of rare and native tree species.

Materials and Methods

Sample collection

Our study area was the Tropical Dry Evergreen Forest (TDEF) of India, which is part of the costal bio-geographic zone. It is narrowly confined to the East coast (9° 22′ –17° 36′ N latitude and 78° 49′ –82° 56′ E longitude) between Visakhapatnam in Andhra Pradesh and Ramanathapuram in Tamil Nadu (Figure 1). The forests have three sub-classifications: sandy coast, interior coastal plains with red lateritic soil, and isolated hillocks wherein dense forest thickets are formed with evergreen and deciduous small

Figure 1. Map of India showing the distribution of Tropical Dry Evergreen Forest (*Painted in green colour) distributed between Visakhapatnam in Andhra Pradesh State and Ramanathapuram in Tamil Nadu State.

trees and thorny shrubs. The TDEF receives an annual rainfall of 900 mm to 1200 mm. Depending on the geographical location, the dry season may extend from January to March or from December to May [37].

We sampled 429 trees representing 143 species (114 genera, 42 families and 19 orders) from different sites within TDEF, and their GPS coordinates are provided in Table S1. Out of the 143 tree species collected, 16 species are on the IUCN red list of threatened species as searched in the website http://www.redlist.org. All the samples were collected for research purpose only from cultivated sources, gardens and open forests which are accessible to any public, hence no permission was required. Voucher specimens from all the collections were professionally identified using local floras. They were mounted on standard herbarium sheets, and were deposited to the SRM University Herbarium. Leaves from each accession were air-dried, stored at room temperature, and later used for DNA extraction and barcoding. In addition, sap woods from 25 freshly logged trees were collected from timber shops at five different locations.

DNA isolation

Genomic DNA was isolated by following the protocol of Saghai-Maroof *et al.* [38] with minor modifications. About 100 mg of leaf tissue was taken for genomic DNA isolation and ground using mortar and pestle by adding 500 μl of CTAB buffer (100 mM Tris-HCl, 1.4 M NaCl, 20 mM EDTA, 1% beta-mercaptoethanol, 2% CTAB). The samples were transferred to 1.5 ml centrifuge tubes, incubated in water bath at 55°C for 30 minutes, and then extracted with equal volume of chloroform. The samples were centrifuged at 10,000 rpm for 10 minutes, and the aqueous phase was transferred to fresh 1.5 ml centrifuge tubes. The DNA was precipitated by adding equal volume of ice-cold isopropanol, and centrifuged at 10,000 rpm for 10 minutes. The DNA pellet was washed with 70% ethanol, air-dried at room temperature, and dissolved in 100 μl TE buffer. In case of wood samples, genomic DNA was isolated by following the same protocol except that 2% PVP was included in the CTAB buffer, and the samples were incubated at 55°C for 10 hours.

PCR amplification and DNA sequencing

PCR amplification of DNA barcode markers was done using 50 ng of total genomic DNA as template and the commonly used primers for *matK* (*matK*-1RKIM-F and *matK*-3FKIM-R, Ki-Joong Kim, School of Life Sciences and Biotechnology, Korea University, Korea, unpublished), and *rbcL* (*rbcL*a-F, *rbcL*ajf634-R) [39,40]. PCR reaction mixture (30 μl) contained 1X buffer with 1.5 mM MgCl$_2$, 200 μM dNTPs, 5 pmol primers, and 1 unit *Taq* DNA polymerase. PCR was done in a thermal cycler (Eppendorf, Germany) using the following protocol: initial denaturation at 95°C for 5 minutes, 30 cycles of denaturation at 95°C for 30 seconds, annealing at 55°C for 30 seconds, and extension at 72°C for 1 minutes, final extension at 72°C for 5 minutes, and hold at 16°C. The PCR products were checked by agarose gel electrophoresis, and purified using EZ-10 Spin Column PCR Purification Kit (Bio Basic Inc. Ontario, Canada). The purified PCR products were sequenced from both ends using the same PCR primers in 3130×l Genetic analyzer (Applied Biosystems, CA, USA). The sequences were manually edited using Sequence Scanner Software v. 1.0 (Applied Biosystems, CA, USA) and full length sequences were assembled.

Data analyses

The fully edited sequences with original trace files for *rbcL* and *matK* markers were submitted to Barcode of Life Database (BOLD Systems v.3.) under the project name "TDEF Project 1" with process IDs TDEF001-12 to TDEF429-12. The details of the 429 samples that were used in the present study, their process IDs in BOLD database, PCR success and length of *rbcL* and *matK* sequences obtained are given in Table S2. These sequences were also used to create a TDEF reference barcode library. Pairwise divergence was calculated in BOLD Systems v. 3 using Kimura 2 parameter distance model and MUSCLE program [41]. Database search for species identification were done using Basic Local Alignment Search Tool (BLAST) against non-redundant nucleotide database at NCBI (www.blast.ncbi.nlm.nih.gov/Blast.cgi). We assessed the species resolution of the two DNA barcodes using three different methods; the best match method [42], phylogenetic method [43], and Characteristic Attribute Organization System (CAOS) [44].

Best match method for species identification was carried out using TaxonDNA version 1.6.2, [42] which is available at http://taxondna.sf.net/. In this method, each sequence was queried against TDEF reference barcode library to identify the species associated with its closest match based on the genetic distance. The query identification was considered a "success" when the two sequences were from the same species, "ambiguous" when it matched with more than one species at the same genetic distance, and "failure" when the two sequences were from mismatched species.

Phylogenetic tree was constructed after combining the *rbcL* and *matK* barcode sequences. Genetic distances were calculated by K2P distance model and phylogenetic trees were constructed by Neighbor-Joining (NJ) method using ClustalW in MEGA v. 5.1 [43]. Bootstrap support was analyzed with 1,000 replications. All positions containing gaps and missing data were eliminated from the analysis. Species were distinguished based on genetic distance and monophyly.

Characteristic Attribute Organization System (CAOS) was used to identify diagnostic characteristic attributes (CAs) for species identification [44,45]. Sequence data matrix and tree file were generated using the program MESQUITE v. 2.6 [46]. The resulting NEXUS file which consists of a non-interleaved DNA data matrix, a translate block (converts the taxon names to higher values in the tree representation) and a Newick tree file with collapsing nodes relative to the taxonomic groupings of interest was used in CAOS in accordance with the manual (www.boli.uvm.edu/casos-workbench/manual). First, it was used in the P-Gnome program to determine diagnostic positions at each major taxonomic grouping. Then, new sequences were classified into taxonomic groupings using the P-Elf program. Finally, the most variable sites that distinguish all the taxa were chosen. The character states at these nucleotide positions were listed and unique combinations of CAs were identified.

Results and Discussion

PCR amplification and bidirectional sequencing of *rbcL* and *matK* markers

Success of PCR amplification and sequence recoverability is an important criterion for assessing the utility of DNA barcodes. In our study, *rbcL* and *matK* barcode markers were amplified using universal primer pairs and standard protocols for most of our samples, despite the fact that these plant samples represented 42 diverse families. The *rbcL* marker was successfully amplified from all the samples, whereas the *matK* marker was amplified only in 75.8% of the samples. There was no variation in sequence length for *rbcL*; bidirectional sequencing recovered the 607 bp target sequence for all the PCR amplicons. Bidirectional sequencing was successful in 98% of the *matK* PCR amplicons, and there was considerable variation in the sequence length. Length of the *matK* sequence (Q value >40) varied between 508 bp and 867 bp with an average of 803 bp (500 bp is acceptable for the submission to BOLD database). Our results support earlier studies that report no variation in sequence length for *rbcL* along with high PCR amplification and sequencing success [25,47], which in some studies reaches 100% [48,49]. Previous researches suggest that *matK* PCR success rate is highly variable, ranging from 40% to 97% [39,48]. Although we did not record any repeat sequences in *matK* as documented in other studies [50] in which it impacted the sequencing quality and success; repeat sequences in *matK* are not as common as those found in *trnH-psbA* [51].

Intra/inter-specific divergence

Intra-specific and inter-specific divergence are useful for assessing DNA barcodes [29,52,53]. We calculated divergence among the individuals of the same species (intra-specific divergence) as well as the species of individual genus (inter-specific divergence) wherever multiple species in a genus where included in the study. Intra-specific divergence varied from 0.0% to 0.33% and 0.0% to 0.49% for *rbcL* and *matK*, respectively. Inter-specific divergence varied from 0.0% to 1.8% for *rbcL*, and 0.0% to 2.6% for *matK*. Our study included 44 congeneric species from 15 genera for which pairwise divergences were considered for their ability to differentiate the species. The number of congeneric species per genus varied between 2 and 7 species, and they formed 63 congeneric species pairs. Data from *rbcL* was available for all the pairs, and it differentiated 28 (44%) species when cut-off for intra-specific divergence was set at 0.5% (Table S3). At this cut-off level, *matK* differentiated 35 (92%) species (Table S4). We defined barcoding gaps as the difference between minimum inter-specific and maximum intra-specific divergence, as calculated for the congeneric species. Barcoding gap was observed in 11 genera, and it varied from 0.16% to 0.66% and 0.38% to 1.55% for *rbcL* and *matK* marker, respectively. In general, the barcoding gap is narrow due to the existence of closely related congeneric species. There was a large overlap between intra-specific and inter-specific pairwise distances among the congeneric species of deciduous trees of which the observed barcoding gap ranged between 0.2% and 0.9% [54]. Comparable levels of the barcoding gap were reported in *Agalinis* that ranged between 0.44% and 0.76% [55]. If pairwise divergence across all the species (non-congeneric) is considered, *rbcL* and *matK* differentiated 45.14% and 90% of the species, respectively. Previous researches have reported *matK* to have only slightly more discriminatory power than *rbcL* [27,28]. We report a considerably larger difference, but this may be attributed to the fact that 24% of our samples are from Fabaceae, and *matK* was shown to have more than 80% species differentiation in this family [56].

Barcode species resolution

It is estimated that the TDEF in India has ca. 1,500 species of which ca. 300 species are trees. Therefore, the TDEF represents about 11.5% of the 2,560 tree species found in India [57]. We have generated TDEF reference barcode library for the first time with 429 *rbcL* and 318 *matK* barcodes that were derived from 143 tree species.

Best match method for species ID

The best match method is the simplest method for species identification [42]. It assigns the query sequence to a species with which it shows the smallest genetic distance. The *rbcL* and *matK* barcode sequences from individual samples were queried against sequences in the TDEF reference barcode library. The *rbcL* marker correctly identified 129 out of 143 species (90.2%) with the smallest genetic distance among all the species. Species identification for the remaining samples was 'ambiguous' because they showed same genetic distance with more than one species. The *matK* marker correctly identified the samples from 113 out of 117 species (96.5%). The strict combined marker (*rbcL+matK*) approach correctly identified the samples from 115 out of 117 species (98.3%) (Table 1). The tiered approach (1st tier *rbcL*; 2nd tier *matK*) correctly identified the samples from 136 out of 143 species (95%). The distance based methods have been criticized because it is extremely difficult to determine a single universal threshold genetic distance for distinguishing taxonomic groups [58,59]; this is supported by the fact that the barcode gap can vary greatly across the groups [60]. Assigning group-specific thresholds either by following the "10X rule" of Herbert *et al.* [61] or otherwise is also not reliable when the estimated intra-group divergence does not represent the entire range of the distribution.

Phylogenetic method for species ID

Phylogenetic tree based analyses are useful for evaluating discriminatory power by calculating the proportion of monophyletic species. A monophyletic clade includes the ancestor and all of its descendants that can be identified by the ability to remove it from the rest of the phylogenetic tree with a single cut. In our study, we constructed phylogenetic tree using the neighbor-joining method, which has been adopted by many floristic barcoding studies [33,62]. Combined data for both *rbcL* and *matK* marker was available for 117 species belonging to 34 families. In the phylogentic tree, 30 families formed monophyletic groups, and 27 of them had bootstrap value between 70% and 100%. (Figure 2). The largest family that we studied was the Fabaceae, which included 23 genera and 34 species. Among the three subfamilies in Fabaceae, *Faboideae* was monophyletic while *Caesalpinioideae* was paraphyletic with respect to *Mimosoideae* (Figure 3). This is supported by the earlier phylogenetic report based on *rbcL* sequences as well as morphological characters [63–65]. Among the four tribes studied in *Faboideae*, *Dalbergieae* and *Robinieae* were monophyletic while *Millettieae* and *Phaseoleae* were not monophyletic (Figure 3). Polyphyly relationship between *Millettieae* and *Phaseoleae* was reported before based on morphological characters [66], chloroplast *rbcL* sequences [67], and nuclear phytochrome gene sequences [68]. *Caesalpinieae* in *Caesalpinioideae* as well as *Acacieae* and *Mimoseae* in *Mimosoideae* were not monophyletic (Figure 3). Earlier studies based on morphological as well as *rbcL* data have shown that *Mimoseae* is paraphyletic [69,70]. In the genus level, all except *Acacia* and *Albizia* formed monophyletic groups. The non-monophyletic clade formed two branches: one branch contained only the species of *Acacia*; the other branch was shared by the species of *Acacia*, *Albizia*, *Enterolobium* and *Pithecelobium*. While *Acacia* belongs to tribe *Acacieae*, *Albizia*, *Enterolobium* and *Pithecelobium* belong to tribe *Ingeae*. Based on *matK* and *trnK* chloroplast sequences, it has been reported that the genus *Acacia* is not monophyletic [71,72]. We also found a non-monophyletic clade outside the Fabaceae that was formed by *Pamburus* and *Aegle*, which belong to tribe *Aurantieae* of Rutaceae.

Our phylogenetic trees can also be used for differentiating species. We could differentiate 90.2%, 95.7%, and 98.3% of the species from the tree constructed using *rbcL*, *matK*, and *rbcL+matK*, respectively (Figure 2, Figure S1, Figure S2, and Table 1). The species that could not be differentiated based on *rbcL* marker included four species of *Acacia*, six species of *Ficus* and two species of *Annona*. In addition, the monotypic species *Aegle marmelos* could not be differentiated from *Pamburus missionis*. However, *matK* differentiated the two species each from *Acacia* and *Annona*, *P. missionis* and *A. marmelos* that could not be differentiated by *rbcL*. *Manilkara hexandra*, *M. zapota* and *Madhuca longifolia* of Sapotaceae were not distinguished by *matK* but were distinguished by *rbcL* albeit with very low genetic distance. It is reported that plastid markers perform poorly in recovering monophyletic species in Sapotaceae [13]. However, by combining the data from *rbcL* and *matK*, we could differentiate all except two species (*Acacia chundra* and *A. ferruginea*). Phylogenetic tree based methods have been criticized because they are not able to make use of low level of divergence, which is sufficient for differentiating groups but not for building phylogenetic relationships [33,60].

Table 1. Performance of DNA barcodes in sequence recovery and species identification success.

| Barcodes | Sequence recovery | | Species identification success | | |
	No. of accessions	No. of species	Best match method	Phylogenetic method	CAOS method
rbcL	429	143	129 (90.2%)	129 (90.2%)	129 (90.2%)
matK	318	117	113 (96.5%)	112 (95.7%)	113 (96.5%)
rbcL+matK	351	117	115 (98.3%)	115 (98.3%)	115 (98.3%)

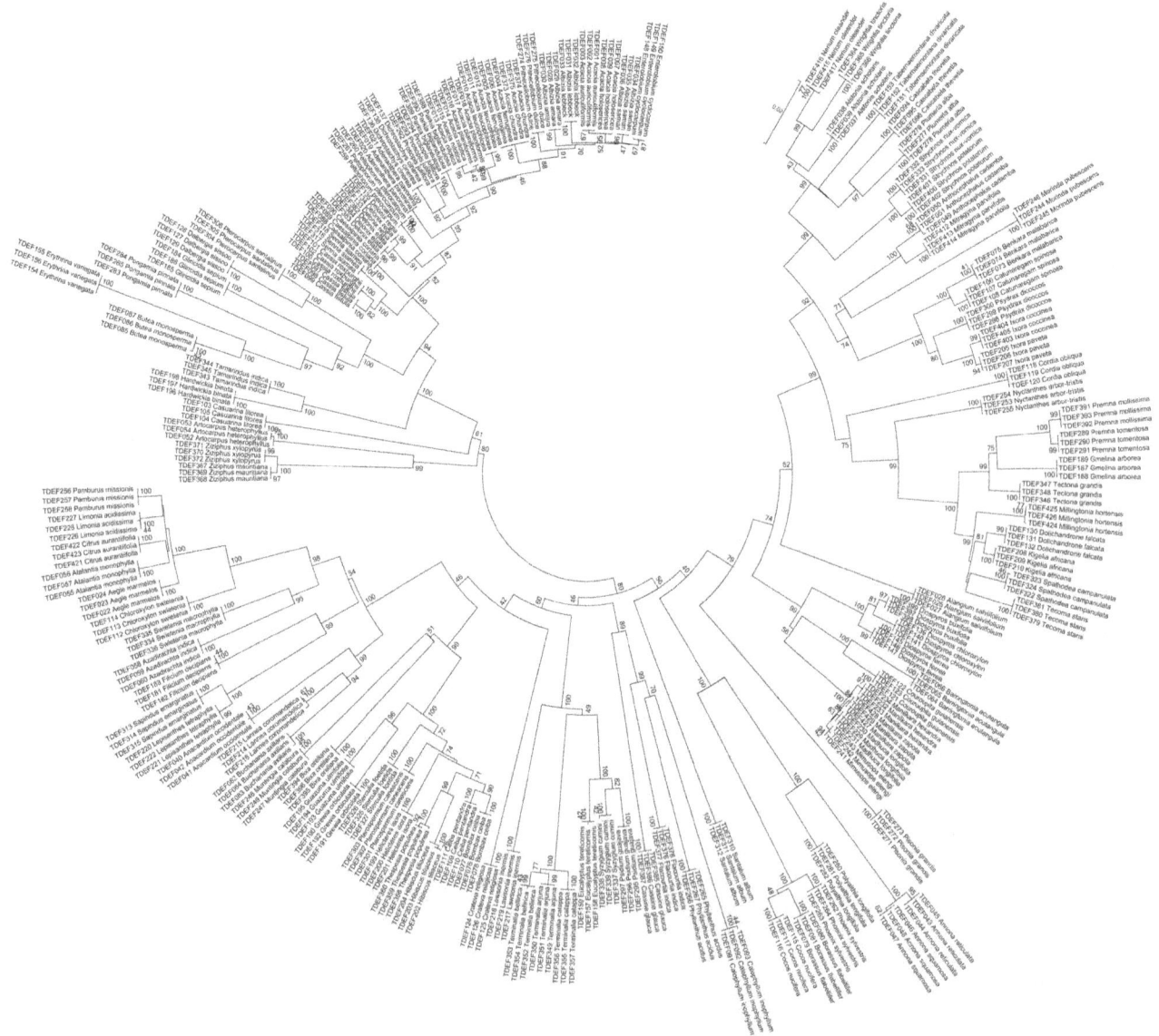

Figure 2. NJ tree of TDEF reference barcode library for *rbcL+matK* marker from 117 tree species.

Characteristic attribute organization system (CAOS) method for species ID

The CAOS method identifies a combination of characteristic attributes (CAs) that is diagnostic to a particular group [73]. This method is based on the concept that members of a taxonomic group share characteristic attributes (CAs) that are absent in comparable groups. The CAOS algorithm thus identifies CAs for every clade at each branching node within a guide tree that is first produced from a data set. The resulting combination of diagnostics CAs can be used for subsequent classification of new data into the taxonomic groupings represented by the guide tree [73,74]. This method has been used for DNA barcoding in animals [75,76] and plants [55,77]. Here we have employed CAOS method using *rbcL* and *matK* markers as character states, and stringently considered only single pure CAs (sPu), which are present in all member of one clade but absent in the other clades. We have found at least one sPu in 90.2%, 96.5%, and 98.3% of

the species with *rbcL*, *matK* and *rbcL+matK*, respectively (Table 1). The number of sPu in individual species varied from 1 to 25 (average 6.5) and 1 to 58 (average 18) for *rbcL* and *matK*, respectively (Table 2, Table S5 and Table S6).

Accuracy and applications of the TDEF reference barcode library

Species resolution from our study within the TDEF in India was 90.2% (*rbcL*) and 96% (*matK*) as estimated using three different methods of analysis. This estimate is much higher when compared with less than 72% species discrimination that is generally reported for *rbcL* and *matK* markers at a global scale [27,28], but very similar to studies at a regional scale [20]. The high species resolution estimates from our study is likely attributed to the fact that the current TDEF reference barcode library is made of highly diverse species; 143 species representing 114 genera, and 42 (36.8%) of them are monotypic to the TDEF in India. In general,

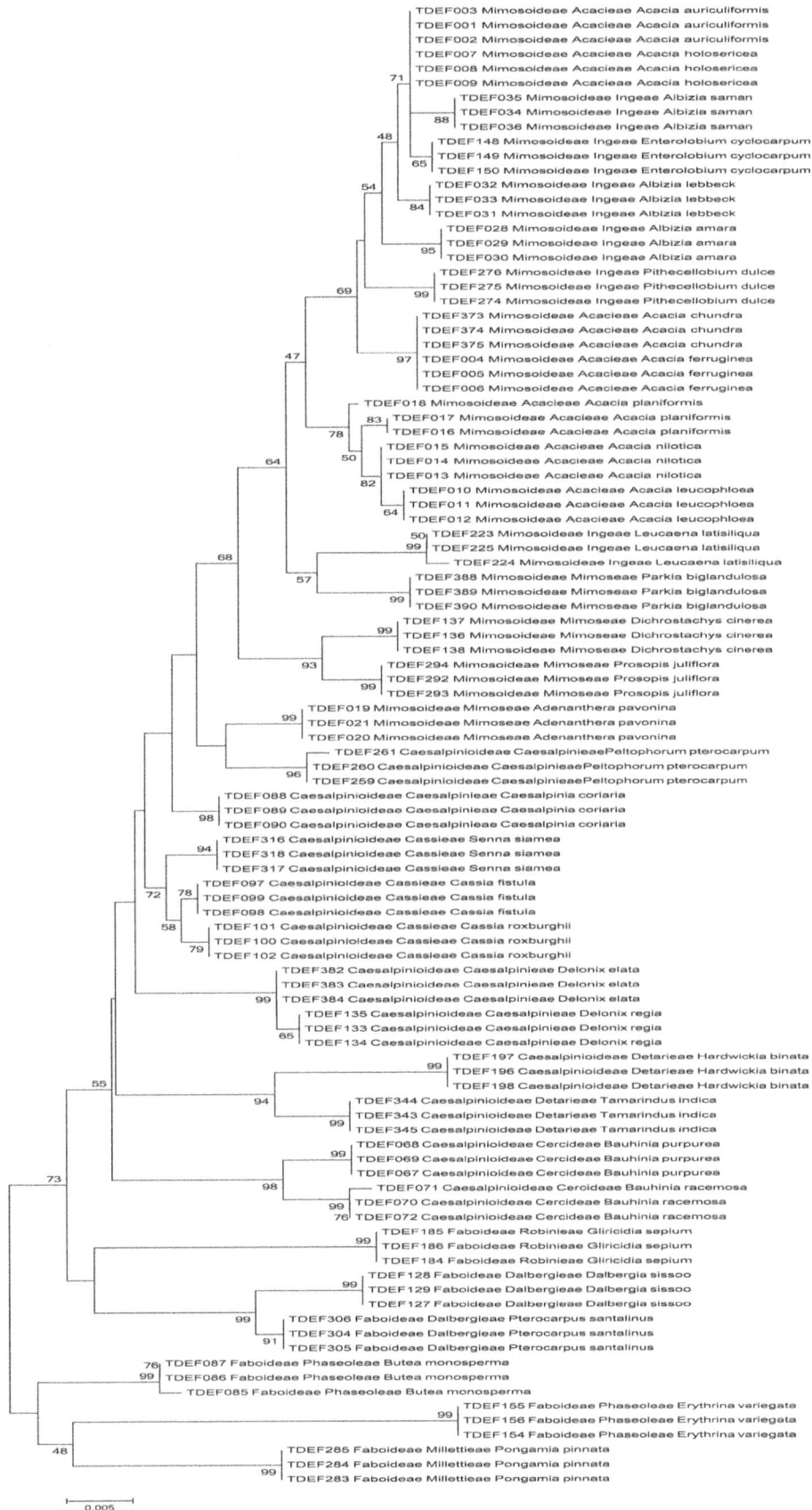

Figure 3. NJ tree of Fabaceae in the TDEF reference barcode library for *rbcL+matK* marker.

Table 2. Number of diagnostic characters (sPu) for TDEF tree species from *rbcL* and *matK* markers.

S.No	No. of sPu	No. of species	
		rbcL	*matK*
1	0	14	4
2	1–5	67	20
3	6–10	37	22
4	11–15	17	16
5	16–20	6	17
6	21–25	2	8
7	26–30	0	5
8	31–35	0	5
9	36–40	0	6
10	41–45	0	7
11	46–50	0	3
12	51–55	0	2
13	55–58	0	2

approximately 20% of the species in the TDEF in India are monotypic. Similar results were reported when DNA barcoding were applied on a regional scale in Barro Colorado Island of Panama and Northeast Puerto Rican forest [34,35] and tropical rain forest of French Guiana [13]. Although the standard barcode markers recommended by CBOL were sufficient to resolve most of

Table 3. Species identification of the logged timbers using TDEF reference barcode library.

S. No.	Sample ID	Vernacular name on the label	Scientific name	Species ID by DNA barcoding
1	TBW001	Poovarasu	*Thespesia populnea*	*Thespesia populnea*
2	TBW002	Sensandhanam	*Pterocarpus santalinus*	*Pterocarpus santalinus*
3	TBW003	Konravagai	*Peltophorum pterocarpum*	*Peltophorum pterocarpum*
4	TBW004	Anikundumani	*Adenanthera pavonina*	*Adenanthera pavonina*
5	TBW005	Cimaivagai	*Albizia saman*	*Albizia sp.*
6	TBW006	Nuna	*Morinda pubescens*	*Morinda pubescens*
7	TBW007	Mahogany	*Swietenia macrophylla*	*Swietenia macrophylla*
8	TBW008	Puliyamaram	*Tamarindus indica*	*Tamarindus indica*
9	TBW009	Saundal	*Leucaena latisiliqua*	*Leucaena latisiliqua*
10	TBW010	Pongum	*Pongamia pinnata*	*Pongamia pinnata*
11	TBW011	Thailam	*Eucalyptus tereticornis*	*Eucalyptus tereticornis*
12	TBW012	Theku	*Tectona grandis*	*Tectona grandis*
13	TBW013	Manja Kondrai	*Senna siamea*	*Senna siamea*
14	TBW014	Vembu	*Azadirachta indica*	*Azadirachta indica*
15	TBW015	Kattu Vaagai	*Albizia lebbeck*	*Albizia sp.*
16	TBW016	Mara Kumizh	*Gmelina arborea*	*Gmelina arborea*
17	TBW017	Pala	*Artocarpus heterophyllus*	*Artocarpus heterophyllus*
18	TBW018	Netilingam	*Polyalthia longifolia*	*Polyalthia longifolia*
19	TBW019	Purasu	*Butea monosperma*	*Butea monosperma*
20	TBW020	Iluppai	*Madhuca longifolia*	*Madhuca longifolia*
21	TBW021	Vilam Palam	*Limonia acidissima*	*Feronia limonia*
22	TBW022	Naval	*Syzygium cumini*	*Syzygium cumini*
23	TBW023	Velvel	*Acacia leucophloea*	*Acacia sp.*
24	TBW024	Sisu	*Dalbergia sissoo*	*Dalbergia sissoo*
25	TBW025	Karuvel	*Acacia nilotica*	*Acacia sp.*

the species, we suggest the addition of a supplementary marker such as ITS2 to increase species resolution based on evidence from other studies [21,62]. It appears that floristic barcode surveys at regional levels that use a local barcode library may provide an excellent tool for quick and reliable species identification. This includes many examples such as biodiversity monitoring, identification of plants that are prohibited from trading, authentication of medicinal plants collected from a region or auditing timber for illegal substitution with rare species of trees.

Commercial harvesting of timber is one of the major threats for its biodiversity in the TDEF of India. The threat is more prominent in case of the rare tree species that are listed in CITES Appendix II [78]. Though trading them within or outside the country is banned, their commercial value does attract illegal trading, which is well documented in the TDEF. Currently, it is very difficult to gather evidence and prosecute illegal trade of rare tree species. For example, the wood of *Santalum album* and *Osyris lanceolata* are anatomically similar which could not be distinguished easily [79]. A DNA barcode could serve as legal evidence of species identity from the traded parts of the plants, which is critical for supporting legal action against fraudulent or illegal trading. We utilized the TDEF reference barcode library that was developed from the current study to identity wood samples from commercial timber operations in the TDEF. We were able to identify 21 of timber samples at the species level, and the remaining 4 were identified at the genus level (Table 3). Although we only provide here a small case study, this does provide proof in principle that the TDEF reference barcode library could be used to more thoroughly audit timber operations throughout the TDEF. The 16 threatened species in the TDEF that are on the IUCN red list could be monitored using our TDEF reference barcode library, which provides legal evidence of enforcing conservation measures in the TDEF. This barcode library could be used to address the unsustainable and indiscriminate collection of plants from the wild for their medicinal value; 77 out of 143 tree species are traded as herbal remedies of which 28 are in high demand because they are highly effective in the commonly used traditional remedies [80]. In this case, the TDEF reference barcode library would also be useful for the authentication of commercial medicinal plant products, which are often adulterated (product substitution or contamination) with other species [81].

Supporting Information

Figure S1   NJ tree of TDEF reference barcode library for *rbcL* marker from 143 tree species.

Figure S2   NJ tree of TDEF reference barcode library for *matK* marker from 117 tree species.

Table S1   Collection sites from Tropical Dry Evergreen Forest

Table S2   Details of the 429 samples collected for the present study, their process IDs in BOLD, PCR success and length of *rbcL* and *matK* sequences obtained.

Table S3   Inter-specific divergence between congeneric species pairs for *rbcL* marker.

Table S4   Inter-specific divergence between congeneric species pairs for *matK* marker.

Table S5   Number and positions of diagnostics single pure CAs (sPu) for the TDEF tree species from *rbcL* marker.

Table S6   Number and positions of diagnostics single pure CAs (sPu) for the TDEF tree species from *matK* marker.

Acknowledgments

The authors are thankful to Department of Science and Technology (Government of India) for the fellowship provided to SN, Department of Biotechnology (Government of India) for DBT-CREST award fellowship provided to MP, Biodiversity Institute of Ontario, University of Guelph (Canada) for partial funding of *matK* sequencing, and SRM University, India for providing major funding for this work.

Author Contributions

Conceived and designed the experiments: MP SGN. Performed the experiments: SN DK SLV MP. Analyzed the data: SN MP SGN SR. Contributed reagents/materials/analysis tools: MP SGN SR. Contributed to the writing of the manuscript: SN MP SGN.

References

1. Kumar NK, Raghunath TP, Jayaraj RSC, Anandalakshmi R, Warrier RR, editors (2012) State of forest genetic resources in India: A Country report. Institute of Forest Genetics and Tree Breeding, Indian Council of Forestry Research and Education, Coimbatore.

2. Goyal AK, Sujata A, editors (2009) India's fourth national report to convention on biological diversity. New Delhi: Ministry of Environment and Forests, Government of India.

3. Rodgers WA, Panwar SH (1988) Biogeographical classification of India. New Forest, Dehra Dun.

4. Ministry of Environment and Forests (2005) State of Forest Report. Dehra Dun: Forest survey of India.

5. Champion HG, Seth SK (1968) A revised survey of the forest types in India. Manager Government of India Press, Nasik.

6. Reid WR (1992) How many species will there be? Tropical deforestation and species extinction. Whitmore TC, Sayer JA, editors. New York: Chapman and Hall, Chapter 3. Avaliable: http://www.ciesin.columbia.edu/docs/002-252a/002-252.html.

7. Venkateswaran R, Parthasarathy N (2003) Tropical dry evergreen forests on the Coromandel coast of India: Structure, composition and human disturbance. Ecotropica 9(1–2): 45–58.

8. Parthasarathy N, Selwyn MA, Udayakumar M (2008) Tropical dry evergreen forests of peninsular India: Ecology and conservation significance. Tropical Conserv Sci 1(2): 89–110.

9. National Environment Policy (2006) Ministry of Environment and Forests. New Delhi: Government of India.

10. Mace GM (2004) The role of taxonomy in species conservation. Phil Trans R Soc B: Biological Sciences 359: 711–719.

11. Parmentier I, Duminil J, Kuzmina M, Philippe M, Thomas DW, et al. (2013) How effective are DNA barcodes in the identification of African rainforest trees? PLoS ONE 8(4): e54921.

12. Costion C, Ford A, Cross H, Crayn D, Harrington M, et al. (2011) Plant DNA barcodes can accurately estimate species richness in poorly known floras. PLoS ONE 6: e26841.

13. Gonzalez MA, Baraloto C, Engel J, Mori SA, Pétronelli P, et al. (2009) Identification of Amazonian trees with DNA barcodes. PLoS ONE 4(10): e7483.

14. Margules CR, Austin M, Mollison D, Smith F (1994) Biological models for monitoring species decline: The construction and use of data bases [and discussion]. Phil Trans R Soc B: Biological Sciences 344: 69–75.

15. de Carvalho MR, Bockmann FA, Amorim DS, Brandao CRF, de Vivo M, et al. (2005) Revisiting the taxonomic impediment. Science 307: 353.

16. Schuster SC (2008) Next-generation sequencing transforms today's biology. Nat Methods 5: 16–18.

17. Newmaster SG, Fazekas AJ, Ragupathy S (2006) DNA barcoding in land plants: evaluation of *rbcL* in a multigene tiered approach. Botany 341: 335–341.

18. De Mattia F, Gentili R, Bruni I, Galimberti A, Sgorbati S, et al. (2012) A multi-marker DNA barcoding approach to save time and resources in vegetation surveys. Bot J Linn Soc 169: 518–529.

19. Lahaye R, van der Bank M, Bogarin D, Warner J, Pupulin F, et al. (2008) From the Cover: DNA barcoding the floras of biodiversity hotspots. Proc Natl Acad Sci 105: 2923–2928.

20. Burgess KS, Fazekas AJ, Kesanakurti PR, Graham SW, Husband BC, et al. (2011) Discriminating plant species in a local temperate flora using the *rbcL+matK* DNA barcode. Methods Ecol Evol 2: 333–340.

21. Thomson KA, Newmaster SG (2014) Molecular taxonomic tools provide more accurate estimates of species richness at less cost than traditional morphology-based taxonomic practices in a vegetation survey. Biodiversity and Conservation 23: 1411–1424.

22. Hebert PDN, Cywinska A, Ball SL, deWaard JR (2003) Biological identifications through DNA barcodes. Proc R Soc Lond B: Biological Sciences 270: 313–321.

23. Rieseberg LH, Wood TE, Baack EJ (2006) The nature of plant species. Nature 440: 524–527.

24. Fazekas AJ, Kesanakurti PR, Burgess KS, Percy DM, Graham SW, et al. (2009) Are plant species inherently harder than animal species using DNA barcoding markers? Mol Ecol Res 9: 130–139.

25. Kress WJ, Wurdack KJ, Zimmer EA, Weigt LA, Janzen DH (2005) Use of DNA barcodes to identify flowering plants. Proc Natl Acad Sci 102: 8369–8374.

26. China Plant BOL Group, Li DZ, Gao LM, Li HT, Wang H, et al. (2011) Comparative analysis of a large dataset indicates that internal transcribed spacer (ITS) should be incorporated into the core barcode for seed plants. Proc Natl Acad Sci 108: 19641–19646.

27. CBOL PWG, Hollingsworth PM, Forrest LL, Spouge JL, Hajibabaei M, et al. (2009) A DNA barcode for land plants. Proc Natl Acad Sci 106: 12794–12797.

28. Hollingsworth P, Graham S, Little D (2011) Choosing and using a plant DNA barcode. PLoS ONE 6: e19254.

29. Newmaster SG, Fazekas AJ, Steeves RAD, Janovec J (2008) Testing candidate plant barcode regions in the Myristicaceae. Mol Eco Resour 8: 480–490.

30. Purushothaman N, Newmaster SG, Ragupathy S, Stalin N, Suresh D, et al. (2014) A tiered barcode authentication tool to differentiate medicinal *Cassia* species in India. Genet Mol Res 13: 2959–2968.

31. Chase MW, Salamin N, Wilkinson M, Dunwell JM, Kesanakurthi RP, et al. (2005) Land plants and DNA barcodes: short-term and long-term goals. Phil Trans R Soc Lond B: Biological Sciences 360: 1889–1895.

32. Tripathi AM, Tyagi A, Kumar A, Singh A, Singh S, et al. (2013) The internal transcribed spacer (ITS) region and *trnH-psbA* are suitable candidate loci for DNA barcoding of tropical tree species of India. PLoS ONE 8(2): e57934.

33. Saarela JM, Sokoloff PC, Gillespie LJ, Consaul LL, Bull RD (2013) DNA barcoding the Canadian Arctic Flora: core plastid barcodes (*rbcL+matK*) for 490 Vascular Plant Species. PLoS ONE 8: 36.

34. Kress WJ, Erickson DL, Jones FA, Swenson NG, Perez R, et al. (2009) Plant DNA barcodes and a community phylogeny of a tropical forest dynamics plot in Panama. Proc Natl Acad Sci 106: 18621–18626.

35. Kress WJ, Erickson DL, Swenson NG, Thompson J, Uriarte M, et al. (2010) Advances in the use of DNA barcodes to build a community phylogeny for tropical trees in a Puerto Rican forest dynamics plot. PLoS ONE 5(11): e15409.

36. Pei N, Lian J-Y, Erickson DL, Swenson NG, Kress WJ, et al. (2011) Exploring tree-habitat associations in a Chinese subtropical forest plot using a molecular phylogeny generated from DNA Barcode Loci. PLoS ONE 6(6): e21273.

37. Meher-Homji VM (1974) On the origin of the tropical dry evergreen forest of south India. Int J Ecol Environ Sci 1: 19–39.

38. SaghaiMaroof MA, Soliman KM, Jorgensen RA, Allard RW (1984) Ribosomal DNA spacer–length polymorphism in barley: Mendelian inheritance, chromosomal location, and population dynamics. Proc Natl Acad Sci 81: 8014–8019.

39. Kress WJ, Erickson DL (2007) A two-locus global DNA barcode for land plants: the coding *rbcL* gene complements the non-coding *trnH-psbA* spacer region. PLoS ONE 2: e508.

40. Fazekas AJ, Burgess KS, Kesanakurti PR, Graham SW, Newmaster SG, et al. (2008). Multiple multi locus DNA barcodes from the plastid genome discriminate plant species equally well. PLoS ONE 3: e2802.

41. Edgar RC (2004) MUSCLE: multiple sequence alignment with high accuracy and high throughput. Nucl Acids Res 32: 1792–1797.

42. Meier R, Shiyang K, Vaidya G, Peter KLN (2006) DNA Barcoding and taxonomy in Diptera: a tale of high intraspecific variability and low identification success. Syst Biol 55: 715–728.

43. Tamura K, Peterson D, Peterson N, Stecher G, Nei M, et al. (2011) MEGA5: Molecular evolutionary genetics analysis using maximum likelihood, evolutionary distance, and maximum parsimony methods. Mol Biol Evol 28: 2731–9.

44. Sarkar IN, Planet PJ, Desalle R (2008) CAOS software for use in character-based DNA barcoding. Mol Ecol Resour 8: 1256–1259.

45. Bergmann T, Hadrys H, Breves G, Schierwater B (2009) Character-based DNA barcoding: a superior tool for species classification. Berl Munch Tierarztl Wochenschr 122: 446–450.

46. Maddison WP, Maddison DR (2009) MESQUITE: A modular system for evolutionary analysis (Counter website. Available: http://mesquiteproject.org.).

47. Roy S, Tyagi A, Shukla V, Kumar A, Singh UM, et al. (2010) Universal plant DNA barcode loci may not work in complex groups: A case study with Indian *Berberis* Species. PLoS ONE 5(10): e13674.

48. Zhang CY, Wang FY, Yan HF, Hao G, Hu CM, et al. (2012) Testing DNA barcoding in closely related groups of Lysimachia L. (Myrsinaceae). Mol Ecol Resour 12: 98–108.

49. Maia VH, Mata CS, Franco LO, Cardoso MA, Cardoso SRS, et al. (2012) DNA barcoding Bromeliaceae: achievements and pitfalls. PLoS ONE 7(1): e29877.

50. Wang NA, Jacques FMB, Milne RI, Zhang CQ, Yang JB (2012) DNA barcoding of *Nyssaceae* (Cornales) and taxonomic issues. Bot Stud 53(2): 265–274.

51. Fazekas AJ, Steeves R, Newmaster SG (2010) Improving sequencing quality from PCR products containing long mononucleotide repeats. Biotechniques 48: 277–281.

52. Yu H, Wu K, Song K, Zhu Y, Yao H, et al. (2014) Expedient identification of Magnoliaceae species by DNA barcoding. POJ 7(1): 47–53.

53. Puillandre N, Lambert A, Brouillet S, Achaz G (2012) ABGD, Automatic barcode gap discovery for primary species delimitation. Mol Ecol Resour 21(8): 1864–1877.

54. Arca M, Hinsinger DD, Cruaud C, Tillier A, Bousquet J, et al. (2012) Deciduous trees and the application of universal DNA barcodes: A case study on the Circumpolar *Fraxinus*. PLoS ONE 7(3): e34089.

55. Pettengill JB, Neel MC (2010) An evaluation of candidate plant DNA barcodes and assignment methods in diagnosing 29 species in the genus *Agalinis* (Orobanchaceae). Am J Bot 97: 1391–1406.

56. Gao T, Sun Z, Yao H, Song J, Zhu Y, et al. (2011) Identification of Fabaceae plants using the DNA barcode *matK*. Planta Med 77: 92–94.

57. Rao RR, editor (1994) Biodiversity in India (Floristic Aspects). Dehra Dun: Bisen Singh Mahendra Pal Singh.

58. Ferguson JWH (2002) On the use of genetic divergence for identifying species. Biol J Linn Soc 75: 509–516.

59. Little DP, Stevenson DW (2007) A comparison of algorithms for the identification of specimens using DNA barcodes: Examples from gymnosperms. Cladistics 23: 1–21.

60. Fazekas A, Burgess KS, Kesanakurti PR, Percy DM, Hajibabaei M, et al. (2008) Assessing the utility of coding and non-coding genomic regions for plant DNA barcoding. PLoS ONE 3: 1–12.

61. Hebert PDN, Penton EH, Burns JM, Janzen DH, Hallwachs W (2004) Ten species in one: DNA barcoding reveals cryptic species in the neotropical skipper butterfly *Astraptes fulgerator*. Proc Natl Acad Sci 101: 14812–14817.

62. Kuzmina ML, Johnson KL, Barron HR, Hebert PDN (2012) Identification of the vascular plants of Churchill, Manitoba, using a DNA barcode library. BMC Ecol 12: 25.

63. Kass E, Wink M (1995) Molecular phylogeny of the Papilionoideae (Family Fabaceae): *rbcL* gene sequences versus chemical taxonomy. Bot Acta 108: 149–162.

64. Bruneau A, Forest F, Herendeen PS, Klitgaard BB, Lewis GP (2001) Phylogenetic relationships in the Caesalpinioideae (Leguminosae) as inferred from chloroplast *trnL* intron sequences. Syst Bot 26(3): 487–514.

65. Simpson MG (2010) Plant systematics. Academic Press - An imprint of Elsevier.

66. Geesink R (1984) Scala Millettiearum. E. J. Brill/Leiden University Press, Leiden, The Netherlands.

67. Doyle JF, Doyle JL, Ballenger JA, Dickson EE, Kajita T, et al. (1997) A phylogeny of the chloroplast gene *rbcL* in the Leguminosae: taxonomic correlations and insights into the evolution of nodulation. Am J Bot 84(4): 541–554.

68. Lavin M, Eshbaugh E, Hu JM, Mathews S, Sharrock RA (1998) Monophyletic subgroups of the tribe Millettieae (Leguminosae) as revealed by phytochrome nucleotide sequence data. Am J Bot 85: 412–433.

69. Chappill JA, Maslin BR (1995) A phylogenetic assessment of tribe Acacieae. In: Crisp MD, Doyle JJ, ed. Advances in legume systematic. Phylogeny Royal Botanic Gardens Kew. 77–99.

70. Sulaiman SF, Culham A, Harborne JB (2003) Molecular phylogeny of Fabaceae based on *rbcL* sequence data: with special emphasis on the tribe Mimoseae (Mimosoideae). Asia Pac J Mol Biol 11(1): 9–35.

71. Miller JT, Bayer RJ (2000) Molecular phylogenetics of *Acacia* (Fabaceae: Mimosoideae) based on chloroplast *trnK/matK* and nuclear histone H3–D sequences. In: Herendeen P, Bruneau A, editors. Adv Legume Syst 9. Royal Botanic Gardens Kew: London. 180–200.

72. Miller JT, Bayer RJ (2001) Molecular phylogenetics of *Acacia* (Fabaceae: Mimosoideae) based on the chloroplast *matK* coding sequence and flanking *trnK* intron spacer region. Am J Bot 88: 697–705.

73. DeSalle R, Egan MG, Siddall M (2005) The unholy trinity: taxonomy, species delimitation and DNA barcoding. Phil Trans R Soc B 360: 1905–1916.

74. Rach J, DeSalle R, Sarkar IN, Schierwater B, Hadrys H (2008) Character-based DNA barcoding allows discrimination of genera, species and populations in Odonata. Proc R Soc Lond B 275: 237–247.

75. Lowenstein JH, Amato G, Kolokotronis SO (2009) Therealmaccoyii: identifying Tuna Sushi with DNA barcodes-contrasting characteristic attributes and genetic distances. PLoS ONE 4: e7866.

76. Kvist S, Sarkar IN, Erseus C (2010) Genetic variation and phylogeny of the cosmopolitan marine genus Tubificoides (Annelida: Clitellata: Naididae: Tubificinae). Mol Phylogenet Evol 57: 687–702.

77. Morejon NF, Silva VF, Astorga GJ, Stevenson DW (2011) Character-based, population-level DNA barcoding in Mexican species of *Zamia* L. (Zamiaceae: Cycadales). Mitochondrial DNA 21: 51–59.

78. Convention on International Trade in Endangered Species of Wild Fauna and Flora (2013). Appendices I, II and III. http://www.cites.org.

79. Rao RV, Hemavathi TR, Sujatha M, Chauhan L, Raturi RD (1988) Stem wood and root wood anatomy of *Santalum album* L. and the problem of wood adulteration in Sandal and its Products. ACIAR Proceedings No. 84. Canberra Australia. 101–110.

80. Ved DK, Goraya GS (2008) Demand and supply of medicinal plants in India (FRLHT, Bangalore). Dehra Dun: Bishen Singh Mahendra Pal Singh.

81. Newmaster SG, Grguric M, Shanmughanandhan D, Ramalingam S, Ragupathy S (2013) DNA barcoding detects contamination and substitution in North American herbal products. BMC Medicine 11: 222.

11

Positive Effects of Plant Genotypic and Species Diversity on Anti-Herbivore Defenses in a Tropical Tree Species

Xoaquín Moreira[1][*][¤], Luis Abdala-Roberts[1], Víctor Parra-Tabla[2], Kailen A. Mooney[1]

1 Department of Ecology and Evolutionary Biology, University of California Irvine, Irvine, California, United States of America, **2** Departamento de Ecología Tropical, Campus de Ciencias Biológicas y Agropecuarias, Universidad Autónoma de Yucatán, Mérida, Yucatán, México

Abstract

Despite increasing evidence that plant intra- and inter-specific diversity increases primary productivity, and that such effect may in turn cascade up to influence herbivores, there is little information about plant diversity effects on plant anti-herbivore defenses, the relative importance of different sources of plant diversity, and the mechanisms for such effects. For example, increased plant growth at high diversity may lead to reduced investment in defenses via growth-defense trade-offs. Alternatively, positive effects of plant diversity on plant growth may lead to increased herbivore abundance which in turn leads to a greater investment in plant defenses. The magnitude of trait variation underlying diversity effects is usually greater among species than among genotypes within a given species, so plant species diversity effects on resource use by producers as well as on higher trophic levels should be stronger than genotypic diversity effects. Here we compared the relative importance of plant genotypic and species diversity on anti-herbivore defenses and whether such effects are mediated indirectly via diversity effects on plant growth and/or herbivore damage. To this end, we performed a large-scale field experiment where we manipulated genotypic diversity of big-leaf mahogany (*Swietenia macrophylla*) and tree species diversity, and measured effects on mahogany growth, damage by the stem-boring specialist caterpillar *Hypsipyla grandella*, and defensive traits (polyphenolics and condensed tannins in stem and leaves). We found that both forms of plant diversity had positive effects on stem (but not leaf) defenses. However, neither source of diversity influenced mahogany growth, and diversity effects on defenses were not mediated by either growth-defense trade-offs or changes in stem-borer damage. Although the mechanism(s) of diversity effects on plant defenses are yet to be determined, our study is one of the few to test for and show producer diversity effects on plant chemical defenses.

Editor: Andrew Hector, University of Oxford, United Kingdom

Funding: This research was supported by a UC MEXUS-CONACyT grant for UC postdocs to XM (UCM-101443), a CONACyT grant to VPT (128856) and a UC MEXUS-CONACyT collaborative grant to all the authors (UCM-55592). XM received financial support from Postdoctoral Fulbright/Spanish Ministry of Education grant program. LAR was funded by a GAANN fellowship and a UC MEXUS-CONACyT scholarship. The funders had no role in study design, data collection and analysis, decision to publish, or preparation of the manuscript.

Competing Interests: The authors have declared that no competing interests exist.

* Email: xmoreira1@gmail.com

¤ Current address: Institute of Biology, Laboratory of Evolutive Entomology, University of Neuchâtel, Neuchâtel, Switzerland

Introduction

Ecological research conducted over the last decade has shown that plant intra- and inter-specific diversity have large effects on ecosystem processes, such as decomposition rates and productivity [1–6], as well as on the structure of associated communities of consumers [1–3,7–11]. Specifically, numerous studies have found that plant diversity increases plant biomass production due to niche partitioning and more efficient resource use among species or genotypes within a given species (i.e., complementarity effect) [12,13]. Such increases in plant biomass, as well as greater habitat complexity may in turn cascade up to influence consumers, particularly in the case of arthropods associated with plant canopies [1,2,8,14,15].

Although plant diversity effects on plant biomass (via resource use) and consumers are well-documented, little information is available about the effects of plant diversity on anti-herbivore defenses. Diversity effects on plant defenses are extremely important because they might influence herbivory and explain

over-yielding [7], alter community structure at higher trophic levels (e.g. via effects on herbivores) [16], as well as mediate ecosystem processes (e.g. food web dynamics, decomposition) [17].

There are two possible mechanisms by which producer diversity may influence plant defenses. First, plant diversity is known to increase plant growth via more efficient resource use [12]; assuming that the production of anti-herbivore defenses is costly for plants [18–20], then greater plant growth at high diversity may lead to reduced investment in defenses via growth-defense trade-offs. To date, only one study has tested this hypothesis (indirectly) and found a trade-off between complementarity for increased plant productivity and resistance to herbivory at high diversity [21], suggesting that growth-defense trade-offs may arise due to greater allocation to plant growth. Second, positive effects of plant diversity on producer biomass frequently lead to increased herbivore loads [1,2,8,14,15] and damage [22,23], which in turn might lead to greater investment in plant defenses. Alternatively, high diversity might lead to reduced herbivore abundance (and damage) due to mechanisms of associational resistance such as

Figure 1. *Hypsipyla grandella* **damage.** Damage caused by *Hypsipyla grandella* (Lepidoptera: Pyralidae), a stem-boring caterpillar specializing on tree species of the neotropical family Meliaceae. The images show a fourth-instar larva inside a terminal shoot of a big-leaf mahogany (*Swietenia macrophylla* King, Meliaceae) sapling and the damage caused. Photo credits: Luis Abdala-Roberts.

reduction in host plant density (i.e. resource concentration effects) [24,25], and in turn reduced investment in plant defenses. This latter mechanism is predicted for specialist insect herbivores which are more sensitive to changes in the density of specific host plants [26,27].

Despite these appealing predictions, few studies have directly evaluated the effects of plant diversity on anti-herbivore defenses [8,28] and the previously described mechanisms for such effects have not been tested. For example, Mraja and colleagues [28] found mixed evidence for diversity effects on plant defenses as *Plantago lanceolata* plants growing in patches of high species diversity exhibited a lower concentration of foliar aucubin and total iridoid glycoside, but a greater concentration of catalpol, another important defensive compound. In addition, Moreira and colleagues [8] found that host-pine species diversity increased pine growth and herbivore density but did not significantly affect the concentration of chemical defenses in pine seedlings. However, in this work the presence of predatory ants (which were more abundant in diverse patches) may have resulted in (indirect) defense against herbivores and this could have influenced patterns of allocation to chemical defenses by plants across levels of diversity [8].

Importantly, the mechanisms by which producer diversity influences plant defenses may vary depending on the source of plant diversity. For instance, the magnitude of trait variation underlying diversity effects is usually greater among species than among genotypes within a given species, therefore we would expect that plant species diversity effects on resource use by producers as well as on higher trophic levels should be stronger than genotypic diversity effects [14,29]; but see Crawford & Rudgers [30] as counter-example. Accordingly, greater trait variation among plant species would be expected to lead to increased niche partitioning and stronger effects on plant growth (relative to genotypic diversity effects), with this in turn causing a stronger reduction in plant defenses via growth-defense trade-offs. Alternatively, greater trait variation among plant species may lead to stronger positive (via greater plant biomass) [1,14,15,22] or negative (e.g. via decreased host plant density or apparency) [24,31,32] effects on herbivore abundance and damage (as compared to genotypic diversity), and thus stronger effects on plant defenses compared to effects of plant genotypic diversity. Nonetheless, these predictions and their mechanisms have not been tested yet as plant intra- and inter-specific diversity effects have usually been studied separately.

The aim of this study was to evaluate the effects of plant species and genotypic diversity on producer anti-herbivore defenses. In particular, we were interested in (a) comparing the relative importance of these two source of plant diversity (following the prediction of stronger effects of species diversity), and (b) evaluating the mechanisms for such effects, namely if diversity effects were mediated via changes in plant growth (due to underlying growth-defense trade-offs) or herbivore damage. To address this, we performed a large-scale field experiment where we manipulated genotypic diversity of the tropical tree big-leaf mahogany (*Swietenia macrophylla*) as well as tree species diversity, and evaluated the effects of diversity on mahogany growth (total height), herbivore damage (caused by the specialist insect *Hypsipyla grandella*), and chemical defensive traits (polyphenolics and condensed tannins in stem and leaves). We focused on *H. grandella* because it is the most important herbivore (in terms of abundance and amount of damage inflicted) of mahogany in our system (Abdala-Roberts et al., unpublished data). Larvae of this herbivore carve tunnels through the terminal shoots of juvenile plants (Fig. 1) which in turn result in die-off of large portions of the plant, stem deformation, and reduced growth [33]. Overall, the present study is one of the few to test for plant diversity effects on anti-herbivore defenses, and uniquely compares such effects among sources of diversity while addressing the mechanisms for such effects. In so doing, we move beyond studying the effects of plant diversity on resource use to understanding how diversity influences plant secondary chemistry, an important but largely ignored suite of plant traits influencing herbivores and associated food webs.

Materials and Methods

Ethics Statement

The research did not involve manipulations of humans or animals. No specific permissions were required for our field work. The plant material used for this study was only sampled at a very limited scale and therefore had negligible effects on broader ecosystem functioning. The location is not privately-owned or protected in any way. The field studies did not involve endangered or protected species.

Natural history

Big-leaf mahogany (*Swietenia macrophylla* King, Meliaceae) is a large-statured tree with a patchy distribution throughout much of southern Mexico, Central America, and South America [34]. In forests, big-leaf mahogany is fed upon by a suite generalist and specialist insect leaf chewers [35,36], leaf-miners [37], and rodents [38]. The most relevant herbivore attacking mahogany in forest plantations, and particularly at our study site (Abdala-Roberts et al., unpublished data), is *H. grandella* (Lepidoptera: Pyralidae) [37,39], a specialist stem-boring caterpillar that feeds only on a few species of Meliaceae (Fig. 1). Larvae of this herbivore carve tunnels through the terminal shoots and poles of saplings and juvenile plants (Fig. 1), resulting in severe deformations of the main stem and shunted growth [33]. Because this herbivore is a stem feeder, we expected a stronger link between damage and chemical defenses in the stem relative to defenses in other types of tissue (e.g. leaves).

In tropical forests of the Yucatan Peninsula, big-leaf mahogany co-occurs with five other tree species [40] that are the subject of this experiment, namely: *Tabebuia rosea* (Bertol.) DC. (Bignonaceae), *Ceiba pentandra* (L.) Gaertn. (Malvaceae), *Enterolobium cyclocarpum* (Jacq.) Griseb. (Fabaceae), *Piscidia piscipula* (L.) Sarg. (Fabaceae), and *Cordia dodecandra* A. DC. (Boraginaceae). These species are long-lived, deciduous, and adult trees reach from 20 m (*P. piscipula*) to 40 m (*C. pentandra*) [34], and are distributed from central México to Central and South America [34].

Experimental design

In December 2011, we established 74 plots of 21×21 m each at a planting density of 64 plants per plot, 3 m spacing among trees, for a total of 4780 plants. Isles between plots were 6-m wide, and the experiment covered a total area of 7.2 ha Mahogany was planted in 59 of these plots which were classified into four types, depending on the diversity treatment combination: a) mahogany monocultures of one genotype (12 plots, two replicate plots/genotype; hereafter "genotypic monocultures"), b) mahogany monocultures of four genotypes (20 plots; hereafter "genotypic polycultures"), c) polycultures of four species within which all mahogany saplings were of one genotype (12 plots, two plots/genotype), and d) polycultures of four species within which mahogany plants were represented by four genotypes (15 plots). For these four plot types, treatments of both species and genotypic diversity included equal numbers of individuals of four species (one

Table 1. Diversity effect on growth and herbivore damage.

	Plant height		Damage by *H. grandella*	
a) Genotypic diversity	$F_{1,218}$	P	$F_{1,218}$	P
Diversity	1.03	0.311	1.47	0.226
b) Species diversity	$F_{1,166}$	P	$F_{1,167}$	P
Diversity	0.04	0.849	0.90	0.344

Summary of results from generalized linear mixed models testing for the effects of (a) big-leaf mahogany (*Swietenia macrophylla* King, Meliaceae) genotypic diversity and (b) tree species diversity on mahogany growth and damage by the specialist *Hypsipyla grandella* (number of attack sites per plant). We tested the effect of mahogany genotypic diversity by comparing genotypic monocultures to genotypic polycultures, whereas to test for a species diversity effect we compared genotypic monocultures and species polycultures. F-values and associated significance levels (*P*) are shown, as well as numerator and denominator degrees of freedom (subscripts).

of which was always mahogany) and four mahogany genotypes drawn from pools of six species or six genotypes, respectively. All non-mahogany species were equally represented across polycultures (each species present in six polyculture plots). Likewise, mahogany genotypes were represented in a similar number of mahogany monocultures of four genotypes (8–9 plots per genotype), and also in a similar number of species polycultures where mahogany plants were of four genotypes (9–10 plots per genotype).

In this study, we sampled a subset of these 59 plots where mahogany was present (see following two sections) and within these plots we restricted our sampling only to mahogany.

Seed sources and measurements of plant growth and herbivory

For measurements of growth and stem borer damage, we randomly selected eight mahogany plants allocated across plot types as follows (N = 352 plants, 44 plots): 12 genotypic monoculture plots, 20 genotypic polyculture plots, and 12 species polyculture plots where mahogany was represented by one genotype; we did not sample species polyculture plots with four genotypes. From January 2011 to March 2011, we collected seeds of each tree species from adult plants located in southern Quintana Roo (SE México), and germinated at the INIFAP (Instituto Nacional de Investigaciones Forestales Agrícolas y Pecuarias) in Mocochá (21°06′N, 89°26′W, Yucatan, SE México). For all species, we collected seed from six mother trees, and distance among trees ranged from 0.5 to 50 km. In the case of mahogany, distance among mother trees ranged from 3 to 50 km which is within the distance range used by previous studies to define genetically distinct populations of this species [41,42]. In December 2011, we established the experiment near the locality of Muna (20°24′44′′N, 89°45′13′′W, Yucatan, SE Mexico) by planting four-month old seedlings. After planting, saplings were fertilized in January 2012 with N, P, K (20:30:10) and irrigated with 2 l of water three times per week from January 2012 until June 2012, and from January 2013 until June 2013. The rainy season typically spans from June until October and therefore artificial irrigation was unnecessary during this time period.

In late July 2013, we recorded mahogany growth by measuring total height, as well as *H. grandella* damage by examining the apical and axilar meristems of plants in search of attack sites (easily identified by the presence of frass) [43] and recorded the number of *H. grandella* attacks per plant.

Chemical analyses of plant anti-herbivore defenses

As a proxy of quantitative chemical defences, we measured polyphenolics and condensed tannins in stems and leaves. In late July 2013, we randomly sampled four out of the eight previously mentioned mahogany plants sampled per plot for growth and stem borer attack (N = 144 plants, 36 plots), allocated across plot types as follows: 12 (i.e. all) genotypic monocultures, 12 (out of the 20) genotypic polycultures, and 12 (i.e. all) species polycultures; again, we did not sample polycultures with four mahogany genotypes. Polyphenolics are carbon-based compounds, non-nutritious, and unpalatable for herbivores because they inhibit herbivore digestion by binding to consumed plant proteins [44]. Condensed tannins are one of the most abundant phenolic compounds in plant tissues and have strong negative effects on food digestibility and herbivore performance [44,45]. Both defensive traits are generally recognized as herbivore feeding deterrents [44,46], and have been shown to reduce *H. grandella* larval performance and survival [47,48]. For each plant, we collected a 10 cm-long segment of the stem and three fully expanded, undamaged terminal leaves of one branch. Samples were transported to the laboratory on ice, immediately weighed, oven-dried (45°C to constant weight), and subsequently ground manually in a mortar with liquid nitrogen for subsequent determination of polyphenolic and condensed tannin concentration.

Polyphenolics were extracted and analysed as described by Moreira et al. [49]. Briefly, polyphenolics were extracted from 300 mg of plant tissue with aqueous methanol (1:1 vol:vol) in an ultrasonic bath for 15 min, followed by centrifugation and subsequent dilution of the methanolic extract. Polyphenolic concentration was determined colorimetrically by the Folin-Ciocalteu method in a Biorad 650 microplate reader (Bio-Rad Laboratories Inc., Philadelphia, PA, USA) at 740 nm, using tannic acid as standard, and concentrations were based on dry weights (d.w.).

Condensed tannins were determined by the acid butanol method [50] in the same 50% aqueous methanol extract used for polyphenolics. A mixture of an aliquot of methanol extract and acid butanol (950 ml of n-butanol mixed with 50 ml of concentrated HCl) and iron (0.5 g of 2% ferric ammonium sulphate in 2N HCl) reagents was placed in a boiling water bath for 50 min and then cooled rapidly to 0°C on ice. Condensed tannins were determined colorimetrically in a Biorad 650 microplate reader at 550 nm using a commercial quebracho tannin extract (72.0% condensed tannins) as standard [51].

Statistical analyses

We performed general linear mixed models to test for plant diversity effects on plant growth, herbivore damage (number of *H.*

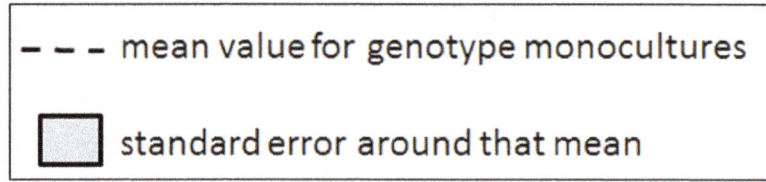

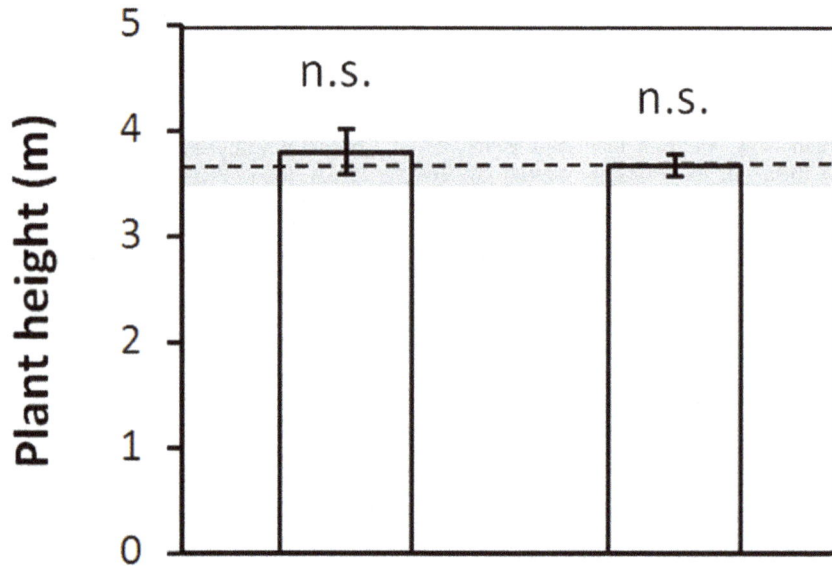

(A)

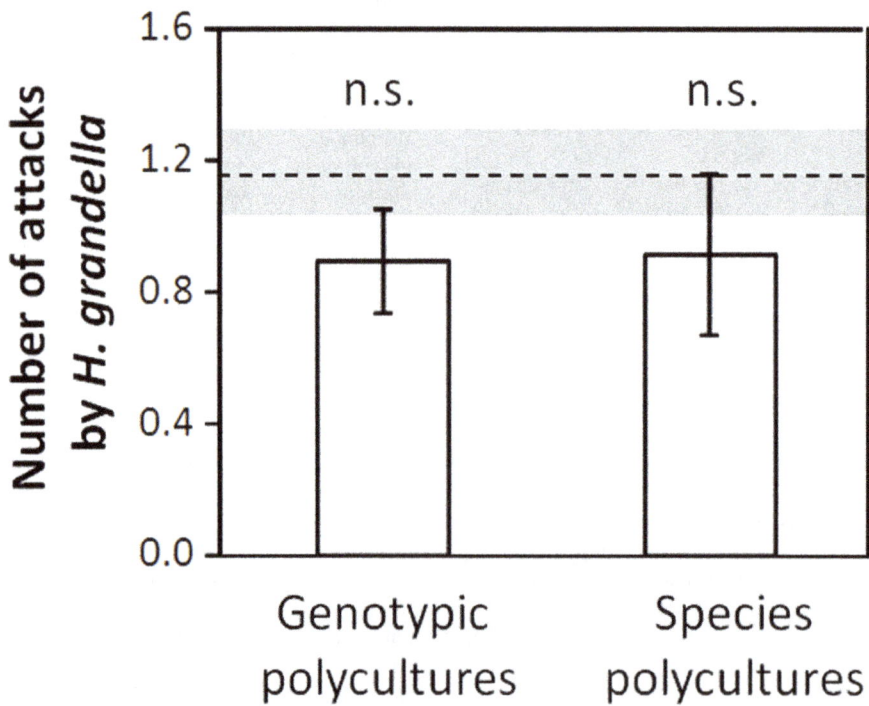

(B)

Figure 2. Diversity effect on growth and herbivore damage. Effect of mahogany genotypic and tree species diversity on: (A) mahogany sapling height and (B) the mean number of attack sites per plant by the specialist stem-boring insect *Hypsipyla grandella*. The dashed line represents the mean value for genotype monocultures (N = 12) and the shaded area represents the standard error around that mean. Least-square means ± S.E. (N = 20 genotypic polycultures and N = 12 species polycultures). "(n.s.)" in the figures indicates non-significant differences (*P*<0.05) between a given diversity treatment and the genotypic monoculture treatment.

grandella attacks per plant), and chemical defenses (polyphenolics and condensed tannins, separately for leaves and stems). For each response variable, we performed two separate models: First, to test the effect of mahogany genotype diversity, we compared genotypic monocultures and genotypic polycultures, and second, to test for a species diversity effect, we compared genotypic monocultures and species polycultures (hereafter "Model 0" in each case). Then, we departed from these initial models and constructed "mechanistic" models which included additional covariates aimed at testing if each source of diversity influenced mahogany defenses via effects on mahogany growth and *H. grandella* attack: (i) "Model 1" tested an effect of genotypic or species diversity on mahogany defenses via trade-offs between growth and defense by including plant height as a covariate. If diversity effects on defenses are mediated by growth-defense trade-offs via increased growth, then a significant effect of diversity on defenses should become non-significant once plant size is accounted for in the model; (ii) "Model 2" tested an effect of genotypic or species diversity via changes in herbivory by including *H. grandella* attack as a covariate in the model. If diversity effects on defenses are mediated by higher or lower attack of *H. grandella* at high diversity, then a significant effect of diversity on defenses should become non-significant once herbivory is accounted for in the model. Normality was met in all cases, and we report least square means ± standard errors as descriptive statistics. All models were performed with PROC MIXED in SAS (SAS 9.2 System, SAS, Cary, NC) using plot and genotype as random effects. The former accounted for non-independence of plants sampled from the same plot, while the latter controlled for variation in growth and defenses among maternal sources. We also performed models using plot as the level of replication (results not shown) and all results were qualitatively idential.

Results

Consequences of plant genotypic and species diversity on mahogany growth and herbivory

Eighteen months after planting, the mean size of mahogany plants was 3.65±0.11 m, 3.79±0.09 m, and 3.67±0.11 m for genotypic monocultures, genotypic polycultures, and species polycultures, respectively. We found that neither genotypic diversity nor species diversity had significant effects on plant height (Table 1, Fig. 2A). On the other hand, we found that 48% of all mahogany saplings were attacked by *H. grandella*. Although attack was lower in genotypic and species polycultures (30% in both cases) relative to genotypic monocultures, we were not able to detect a significant effect of either source of plant diversity on attack by this herbivore (Table 1, Fig. 2B).

Consequences of plant genotypic and species diversity on mahogany defenses

Genotypic diversity had a significant effect on the concentration of chemical defenses in stems (Model 0, Table 2a). Specifically, we found that the concentration of polyphenolics and condensed tannins in stems were 40% and 60% greater, respectively, in genotypic polycultures than in genotypic monocultures (Fig. 3A, 3B). For both types of defenses, this positive effect of genotypic

diversity remained significant after accounting for *H. grandella* attack (Model 1, Table 2a). Similarly, the genotypic diversity effect on stem polyphenolics and tannins remained significant after accounting for plant height (Model 2, Table 2a). Herbivore attack had only significant effects on stem tannins (Model 2, Table 2a), whereas plant size had no significant effects on stem defenses (Model 1, Table 2a). Similarly, we found that species diversity also had a significant effect on stem defenses (Model 0, Table 2b), with the concentration of polyphenolics and condensed tannins in stems being 36% and 42% greater, respectively, in species polycultures compared with genotypic monocultures (Fig. 3A, 3B). Such effect remained significant after accounting for plant height (Model 1, Table 2b) or *H. grandella* attack (Model 2, Table 2b). Plant height and herbivory damage effects were non-significant in these models (Models 1 and 2, Table 2a, 2b), suggesting that Model 0 is the most appropriate.

By contrast, genotypic and species diversity did not have significant effects on the concentration of polyphenolics or condensed tannins in leaves (Model 0, Tables 2a, 2b; Fig. 3C, 3D). The effect of plant genotypic and species diversity on leaf polyphenolics and condensed tannins remained non-significant after accounting for plant height (Model 1, Tables 2a, 2b) or *H. grandella* attack (Model 2, Tables 2a, 2b), and neither of these covariates influenced leaf defensive traits.

Discussion

Our results showed that plant genotypic and species diversity had strong positive effects on stem (but not leaf) anti-herbivore defenses in big-leaf mahogany, namely the concentration of stem polyphenolics and condensed tannins. In addition, the effects of these two sources of plant diversity on defenses were similar in magnitude. Contrarily to our expectations, positive effects from both sources of plant diversity were not mediated by the effects of diversity on plant growth or herbivore damage. Although the mechanism(s) of diversity effects on plant defenses in this study are yet to be determined, our study is one of the few to test for and show producer diversity effects on plant chemical defenses (i.e. secondary compounds). Such effects were particularly strong and are likely to play an important role in mediating plant diversity effects on higher trophic levels and ecosystem function.

To the best of our knowledge, there are no previous studies testing the relative contribution of plant intra- and inter-specific diversity on plant defense allocation patterns. Our results showed that plant genotypic and species diversity effects on mahogany anti-herbivore defenses tended to be similar in magnitude, despite the presumption that greater magnitude of trait variation among species than among genotypes should lead to stronger effects of species diversity (relative to genotypic diversity) on producers. Moreover, for stem tannins we even observed a stronger effect of genotypic diversity relative to species diversity. Accordingly, these findings agree with recent work showing that plant intra- and inter-specific diversity effects on ecosystem function and arthropod communities can be of similar importance [2,14,52]. For example, Crawford & Rudgers [52] found effects of similar magnitude from species and genotypic diversity on biomass in *Ammophila breviligulata*, a dominant species in dune ecosystems. Similarly, Cook-Patton et al. [14] found equivalent increases in aboveground

Table 2. Diversity effect on plant defenses.

	Stem phenolics		Leaf phenolics		Stem tannins		Leaf tannins	
	F	P	F	P	F	P	F	P
a) Genotypic diversity								
Model 0								
Diversity [(1,63)]	4.33	**0.041**	2.07	0.155	5.04	**0.028**	0.94	0.335
Model 1								
Diversity [(1,62)]	4.16	**0.046**	1.99	0.164	4.44	**0.039**	0.98	0.326
Height [(1,62)]	0.00	0.978	0.00	0.980	0.42	0.520	0.06	0.813
Model 2								
Diversity [(1,62)]	4.39	**0.040**	2.06	0.156	5.21	**0.026**	0.93	0.338
Damage by *H.grandella* [(1,62)]	2.60	0.112	0.52	0.473	4.37	**0.041**	0.03	0.873
b) Species diversity								
Model 0								
Diversity [(1,67)]	4.16	**0.045**	0.03	0.874	4.27	**0.043**	0.04	0.841
Model 1								
Diversity [(1,66)]	4.07	**0.048**	0.04	0.876	4.32	**0.042**	0.04	0.843
Height [(1,66)]	0.05	0.819	0.04	0.847	0.19	0.668	0.00	0.945
Model 2								
Diversity [(1,66)]	4.20	**0.044**	0.03	0.852	4.15	**0.046**	0.06	0.801
Damage by *H.grandella* [(1,66)]	0.13	0.721	0.34	0.563	0.29	0.590	1.17	0.283

Summary of results from generalized linear mixed models testing for the effects of (a) big-leaf mahogany genotypic diversity and (b) plant species diversity on mahogany chemical defenses (polyphenolics and condensed tannins in the stem and leaves). To test the effect of mahogany genotype diversity we compared genotypic monocultures and genotypic polycultures, whereas to test for a species diversity effect we compared genotypic monocultures and species polycultures. Initial models testing only for an effect of diversity are designed as "Model 0". For models labelled "Model 1", we tested for an effect of diversity on defenses via growth-defense trade-offs by including plant height (proxy of growth) as a covariate. Finally, for models labelled "Model 2" we tested an effect of diversity via changes in herbivory by including *H. grandella* attack as a covariate. F-values and associated significance levels (*P*) are shown, as well as degrees of freedom (as subscripts in parenthesis). Significant (*P*<0.05) and marginal (0.05<*P*<0.10) *P* values are typed in bold.

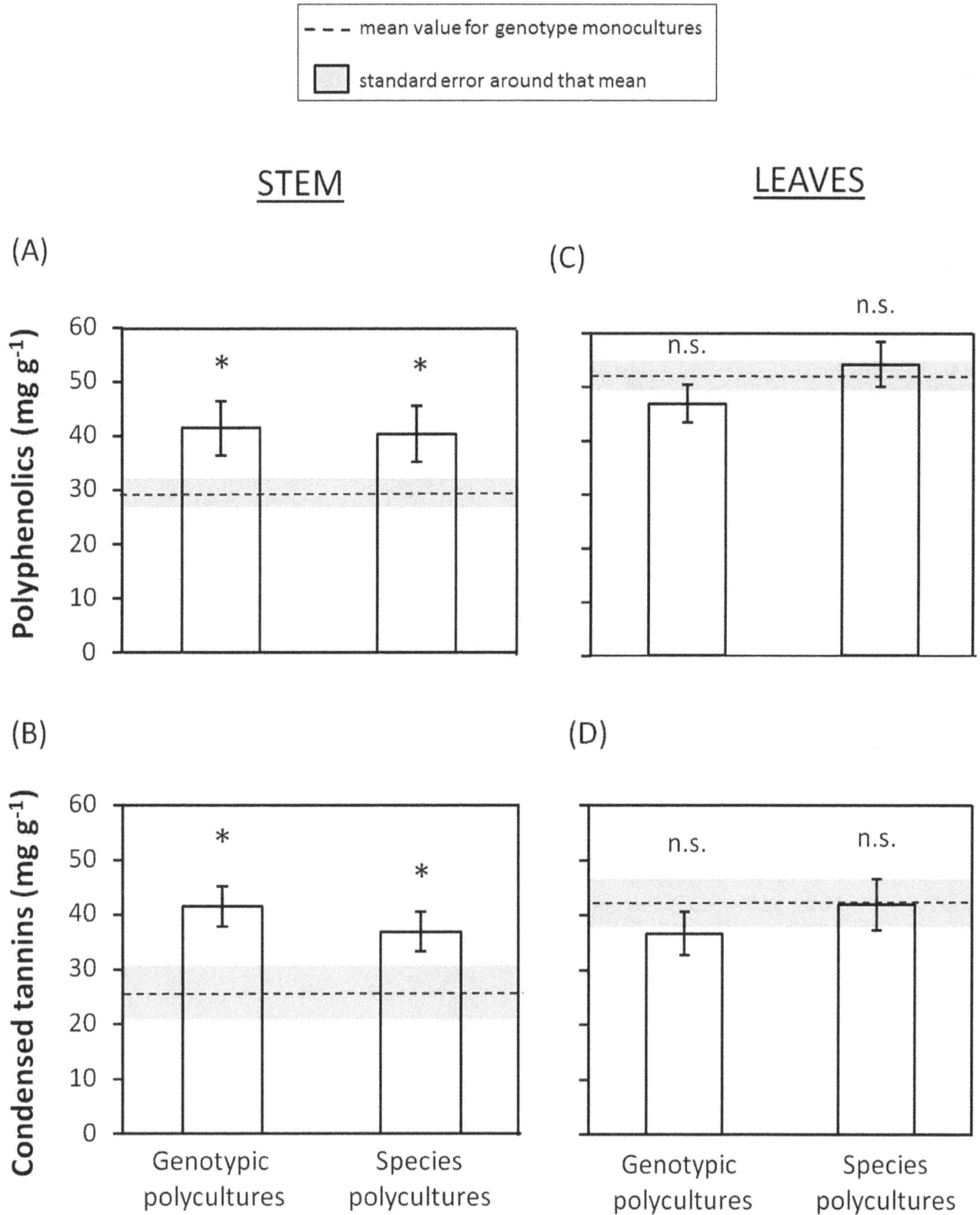

Figure 3. Diversity effect on plant defenses. Effect of mahogany genotypic and species diversity on the concentration of: (A) mahogany stem polyphenolics, (B) stem tannins, (C) leaf polyphenolics, and (D) leaf tannins. The dashed line represents the mean value for mahogany genotypic monocultures (N = 12) and the shaded area represents the standard error around that mean. Least-square means ± S.E. (N = 12 for genotypic polycultures and N = 12 for species polycultures). Significant differences ($P<0.05$) between a given diversity treatment and the genotypic monoculture treatment are indicated by an asterisk.

biomass and arthropod species richness due to genotypic and species diversity for the common evening primrose (*Oenothera biennis*). Nonetheless, further research in other systems comparing the effects of plant intra- and inter-specific diversity is necessary in order to assess the relative importance and mechanisms by which different forms of plant diversity shape anti-herbivore defenses in plants, and in turn potentially explain differential effects on arthropod faunas.

Our finding that diversity did not influence plant defenses via growth-defense trade-offs is not surprising, given that neither source of diversity influenced mahogany growth. Contrary to our findings, results from a previous study by McArt & Thaler [21] are consistent with the idea of growth-defense trade-offs act as a mechanism of diversity effects on plant defenses. Specifically, they found a trade-off where increased productivity of *Oenothera biennis* (via complementarity effects) at high genotypic diversity resulted in reduced resistance to an exotic leaf herbivore [21]. In addition, our results also run counter with the idea that producer diversity effects on plant defenses are mediated through effects on herbivores. In this case, a previous study by Mraja et al. [28] agrees with our findings, as they found that diversity effects on *P. lanceolata* were only weakly related to changes in herbivory across levels of diversity. Our results showed a tendency for a negative effect of plant diversity on *H. grandella* attack, suggesting that some unknown mechanism drove an increase in plant defenses, and that such effect in turn could have negatively influenced stem borer attack. However, this argument remains speculative as reductions in attack by this specialist herbivore may have also responded to resource concentration effects with increasing genotypic or species diversity, as suggested by previous findings in this system (Abdala-Roberts et al. unpublished data). Regardless of the mechanism at work, evidence for diversity effects on plant defenses remains limited and further research is needed to derive general patterns and determine which mechanisms and defensive compounds are more important and under what conditions.

Although we found that plant diversity effects on anti-herbivore defenses were not mediated by diversity effects on *H. grandella* attack or via growth-defense trade-offs, these mechanisms cannot be entirely discarded. First, it is possible that our measure of *H. grandella* abundance (i.e. number of attack sites per plant) was not predictive of the amount of damage (and thus defense induction) experienced by each plant. Second, and related also to the effects of diversity on herbivory, previous work in our system (during 2012) has shown that attack by *H. grandella* was significantly lower in species polycultures at the middle of the rainy season (early September), but this pattern reversed towards the end of the rainy season (late October) (Abdala-Roberts et al., unpublished data). Therefore, whereas we associated plant defenses with current patterns of herbivory (tissue samples taken during the same month as herbivore measurements), it is possible that stem-borer attack levels during previous months would have been a better predictor of defensive investment. Third, although results suggest that diversity effects on defenses via growth-defense trade-offs are not occurring at present in this system, our measurements were conducted at an early time point in the experiment. Accordingly, it is possible that such trade-offs may arise subsequently, once

diversity effects on plant growth presumably become stronger [13,53].

It is important to note that some authors have suggested that producer diversity may influence plant defenses in the absence of effects on plant growth and growth-defense trade-offs [28,54,55]. For example, plant species diversity may decrease light availability (due to architecture differences among species or genotypes) and such effect is predicted to reduce the concentration of carbon-based defenses such as phenolics, which are involved in photo-protection [56]. Accordingly, we previously found that low light availability reduces the concentration of polyphenolics and tannins in leaves and stems of mahogany [57]. However, in the present study we instead found a positive effect of species diversity on defenses which runs counter the argument that diversity effects on polyphenolics and tannins were mediated by light availability. Alternatively, as suggested by other studies, it is possible that plant species diversity improves nutrient acquisition (e.g. nitrogen) and use by plants which in turn increased nutrient concentrations in tissues [28,55,58] as well as the concentration of nitrogen-based defenses [28]. Finally, there is evidence that plants are able to recognize conspecific vs. hetero-specific individuals, as well as more closely vs. more distantly related individuals of their species and modulate the release of volatiles associated with defense [59]. Based on this, the prediction would be increased expression of defenses in monocultures (species or genotypic) as there would be a greater density of con-specifics or individuals of the same genotype emitting volatiles. However, we found the opposite pattern, which suggests that this mechanism was not at work.

In summary, the results from this study demonstrated that both plant inter- and intra-specific diversity can cause important changes in allocation to chemical defenses by plants, but that such effects were not driven by changes in plant growth (and thus growth-defense trade-offs) or herbivore damage. We suggest that additional abiotic factors (e.g. nutrient availability and uptake) and mechanisms should be considered in future studies in order to fully understand the observed patterns. Understanding the mechanisms by which plant diversity influences traits of importance to herbivores, in particular plant secondary metabolites, will contribute to a better understanding of plant diversity effects on consumers and ecosystem function.

Acknowledgments

We thank Nicolás Salinas-Peba, Luis Salinas-Peba, Teresa Quijano-Medina, María José Campos-Navarrete and Alejandra González-Moreno for their technical assistance in the experimental setup, assessments and plant sampling. We also thank Raúl Mena, Virginia Solís, Luis Esquivel and Denis Marrufo for laboratory assistance, and Luz Pato for help with chemical analyses.

Author Contributions

Conceived and designed the experiments: XM LAR KAM VPT. Performed the experiments: XM LAR. Analyzed the data: XM. Contributed reagents/materials/analysis tools: XM LAR KAM VPT. Contributed to the writing of the manuscript: XM LAR KAM VPT. Wrote the first draft of the manuscript: XM LAR. Contributed substantially to manuscript revisions: KAM VPT.

References

1. Crutsinger GM, Collins MD, Fordyce JA, Gompert Z, Nice CC, et al. (2006) Plant genotypic diversity predicts community structure and governs an ecosystem process. Science 313: 966–968.

2. Johnson MT, Lajeunesse MJ, Agrawal AA (2006) Additive and interactive effects of plant genotypic diversity on arthropod communities and plant fitness. Ecol Lett 9: 24–34.

3. Haddad NM, Crutsinger GM, Gross K, Haarstad J, Knops JMH, et al. (2009) Plant species loss decreases arthropod diversity and shifts trophic structure. Ecol Lett 12: 1029–1039.

4. Tilman D, Lehman CL, Thomson KT (1997) Plant diversity and ecosystem productivity: Theoretical considerations. Proc Natl Acad Sci USA 94: 1857–1861.

5. Cardinale BJ, Wright JP, Cadotte MW, Carroll IT, Hector A, et al. (2007) Impacts of plant diversity on biomass production increase through time because of species complementarity. Proc Natl Acad Sci USA 104: 18123–18128.

6. Hughes AR, Inouye BD, Johnson MTJ, Underwood N, Vellend M (2008) Ecological consequences of genetic diversity. Ecol Lett 11: 609–623.

7. Haddad NM, Crutsinger GM, Gross K, Haarstad J, Tilman D (2011) Plant diversity and the stability of foodwebs. Ecol Lett 14: 42–46.

8. Moreira X, Mooney KA, Zas R, Sampedro L (2012) Bottom-up effects of host-plant species diversity and top-down effects of ants interactively increase plant performance. Proc R Soc Lond B 279: 4464–4472.

9. Moreira X, Mooney KA (2013) Influence of plant genetic diversity on interactions between higher trophic levels. Biol Lett 9: 20130133.

10. Castagneyrol B, Lagache L, Giffard B, Kremer A, Jactel H (2012) Genetic diversity increases insect herbivory on oak saplings. PLoS ONE 7: e44247.

11. Parker JD, Salminen JP, Agrawal AA (2010) Herbivory enhances positive effects of plant genotypic diversity. Ecol Lett 13: 553–563.

12. Loreau M, Hector A (2001) Partitioning selection and complementarity in biodiversity experiments. Nature 412: 72–76.

13. Cardinale B, Matulich KL, Hooper DU, Byrnes JE, Duffy E, et al. (2011) The functional role of producer diversity in ecosystems. Am J Bot 98: 572–592.

14. Cook-Patton SC, McArt SH, Parachnowitsch AL, Thaler JS, Agrawal AA (2011) A direct comparison of the consequences of plant genotypic and species diversity on communities and ecosystem function. Ecology 92: 915–923.

15. McArt SH, Cook-Patton SC, Thaler JS (2012) Relationships between arthropd richness, evenness, and diversity are altered by complementarity among plant genotypes. Oecologia 168: 1013–1021.

16. Ohgushi T (2005) Indirect interaction webs: Herbivore-induced effects through trait change in plants. Annu Rev Ecol Evol Syst 36: 81–105.

17. Van der Putten WH (2003) Plant defense belowground and spatiotemporal processes in natural vegetation. Ecology 84: 2269–2280.

18. Coley PD, Bryant JP, Chapin FS (1985) Resource availability and plant antiherbivore defense. Science 230: 895–899.

19. Herms DA, Mattson WJ (1992) The dilemma of plants: to grow or defend. Q Rev Biol 67: 283–335.

20. Mooney KA, Halitschke R, Kessler A, Agrawal AA (2010) Evolutionary trade-offs in plants mediate the strength of trophic cascades. Science 237: 1642–1644.

21. McArt SH, Thaler JS (2013) Plant genotypic diversity reduces the rate of consumer resource utilization. Proc R Soc Lond B 280: 20130639.

22. Koricheva J, Mulder CPH, Schmid B, Joshi J, Huss-Danell K (2000) Numerical responses of different trophic groups of invertebrates to manipulations of plant diversity in grasslands. Oecologia 125: 271–282.

23. Scherber C, Mwangi PN, Temperton VM, Roscher C, Schumacher J, et al. (2006) Effects of plant diversity on invertebrate herbivory in and experimental grassland. Oecologia 147: 489–500.

24. Otway S, Hector A, Lawton JH (2005) Resource dilution effects on specialist insect herbivores in a grassland biodiversity experiment. J Anim Ecol 74: 234–240.

25. Root RB (1973) Organization of a plant-arthropod association in simple and diverse habitats: The fauna of collards (Brassica oleracea). Ecol Monogr 43: 95–124.

26. Kareiva P (1983) Influence of vegetation texture on herbivore populations: resource concentration and herbivore movement. In: Denno RF, McClure MS, editors. Variable plants and herbivores in natural and managed systems. New York: New York Academic Press. pp.259–289.

27. Vehviläinen H, Koricheva J, Ruohomäki K (2007) Tree species diversity influences herbivore abundance and damage: meta-analysis of long-term forest experiments. Oecologia 152: 287–298.

28. Mraja A, Unsicker S, Reichelt M, Gershenzon J, Roscher C (2011) Plant community diversity influences allocation to direct chemical defense in Plantago lanceolata. PLoS ONE 6: e28055.

29. Fridley JD, Grime PJ (2010) Community and ecosystem effects of intraspecific genetic diversity in grassland microcosms of varying species diversity. Ecology 91: 2272–2283.

30. Crawford KM, Rudgers JA (2013) Genetic diversity within a dominant plant outweighs plant species diversity in structuring an arthropod community. Ecology 94: 1025–1035.

31. Hämback PA, Ågren J, Ericson L (2000) Associational resistance: insect damage to purple loosestrife reduced in thickets of sweet gale. Ecology 81: 1784–1794.

32. Castagneyrol B, Giffard B, Pére C, Jactel H (2013) Plant apparency, and overlooked driver of associational resistance to insect herbivory. J Ecol 101: 418–429.

33. Grijpma P, Ramalho R (1973) Toona spp., posibles alternativas para el problema del barrenador Hypsipyla grandella de las Meliaceae en América Latina. In: Grijpma P, editor. Studies on the shootborer Hypsypla grandella (Zeller) Lep: Pyralidae. Costa Rica: IICA Miscellaneous Publication No. 101, v. 1. pp.3–17.

34. Pennington T, Sarukhán J (2005) Árboles Tropicales de México. México: Fondo de Cultura Económica. 523 p.

35. Norghauer JM, Malcolm JR, Zimmerman BL (2008) Canopy cover mediates interactions between a specialist caterpillar and seedlings of a neotropical tree. J Ecol 96: 103–113.

36. Norghauer JM, Grogan J, Malcolm JR, Felfili JM (2010) Long-distance dispersal helps germinating mahogany seedlings escape defoliation by a specialist caterpillar. Oecologia 162: 405–412.

37. Mayhew JE, Newton AC (1998) The silviculture of mahogany. Wallingford: CABI Publishing. 240 p.

38. Grogan J, Galvão J, Simões L, Veríssimo A (2003) Regeneration of big-leaf mahogany in closed and logged forests of southeastern Pará, Brazil. In: Lugo AE, Figueroa-Colón JC, Alayón M, editors. Big-leaf mahogany: Genetics, ecology, and management. New York: Springer-Verlag. pp.193–208.

39. Nair KSS (2007) Tropical forest insect pests: ecology, impact and management. Cambridge: Cambridge University Press. 404 p.

40. CICY (2010) Centro de Investigación Científica de Yucatán (CICY). Flora de la Península de Yucatán. Catálogo de Flora. Yucatán, México. Available at: http://www.cicy.mx/sitios/flora%20digital/index.php.

41. Gillies ACM, Navarro C, Lowe AJ, Newton AC, Hernández M, et al. (1999) Genetic diversity in Mesoamerican populations of mahogany (Swietenia macrophylla), assessed using RAPDs. Heredity 83: 722–732.

42. Loveless MD, Gullison RE (2003) Genetic variation in natural mahogany populations in Bolivia. In: Lugo AE, Figueroa-Colón JC, Alayón M, editors. Big-leaf mahogany Genetics, ecology, and management. New York: Springer-Verlag. pp.9–28.

43. Taveras R, Hilje L, Hanson P, Mexzón R, Carballo M, et al. (2004) Population trends and damage patterns of Hypsipyla grandella (Lepidoptera: Pyralidae) in a mahogany stand, in Turrialba, Costa Rica. Agric For Entomol 6: 89–98.

44. Salminen J-P, Karonen M (2011) Chemical ecology of tannins and other phenolics: we need a change in approach. Funct Ecol 25: 325–338.

45. Heil M, Baumann B, Andary C, Linsenmair EK, McKey D (2004) Extraction and quantification of "condensed tannins" as a measure of plant anti-herbivore defense? Revisiting an old problem. Naturwissenschaften 89: 519–524.

46. Moreira X, Mooney KA, Rasmann S, Petry WK, Carrillo-Gavilán A, et al. (2014) Trade-offs between constitutive and induced defenses drive geographical and climatic clines in pine chemical defenses. Ecol Lett 17: 537–546.

47. Pérez-Flores J, Eigenbrode SD, Hilje-Quiroz L (2012) Alkaloids, limonoids and phenols from Meliaceae species decrease survival and performance of Hypsipyla grandella larvae. Am J Plant Sci 3: 988–994.

48. Newton AC, Watt AD, Lopez F, Cornelius JP, Mesén JF, et al. (1999) Genetic variation in host susceptibility to attack by the mahogany shoot borer, Hypsipyla grandella (Zeller). Agric For Entomol 1: 11–18.

49. Moreira X, Zas R, Sampedro L (2012) Differential allocation of constitutive and induced chemical defenses in pine tree juveniles: a test of the optimal defense theory. PLoS ONE 7: e34006.

50. Porter LJ, Hrstich LN, Chan BG (1986) The conversion of procyanidins and prodelphinidins to cyanidin and delfinidin. Phytochem 25: 223–230.

51. Sampedro L, Moreira X, Zas R (2011) Costs of constitutive and herbivore-induced chemical defenses in pine trees emerge only under low resources availability. J Ecol 99: 818–827.

52. Crawford KM, Rudgers JA (2012) Plant species diversity and genetic diversity within a dominant species interactively affect plant community biomass. J Ecol 100: 1512–1521.

53. Tilman D, Knops J, Wedin D, Reich P (2002) Plant diversity and composition: effects on productivity and nutrient dynamics of experimental grasslands. In: Loreau M, Naeem S, Inchausti P, editors. Biodiversity and ecosystem functioning: Synthesis and perspectives. Oxford: Oxford University Press. pp.21–35.

54. Spehn EM, Joshi J, Schmid B, Diemer M, Körner C (2000) Above-ground resource use increases with plant species richness in experimental grassland ecosystems. Funct Ecol 14: 326–337.

55. Roscher C, Kutsch WL, Schulze ED (2011) Light and nitrogen competition limit Lolium perenne in experimental grasslands of increasing plant diversity. Plant Biol 13: 134–144.

56. Close DC, McArthur C, Paterson S, Fitzgerald H, Walsh A, et al. (2003) Photoinhibition: a link between effects of the environment on eucalypt seedling leaf chemistry and herbivory. Ecology 84: 2952–2966.

57. Abdala-Roberts L, Moreira X, Cervera JC, Parra-Tabla V (2014) Light availability influences growth-defense trade-offs in big-leaf mahogany (Swietenia macrophylla King). Biotropica in press.

58. Lang AC, von Oheimb G, Scherer-Lorenzen M, Yang B, Trogisch S, et al. (2014) Mixed afforestation of young subtropical trees promotes nitrogen acquisition and retention. J Appl Ecol 51: 224–233.

59. Karban R, Yang LH, Edwards KF (2014) Volatile communication between plants that affects herbivory: a meta-analysis. Ecol Lett 17: 44–52.

12

Trends in Extinction Risk for Imperiled Species in Canada

Brett Favaro[1,2,3]*, **Danielle C. Claar**[1], **Caroline H. Fox**[4,5], **Cameron Freshwater**[1], **Jessica J. Holden**[1], **Allan Roberts**[6], **UVic Research Derby**[1,4,7]¶

1 Department of Biology, University of Victoria, Victoria, Canada, 2 Centre for Sustainable Aquatic Resources, Fisheries and Marine Institute of Memorial University of Newfoundland, St John's, Canada, 3 Department of Ocean Sciences, Memorial University of Newfoundland, St. John's, Canada, 4 Department of Geography, University of Victoria, Victoria, Canada, 5 Raincoast Conservation Foundation, Sidney, Canada, 6 Bamfield Marine Sciences Centre, Bamfield East, Canada, 7 School of Environmental Studies, University of Victoria, Victoria, Canada

Abstract

Protecting and promoting recovery of species at risk of extinction is a critical component of biodiversity conservation. In Canada, the Committee on the Status of Endangered Wildlife in Canada (COSEWIC) determines whether species are at risk of extinction or extirpation, and has conducted these assessments since 1977. We examined trends in COSEWIC assessments to identify whether at-risk species that have been assessed more than once tended to improve, remain constant, or deteriorate in status, as a way of assessing the effectiveness of biodiversity conservation in Canada. Of 369 species that met our criteria for examination, 115 deteriorated, 202 remained unchanged, and 52 improved in status. Only 20 species (5.4%) improved to the point where they were 'not at risk', and five of those were due to increased sampling efforts rather than an increase in population size. Species outcomes were also dependent on the severity of their initial assessment; for example, 47% of species that were initially listed as special concern deteriorated between assessments. After receiving an at-risk assessment by COSEWIC, a species is considered for listing under the federal Species at Risk Act (SARA), which is the primary national tool that mandates protection for at-risk species. We examined whether SARA-listing was associated with improved COSEWIC assessment outcomes relative to unlisted species. Of 305 species that had multiple assessments and were SARA-listed, 221 were listed at a level that required identification and protection of critical habitat; however, critical habitat was fully identified for only 56 of these species. We suggest that the Canadian government should formally identify and protect critical habitat, as is required by existing legislation. In addition, our finding that at-risk species in Canada rarely recover leads us to recommend that every effort be made to actively prevent species from becoming at-risk in the first place.

Editor: Clinton N. Jenkins, Instituto de Pesquisas Ecológicas, Brazil

Funding: BF was supported by a Liber Ero postdoctoral fellowship (http://liberero.ca/). CHF was supported by the Raincoast Conservation Foundation (http://www.raincoast.org/) and the National Sciences and Engineering Research Council of Canada (http://www.nserc-crsng.gc.ca/). DCC and CF were supported by University of Victoria Graduate Fellowships. Catering for the UVIC Research Derby event was provided by the University of Victoria Environmental Science and Biology departments. The funders had no role in study design, data collection and analysis, decision to publish, or preparation of the manuscript.

Competing Interests: The authors have declared that no competing interests exist.

* Email: brett.favaro@mi.mun.ca

¶ Membership of the UVic Research Derby is provided in the Acknowledgments.

Introduction

Unsustainable exploitation, climate change, ocean acidification and other anthropogenic impacts have resulted in a global extinction rate that is as much as 1000 times the historic background rate [1–3]. Given the irreversibility of extinctions, preventing or reversing the continuing decline of at-risk species is a major focus of conservation [4]. Preserving global biodiversity is also considered essential for human well-being and the maintenance of ecosystem processes [3,5–7].

Many countries have legislation that explicitly protects species at risk of extinction. In general, such legislation is designed to identify vulnerable taxa, establish recovery plans, prevent further declines, and promote recovery [8]. Recognizing that habitat loss is the leading cause of extinction [9–12], the identification and preservation of habitat is often required, contributing to the stabilization and recovery of threatened species [13]. Despite the

implementation of laws and conservation programs, global biodiversity continues to decline [3,5,14].

In Canada, species at risk are identified and protected in a multi-step process. The process begins with the Committee on the Status of Endangered Wildlife in Canada (COSEWIC), an independent scientific body formed in 1977. COSEWIC assesses the status of candidate species that are potentially at risk of extinction or extirpation [15]. This body only considers scientific evidence relevant to a species' recovery potential, and ignores socioeconomic costs or benefits of protection [15]. Their assessments are based on the Canadian extent of the species' range, even if the species occurs in the United States or elsewhere [15]. Species assessed as at-risk by COSEWIC do not automatically secure legal protection from the committee's decision. Rather, protection is issued under the Species at Risk Act (SARA), which was passed in 2003, and which formalized the use of COSEWIC assessments as the scientific basis for listing decisions [16]. Upon receipt of a COSEWIC assessment, the Minister of the

Environment must issue a response statement indicating whether the species will be listed at the status recommended by the assessment, or whether more information is required [17]. The ultimate decision whether to list may also incorporate the socio-economic impacts of listing. If listed, it becomes illegal to kill or harm individuals of that species and the species' critical habitat must be identified and protected to the extent possible [18]. Protections derived from SARA automatically apply to federal lands (including oceans), but for these measures to apply outside of these areas (e.g. in provincial or private land) complementary protections must be implemented that applies to the species and its habitat [19]. If the Minister of the Environment decides that provincial or territorial law is insufficient to effectively protect a listed species, the federal government has the power to apply SARA's protections to provincial or territorial lands [20]. Recovery strategies and action plans (required by SARA) are mandated to be released within a specified timeframe, the length of which depends on the species' designation and when it was listed [21,22]. Furthermore, SARA requires that management plans be produced for species that are not at immediate risk of extinction or extirpation, but are nevertheless of 'special concern' since they may become at-risk in the future [23]. Species of special concern do not receive the full protections offered to species listed under SARA as threatened, endangered, or extirpated.

Previous studies have identified biases, limitations, and a general lack of implementation associated with at-risk species legislation in Canada (Table 1). For instance, harvested species and those found in northern territories are less likely to be listed under SARA [24,26], as are species threatened by biological resource use, including those that are unintentionally harvested [27]. In addition, recovery strategies and action plans are often not completed within deadlines established by SARA [22], leading to concerns that species are not receiving timely protections, despite being listed. This finding was confirmed in a recent federal court decision [28]. Moreover, COSEWIC itself has been criticized for its assessment criteria and apparent biases [31–32]. For example, species with low information availability tend to receive less severe assessments than would be suggested under the precautionary principle [30].

In this study, we examine trends in the status of species repeatedly assessed by COSEWIC, used here as a proxy for Canada's effectiveness in species conservation. Using records obtained from COSEWIC [33] and the SARA Public Registry [34], we analyze trends in the designations of species with two or more assessments to determine whether species are, on average, improving, deteriorating, or remaining stable in status. In addition, we examine outcome differences across taxonomic groups, and whether the basic obligation to identify critical habitat for listed species has been met. Finally, we assess whether species listed for longer have better outcomes. While this study builds on previous work reviewing the process of listing species under SARA (Table 1), it is the first to assess the overall trends in status of at-risk species that have been assessed more than once by COSEWIC.

Methods

COSEWIC Wildlife Species Database

Species, subspecies, and populations (hereafter 'species') that are assessed by COSEWIC are placed into one of five categories on a scale of increasing risk of extinction or extirpation [32]: not at risk, special concern, threatened, endangered, or extinct or extirpated (hereafter extirpated). A sixth category, data deficient, is used when there are insufficient data to classify a species. The criteria

for categorization depends on a variety of factors including changes in total numbers of mature individuals, whether a species has a small and declining population size, and the changes in size of its range [35]. COSEWIC's assessment criteria were updated in 2001 [35]. Although the same assessment categories were employed after the revision in 2001, species required a more severe decline in total population to qualify as either endangered or threatened. For example, prior to 2001 the criteria for an Endangered assessment (Criteria A1: [35]) required a reduction of ≥50% in the total number of mature individuals over the last 10 years or 3 generations. As of 2001, this same category required a ≥70% decline [36].

The definition of full recovery that we employ here is a change in COSEWIC assessment status to 'not at risk.' Although this may not equate to a full ecological recovery (e.g. [37,38]) it allows us to broadly assess trends with a consistent metric.

Data collection

We identified all species that COSEWIC has assessed more than once using the Wildlife Species Search Engine on the COSEWIC website [33]. Assessments occurred between 1977 (COSEWIC's establishment) and December 2013 (when we conducted our search). We recognize that our analysis may be limited by biases associated with which species have been assessed multiple times and that COSEWIC listing criteria have changed since 2001. However, because the 2001 revision tightened criteria for more-severe listings, the expected bias introduced by the criteria change would be a greater proportion of less severe species listings. We used the 'change in status' tab on the website to include any species 'in a higher risk category' (N = 81); 'in a lower risk category' (N = 36); 'no longer at risk' (N = 21); 'changed' (N = 20); 'reassigned' (N = 69); and 'no change' (N = 272); populating our database with a total of 499 species. In our analysis, we excluded any species that had ever been assessed as 'data deficient' by COSEWIC (N = 30 species), or that had experienced some form of reassignment (e.g. a species split into multiple designatable units, N = 106 species), because changes in status for these species cannot be interpreted as a true change in extinction risk. We excluded six species that were both data deficient and reclassified, leaving a total of 369 species in our analysis.

We collected the history of COSEWIC assessments for each species on its respective species summary page. We recorded the date and status designation of each COSEWIC assessment for each species, and the taxonomic group to which that species belongs (due to small sample sizes, we combined molluscs and arthropods into a single 'invertebrates' category, and combined lichens and mosses). We then used the SARA public registry [34] to record whether species were SARA listed, and to access COSEWIC status reports for each species. To account for differences between species' life histories, COSEWIC and the IUCN place rates of decline into a biological context by scaling assessments by the generation time (GT) of each species [35,39]. Therefore we also recorded each species' generation time, if available. If the species was SARA-listed as extirpated, endangered, or threatened, we recorded whether a recovery strategy had been completed. If the recovery strategy was completed, we recorded whether the strategy indicated that critical habitat had been fully, partially, or not identified. If a recovery strategy was not complete for the species, we recorded that critical habitat had not been identified. For SARA-listed species, Environment Canada provided us with the dates when species were listed.

For some species, the COSEWIC summary explained that an apparent improvement in status was due to the discovery of new populations through increased sampling, and not because of an

Table 1. Summary of review papers related to endangered species assessment and legislation in Canada.

Publication	Reference #	Primary findings
Vanderzwaag and Hutchings (2005)	[51]	Review of SARA implementation related to marine fish. Paper advocates for biodiversity preservation by implementing marine protected areas and modernizing Fisheries Act.
Mooers et al. (2007)	[52]	First identification of taxonomic and regional biases in SARA listing. Northern species and marine fish and terrestrial mammals unlikely to receive SARA-listing.
Findlay et al. (2009)	[24]	Commercially harvested species, species managed by DFO, and species that occur entirely within Canada are less likely to receive listing.
Lukey et al. (2009)	[29]	Changes in assessment status from 'endangered' to 'threatened' often occur without sufficient justification for the change. Assessment criteria are not always applied consistently.
Lukey et al. (2010)	[30]	COSEWIC assessments do not follow the precautionary principle – lack of information is associated with assessments of species to lower risk categories.
Mooers et al. (2010)	[25]	Most SARA-listed species lack recovery plans. Scientific advice is insufficiently reflected in conservation policy for at-risk species. First review of changes in COSEWIC status across assessments.
Powles (2011)	[31]	General overview of marine fish assessed by COSEWIC.
Taylor and Pinkus (2013)	[53]	Only 17% of recovery strategies led by DFO included critical habitat, as opposed to 63% for those led by Environment Canada. Recovery strategies written after court judgments related to SARA were more likely to identify critical habitat.
Waples et al. (2013)	[8]	Comparison of United States' Endangered Species Act (ESA) with SARA. ESA should adopt a single national scientific body, while SARA should adopt strict deadlines for listing action. The emphasis on socioeconomic factors should also be reduced.
McCune et al. (2013)	[27]	Human disturbance, invasive species, residential development, and ultimately loss of habitat are major threats to the majority of SARA-listed species. Threats differ by taxonomic grouping.
Schultz et al. (2013)	[26]	Imperiled marine fish are unlikely to receive SARA listing if the forecasted cost of listing exceeds $90,000 per decade. The threshold for freshwater fish is $5,000,000. Rationale used in the decision to list was inconsistent between freshwater and marine fish.
Auditor General of Canada (2013)	[22]	All three government agencies involved with species at risk (Environment Canada, Fisheries and Oceans Canada, and Parks Canada) are not meeting obligations to complete recovery strategies, action plans, or management plans. Only seven of 97 required action plans were complete.

actual recovery in population. We made note of these cases (N = 20) so as to distinguish them from improvements due to conservation action.

Data analysis

We assigned each COSEWIC status a numerical value in descending order of severity (5 = not at risk, 4 = special concern, 3 = threatened, 2 = endangered, 1 = extirpated). We then calculated the overall change for each species, between the first and most recent COSEWIC assessment. For example, a species that was initially classified as special concern, and then deteriorated to endangered in its most recent assessment would receive a score of −2. For species requiring critical habitat designation (extirpated, endangered and threatened), we performed a Pearson's chi-squared test to determine whether critical habitat identifications (full, partial, or not identified) across the ten taxonomic groups differed from the proportions across all species. We used the statistical software R for computations and data plots [40].

We identified species whose most recent COSEWIC assessment occurred after the species had been listed under SARA for at least three of that species' GT, to identify species for which some recovery may be biologically possible. We then employed a cumulative link mixed model using the "ordinal" package in R [41] to test whether the number of GTs since SARA-listing was associated with a change in final COSEWIC status across species. Cumulative link mixed models test the influence of one or more independent variables (herein: number of generation times since listing) on an ordinal dependent variable (change in COSEWIC status), and are analogous to generalized linear mixed effects models in that they allow for the incorporation of a random effect, or 'grouping' factor (taxonomic order) [41–43].

Results

There were a total of 369 species across ten taxonomic groups that met our criteria for inclusion (Table S1, Fig. 1). For all taxonomic groups (except marine fish), the majority of species declined or remained the same in status (Table S1, Fig. 2). However, species trajectories varied substantially based on their initial assessment (Fig. 3). For species initially classified as not at risk, endangered, or extirpated, the most common outcome was to remain at their initially assessed status. However, for species initially classified as special concern or threatened, deterioration in status was the most common outcome. For example, 47% of species listed as special concern deteriorated in status (Fig. 3). Only 20 species (5.4%) received a 'not at risk' assessment after previously being listed in an at-risk category (three terrestrial mammals, nine birds, four freshwater fish, one marine fish, two vascular plants, and one lichen). Five of these cases were not due to conservation action, but were instead due to increased sampling effort. A total of 221 species are listed under SARA as threatened, endangered, or extirpated (Fig. 4), and therefore their critical habitat should be identified. Overall, and for all taxonomic groups, critical habitat has not been fully identified for more than half of SARA-listed species (Fig. 4). Further, critical habitat identifications (full, partial, or not identified) significantly varied across the ten taxonomic groups ($x^2 = 51.90$, df = 18, p<0.001).

There were 163 species that met our criteria for inclusion in the Generation Time analysis (listed as extirpated, endangered, or

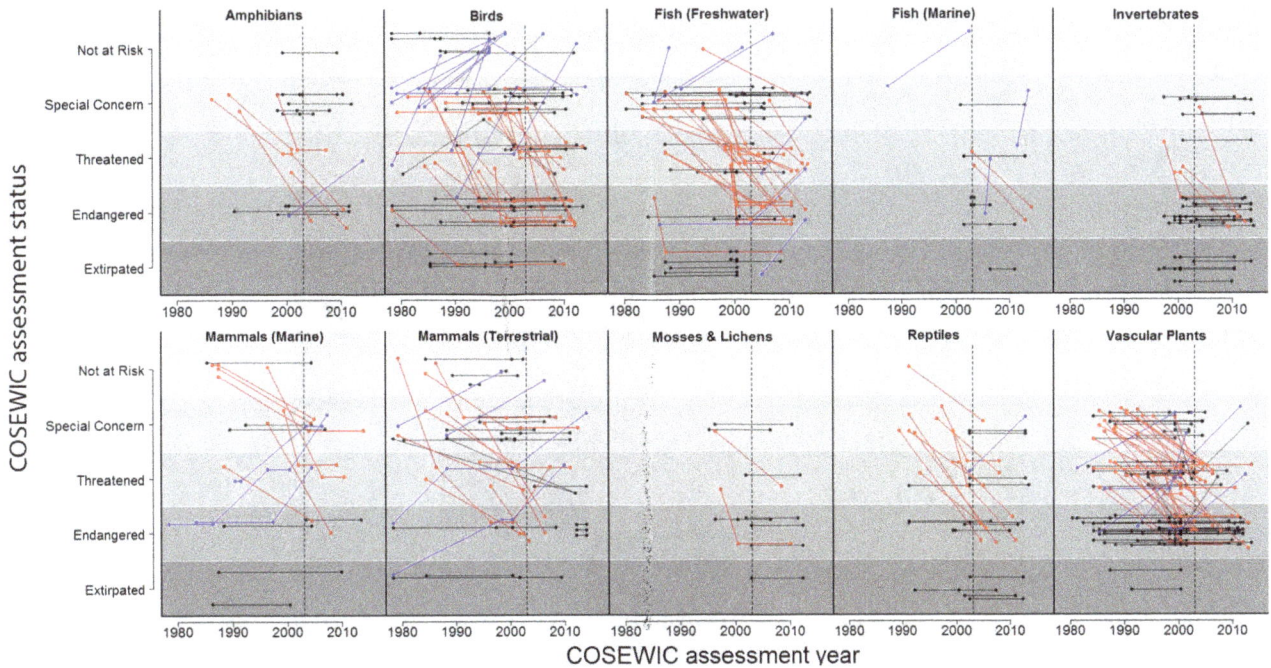

Figure 1. Overview of COSEWIC assessment statuses for all species that have been assessed more than once, and that have never been data deficient or taxonomically reassigned. Each species is represented by points (assessment date and outcome) connected by lines. Species that deteriorated in status from their first to final COSEWIC assessments are red, species that have improved are blue, and species that have remained constant are black. The vertical dotted line indicates 2003, or the passing of SARA. Species whose apparent recovery was due to increased sampling effort, and not biological recovery (N = 20) are not shown.

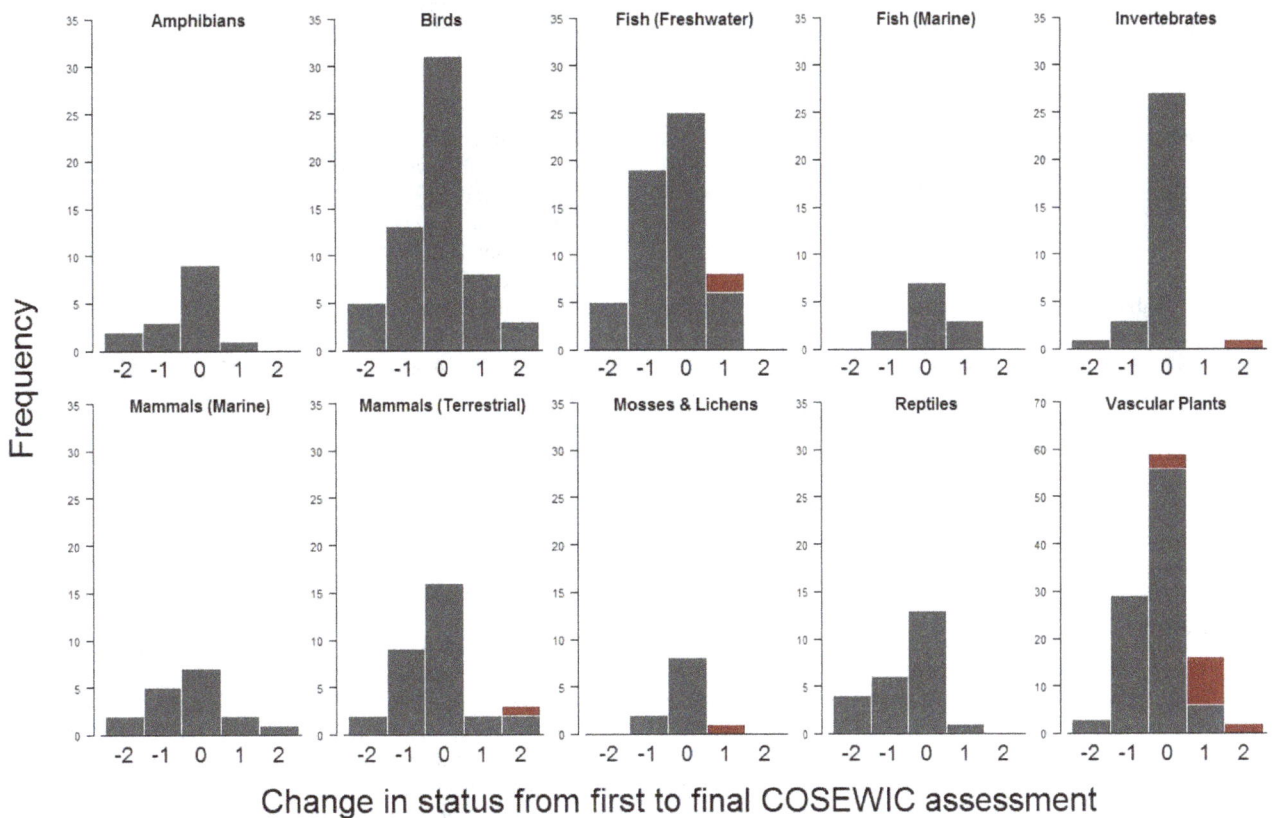

Figure 2. Frequencies of occurrence of change in COSEWIC assessed status for each taxonomic group. Positive numbers indicate improvement (e.g. a transition from endangered to threatened would be +1, while endangered to special concern would be +2), negative numbers indicate deterioration, and zero indicates no change across assessments. Red bars indicate apparent recoveries due to increased sampling efforts. Note that the y-axis for vascular plants is scaled differently from other taxonomic groups.

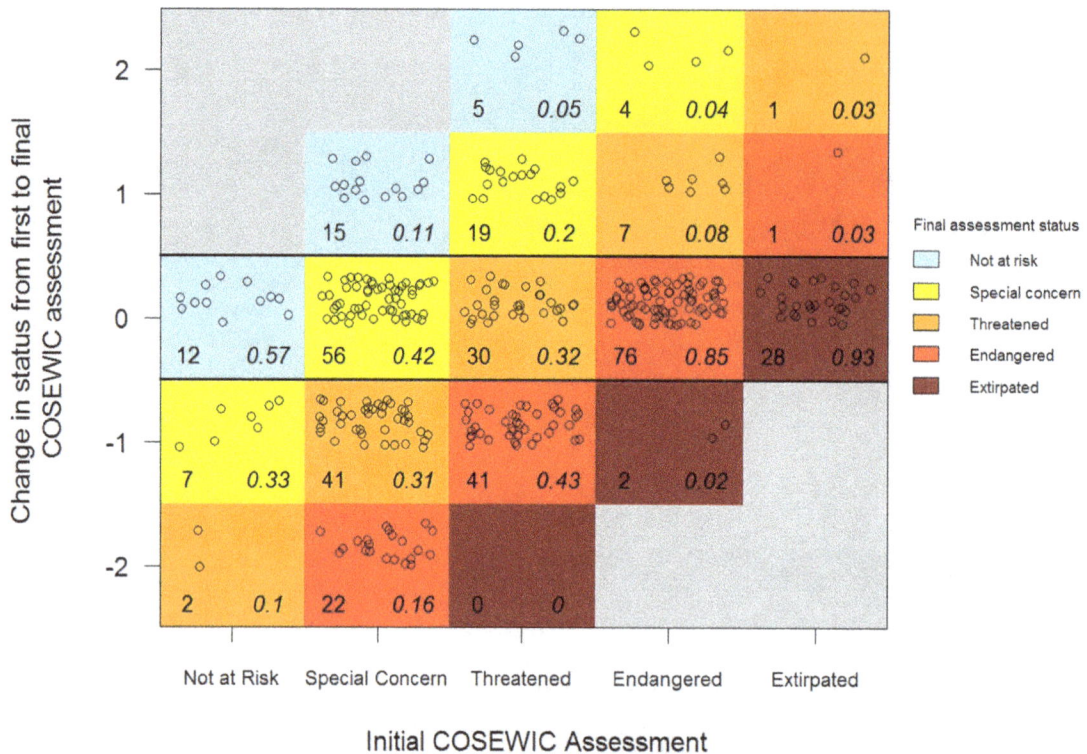

Figure 3. Trends in COSEWIC statuses for each species, grouped by initial assessments. Proportions are grouped by initial assessment status (column). Change from first to final assessment is indicated on the Y axis, and colours indicate the final assessment status. For example, of species that were initially assessed as threatened (third column), 19 of them (or a proportion of 0.2) improved by one status level (+1 on Y-axis), ultimately placing them into the special concern category (yellow).

threatened; GT reported; schedule 1), and 69 that have been listed for three or more GT. The number of GT since listing ranged from 0.11 to 35.8 (mean ±1 S.D., 3.7±4.3, Figure 5). If a species has been adequately protected, the probability of a species improving in status should increase with the number of generations since listing. However, there was an estimated decline in the probability of a species improving as the number of generations since listing increased ($\hat{\beta} = -0.05$, $\{\hat{\theta}_j\} = \{-2.40, -0.18, 2.44, 4.31\}$, 95% C.I. $= -0.124$ to 0.008), but it was not significant (Wald p = 0.084). This model had a small absolute gradient ($7.4*10^{-6}$) and a condition number of the Hessian of 720, indicating that the model was able to converge and was well-defined, respectively [42]. The cumulative odds ratio was between 0.88 and 1.01 (95% C.I.), meaning that for each unit increase in GT since listing, the odds of a species improving in status ranged from increasing by 1% to decreasing by 12%.

Discussion

For species that have been assessed more than once by COSEWIC, improvement was rare, and recovery to a 'not at risk' status occurred in only 5.7% of cases. Moreover, one quarter of those were apparent recoveries due to increased sampling, rather than increases in population size driven by conservation action. Contrary to the intent of endangered species legislation in Canada, the probability of a species improving in status did not increase with the number of generations since initial listing under SARA. Moreover, species had a greater probability of deteriorating in status with the number of generations since listing, although this relationship was marginally statistically non-significant (Wald

p = 0.084). In contrast, species recovery in the United States was strongly correlated with the number of years of protection under the Endangered Species Act [44], indicating that endangered species legislation can be effective. These results suggest a potential failure of Canadian legislation, its subsequent implementation, or both. While COSEWIC's assessment criteria were modified in 2001 [35], the revisions made it harder to qualify for more-severe categories. Therefore we have no evidence to suggest that observed declines in status were an artifact of changes in criteria.

The lack of observed recovery for SARA-listed species may be due to the lack of implementation of the law. For example, for those species without identified critical habitat, the habitat protection provisions of SARA cannot be fully implemented. Overall, the proportion of species with critical habitat identified was higher than what has been reported in a previous study [25], although our study focused on a subset of species that have been assessed multiple times, and that have never been data deficient. These are the species for which it should have been most feasible to identify critical habitat, as they have been scrutinized for a relatively long time and have sufficient data to complete an accurate assessment. The fact remains that despite the importance of critical habitat identification, the proportion of species whose critical habitat was fully identified remained low, and varied significantly across the ten taxonomic groups. This pattern suggests that there are considerable differences between the protections that species should receive under SARA, and what is actually achieved.

Since our data demonstrate that it is rare for at-risk species to recover in Canada, it is essential that substantial efforts be made to prevent species from becoming at-risk in the first place. Given that

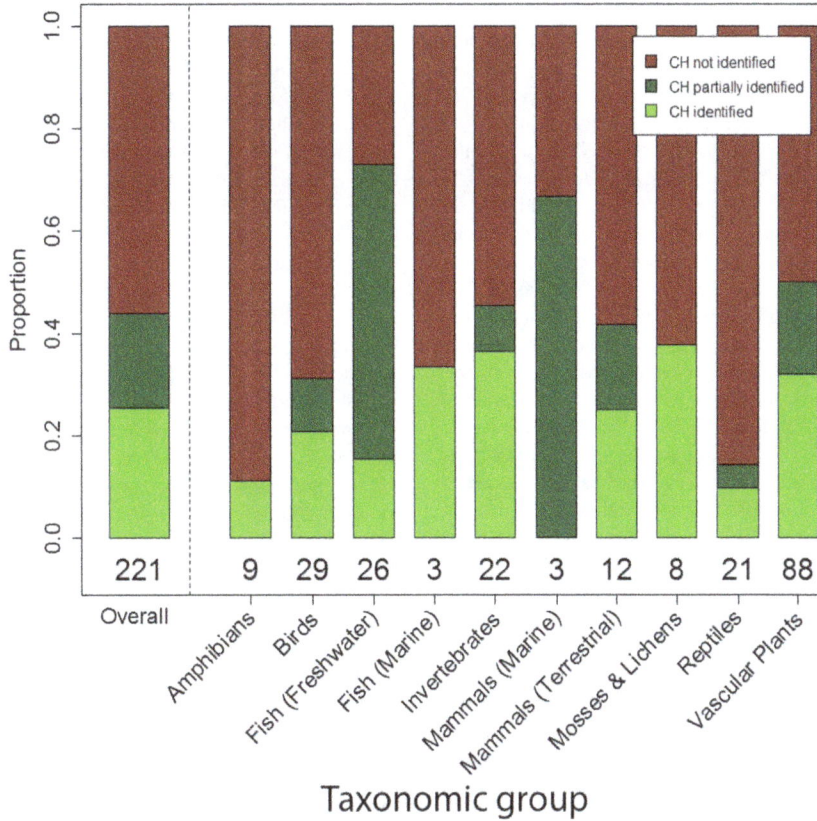

Figure 4. Proportion of SARA-listed species (listed as extirpated, endangered, or threatened) that have critical habitat (CH) fully identified (light green), partially identified (dark green), or not at all identified (dark red), overall and by taxonomic group. Values under bars indicate the number of species in each taxonomic category.

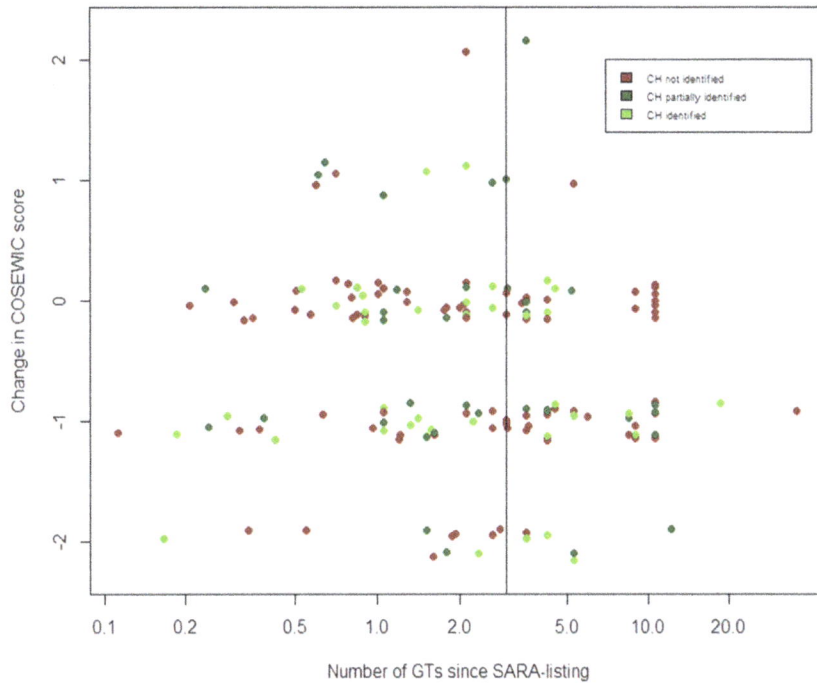

Figure 5. Change in COSEWIC assessment statuses versus the number of GT that have passed since initial SARA listing for all schedule 1 species listed as threatened, endangered, or extirpated. Dot colour indicates whether CH has been fully identified (green), partially identified (dark green), or not identified (red). The black vertical line indicates three GT.

imperiled species are usually threatened by loss of habitat [45], recent weakening of federal laws that protect habitat in Canada [25,46,47,48] will be unhelpful in the long term, as it may result in additional species declining to the point where they receive an at-risk designation. Habitat protection should not be limited to critical habitat – it should be managed appropriately for all habitat such that protection does not become critical. The experience in the United States demonstrates that protection of critical habitat is associated with improved conservation outcomes relative to species without such protection [44,47]. In addition, since we found that the most common outcome for a species assessed as special concern was to deteriorate, an assessment of special concern does not currently result in sufficient protection to promote recovery.

The single most common outcome across taxonomic groups was for species to remain at the same status across assessments, and the number of species that declined outnumbered those that recovered by a ratio of approximately 2:1. These findings are alarming for three reasons. First, by definition, a species that has been assessed as anything other than 'not at risk' is at elevated risk of extinction or extirpation given current conditions, and therefore maintaining a species at a threatened status should not be interpreted as a conservation victory. Second, it takes a substantial decline in population size or range size to trigger a change in assessment status, so real declines (or increases) could still occur within species held at the same COSEWIC threat level across assessments. Third, a 'not at risk' designation only means that the species is not at elevated risk of extinction or extirpation – it does not imply that the population has recovered to historical levels. Even species that are classified 'not at risk' can be heavily depleted and unable to serve their historic roles in ecosystem structure or function. This has implications for managers in the United States as well as Canada, because many at-risk species have ranges that extend into both countries. At the very least, a successful species at risk program should demonstrate species recovering to a point where they are not at risk of extinction or extirpation given current conditions. Currently, this goal is not being achieved in Canada for the overwhelming majority of species.

Recommendations

Our results lead us to make three core recommendations for at-risk species in Canada. First, given that it was much more common for species of special concern to deteriorate than to improve, we should recognize that a special concern listing warns of a coming deterioration, and we therefore suggest that the protections

associated with this listing should be strengthened. Second, given the poor outcomes of at-risk species in Canada, it should be a policy priority to prevent species from becoming at-risk in the first place. The importance of critical habitat indicates that future legislation should be underpinned by a strong mandate to conserve habitat and we recommend that any legislative changes that may reduce habitat protection (e.g. the Fisheries Act [48]) should be reconsidered. Third, to experience conservation benefits from SARA, this law must be fully implemented. Implementation requires that critical habitat be fully identified and subsequently protected for SARA-listed species. The federal government should also be prepared to enact its 'safety net' provision, in the event that species in these regions are not receiving adequate protection to enable recovery (as it did with greater sage-grouse, *Centrocercus urophasianus* [49]).

Finally, even if these recommendations were accepted and put into effect, recovery takes time. Effective management requires that conservation measures be sustained over the long term, even if positive outcomes are not immediately observed.

Supporting Information

Table S1   Summary counts of species across taxonomic groups, for all species assessed more than once by COSEWIC. Values for species listed for >3 generation times (GT) refer to the subset of species for which GTs were reported.

Acknowledgments

We thank Environment Canada for providing us with dates that species were listed under SARA. The idea for this paper originated within a Research Derby conducted at the University of Victoria [50]. The 2013 UVic Research Derby author group consisted of the authors of this paper, as well as A.F. Martin, K. Pawluk, D. Roberts, and J. Robinson, and can be contacted through B. Favaro (brett.favaro@mi.mun.ca). We thank S. Pinkus and three anonymous reviewers for providing useful comments.

Author Contributions

Conceived and designed the experiments: BF DCC CHF CF JJH. Performed the experiments: BF DCC CHF CF JJH. Analyzed the data: BF DCC CHF CF JJH AR. Wrote the paper: BF DCC CHF CF JJH AR.

References

1. Pimm SL, Russell GJ, Gittleman JL, Brooks TM (1995) The future of biodiversity. Science 269: 347–350.

2. Chapin III FS, Zavaleta ES, Eviner VT, Naylor RL, Vitousek PM, et al. (2000) Consequences of changing biodiversity. Nature 6783: 234–242.

3. Rands MRW, Adams WM, Bennun L, Butchart SHM, Clements A, et al. (2010) Biodiversity conservation: Challenges beyond 2010. Science 329: 1298–1303.

4. Arponen A (2012) Prioritizing species for conservation planning. Biodivers Conserv 21: 875–893.

5. Hassan RM, Scholes R, Ash N, editors. (2005) Ecosystems and human well-being: Current state and trends. Findings of the condition and trends working group millennium ecosystem assessment series. Washington DC: Island Press.

6. Hooper DU, Chapin III FS, Ewel JJ, Hector A, Inchausti P, et al. (2005) Effects of biodiversity on ecosystem functioning: A consensus of current knowledge. Ecol Monogr 75: 3–35.

7. Barton J, Pretty J (2010) What is the best dose of nature and green exercise for improving mental health? A multi-study analysis. Environ Sci Tech 44: 3947–3955.

8. Waples RS, Nammack M, Cochrane JF, Hutchings JA (2013) A tale of two acts: Endangered species listing practices in Canada and the United States. Bioscience 63: 723–734.

9. Brooks TM, Mittermeier RA, Mittermeier CG, Da Fonseca GAB, Rylands AB, et al. (2002) Habitat loss and extinction in the hotspots of biodiversity. Conserv Biol 16: 909–923.

10. Kareiva P, Wennergren U (1995) Connecting landscape patterns to ecosystem and population processes. Nature 373: 299–302.

11. Wilcove DS, Rothstein D, Jason Dubow, Phillips A, Losos E (1998) Quantifying threats to imperiled species in the United States. Bioscience 48: 607–615.

12. Homan RN, Windmiller BS, Reed JM (2004) Critical thresholds associated with habitat loss for two vernal pool-breeding amphibians. Ecol App 14: 1547–1553.

13. Taylor MFJ, Sattler PS, Evans M, Fuller RA, Watson JEM, et al. (2011) What works for threatened species recovery? An empirical evaluation for Australia. Biodivers Conserv 20: 767–777.

14. "Secretariat of the Convention on Biological Diversity." (2010) Global biodiversity outlook 3. Montreal. 94 p.

15. COSEWIC (2013) Terms of reference: Committee on the status of endangered wildlife in Canada (COSEWIC), revised and approved by COSEWIC November 2011.

16. *SARA (Species at Risk Act)* (2002) An act respecting the protection of wildlife species at risk in Canada. SC 2002, c 29, s 15. Available: http://Laws-lois.justice.gc.ca/PDF/S-15.3.pdf. Accessed 2014 August 11.

17. *SARA (Species at Risk Act)* (2002) An act respecting the protection of wildlife species at risk in Canada. SC 2002, c 29, s 27. Available: http://Laws-lois.justice.gc.ca/PDF/S-15.3.pdf. Accessed 2014 August 11.

18. *SARA (Species at Risk Act)* (2002) An act respecting the protection of wildlife species at risk in Canada. SC 2002, c 29, ss 32–36. Available: http://Laws-lois.justice.gc.ca/PDF/S-15.3.pdf. Accessed 2014 August 11.

19. *SARA (Species at Risk Act)* (2002) An act respecting the protection of wildlife species at risk in Canada. SC 2002, c 29, ss 34(1)-35(1,2). Available: http://Laws-lois.justice.gc.ca/PDF/S-15.3.pdf. Accessed 2014 August 11.

20. *SARA (Species at Risk Act)* (2002) An act respecting the protection of wildlife species at risk in Canada. SC 2002, c 29, ss 3435. Available: http://Laws-lois.justice.gc.ca/PDF/S-15.3.pdf. Accessed 2014 August 11.

21. *SARA (Species at Risk Act)* (2002) An act respecting the protection of wildlife species at risk in Canada. SC 2002, c 29, ss 41–45. Available: http://Laws-lois.justice.gc.ca/PDF/S-15.3.pdf. Accessed 2014 August 11.

22. Auditor General of Canada (2013) Chapter 6: Recovery planning for species at risk. In: Anonymous Report of the commissioner of the environment and sustainable development. Office of the Auditor General of Canada. pp. 1–22. Available: http://www.oag-bvg.gc.ca/internet/English/parl_cesd_201311_06_e_38676.html. Accessed 2014 August 11.

23. *SARA (Species at Risk Act)* (2002) An act respecting the protection of wildlife species at risk in Canada. SC 2002, c 29, ss 65–72. Available: http://Laws-lois.justice.gc.ca/PDF/S-15.3.pdf. Accessed 2014 August 11.

24. Findlay C, Elgie S, Giles B, Burr L (2009) Species listing under Canada's Species at Risk Act. Conserv Biol 23: 1609–1617.

25. Mooers AO, Doak DF, Findlay CS, Green DM, Grouios C (2010) Science, policy, and species at risk in Canada. Bioscience 60: 843–849.

26. Schultz JA, Darling ES, Côté IM (2013) What is an endangered species worth? Threshold costs for protecting imperilled fishes in Canada. Mar Policy 42: 125–132. http://dx.doi.org/10.1016/j.marpol.2013.01.021.

27. McCune JL, Harrower WL, Avery-Gomm S, Brogan JM, Csergő A, et al. (2013) Threats to Canadian species at risk: An analysis of finalized recovery strategies. Biol Conserv 166: 254–265. http://dx.doi.org/10.1016/j.biocon.2013.07.006.

28. *Western Canada Wilderness Committee, David Suzuki Foundation, Greenpeace Canada, Sierra Club of British Columbia Foundation, and Wildsite v. Minister of Fisheries and Oceans and Minister of the Environment* (2014) Canada Federal Court, 2014 FC 148, T-1777-12. Available: http://cas-ncr-nter03.cas-satj.gc.ca/rss/T-1777-12%20SARA%20decision%2014-02-2014%20ENG.pdf. Accessed 2014 August 11.

29. Lukey JR, Crawford SS (2009) Consistency of COSEWIC species at risk designations: Freshwater fishes as a case study. Can J Fish Aquat Sci 66: 959–971.

30. Lukey JR, Crawford SS, Gillis D (2010) Effect of information availability on assessment and designation of species at risk. Conserv Biol 24: 1398–1406.

31. Powles H (2011) Assessing risk of extinction of marine fishes in Canada - the COSEWIC experience. Fisheries 36: 231–246.

32. Kowalchuk KA, Kuhn RG (2012) Mammal distribution in Nunavut: Inuit harvest data and COSEWIC's species at risk assessment process. Ecol Soc 17: 4.

33. COSEWIC (2013) Database of wildlife species assessed by COSEWIC. Available: http://www.cosewic.gc.ca/eng/sct1/searchform_e.cfm. Accessed 2014 August 11.

34. Government of Canada (2014) Species at risk act public registry. Available: http://www.sararegistry.gc.ca/default_e.cfm. Accessed 2014 August 11.

35. COSEWIC (2011) COSEWIC's Assessment Process and Criteria. Available: http://www.cosewic.gc.ca/pdf/Assessment_process_and_criteria_e.pdf. Accessed 2014 August 11.

36. COSEWIC (2005) Original criteria and definitions used in the status assessment of species from October 1999 to May 2001. Available: http://www.cosewic.gc.ca/eng/sct0/original_criteria_e.cfm. Accessed 2014 August 11.

37. Redford KH, Amato G, Baillie J, Beldomenico P, Bennet EL, et al. (2011) What does It mean to successfully conserve a (vertebrate) species? BioScience 61:39–48.

38. Westwood A, Reuchlin-Hugenholtz E, Keith DM (2014) Re-defining recovery: A generalized framework for assessing species recovery. Biological Conservation 172:155–162.

39. IUCN Standards and Petitions Subcommittee (2014) Guidelines for using the IUCN Red List categories and criteria. Version 11. Available: http://www.iucnredlist.org/documents/RedListGuidelines.pdf. Accessed 2014 August 11.

40. R Core Team (2013) R: A language and environment for statistical computing. R Foundation for Statistical Computing, Vienne, Austria. http://www.R-project.org.

41. Christensen RHB (2013) Ordinal – Regression models for Ordinal Data. R package version 2013.9-30. Available: http://www.cran.r-project.org/package=ordinal/.

42. Tutz G, Hennevogl W (1996) Random effects in ordinal regression models. Comput Stat Data An 22: 537–557.

43. Christensen RHB (2013) A tutorial on fitting Cumulative Link Models with the ordinal package. Available: ftp://ftp.stat.math.ethz.ch/R-CRAN/web/packages/ordinal/vignettes/clm_tutorial.pdf.

44. Taylor MFJ, Suckling KF, Rachlinski JJ (2005) The effectiveness of the endangered species act: A quantitative analysis. Bioscience 55: 360–367.

45. Venter O, Brodeur NN, Nemiroff L, Belland B, Dolinsek IJ, et al. (2006) Threats to endangered species in Canada. Bioscience 56: 903–910.

46. Favaro B, Reynolds JD, Côté IM (2012) Canada's weakening aquatic protection. Science 337: 154.

47. Kirchhoff D, Tsuji LJS (2014) Reading between the lines of the 'Responsible resource development' rhetoric: The use of omnibus bills to 'streamline' Canadian environmental legislation. Impact Assessment and Project Appraisal: 1–13. DOI: 10.1080/14615517.2014.894673.

48. Hutchings JA, Post JR (2013) Gutting Canada's fisheries act: No fishery, no fish habitat protection. Fisheries 38: 497–501.

49. Canada Gazette (2013) Emergency Order for the Protection of the Greater Sage-Grouse. Species At Risk Act. Government of Canada. P.C. 2013–1245. Available: http://canadagazette.gc.ca/rp-pr/p2/2013/2013-12-04/html/sup-eng.php Accessed 2014 July 22.

50. Favaro B, Braun DC, Earth2Ocean Research Derby (2013) The 'research derby': a pressure cooker for creative and collaborative science. Ideas Ecol Evol 6:40–66.

51. Vanderzwaag DL, Hutchings JA (2005) Canada's marine species at risk: Science and law at the helm, but a sea of uncertainties. Ocean Dev Int Law 36: 219–259.

52. Mooers AO, Prugh LR, Festa-Bianchet M, Hutchings JA (2007) Biases in legal listing under Canadian endangered species legislation. Conserv Biol 21: 572–575.

53. Taylor EB, Pinkus S (2013) The effects of lead agency, nongovernmental organizations, and recovery team membership on the identification of critical habitat for species at risk: Insights from the Canadian experience. Environ Rev 21: 93–102.

13

Interspecific Neighbor Interactions Promote the Positive Diversity-Productivity Relationship in Experimental Grassland Communities

Yuhua Zhang, Yongfan Wang*, Shixiao Yu

Department of Ecology, School of Life Sciences/State Key Laboratory of Biocontrol, Sun Yat-sen University, Guangzhou, China

Abstract

Because the frequency of heterospecific interactions inevitably increases with species richness in a community, biodiversity effects must be expressed by such interactions. However, little is understood how heterospecific interactions affect ecosystem productivity because rarely are biodiversity ecosystem functioning experiments spatially explicitly manipulated. To test the effect of heterospecific interactions on productivity, direct evidence of heterospecific neighborhood interaction is needed. In this study we conducted experiments with a detailed spatial design to investigate whether and how heterospecific neighborhood interactions promote primary productivity in a grassland community. The results showed that increasing the heterospecific: conspecific contact ratio significantly increased productivity. We found there was a significant difference in the variation in plant height between monoculture and mixture communities, suggesting that height-asymmetric competition for light plays a central role in promoting productivity. Heterospecific interactions make tall plants grow taller and short plants become smaller in mixtures compared to monocultures, thereby increasing the efficiency of light interception and utilization. Overyielding in the mixture communities arises from the fact that the loss in the growth of short plants is compensated by the increased growth of tall plants. The positive correlation between species richness and primary production was strengthened by increasing the frequency of heterospecific interactions. We conclude that species richness significantly promotes primary ecosystem production through heterospecific neighborhood interactions.

Editor: Andrew Hector, University of Oxford, United Kingdom

Funding: The research was funded by the National Natural Science Foundation of China (Project 30970472, 31230013) and by the Zhang-Hongda Science Foundation in Sun Yat-sen University. The funders had no role in study design, data collection and analysis, decision to publish, or preparation of the manuscript.

Competing Interests: The authors have declared that no competing interests exist.

* Email: lsswyf@mail.sysu.edu.cn

Introduction

Understanding the role of biodiversity in promoting ecosystem functions, such as primary production, is critically important to biodiversity conservation and ecosystem management. Empirical studies showed that diversity-productivity relationships can take various forms [1,2,3,4]. However, positive relationships, the productivity increases with diversity, have been overwhelmingly documented by many manipulative biodiversity experiments [5,6,7,8,9,10,11].

Complementarity and selection effects are the two primary mechanisms used to interpret the positive diversity-productivity relationships [12]. The complementarity effect results from resource partitioning, natural enemy regulation or facilitative interactions between species in mixture communities [7,13,14,15], while the selection effect is due to shifts in dominance driven by heterospecific competition [16,17,7]. Although the selection effect does play a role, its effect is variable and its importance often tends to decrease over time, leaving the complementarity effect as the main factor explaining the positive effect of biodiversity on ecosystem functioning [12,13,8,9,18,19]. A critical, but largely unresolved question is: what are the biological mechanisms that drive the complementarity effect and how do they enhance productivity in mixture communities?

The main difference between mixtures and monocultures is that the former are subject to heterospecific interaction effects while the latter experience only conspecific competition. Heterospecific interactions must produce more improvements on average than conspecific interactions when generating positive effects on ecosystem functioning. They must also ensure the coexistence of the species present, which links the biodiversity–ecosystem functioning and species coexistence theories, albeit in complex ways [20,15]. Therefore, resource partitioning and facilitation have often been used to explain the positive effects of biodiversity on ecosystem functioning, whether through belowground processes that lead to enhanced soil nutrient utilization [21,22,23] or by reduced competition for light [24,25]. Variation in traits, such as plant height, plant architecture effects on light competition, and rooting depth and spread for water or nutrient utilization, are essential complementarity effects that can produce overyielding in mixtures.

Heterospecific competition can further increase heterospecific trait variation in mixtures, either through niche shifts in the presence of heterospecific competitors or by differential growth, thereby increasing the potential for complementarity between species and overyielding in mixtures. For instance, under light competition, short plants may become shorter while tall plants become taller in mixtures, compared to monocultures, so that the loss of biomass by the inferior species is compensated by the biomass gain in the superior species. Consequently, both negative (mainly for short species) and positive (mainly for tall species) neighborhood interspecific interactions can be observed in a mixture, e.g., as in the case of soil nitrate competition [21].

Experiments have been conducted to test for the effect of interspecific interactions on biomass [26,27,21,25,28]. Such experiments are commonly designed to compare biomass in plots where the seeds have been sown by random broadcasting versus that of aggregated sowing. Most results have shown that biomass in randomly sown plots is higher than that in aggregated sown plots due to the more efficient use of soil nutrients or light in the former case, presumably due to the higher frequency of heterospecific interactions [24,21,23,25,28,29]. However, in these studies, the plots are considered the basic experimental unit and sowing method (dispersed versus aggregated) as a treatment factor. No data on the frequency of heterospecific neighbor interactions within the plots were collected. Such data are essential for directly inferring the effects of heterospecific interactions [6].

We conducted an intensive study to investigate the effect of interspecific interactions on plant growth in grassland communities and to show how that gives rise to a positive diversity-productivity relationship. The study consisted of two complementary experiments. The first experiment was designed to measure the growth of plants in each grid within a series of plots that varied in sowing density, species richness and frequency of heterospecific interactions. The second experiment enhanced the first one by adding more diversity levels. Based on these, we first modeled the effects of species richness and the frequency of heterospecific interactions on plant growth across the plots. We then tested the hypothesis that the positive effect of heterospecific interactions on productivity in grassland communities was driven by height-asymmetric competition for light in mixture plots.

Materials and Methods

Ethics Statement

The study site is maintained by Heerkou Town, Fengkai County, Guangdong Province, China. The site was leased to Sun Yat-sen University from 2009 to 2013 for conducting the grassland experiments reported in this study. Our study did not involve any damage to land resources and no specific permissions were required for this research. Our experiments didn't contain any treatments with chemical addition. The seeds of all plant species used in our experiments were collected from the area around the experimental site, and these plants were neither endangered nor protected species.

Experimental Design

The experimental site was located in a subtropical arable field near the Heishiding Nature Reserve (111°53′ E, 23°26′ N), Guangdong Province, China [30]. The soil was a ferralosol. The first experiment was established in 2009 and comprised of nine blocks involving eight species (Fig. 1a). Each block consisted of 36 plots (each 1 m×1 m) that varied in sowing density, number of species and spatial pattern. Sowing density across plots within a block varied from low (64 grids/plot), to medium (144 grids/plot) to high (256 grids/plot) (Fig. 1a). There were two diversity levels (monoculture and 8 species mixtures). Each of the eight species had three monocultures with densities varying from low, medium

to high in each of the nine blocks, which made a total of 24 monoculture plots per block. In addition, there were two spatial patterns (aggregated and dispersed) for each density level (Fig. 1a) and the aggregated and dispersed patterns were replicated twice, making a total of 12 plots (i.e., 2 spatial patterns×3 densities×2 replicates). The dispersed spatial pattern was designed to maximize heterospecific interactions. In total, there were 324 plots across the nine blocks for this experiment. All of the 324 plots were hand-seeded in February 2009 and 10 seeds of each species were sown in each grid of a plot. In total, 640, 1440, 2560 seeds, respectively, were sown in each plot in order to represent the three density treatment levels. The eight species used in this experiment were all native species and were randomly selected from the local area. They were *Ambrosia artemisiifolia* (Compositae), *Urena lobata* (Malvaceae), *Triumfetta rhomboidea* (Tiliaceae), *Bidens pilosa* (Compositae), *Mosla dianthera* (Labiatae), *Pennisetum alopecuroides* (Gramineae), *Epimeredi indica* (Labiatae) and *Corchorus capsularis* (Tiliaceae), which were respectively referred to as A, B, C, D, E, F, G, H (Fig. 1a).

The second experiment was established in a nearby arable field in 2010. It had five blocks. Each block comprised of 32 plots, including eight monocultures (each for one of the eight species) and six mixtures that had two spatial patterns (aggregated and dispersed) and three diversity levels (two, four and eight species) (Fig. 1b). Each of the six mixtures was replicated four times. In total, there were 160 plots across the five blocks for this experiment. In contrast to the first experiment, the sowing density per plot was fixed at 64 grids, while the diversity levels were either one, two, four or eight species (Fig. 1b). Another difference in this experiment was that the eight species in each plot were not fixed but were randomly selected from a species pool of 15 species. The species pool included all the eight species included in the first experiment plus seven new species: *Chenopodium ambrosioides* (Chenopodiaceae), *Cassia occidentalis* (Leguminosae), *Cassia tora* (Leguminosae), *Keiskea australis* (Labiatae), *Lespedeza cuneata* (Leguminosae), *Elsholtzia ciliata* (Labiatae) and *Sida acuta* (Malvaceae). All 160 plots were hand-seeded in April 2010 and 10 seeds per species were sown in each plot grid.

Biomass Harvesting

The plots were weeded monthly during the study, but the weeds were not measured and were thus excluded from the study. The plots were harvested at the end of the growing season (late September in 2009 and 2010 for the respective experiments). For experiment I, the height of the tallest plant in each plot grid was measured. The plants in each grid were then gently removed, and the soil was shaken and washed off. The roots were separated from the shoots. The clipped shoots for each grid were weighed after oven drying at 80°C for 48 hours. The clipped roots for each species were combined across all the grids for each plot and weighed after oven drying at 80°C for 48 hours. This meant that in this experiment both the aboveground and belowground biomasses were measured for each plot. This experiment consisted of two diversity levels (monocultures versus eight species mixtures). All plants in one plot were dead at before harvesting and this plot was excluded from analysis. At the end, experiment I only had 323 plots (see data in Table S1).

For experiment II, aboveground biomass for each species was harvested for each plot (belowground biomass was not measured). The plants of each species in each plot were weighed after oven drying at 80°C for 48 hours. This experiment had four species richness levels (one, two, four, and eight species) compared to the first experiment that had two species richness levels (one and eight species). This experiment complemented the first one by showing

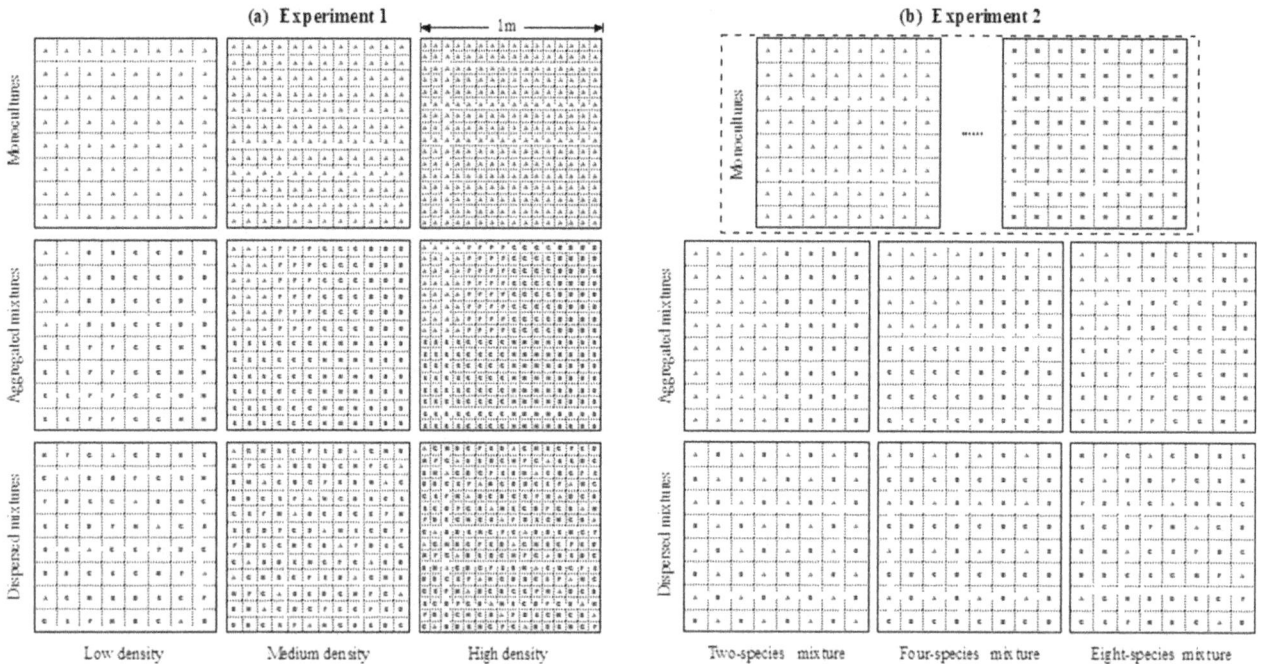

Figure 1. Experimental esign of the two experiments. (*a*) The first experiment comprised 9 blocks involving 8 species. Each block consisted of 36 plots (1×1 m in size). Illustrated here is one block but only shows 9 plots. The other 27 plots are not shown, including 21 monoculture plots for other 7 species (each being sown at low, medium and high density) and six mixcultures (i.e., the two spatial configurations for the 3 density levels, the same as the second and third rows). The aggregated and dispersed mixcultures both had the same 8 species but different spatial configurations. (*b*) The second experiment comprised 5 blocks involving 8 species. Each block consisted of 32 1×1 m plots. Illustrated here is one block but only shows 8 plots. The other 24 plots are not shown, including 6 monoculture plots for the other 6 species and 3 replicates for each of the six mixcultures (i.e., the two spatial configurations for the 3 diversity level). In both experiments, blocks were separated by 2 m walkways, and plots were separated 1 m apart.

how productivity changed with species richness at different levels of heterospecific interactions in the aggregated versus dispersed plot designs.

Data Analysis

We modeled the effects of species richness and the frequency of heterospecific neighbor interactions on plant growth. The analysis was based on the biomass data measured at the plot level for both experiment I (for above and belowground biomass) and experiment II (for aboveground biomass only). For experiment I (Fig. 1*a*), there were three density levels (64, 144 and 256 grids) and two diversity levels (one versus eight species). Spatial patterns in this experiment were either monocultures, aggregated mixtures or dispersed mixtures. Monocultures had no heterospecific interactions, aggregated mixtures had an intermediate heterospecific interaction frequency and dispersed mixtures had the greatest heterospecific interaction frequency.

Multiple linear regression (i.e., ANOVA) can be used to model biomass by considering the frequency of heterospecific interactions, species richness and plant density as explanatory variables. A more efficient approach is to treat plant density as a grouping factor and use a mixed effect model to show biomass changes (although both models led to the same conclusions). The mixed effect model was:

$$y_{ij} = \beta_0 + \beta_1 \text{richness}_{ij} + \beta_2 \text{heterospecific}_{ij} + d_i, \quad d \sim N(0,\sigma^2) \quad (1)$$

where y_{ij} is biomass at density i and plot j, heterospecific$_{ij}$ is the average number of heterospecific neighbors around a focal plant and d_i is plant density as a random effect. The experimental design

does not permit modeling the interactive term between *richness* and *interspecific* because when *richness* = 1, no aggregated or dispersed mixtures exist. This mixed effect model (1) was used to model both the aboveground and belowground biomass in experiment I, in terms of species richness and heterospecific interactions. All the data were Box-Cox transformed to ensure normality (see the transformed data were enclosed in Table S2). Because a multiple regression is a "partial" regression, β_2 describes the effect of heterospecific interactions on biomass, given that the effects of species richness and density are already accounted for.

In a similar manner to experiment I, experiment II had the same three levels of spatial patterns (monocultures, aggregated and dispersed mixtures) (Fig. 1*b*). Different from experiment I, it only had one density level (64 grids) for all 160 plots and four species richness levels (one, two, four or eight species). The model was:

$$y_i = \beta_0 + \beta_1 \text{richness}_i + \beta_2 \text{heterospecific}_i + \varepsilon_i, \quad \varepsilon \sim N(0,\sigma^2), \quad (2)$$

where y_i is biomass in plot i. The biomass was Box-Cox transformed to achieve normality.

To test the hypothesis that height-asymmetric competition for light is responsible for overyielding in mixture plots, we compared the root/shoot biomass ratio, plant height and height variance for monocultures, aggregated mixtures and dispersed mixtures. If competition for light is a determining factor, we should expect the variation in height to increase from monocultures to aggregated and finally to dispersed mixtures, while the average height remains largely unchanged. We should also see that heterospecific competition would make tall plants grow taller and for short plants to become smaller in mixture plots compared to monocultures.

Statistical program R (http://www.r-project.org/) was used to analyze the data. Package "nlme" was used to estimate model (1).

Results

For the first experiment, highly significant positive effects for both diversity and the frequency of interspecific interactions on aboveground biomass were demonstrated with the mixed effect model (1), while the effects of diversity and the frequency of interspecific interaction on belowground biomass were weaker (Table 1). The relationships between species richness and biomass and between the frequency of heterospecific interactions and biomass for experiment I are shown in Fig. 2. It is clear both relationships were positive.

For the second experiment, the results of the multiple linear regression model (2) showed that both diversity and the frequency of heterospecific neighbor interactions significantly affected plot biomass (Table 1). The positive effects of species richness and heterospecific interactions on biomass are shown in Fig. 2. We can also show how biomass changed with species richness for the aggregated and dispersed mixtures by comparing the observed biomass of the mixture plots against the mean biomass averaged from all the monoculture plots (Fig. 3). For the aggregated pattern, there was a positive relationship between species richness and aboveground biomass ($r^2 = 0.05$, $P = 0.026$) (Fig. 3a). This positive relationship became stronger in the dispersed pattern ($r^2 = 0.17$, $P < 0.0001$), although the slope of this relationship is only marginally significantly higher than that of the aggregated pattern ($t = -1.782$, $P = 0.0763$; Fig. 3a). Figure 3b shows the difference between the observed biomass for mixture plots (aggregated and dispersed) and the mean biomass averaged from all the monoculture plots. For the aggregated mixtures, the difference did not significantly increase with species richness ($r^2 = 0.04$, $P = 0.123$) although there is a noticeable trend of increase. For the dispersed mixtures, the difference significantly increased with species richness ($r^2 = 0.19$, $P = 0.0006$). The slope of the linear relationship for the dispersed pattern was significantly (marginally) higher than the slope for the aggregated pattern ($t = -1.917$, $P = 0.0577$; Fig. 3b). These results indicated that increasing the heterospecific interaction frequency strengthened the positive diversity-productivity relationship.

The root to shoot ratio of the plants in experiment I is shown in Fig. 4a. The ratio consistently decreased from monocultures, to aggregated and finally to dispersed mixtures. The ratio for the monoculture plots was significantly higher than that for the aggregated and dispersed mixture plots ($P < 0.05$), although the

latter two were not significantly different. These results indicated that significantly more biomass was allocated to the shoots than to the roots in mixture plots than in monocultures. Although plant height showed no significant difference among the three types of plots, the within-plot variance in height increased substantially in the dispersed plots (Fig. 4b). At the individual species level, it is obvious that short plants became shorter as we move from monocultures to aggregated and finally to dispersed mixtures, while tall plants tended to become taller along the gradient (Fig. 4c).

Discussion

An inevitable outcome of any increase in species diversity is the increase in the frequency of heterospecific interactions. Understanding how heterospecific interactions affect productivity lies at the heart of the diversity-productivity studies. Niche differentiation, facilitation and frequency-dependent growth are among the major mechanisms used to explain the positive effects of biodiversity on ecosystem functioning [24,7,5,13,21,18]. These mechanisms can enhance the utilization of soil nutrients [21,22,23] and reduce competition for resources [24,25]. Central to the understanding of these mechanisms is how heterospecific neighborhood interactions promote the productivity of multispecies communities.

Our results showed that there was a strong residual effect of heterospecific neighbor interactions after the diversity effects had been taken into account and an equally strong residual diversity effect after the number of heterospecific neighbors had been taken into account for aboveground biomass in both experiments (Table 1). Similar results held for belowground biomass in experiment I although the effects of diversity and heterospecific interactions were quite weak. These results suggested that it was important to include both diversity and neighbor interactions when explaining the positive diversity-productivity relationship. For example, for the belowground biomass in experiment I, if the frequency of heterospecific interactions were excluded from model (1), the effect of richness would be extremely significant with P virtually being 0 (Fig. 2). The same interpretations applied to the results for aboveground biomass of both experiments where the effect of species richness on biomass for the analysis without inclusion of the heterospecific interaction term was significantly higher than that with the heterospecific term ($P < 0.001$, log-likelihood ratio test). The inclusion of heterospecific interactions decreased the effect of species richness because species richness and the frequency of heterospecific interactions were highly correlated. Their correlations for experiment I was $R^2 = 0.86$ (high because there were only two species richness levels) and $R^2 = 0.25$

Table 1. Results of the mixed effects model for experiment I and of multiple regression model for experiment II for testing the effects of species richness and interspecific interactions on plot biomass for each experiment.

Experiment	β_0 (intercept)	β_1 (species richness)	β_2 (interspecific interactions)
Experiment I: Above-ground biomass	40.579 ± 2.351***	1.351 ± 0.416***	4.932 ± 1.630***
Experiment I: Below-ground biomass	32.577 ± 1.921***	0.828 ± 0.387*	$2.644 \pm 1.516^{\dagger}$
Experiment II: Above-ground biomass	19.811 ± 0.835***	0.843 ± 0.192***	0.766 ± 0.331*

The Box-Cox transformed biomass were used in the models. The values ($\pm$ SE) are regression coefficients. ***for $P < 0.001$, **for $P < 0.01$, *for $P < 0.05$, and †for $P < 0.1$.

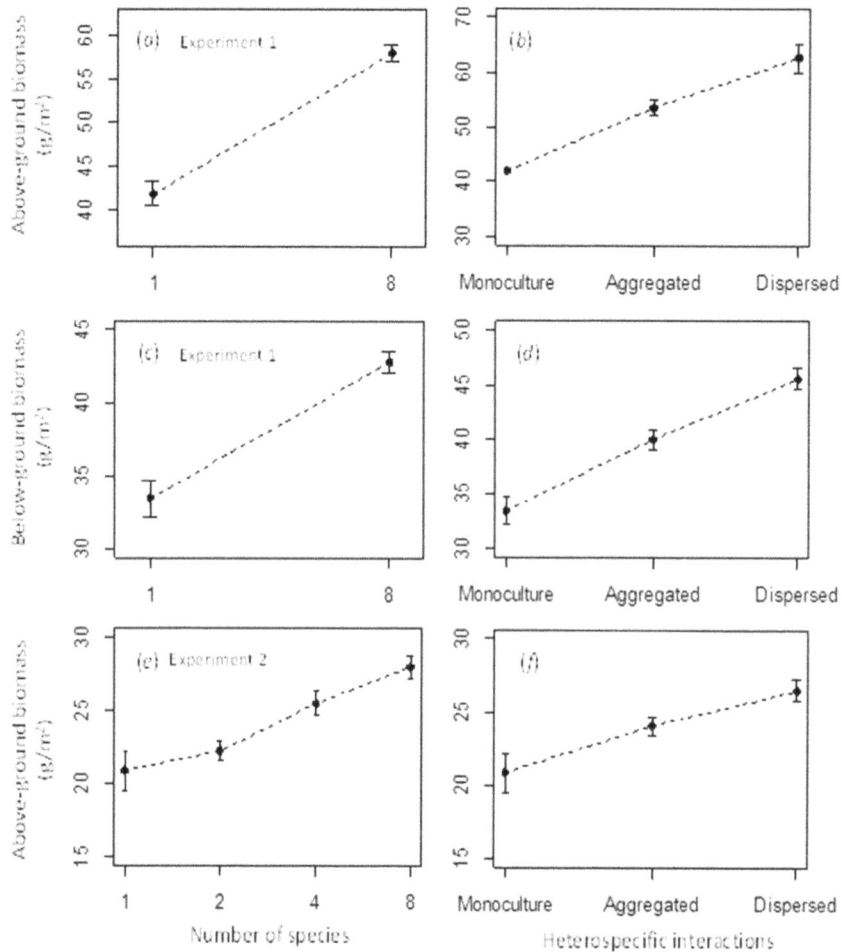

Figure 2. Effects of the number of species and heterospecific interactions on above- and belowbiomass. Mean ± SE for the Box-Cox transformed above-ground biomass (*a* and *b*) and below-ground biomass (*c* and *d*) of experiment I and for the Box-Cox transformed above-ground biomass for experiment II (*e* and *f*). The x-axis label on the left column is the number of species. Experiment I had 2 richness levels (1 and 8 species). Experiment II had four richness levels (1, 2, 4 and 8 species).

for experiment II (four richness levels). It is remarkable to note that the dispersed mixtures significantly enhanced the effect of species richness on productivity compared to the aggregated mixtures (Fig. 3), which provided unequivocal evidence for the importance of heterospecific interactions on productivity.

It is, however, worth noting that the effect of species richness is not entirely explained by the frequency of heterospecific interactions. There are three possible reasons why the richness effect is not amount to the effect of the frequency of heterospecific interactions. First, the frequency of heterospecific interactions depends on the spatial distribution of conspecifics within plot. For a given number of species in a plot, the frequency of heterospecific interactions is small for aggregated conspecifics compared to dispersed conspecifics. Second, because the frequency of heterospecific interactions is a simple count of neighborhoods, it does not necessarily represent the genuine competitive (or facilitative) interactions between heterospecific individuals. To measure the competitive (or facilitative) interactions, distance between neighbors has to be considered. Third, the interactive strength is different between individuals of different species, e.g., the neighborhood interaction between individuals of species A and B is not equivalent to the interaction between the individuals of species B and C. This "diffused" neighborhood interactions

among species could also contribute to the difference in the residual effects between richness and the frequency of heterospecific interactions.

Both theoretical and empirical studies have shown that neighbor interactions play a major role in shaping community structure, species coexistence and individual performance [31,32,33] and thus are central to understanding complementarity effects [24,21,18,25,28,29]. Hille Ris Lambers et al. [21] showed that heterospecific interactions are an important complementarity mechanism that results in a more efficient use of soil nitrate. The effect of heterospecific interactions could be both positive (facilitation by legume species) and negative (competition with C4 species) [21]. Our study also showed that heterospecific interactions that may not be due to belowground heterospecific interactions, but were a result of neighborhood competition for light, play key roles in enhancing community productivity. The results in Table 1 show that, in addition to heterospecific interactions, diversity also significantly contributed to increasing productivity.

Several mechanisms can explain the effect of neighbor interactions on diversity-productivity relationships. Our study suggested that in grassland communities, variation in plant height played an important role. Although there was little change in

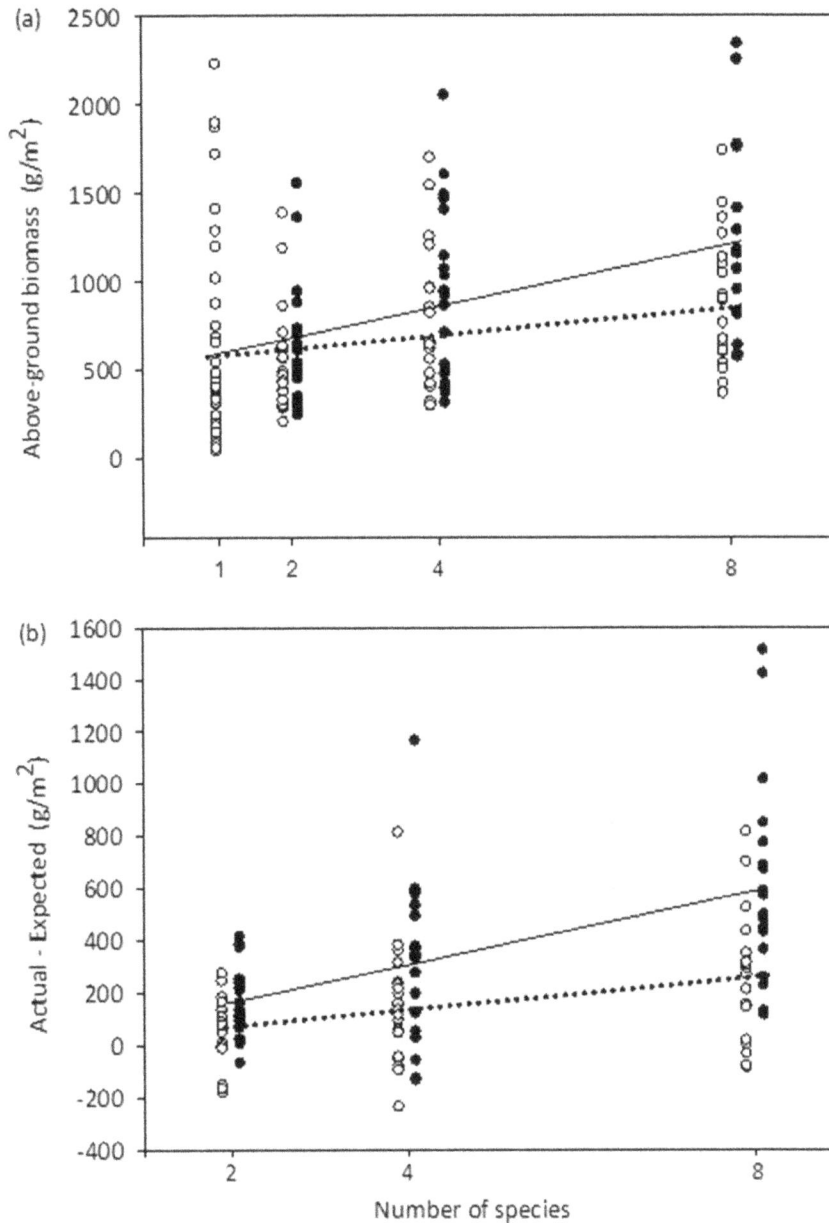

Figure 3. Overyielding and the magnitude of complementary effects. (*a*) Linear relationships between aboveground biomass and species richness for aggregated (dashed lines and open circles) and dispersed mixtures (solid lines and filled circles) of experiment II. (*b*) Difference in aboveground biomass between the observed mixture plots and the mean monoculture biomass of all species across diversity gradients for experiment II. This difference measures overyielding and the degree of difference indicates the degree of complementarity effects. Dashed lines and open circles refer to plots with aggregated mixtures, and solid lines and filled circles refer to dispersed plots.

mean plant height, the variation in height increased substantially from monoculture to aggregated and finally to dispersed mixtures (Fig. 4*b*). This increase in height variation means that the vertical structure from monoculture to aggregated and dispersed mixtures became more complex and heterogeneous, which probably increased the efficient interception and utilization of light. This complexity in complementary space occupancy is predicted by theory and can occur both belowground [34] and aboveground [35]. Overyielding in mixture plots could occur if tall plants become taller while short plants become shorter in more dispersed mixture plots, compared to less dispersed mixtures or monoculture plots, and the reduced growth of the short plants in mixture plots is

compensated by the increased growth of tall plants. This was observed in our experiments (Fig. 4*c*) and means that hetero-specific interactions in mixture plots led to the differentiation in height growth and thus to a more efficient use of light resources. We would further hypothesize that heterospecific neighbor interactions will also make shallow-root species produce shallower roots and deeper root species produce deeper roots, thus optimizing the use of soil water and nutrients. Overyielding is achieved by compensating the growth loss in the shallow root species by the growth gain in the deeper root species. We conclude that no matter what complementarity mechanisms invoke the positive diversity-productivity relationship, a differentiation in

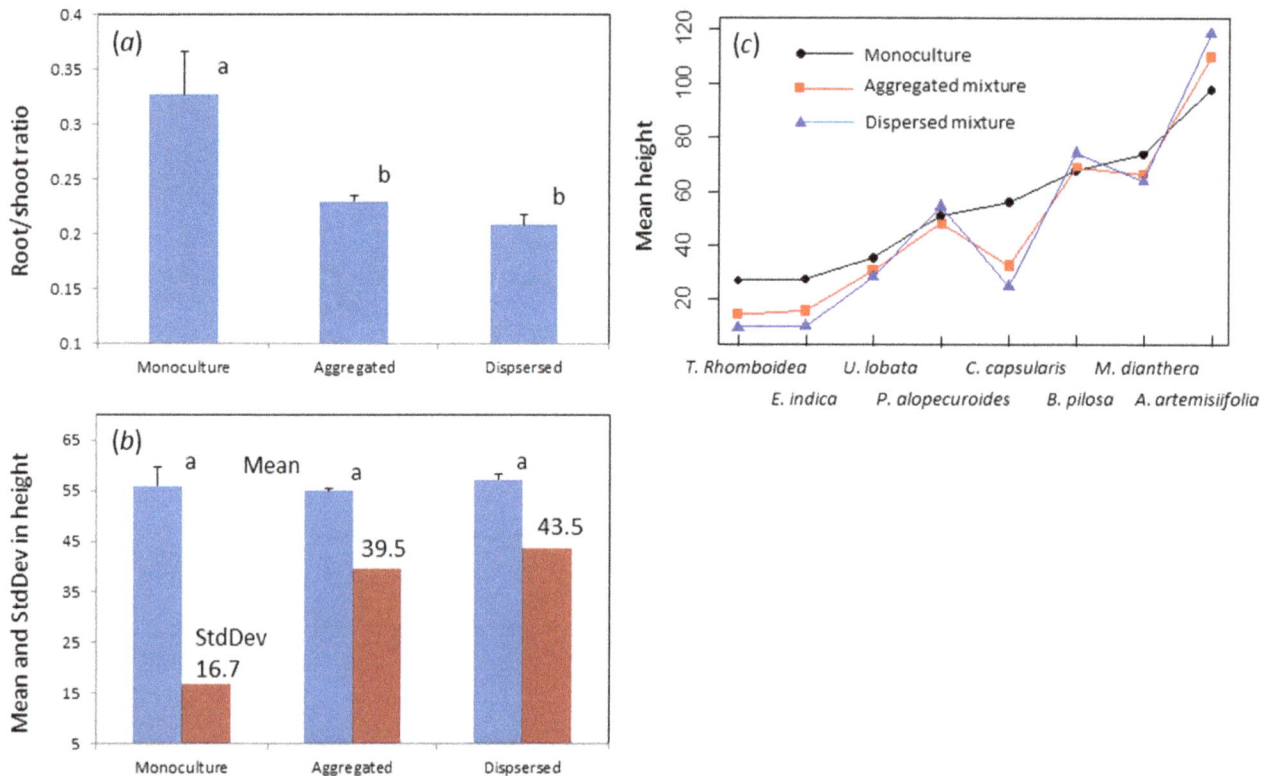

Figure 4. Biomass allocation and plant height variation for experiment I. (*a*) The mean root/shoot ratio varied with heterospecific interactions from monocultures, aggregated to dispersed mixture plots. (*b*) The within-plot mean (left bars) and standard deviation (right bars) in plant height (in cm, calculated over all grids in each plot) versus heterospecific interactions. (*c*) The average height for each of the eight species varied from monocultures, aggregated mixtures to dispersed mixtures. Data bars for the root/shoot ratio in (*a*) and the mean height in (*b*) are mean+1 standard error. Bars with different letters above are significantly different at $P = 0.05$. The values for the standard deviation for height are shown in (*b*).

plant growth (either aboveground or belowground or both) must be seen for the relationship to hold.

Acknowledgments

We are grateful to Weinan Ye, Xianggang Shi, Xubing Liu and Jie Li for their help in the field. Michel Loreau and Fangliang He gave very helpful suggestions on an earlier version of the manuscript.

Author Contributions

Conceived and designed the experiments: YHZ YFW. Performed the experiments: YHZ YFW. Analyzed the data: YHZ YFW SXY. Contributed reagents/materials/analysis tools: YHZ YFW SXY. Contributed to the writing of the manuscript: YHZ YFW SXY.

References

1. Huston MA, Aarssen LW, Austin MP, Cade BS, Fridley JD, et al. (2000) No consistent effect of plant diversity on productivity. Science 289: 1255.

2. Díaz S, Cabido M (2001) Vive la difference: plant functional diversity matters to ecosystem processes. Trends Ecol Evol 16: 646–655.

3. Mittelbach GG, Steiner CF, Scheiner SM, Scheiner SM, Gross KL, et al. (2001) What is the observed relationship between species richness and productivity? Ecology 82: 2381–2396.

4. Whittaker RJ (2010) Meta-analyses and mega-mistakes: calling time on meta-analysis of the species richness-productivity relationship. Ecology 91: 2522–2533.

5. Hector A, Schmid B, Beierkuhnlein C, Caldeira MC, Diemer M, et al. (1999) Plant diversity and productivity experiments in European grasslands. Science 286: 1123–1127.

6. Loreau M, Naeem S, Inchausti P, Bengtsson J, Grime JP, et al. (2001) Biodiversity and ecosystem functioning: current knowledge and future challenges. Science 294: 804–808.

7. Tilman D, Lehman CL, Thomson KT (1997) Plant diversity and ecosystem productivity: theoretical considerations. Proc. Natl Acad Sci USA 94: 1857–1861.

8. Spehn EM, Hector A, Joshi J, Scherer-Lorenzen M, Schmid B, et al. (2005) Ecosystem effects of biodiversity manipulations in European grasslands. Ecol Monogr 75: 37–63.

9. van Ruijven J, Berendse F (2005) Diversity-productivity relationships: initial effects, long-term patterns, and underlying mechanisms. Proc Natl Acad Sci USA 102: 695–700.

10. Maron JL, Marler M, Klironomos JN, Cleveland CC (2011) Soil fungal pathogens and the relationship between plant diversity and productivity. Ecol Lett 14: 36–41.

11. Paquette A, Messier C (2011) The effect of biodiversity on tree productivity: from temperate to boreal forests. Global Ecol and Biogeogr 20: 170–180.

12. Loreau M, Hector A (2001) Partitioning selection and complementarity in biodiversity experiments. Nature 412: 72–76.

13. Tilman D, Reich PB, Knops J, Wedin D, Mielke T, et al. (2001) Diversity and productivity in a long-term grassland experiment. Science 294: 843–845.

14. Hector A (1998) The effect of diversity on productivity: detecting the role of species complementarity. Oikos 82: 597–599.

15. Loreau M, Sapijanskas J, Isbell F, Hector A (2012) Niche and fitness differences relate the maintenance of diversity to ecosystem function: comment. Ecology 93: 1482–1487.

16. Aarssen LW (1997) High productivity in grassland ecosystems: effected by species diversity or productive species? Oikos 80: 182–183.

17. Huston MA (1997) Hidden treatments in ecological experiments: re-evaluating the ecosystem function of biodiversity. Oecologia 110: 449–460.

18. Cardinale BJ, Wright JP, Cadotte MW, Carroll IT, Hector A, et al. (2007) Impacts of plant diversity on biomass production increase through time because of species complementarity. Proc Natl Acad Sci USA 104: 18123–18128.

19. Marquard E, Weigelt A, Temperton VM, Roscher C, Schumacher J, et al. (2009) Plant species richness and functional composition drive overyielding in a six-year grassland experiment. Ecology 90: 3290–3302.

20. Carroll IT, Cardinale BJ, Nisbet RM (2011) Niche and fitness differences relate the maintenance of diversity to ecosystem function. Ecology 92: 1157–1165.

21. Hille Ris Lambers J, Harpole WS, Tilman D, Knops J, Reich PB (2004) Mechanisms responsible for the positive diversity-productivity relationship in Minnesota grasslands. Ecol Lett 7: 661–668.

22. Fargione J, Tilman D (2006) Plant species traits and capacity for resource reduction predict yield and abundance under competition in nitrogen-limited grassland. Funct Ecol 20: 533–540.

23. Fargione J, Tilman D, Dybzinski R, Hille Ris Lambers J, Clark C, et al. (2007) From selection to complementarity, shifts in the causes of biodiversity-productivity relationships in a long-term biodiversity experiment. Proc R Soc Lond B 274: 871–876.

24. Naeem S, Thompson LJ, Lawler SP, Lawton JH, Woodfin RM (1994) Declining biodiversity can alter the performance of ecosystems. Nature 368: 734–737.

25. Potvin C, Dutilleul P (2009) Neighborhood effects and size-asymmetric competition in a tree plantation varying in diversity. Ecology 90: 321–327.

26. Mulder CPH, Uliassi DD, Doak DF (2001) Physical stress and diversity-productivity relationships: The role of positive interactions. Proc Natl Acad Sci USA 98: 6704–6708.

27. Cardinale BJ, Palmer MA, Collins SL (2002) Species diversity enhances ecosystem functioning through interspecific facilitation. Nature 415: 426–429.

28. Lamošová T, Doležal J, Lanta V, Lepš J (2010) Spatial pattern affects diversity-productivity relationships in experimental meadow communities. Acta oecologica 36: 325–332. Bruno JF, Stachowicz JJ, Bertness MD (2003) Inclusion of facilitation into ecological theory. Trends Ecol Evol 18: 119–125.

29. Sapijanskas J, Potvin C, Loreau M (2013) Beyond shading: litter production by neighbors contributes to overyielding in tropical trees. Ecology 94: 941–952.

30. Wang YF, Yu SX, Wang J (2007) Biomass-dependent susceptibility to drought in experimental grassland communities. Ecol Lett 10: 401–410.

31. Goldberg DE, Barton AM (1992) Patterns and consequences of interspecific competition in natural communities: a review of field experiments with plants. Am Nat 139: 771–801.

32. Stoll P, Prati D (2001) Intraspecific aggregation alters competitive interactions in experimental plant communities. Ecology 82: 319–327.

33. Bruno JF, Stachowicz JJ, Bertness MD (2003) Inclusion of facilitation into ecological theory. Trends Ecol Evol 18: 119–125.

34. Loreau M (1998) Biodiversity and ecosystem functioning: A mechanistic model. Proc Natl Acad Sci USA: 5632–5636.

35. Yachi S, Loreau M (2007) Does complementary resource use enhance ecosystem functioning? A model of light competition in plant communities. Ecol Lett 10: 54–62.

14

Modeling Pollinator Community Response to Contrasting Bioenergy Scenarios

Ashley B. Bennett[1], Timothy D. Meehan[2], Claudio Gratton[2], Rufus Isaacs[1]*

1 Department of Entomology and Great Lakes Bioenergy Research Center, Michigan State University, East Lansing, Michigan, United States of America, **2** Department of Entomology and Great Lakes Bioenergy Research Center, University of Wisconsin - Madison, Madison, Wisconsin, United States of America

Abstract

In the United States, policy initiatives aimed at increasing sources of renewable energy are advancing bioenergy production, especially in the Midwest region, where agricultural landscapes dominate. While policy directives are focused on renewable fuel production, biodiversity and ecosystem services will be impacted by the land-use changes required to meet production targets. Using data from field observations, we developed empirical models for predicting abundance, diversity, and community composition of flower-visiting bees based on land cover. We used these models to explore how bees might respond under two contrasting bioenergy scenarios: annual bioenergy crop production and perennial grassland bioenergy production. In the two scenarios, 600,000 ha of marginal annual crop land or marginal grassland were converted to perennial grassland or annual row crop bioenergy production, respectively. Model projections indicate that expansion of annual bioenergy crop production at this scale will reduce bee abundance by 0 to 71%, and bee diversity by 0 to 28%, depending on location. In contrast, converting annual crops on marginal soil to perennial grasslands could increase bee abundance from 0 to 600% and increase bee diversity between 0 and 53%. Our analysis of bee community composition suggested a similar pattern, with bee communities becoming less diverse under annual bioenergy crop production, whereas bee composition transitioned towards a more diverse community dominated by wild bees under perennial bioenergy crop production. Models, like those employed here, suggest that bioenergy policies have important consequences for pollinator conservation.

Editor: Shuang-Quan Huang, Central China Normal University, China

Funding: This work was funded by the Department of Energy Great Lakes Bioenergy Research Center (DOE BER Office of Science DE-FC02-07ER64494), DOE OBP Office of Energy and Renewable Energy (DE-AC05-76RL01830), and by a USDA-NIFA grant to CG and RI(2012-67009-20146). The funders had no role in the study design, data collection and analysis, decision to publish, or preparation of the manuscript.

Competing Interests: The authors have declared that no competing interests exist.

* Email: isaacsr@msu.edu

Introduction

Demand for sustainable sources of energy has spurred increasing interest in bioenergy crops as a fuel source. In the United States, government mandates to increase biofuel production to 36 billion gallons per year by 2022 are advancing research into the production and sustainability of both first- and second-generation biofuels [1,2]. First-generation biofuels are produced from annual row crops such as corn, soybean, and canola, while second-generation cellulosic biofuels can be produced from corn stover, switchgrass, or mixed grasslands, a combination of warm-season grasses and forbs [3]. These contrasting options for biofuel cropping systems have the potential to dramatically alter the types and perenniality of vegetative cover in agricultural landscapes, significantly affecting wildlife [4]. Because policies that promote biofuel production have the potential to cause large scale changes in land use [5,6], identifying how bioenergy crops affect biodiversity will be critical to developing sustainable biofuel policies appropriate for regional implementation across the United States.

A growing body of research suggests that bioenergy crops differentially support biodiversity. For example, bird abundance and richness was consistently higher in perennial grassland biofuel plantings compared to annual biofuel plantings, with expanded grasslands predicted to increase bird richness by 12–207% [4,7]. Similarly, predatory arthropods increased in abundance and diversity in perennial grassland biofuel crops compared to corn monocultures [8,9]. Transitioning the landscape into either annual or perennial biofuel crops will affect species diversity and community composition but also the provisioning of valuable ecosystem services, such as arthropod-mediated predation of crop pests [10,11].

Wild bees, which provide $3.1 billion in pollination services annually to agricultural landscapes in the United States [12], are also expected to be affected by the type of bioenergy crops selected. Research at the field level has found bee abundance to be three to four times higher within perennial grasslands than in corn fields [8], while at the landscape level, pollinators respond positively to increasing amounts of natural area and negatively to landscapes dominated by annual agriculture [13,14,15]. The response of pollinators to land-cover change suggests that the selection of bioenergy crops for large-scale production has the potential to positively or negatively impact these organisms. Expanding production of annual bioenergy crops such as corn or soybeans would further simplify the landscape by increasing the proportion of monoculture plantings across landscapes, reducing the availability of food and nesting resources for pollinators. In contrast, expansion of perennial grassland bioenergy crops could benefit pollinators by increasing landscape heterogeneity and augmenting the amount of resource rich-habitat available for foraging and nesting by bees.

With interest in identifying viable bioenergy crops, predictive models have been employed to investigate the effect of different bioenergy crops on species of conservation concern such as grassland birds [4], as well as a range of ecosystem services including biocontrol, carbon sequestration, and phosphorous loading [11,16,17]. Because pollinators provide a valuable ecosystem service and many are experiencing declining populations [18,19], models have also been developed to explore the effects of landscape composition on bee populations [20,21,22]. The conversion of land into more intensive uses is expected to have significant effects on pollinators [23], yet the use of modeling to predict the effects of intensive large scale land use change on pollinators is limited [24]. Because pollination is a critical service provided to agricultural crops and to natural plant communities [25,26], bioenergy policies should proactively address how bees can benefit from the development of agricultural systems that advance crop production as well as conservation objectives.

In this research, our aim was to explore the potential effects of two different bioenergy crop production scenarios on pollinator communities using a modeling approach. First, using observations from flower-visiting bees across the state of Michigan, U.S.A, we developed models that related bee abundance, diversity, and community composition to the land cover surrounding study sites. The three community metrics we modeled can indirectly provide insights into how pollination services may be affected by changes in bioenergy production. For example, pollination services tend to increase as flower-visiting bees become more abundant and diverse [27,28,29]. Also, shifts in community composition that result from changes in bioenergy production may affect particular species known to be important pollinators or of conservation concern. Next, we used empirical models to predict the effect of different bioenergy production scenarios on bees. The bioenergy production scenarios tested in this study represented opposite extremes of possible future production scenarios, and assumed a transition to annual bioenergy crops or to perennial grassland bioenergy crops on marginal lands. Production scenarios were limited to marginal lands because a sustainable biofuel policy will likely need to minimize competition between lands devoted to food and biofuel production [30]. Because perennial grassland biofuel production would increase landscape diversity and incorporate more resource-rich habitats into the landscape [21], we predicted that bee abundance, diversity, and community composition would benefit from a biofuel policy that increases perennial grassland production.

Methods

Study Sites

Field sampling was conducted with the permission of land owners in 20 soybean fields located across southern Michigan (Fig. 1) during the summer of 2012. We observed bee visitation at sentinel insect-pollinated plants to estimate bee abundance, diversity, and community composition. Sites were at least 3 km apart and varied in the proportion of annual crops and semi-natural habitats in a 1500 m radius surrounding each site. Land use proportions surrounding study sites were calculated using a geographic information system (GIS, ArcGIS, version 10.0, ESRI,

Redlands, CA). The proportion of grassland in the landscape ranged from <1% to 60%, representing a range of possible biofuel production scenarios from sites dominated by annual production to those dominated by perennial grassland production. Bee observations were conducted in soybean fields across a landscape gradient because soybean is a flowering, first-generation biofuel crop that is intensively managed for even plant density and low plant diversity using weed control. This approach reduced variability in bee counts due to variation in flower abundance, plant diversity, and other management practices. However, this approach also requires that inferences about patterns in other crops be viewed with an appropriate degree of caution.

Bee abundance, diversity and community composition

Bee visitation to sentinel sunflowers, *Helianthus annuus*, variety "Sunspot", was measured at each study site to sample the pollinating bee community. Plants were grown in 15.2 cm diameter pots in a greenhouse under 24-h light and a temperature of $26.7 \pm 2°C$. Two sunflowers with open disk flowers were placed at each of two sampling stations located 30 m from field edges and 20 m apart. Bee sampling was conducted simultaneously at each station during a 30-min period with one observer per station. Bees were collected using a hand-held vacuum (Bioquip, Rancho Dominquez, CA) when bee contact was made with disk flowers (i.e., anthers and/or stigmas). Each field (n = 20) was surveyed 3–5 times during the 2012 field season. Unequal sampling across sites was the result of agronomic activities which prevented access to fields on some dates. Sampling occurred on sunny days with temperatures above 24°C, and each field was sampled at least once in the morning and once in the afternoon. Bees were returned to the lab and identified to species using the online key to Bees of Eastern North America at www.discoverlife.org and published species-level keys [31,32].

Bee abundance, diversity, and community composition were quantified for each site. Bee abundance was calculated by averaging the number of bees collected during a 60 min

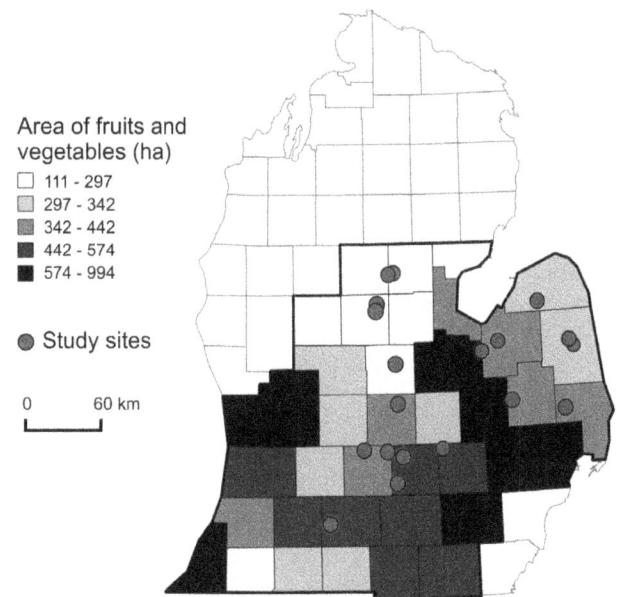

Figure 1. Study sites. Location of study sites sampled for bees across Michigan. Hectares of fruits and vegetables are calculated on a county basis and shown for the lower region of Michigan.

observation period (2 plants/site ×30 min) across all sampling dates for each site, and these values were square-root transformed prior to analysis to improve normality. To avoid obscuring the response of wild bees to landscape change, we excluded *Apis mellifera*, a managed pollinator commonly brought to agricultural landscapes, from calculations of abundance. Because *A. mellifera* was prevalent in agricultural landscapes and influenced composition (see Results), this species was included in calculations of bee diversity and community composition. Bee diversity was quantified for each site using Simpson's diversity index because this index is less sensitive to the degree of sampling effort (Simpson's diversity = 1-D)[33] than species richness and other diversity indices such as Shannon's diversity.

The response of the bee community was then analyzed across sites. First, the abundance of each bee species at each site was averaged across the 2012 season and then square-root transformed and standardized by species and site (i.e., Wisconsin double standardization) [34]. The similarity between sites was then quantified using the Bray-Curtis coefficient. Bee community composition was then ordinated using non-metric multidimensional scaling (NMDS). Next, we explored the relationship between measured landscape variables (e.g., proportions of grassland, forest, wind-pollinated crops, and annual flowering crops; see below) and bee community composition using environmental vector fitting [34]. When viewing the plotted NMDS scores from the community ordination, along with the corresponding environmental vectors (i.e., landscape proportions), we found that most of the variation in bee communities, and most of the association between bee communities and landscape structure, was represented along the first NMDS axis. In order to draw a qualitative link between NMDS scores, bee community composition, and landscape composition, we identified the bee species that contributed most to differences in NMDS scores using a SIMPER (similarity percentages) analysis [34]. Here, the bee community at each site was classified as having a high (>0) or low (<0) NMDS score. The output of the SIMPER analysis elucidated those species that were most highly associated with the two ends of the NMDS spectrum. Given the association between NMDS and landscape gradients, the SIMPER output allowed us to understand how particular bee species responded to landscape composition. Following these preliminary analyses, we used the first NMDS axis values to represent bee composition in subsequent linear regression modeling, described below. NMDS ordinations, environmental vector fitting, and SIMPER analyses were performed using the vegan package [35] written for R statistical computing software [36].

Modeling bee abundance, diversity, and community composition

Multiple linear regression was used to model bee abundance, diversity, and community composition as a function of land cover. Using the 2012 Cropland Data Layer (USDA NASS 2012), the proportion of land cover was calculated in the 1,500 m surrounding sites for nine classes, which accounted for 0.87 ± 0.19 (SD) of land cover: annual wind-pollinated crops, which combined 24 classes of annual crops but was dominated by corn (average proportion of corn = 0.60 in this category, based on all 20 sites); annual flowering crops, which combined 17 classes of annual crops that benefit from pollinators but were dominated by soybean (average proportion of soybean = 0.89); perennial flowering crops (all fruit crops); grasslands (included herbaceous grasslands, old fields, pastures, wildflowers, hayfields, alfalfa fields, and shrublands); forests (combined deciduous, coniferous, mixed forest, and wooded wetlands); wetlands (herbaceous wetlands);

suburbs (areas of low development and open areas dominated by turf); cities (areas of moderate to high development); and other (included water, walnut and Christmas tree farms, and barren). Wind-pollinated crops, annual flowering crops, forests, and grasslands were included as explanatory variables because they accounted for a large proportion of the land cover and (with the exception of forest) are the cover classes that will change under contrasting bioenergy scenarios. Although the proportion of forest does not change under the different bioenergy scenarios, this variable was retained in the model because forests are ecologically important to pollinators, providing nesting habitat and floral resources early in the season [37,38,39].

An information theoretic approach to model selection began with the full model, which included the proportion of wind-pollinated crops, annual flowering crops, grassland, and forest. The full model and all possible subsets of the full model were analyzed using the multimodel inference package, MuMIn, in R [36,40]. The overall best model and all competing models were identified and ranked using bias-corrected Akaike's Information Criterion (AIC$_c$). Because multiple competing models explained bee abundance, diversity, and community composition, we used model-averaged coefficients from the model set to make predictions about changes in bee communities under the different bioenergy scenarios. Model-averaged coefficients were calculated as weighted averages using model coefficients and Akaike weights, where coefficients were set to zero when a variable was not included in a given model (i.e., a shrinkage coefficient) [41]. Spatial autocorrelation in model residuals was assessed with spline correlograms using the ncf package in R [42]. Statistically significant spatial autocorrelation was not detected in model residuals (Text S1).

Projecting bee abundance, diversity, and community composition

Bee abundance, diversity, and community composition were first estimated across the lower peninsula of Michigan under the current landscape scenario. In GIS, proportional land cover maps were calculated separately for each cover type using a moving window approach. The moving window analysis, which was preformed across Lower Michigan, calculates for each pixel the proportion of cover (e.g., grassland, forest, etc.) in a neighborhood with a radius of 1500 m. To generate predicted values for bees under the current land cover scenario, model-averaged coefficients from the empirical models were multiplied by their respective proportional maps. For example, the model-averaged coefficient for grassland was multiplied by the proportion grassland map and then summed with the products of other terms to produce predicted values for each pixel. The equation used to calculate predicted values for bees in GIS was as follows:

$$Y_{i,j} = b_j + g_j G_i + f_j F_i + w_j W_i + a_j A_i \qquad (\text{eq.1})$$

where Y is the prediction for the ith pixel for the jth bee community metric (i.e., bee abundance, diversity, or composition), b_j is the intercept for the jth metric, and g_j, f_j, w_j, and a_j are the metric-specific model-averaged parameter values for the proportions of grassland, G, forest, F, wind-pollinated crops, W, and flowering annual crops, A, in the landscape surrounding the ith pixel. The results of this analysis generated maps that estimated bee abundance, diversity, and community composition under current landscape conditions.

The next step was to model abundance, diversity, and community composition under the two contrasting bioenergy scenarios. Because a sustainable bioenergy policy will need to minimize competition with highly productive agricultural land [30], the bioenergy scenarios we developed were focused on marginal land. The U.S. Department of Agriculture's SSURGO database, which lists land capability classes based on soil quality, erosion potential, and water saturation, was used to identify marginal land. Marginal land was defined as cropland with "severe limitations" to "very severe limitations" (land capability classes 3 and 4, respectively) in addition to other lands consider unsuitable for row crop production (classes 5–8).

Of the marginal land in Lower Michigan, we identified 1,200,352 ha in grassland, 550,750 ha in annual wind-pollinated crops (76% corn), and 290,033 ha in flowering annual crops (86% soybean). To keep the number of hectares converted in each scenario consistent, 600,000 ha of marginal lands were converted in each scenario. In the "perennial grassland bioenergy scenario" approximately 360,000 ha of wind-pollinated crops on marginal land and approximately 240,000 ha of flowering annual crops on marginal land were randomly selected and converted into grassland in GIS. In the "annual bioenergy scenario", 600,000 ha of grassland on marginal land was randomly selected and converted into wind-pollinated crops or flowering annual crops based on the proportion of corn (0.58) and soybean (0.41) in the current landscape (CDL 2012). Once the land-cover conversions were completed in GIS, resulting maps represented land cover change under the perennial grassland bioenergy scenario and the annual bioenergy scenario.

The final step was to predict bee abundance, diversity, and community composition under the annual and perennial bioenergy production scenarios. First, we created proportional land cover maps for each cover type using the annual and perennial bioenergy scenario maps. Then model-averaged coefficients from the empirical models were multiplied by their respective proportional land cover maps (Eq. 1). Finally, we calculated the percent change in bee abundance and diversity between the current landscape $(Y_{i,j,c})$ and each bioenergy scenario $(Y_{i,j,s})$ using the equation: percent change $= ((Y_{i,j,s} - Y_{i,j,c})/Y_{i,j,c}) \times 100$. Because bee composition used NMDS axis scores, which include both positive and negative values, the difference between the bioenergy landscape and the current landscape was calculated using: difference $= Y_{i,j,s} - Y_{i,j,c}$.

We produced aggregate summaries of the percent change in bee communities for each scenario across the study region in two distinct ways. First, we calculated summary statistics for changes in bee community metrics across all grassland and cropland pixels in the study region. Second, we calculated summary statistics for changes in bee community metrics only for changed pixels and their immediate neighbors across the study region. Hereafter, we refer to results from the former calculations as landscape-level results, and from the latter calculations as local-level results.

Results

Empirical data

During 4,800 min (80 hr) of observation on sentinel plants across all study sites, we observed an average of 0 to 7.25 bee visitors per hour. A total of 38 bee species were identified, and species richness ranged from 1 to 16 species per site, which translated into a range of Simpson's diversity values of 0 to 0.89 per site. The NMDS and vector fitting analyses showed that bee community composition changed along a landscape gradient (Fig. 2, two-dimensional stress $= 0.17$), where communities in sites with low proportions of grassland and forest had positive NMDS axis scores, while communities associated with high proportions of

grassland and forest had negative NMDS axis scores. The proportion of grassland and forest were significantly negatively correlated with the first NMDS axis (Fig. 2; Grassland: $R^2 = 0.46$, $P = 0.005$; Forest: $R^2 = 0.43$, $P = 0.007$), while the proportion of wind-pollinated and flowering annual crops were positively correlated with the first NMDS axis (Fig. 2; Wind: $R^2 = 0.51$, $P = 0.003$, Flowering annual: $R^2 = 0.45$, $P = 0.006$). In general, the bee communities in landscapes dominated by annual crops were less diverse while those in grassland and forest dominated landscapes were more diverse (Text S2). The SIMPER analysis indicated that *A. mellifera*, *Augochlorella aurata*, and *Halictus ligatus* contributed to the largest difference: 17%, 8%, and 7%, respectively, among sites along the landscape gradient (Text S2). *A. mellifera* and *A. aurata* contributed more to the bee community in agricultural landscapes while *H. ligatus* was increasingly associated with sites having higher proportions of grassland and forest in the surrounding landscape (Text S2).

Modeling pollinators

Bee abundance was best explained by the proportion of forest and grassland in the surrounding landscape (Table 1). Because several competing models were present, model-averaged coefficients calculated with shrinkage were used to predict and map bee abundance in GIS (Table 1, Fig 3A). Overall, forest and grassland had positive model-average coefficients for bee abundance, while wind-pollinated crops and flowering annual crops had negative model-average coefficients.

The best model for bee diversity included only one variable, the proportion of forest in the landscape (Table 1). Again, multiple

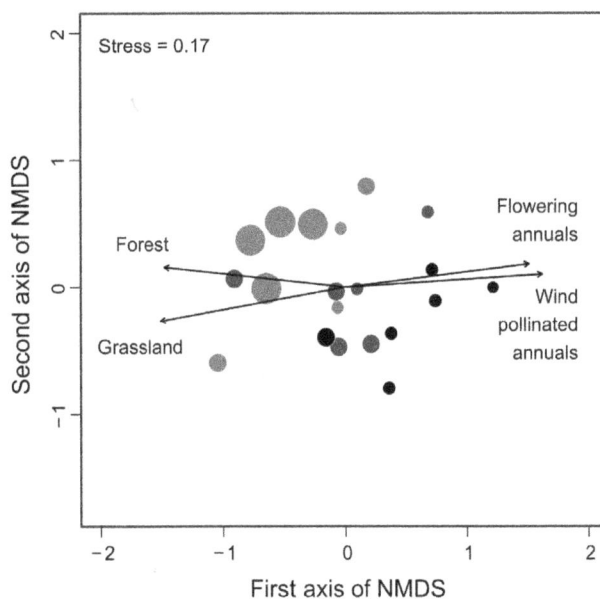

Figure 2. Ordination of bee communities with landscape variables. Ordination of bee community composition using non-metric multidimensional scaling (NMDS) shows that communities change as the proportion of grassland and forest cover increase in a 1500 m radius surrounding sites (stress $= 0.17$). Sites with negative NMDS axis one scores are correlated with grassland ($R^2 = 0.46$, $P = 0.0054$) and forest ($R^2 = 0.43$, $P = 0.0071$), while sites with positive NMDS axis one scores are correlated with wind pollinated crops ($R^2 = 0.52$, $P = 0.0025$) and flowering annual crops ($R^2 = 0.45$, $P = 0.0057$). Increasing circle size represents sites with higher bee abundance while decreasing color intensity (black to light gray) indicates sites with higher levels of bee diversity.

(A)

(B)

(C)

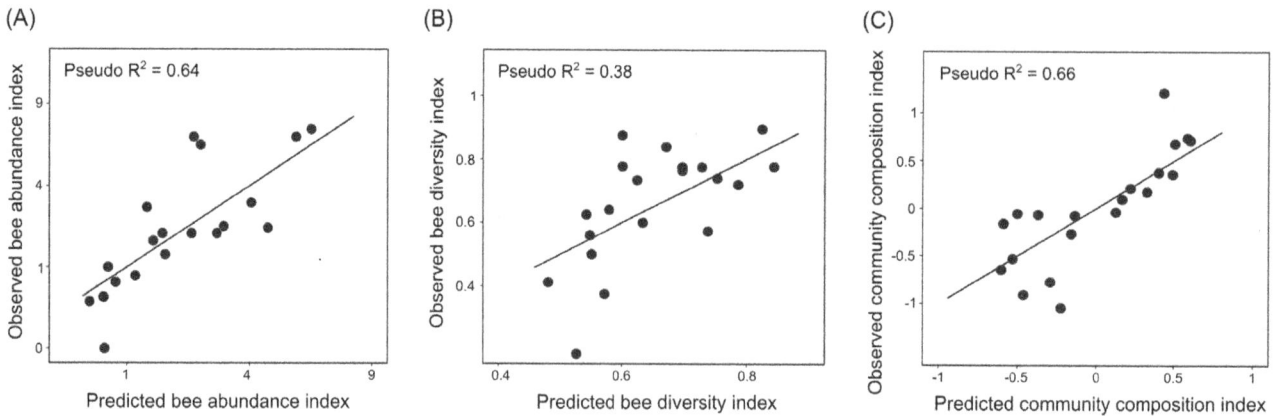

Figure 3. Observed versus predicted bee community metrics. Relationships between observed and predicted values for abundance, diversity, and community composition. Pseudo-R^2 is derived from regressing observed values versus model-averaged predictions. The graph for abundance displays only 19 points because two data points overlap.

competing models were present in the model set for pollinator diversity and model-averaged coefficients calculated with shrink-age were used to estimate bee diversity in GIS (Table 1, Fig 3B). Model-averaged coefficients for bee diversity were also generally positive for grassland and forest and negative for wind-pollinated crops and flowering annual crops.

Bee community composition, as reflected by the first axis in the NMDS ordination, was best explained by the proportion of grassland and wind-pollinated crops in the surrounding landscape (Table 1). Given the large set of competing models, model-averaged coefficients were used to predict community composition in GIS (Table 1, Fig. 3C). Generally, grassland and forest had negative model-averaged coefficients for this factor, while the model-averaged coefficients for wind-pollinated and flowering annual crops were positive. These results agree with results from vector fitting, which are discussed above in the first paragraph of the results, and have an analogous interpretation.

Projected landscape-level effects of biofuel scenarios on bees

Bee abundance responded positively to the perennial bioenergy scenario and negatively to the annual bioenergy scenario. Under a perennial bioenergy scenario, abundance increased from 0 to 600%, with a landscape-level mean increase of 24% (Fig. 4a). The highest predicted increases in abundance occurred where marginal annual cropland was converted into perennial bioenergy crops (Fig. 5a). Under the annual bioenergy scenario, bee abundance was predicted to decline from 0 to 71%, with a landscape-level mean decrease of 6% (Fig. 4b). The largest loss of abundance is projected to occur in central and western Michigan (Fig. 5a). Areas expected to experience little or no change in abundance, such as pockets of central Michigan and the northwest peninsula of Michigan (Fig. 5A), are regions dominated by prime agricultural soils that are not available for conversion into either perennial or annual bioenergy crops. In contrast, bee abundance is expected to change substantially in regions of Michigan where landscapes support grassland and annual crops on marginal lands. For example, in west Michigan wild bee abundance was projected to experience large declines under the annual scenario, whereas increases were predicted for north-central and southwest Michigan under the perennial scenario.

Similar to the results found for abundance, bee diversity was predicted to increase under the perennial bioenergy scenario and

decrease under the annual bioenergy scenario. Bee diversity was predicted to increase from 0 to 53% (Fig. 4c), with a landscape-level mean increase of 10%, under the perennial bioenergy scenario (Fig. 5b). In contrast, bee diversity was predicted to decrease from 0 to 28% (Fig. 4d), with a landscape-level mean decrease in diversity of only approximately 1% under the annual bioenergy scenario (Fig. 5b). Increases in bee diversity under the perennial scenario were predicted for the counties in the peninsula neighboring Lake Michigan, while declines in bee diversity were not predicted for this area of Michigan under the annual scenario. This region of Michigan is an area of intensive agriculture, meaning that areas of annual crops on marginal soils were available for conversion into perennial bioenergy crops, with potential for increasing pollinator diversity. However, areas of marginal grassland were lacking in this region, preventing land conversion into annual bioenergy crops.

Community composition is also predicted to respond to contrasting bioenergy production scenarios. Change in community composition under the perennial scenario, based on NMDS axis one scores, ranged from 0 to −1.73 under the perennial bioenergy scenario (increasing bee abundance and diversity, Fig. 4E), while changes in scores ranged from 0 to 1.37 (decreasing bee abundance and diversity, Fig. 4F) under the annual scenario. Under the perennial scenario, shifts in composition were predicted to occur in north central, western, and south central Michigan (Fig. 5c). Bee composition scores under the annual scenario were expected to change predominantly in west Michigan (Fig. 5c) where bee communities are expected to experience species declines.

Discussion

Biofuel policies set at the national level are expected to expand biofuel crop production [5,43], causing substantial changes in land cover across the Midwest [44,45]. Policies have the potential to influence crop choice as well as crop placement within the landscape, shaping land cover change in ways that are predicted to affect biodiversity and the provisioning of ecosystem services. Our study shows that bee abundance, diversity, and community composition are sensitive to changes in land cover. We found that bee abundance and diversity were greater where there was a greater proportion of grassland and forest in the landscape and were lower where annual agriculture was more prevalent. These results were used to make predictions about how bees in

Table 1. Model selection.

	Intercept	Grassland	Forest	Wind[a]	FA[b]	ΔAICc	Model weight	Model R^2
Abundance								
Competing model								
1	0.66	1.83	2.98			0.00	0.33	0.63
2	1.35		3.23		−1.50	1.28	0.17	0.60
3	0.83		4.37			1.65	0.14	0.52
Average model[c]								0.64
Parameter value	0.95	1.13	2.84	−0.04	−0.50			
SE	0.46	1.20	1.64	0.49	0.88			
Variable weight		0.60	0.86	0.17	0.37			
Diversity								
Competing model								
1	0.52		0.94			0.00	0.23	0.30
2	0.82				−0.70	0.25	0.20	0.29
3	0.68		0.62		−0.40	0.68	0.16	0.38
Average model								0.38
Parameter value	0.68	0.03	0.45	−0.06	−0.30			
SE	0.16	0.18	0.51	0.02	0.36			
Variable weight		0.19	0.56	0.23	0.54			
Composition								
Competing model								
1	−0.20	−1.46		1.64		0.00	0.25	0.62
2	−0.82			1.63	1.34	0.61	0.18	0.61
3	−0.43	−1.03		1.38	0.85	1.94	0.094	0.65
4	−0.69			2.36		1.97	0.093	0.51
Average model								0.66
Parameter value	−0.33	−0.76	−0.42	1.26	0.62			
SE	0.47	0.90	0.94	0.93	0.83			
Variable weight		0.54	0.29	0.76	0.48			

Summary of model selection statistics for the competing models and model-averaged coefficients predicting bee abundance, diversity, and community composition as a function of land cover variables measured in the 1500 m surrounding study sites.

[a]Wind represents the variable wind pollinated annual crops.

[b]FA represents flowering annual crops potentially visited by bees.

[c]For the averaged model the R^2 is a pseudo-R^2 derived from regressing the observed values versus model-averaged predictions.

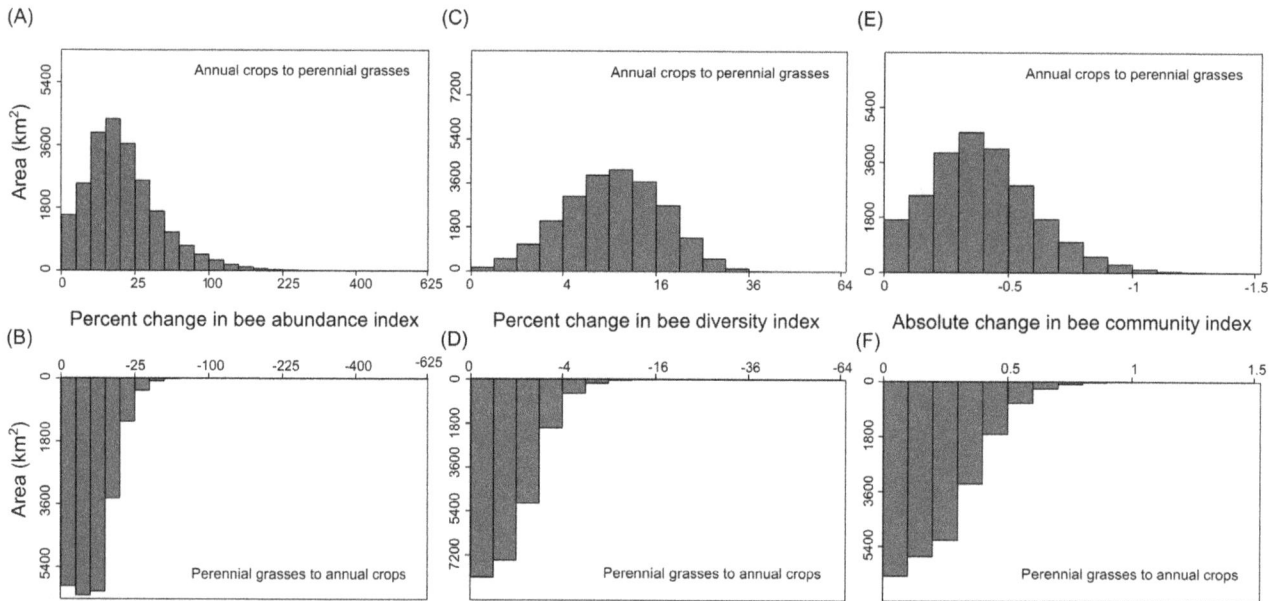

Figure 4. Distribution of percent change for measured bee metrics. The distribution of percent change values calculated under the perennial grassland bioenergy scenario for bee abundance (A), diversity (C), and community composition (E) and under the annual bioenergy scenario for bee abundance (B), diversity (D), and community composition (F). In the annual bioenergy scenario 600,000 ha of marginal grassland was converted into annual bioenergy crops, whereas in the perennial bioenergy scenario 600,000 ha of marginal agricultural land were converted into grassland. Predicted changes in bee communities are only shown for the lower portion of the state where empirical data were collected.

agricultural landscapes would change as annual or perennial bioenergy crops expanded across the region. In a scenario where perennial grassland bioenergy crops are favored, we expect bee abundance and diversity to increase, with shifts to communities that are more dominated by wild bees. In contrast, if policies and markets favor increased adoption of annual bioenergy crops, we predict a reduction in bee abundance and diversity, with community composition moving towards fewer species dominated by generalists such as *A. mellifera*.

In the results reported here, the bee-related metrics were sensitive to landscape-scale land cover change. Mean values reported for increases and decreases in bee abundance and diversity were calculated using all grassland and agricultural pixels across our study region, whether or not they were changed under each scenario. As a result, the magnitudes of the reported percent changes (landscape-level) are smaller than they would be if they were calculated only for the pixels adjacent to those selected for land use change (local-level). At a local level, mean bee abundance increased by 40% under the perennial crop scenario and decreased by 14% under the annual crop scenario. Interestingly, local mean values for bee diversity remained relatively unchanged, with a 9% increase (landscape-scale increase 10%) under the perennial crop scenario and a 2.6% decrease (landscape-scale decrease 1%) under the annual crop scenario. These results suggest that the local effects of bioenergy crop production are more pronounced for bee abundance than for bee diversity.

Forested land played an important role in explaining bee abundance and bee diversity, having the highest variable weight in both sets of models (Table 1). Grassland, however, had the second highest variable weight when explaining bee abundance but the lowest variable weight for bee diversity. The significant relationship between grassland cover and bee abundance may explain why strong local effects are predicted for bee abundance under the perennial and annual land change scenarios but not for bee diversity. The strongest driver of bee diversity was forest cover,

and the proportion of forest in the landscape was not changed under either bioenergy scenario. This might explain why switching energy crops had limited local effects on bee diversity. Although our results suggest that increasing perennial grassland bioenergy production will have local effects on bee abundance, conserving forest habitat will be important for maintaining landscape-level bee diversity.

The response of the bee community to annual and perennial bioenergy scenarios was similar for bee abundance and diversity. In general, the bee community under annual bioenergy production shifted to a community composed of fewer bee species, while the bee community under perennial grassland production transitioned to a more diverse community of wild bees (Text S2). The effect of annual bioenergy crop production on bee community composition is of particular interest in the western counties of Michigan where there is significant production of pollinator-dependent fruit and vegetable crops. Because diverse pollinator communities are expected to provide more reliable pollination services by containing redundant [46] or complimentary [47] pollinator species, a shift to annual biofuel production may lead to a decline in bee diversity, with potential effects on the pollination services provided to fruit growers in this region. The relative location of land used for pollinator-dependent crops and biofuels would be expected to influence the degree to which these changes might affect crop yield.

Changes in bee abundance, diversity, and community composition under the perennial bioenergy scenario highlight where opportunities and challenges for grassland bioenergy production exist across Michigan. Several counties located in north-central and south-east Michigan show little or no change in bee abundance, diversity, or community composition. The lack of change in these counties is due to the presence of fruits, vegetables, corn, and soybeans on prime agricultural soils, yielding few opportunities for perennial grassland bioenergy production on marginal lands. In some cases, small isolated patches of perennial

Perennial grasses to annual crops

Annual crops to perennial grasses

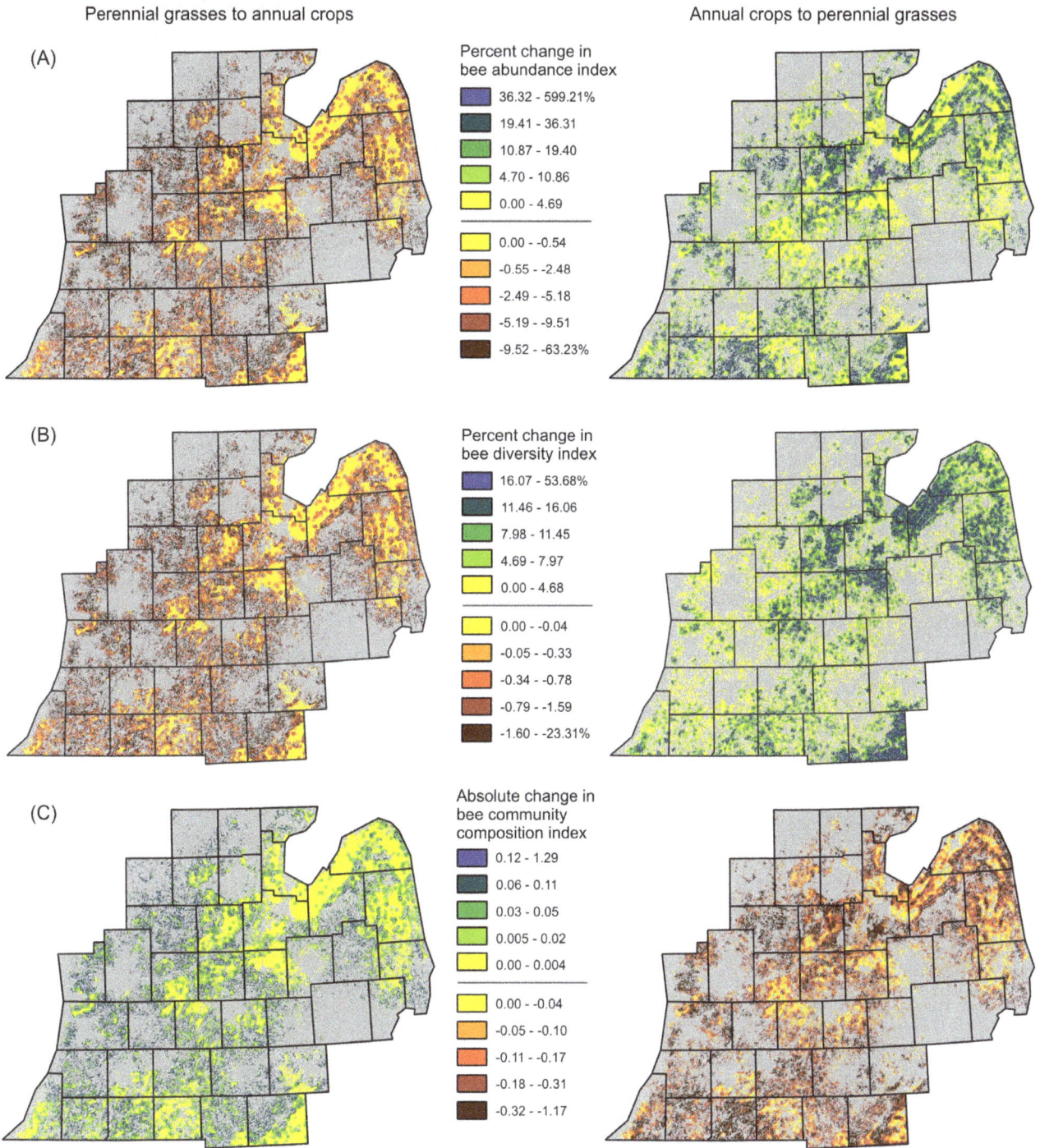

Figure 5. Projected bee metrics. Percent change in bee abundance (A), percent change in bee diversity (B), and difference in community composition (C) predicted for Michigan by an empirical model under annual (left maps) and perennial (right maps) bioenergy production scenarios.

habitats may actually serve as population sinks for bees [48,49], suggesting that one challenge for perennial bioenergy production will be to determine how the size and position of these crops within the landscape will influence pollination services. While placement of bioenergy plantings across the landscape might present challenges, an opportunity exists to target the placement of perennial bioenergy crops near pollinator dependent crops in an effort to increase pollinators and potentially augment pollination services and crop yield.

While the models developed here give insights into the possible effects of future bioenergy production on bees, the resulting maps and our interpretation of these maps depend on several assumptions. First, the models used to predict bee abundance, diversity, and community composition were based on empirical data collected from the lower portion of Michigan and may not extend to other parts of the Midwest. Furthermore, visitation to sunflower measured during the summer may not serve as a good proxy for other pollinator dependent crops, especially early season crops. However, many of the bees we collected, including those in the genus *Bombus* and *Halictus*, are generalists that are present throughout the growing season. Second, conclusions regarding the effect of perennial and annual bioenergy production assume that management practices currently employed for annual crops and perennial grassland habitats do not change substantially with increased bioenergy production. Increasing insecticide use to control emerging pests or annual harvest of bioenergy grasslands could affect bees in ways not reflected in our models. Third, forest was an important variable explaining bee abundance, diversity, and community composition in our study. Under the contrasting bioenergy scenarios developed here, we assumed the proportion of forest remains constant across the landscape, suggesting future loss of forest habitat due to agricultural intensification or urbanization could alter model predictions. Finally, while increasing bee abundance and diversity are generally correlated with higher rates of pollination [28,50,51], we recognize that transitioning the landscape into perennial grassland bioenergy production may not translate into increased pollination services, especially if one effective pollinator species is capable of persisting under both scenarios. However, biologically diverse pollinator communities play an important role in providing stable pollination services [52,53,54], suggesting that land conversion to perennial grassland bioenergy crops can contribute to supporting services in regions where bees have experienced declines.

Conclusions

Using field observations, we generated empirical models and predicted bee abundance, diversity, and community composition across Lower Michigan for two contrasting bioenergy production scenarios. From these analyses, we identified areas where bees are expected to benefit substantially from bioenergy policies that promote perennial grassland production, as well as areas where

further land conversion to annual bioenergy crops is likely to produce significant challenges for the persistence of diverse bee communities. The methods and models developed here have application for the identification of area thresholds required to maximize biodiversity conservation and target areas of the landscape where perennial bioenergy plantings could facilitate pollination services. However, given market values for annual commodity crops, conversion to perennial grassland bioenergy production will likely be limited without policy changes [55]. Policies that acknowledge the value of biodiversity and the services it provides will be necessary for implementing bioenergy production systems that balance trade-offs between crop production and the support of ecosystem services [56].

Acknowledgments

We thank Ashley McNamara, Lindsey Pudlo, Jon Roney, and Laura Maihofer who provided invaluable field assistance, and the landowners who allowed us access to study sites. Thanks to Dr. Jason Gibbs who provided assistance with bee identifications and Dr. David Lusch who provided useful insights into the GIS methodology.

Author Contributions

Conceived and designed the experiments: AB TM CG RI. Performed the experiments: AB TM. Analyzed the data: AB TM. Wrote the paper: AB TM CG RI.

References

1. Tyner WE (2008) The US ethanol and biofuels boom: its origins, current status, and future prospects. Bioscience 58: 646–653.

2. Jarchow ME, Liebman M (2012) Tradeoffs in biomass and nutrient allocation in prairies and corn managed for bioenergy production. Crop Sci 52: 1330–1342.

3. Schubert C (2006) Can biofuels finally take center stage? Nat Biotechnol 24: 777–784.

4. Meehan TD, Hurlbert AH, Gratton C (2010) Bird communities in future bioenergy landscapes of the Upper Midwest. Proc Natl Acad Sci USA 107: 18533–18538.

5. Wright CK, Wimberly MC (2013) Recent land use change in the western Corn Belt threatens grasslands and wetlands. Proc Natl Acad Sci USA 110: 4134–4139.

6. Wiens J, Fargione J, Hill J (2011) Biofuels and biodiversity. Ecol Appl 21: 1085–1095.

7. Robertson BA, Doran PJ, Loomis LR, Robertson JR, Schemske DW (2011) Perennial biomass feedstocks enhance avian diversity. Glob Change Biol Bioenergy 3: 235–246.

8. Gardiner MA, Tuell JK, Isaacs R, Gibbs J, Ascher JS, et al. (2010) Implications of three biofuel crops for beneficial arthropods in agricultural landscapes. Bioenerg Res 3(1): 6–19.

9. Werling BP, Meehan TD, Gratton C, Landis DA (2011) Influence of habitat and landscape perenniality on insect natural enemies in three candidate biofuel crops. Biol Control 59: 304–312.

10. Werling BP, Meehan TD, Robertson BA, Gratton C, Landis DA (2011) Biocontrol potential varies with changes in biofuel-crop plant communities and landscape perenniality. Glob Change Biol Bioenergy 3: 347–359.

11. Meehan TD, Werling BP, Landis DA, Gratton C (2012) Pest-suppression potential of Midwestern landscapes under contrasting bioenergy scenarios. PLOSONE 7(7). E41728. Doi10.1371/journal.pone.0041728.

12. Losey JE, Vaughan M (2006) The economic value of ecological services provided by insects. Bioscience 56: 311–323.

13. Ricketts TH (2004) Tropical forest fragments enhance pollinator activity in nearby coffee crops. Conserv Biol 18: 1262–1271.

14. Ricketts TH, Regetz J, Steffan-Dewenter I, Cunningham SA, Kremen C, et al. (2008) Landscape effects on crop pollination services: are there general patterns? Ecol Lett 11: 499–515.

15. Garibaldi LA, Steffan-Dewenter I, Winfree R, Aizen MA, Bommarco R, et al. (2013) Wild pollinators enhance fruit set of crops regardless of honey bee abundance. Science 339: 1608–1611.

16. Gelfand I, Sahajpal R, Zhang XS, Izaurralde RC, Gross KL, et al. (2013) Sustainable bioenergy production from marginal lands in the US Midwest. Nature 493: 514–ZZZ.

17. Meehan TD, Gratton C, Diehl E, Hunt ND, Mooney DF, et al. (2013) Ecosystem-service tradeoffs associated with switching from annual to perennial energy crops in riparian zones of the US Midwest. PLOSONE 8(11): e89993. doi: 10.137/journal.pone.0080093

18. Potts SG, Biesmeijer JC, Kremen C, Neumann P, Schweiger O, et al. (2010) Global pollinator declines: trends, impacts and drivers. Trends Ecol Evol 25: 345–353.

19. Cameron SA, Lozier JD, Strange JP, Koch JB, Cordes N, et al. (2011) Patterns of widespread decline in North American bumble bees. Proc Natl Acad Sci U S A 108: 662–667.

20. Lonsdorf E, Kremen C, Ricketts T, Winfree R, Williams N, et al. (2009) Modelling pollination services across agricultural landscapes. Ann Bot103: 1589–1600.

21. Kennedy CM, Lonsdorf E, Neel MC, Williams NM, Ricketts TH, et al. (2013) A global quantitative synthesis of local and landscape effects on wild bee pollinators in agroecosystems. Ecol Lett 16: 584–599.

22. Schulp CJE, Lautenbach S, Verburg PH (2014) Quantifying and mapping ecosystem services: demand and supply of pollination in the European Union. Ecol Indic 36: 131–141.

23. Kremen C, Williams NM, Aizen MA, Gemmill-Herren B, LeBuhn G, et al. (2007) Pollination and other ecosystem services produced by mobile organisms: a conceptual framework for the effects of land-use change. Ecol Lett 10: 299–314.

24. Ricketts TH, Lonsdorf E (2013) Mapping the margin: comparing marginal values of tropical forest remnants for pollination services. Ecol Appl 23: 1113–1123.23.

25. Ollerton J, Winfree R, Tarrant S (2011) How many flowering plants are pollinated by animals? Oikos 120: 321–326.

26. Nicholls CI, Altieri MA (2013) Plant biodiversity enhances bees and other insect pollinators in agroecosystems. A review. Agron Sustain Dev 33: 257–274.

27. Hoehn P, Tscharntke T, Tylianakis JM, Steffan-Dewenter I (2008) Functional group diversity of bee pollinators increases crop yield. Proc R Soc B-Biol Sci 275: 2283–2291.

28. Frund J, Dormann CF, Holzschuh A, Tscharntke T (2013) Bee diversity effects on pollination depend on functional complementarity and niche shifts. Ecol 94: 2042–2054.

29. Garibaldi LA, Steffan-Dewenter I, Kremen C, Morales JM, Bommarco R, et al. (2011) Stability of pollination services decreases with isolation from natural areas despite honey bee visits. Ecol Lett 14: 1062–1072.

30. Tilman D, Socolow R, Foley JA, Hill J, Larson E, et al. (2009) Beneficial biofuels-the food, energy, and environment trilemma. Science 325: 270–271.

31. Rehan SM, Sheffield CS (2011) Morphological and molecular delineation of a new species in the *Ceratina dupla* species-group (Hymenoptera: Apidae: Xylocopinae) of eastern North America. Zootaxa 35–50.

32. Gibbs J (2011) Revision of the metallic *Lasioglossum* (*Dialictus*) of eastern North America (Hymenoptera: Halictidae: Halictini). Zootaxa 1–216.

33. Clarke KR, Warwick RM (2001) Change in marine communities: an approach to statistical analysis and interpretation, 2nd ed. Plymouth: PRIMER-E.

34. McCune B, Grace JB (2002) Analysis of ecological communities. MjM software, Gleneden Beach, Oregon, USA: MjM software. 102 p.

35. Oksanen J, Blanchet FG, Kindt R, Legendre P, Minchin PR, et al. (2013) Vegan: community ecology package. R package version 2.0–10. Available: http://CRAN.R-project.org/package=vegan

36. R Development Core Team (2013) R: A language and environment for statistical computing. Vienna, Austria.38

37. Macior LW (1968) *Bombus* (Hymenoptera Apidae) queen foraging in relation to vernal pollination in Wisconsin. Ecol 49: 20–25.

38. Ginsberg HS (1983) Foraging ecology of bees in an old field. Ecol 64: 165–175.

39. Fabian Y, Sandau N, Bruggisser OT, Aebi A, Kehrli P, et al. (2013) The importance of landscape and spatial structure for hymenopteran-based food webs in an agro-ecosystem. J Anim Ecol 82: 1203–1214.

40. Barton K (2013) MuMIn: mulit-model inference. R package version 1.9.5.Avail-.Available: http://CRAN.R-project.org/package=MuMIn

41. Burnham KP, Anderson DR (2002) Model selection and multimodel inference: a practical information-theoretic approach. New York: Springer. 159 p.

42. Bjornstad ON (2013) ncf: spatial nonparametric covariance functions. R package version 1. 1–5. Available: http://CRAN.R-project.org/package=ncf

43. Fargione JE, Cooper TR, Flaspohler DJ, Hill J, Lehman C, et al. (2009) Bioenergy and wildlife: threats and opportunities for grassland conservation. Bioscience 59: 767–777.

44. Wu F, Guan ZF, Yu F, Myers RJ (2013) The spillover effects of biofuel policy on participation in the conservation reserve program. J Econ Dyn Control 37: 1755–1770.

45. Johnston CA (2014) Agricultural expansion: land use shell game in the U.S. Northern Plains. Landscape Ecol 29: 81–95.

46. Winfree R, Kremen C (2009) Are ecosystem services stabilized by differences among species? A test using crop pollination. Proc R Soc B-Biol Sci 276: 229–237.

47. Brittain C, Kremen C, Klein AM (2013) Biodiversity buffers pollination from changes in environmental conditions. Glob Chang Biol 19: 540–547.

48. Fahrig L (2003) Effects of habitat fragmentation on biodiversity. Ann Rev Ecol Evol Syst 34: 487–515.

49. Rosch V, Tscharntke T, Scherber C, Batary P (2013) Landscape composition, connectivity and fragment size drive effects of grassland fragmentation on insect communities. J Applied Ecol 50: 387–394.

50. Albrecht M, Schmid B, Hautier Y, Muller CB (2012) Diverse pollinator communities enhance plant reproductive success. Proc. R Soc B Biol Sci 279: 4845–4852.

51. Holzschuh A, Dudenhoffer JH, Tscharntke T (2012) Landscapes with wild bee habitats enhance pollination, fruit set and yield of sweet cherry. Biol Conserv 153: 101–107.

52. Klein AM, Steffan-Dewenter I, Tscharntke T (2003) Fruit set of highland coffee increases with the diversity of pollinating bees. Proc R Soc B-Biol Sci 270: 955–961.

53. Ebeling A, Klein AM, Schumacher J, Weisser WW, Tscharntke T (2008) How does plant richness affect pollinator richness and temporal stability of flower visits? Oikos 117: 1808–1815.

54. Bartomeus I, Park MG, Gibbs J, Danforth BN, Lakso AN, et al. (2013) Biodiversity ensures plant-pollinator phenological synchrony against climate change. Ecol Lett 16: 1331–1338.

55. James LK, Swinton SM, Thelen KD (2010) Profitability analysis of cellulosic energy crops compared with corn. Agron J 102: 675–687.

56. Landis DA, Werling BP (2010) Arthropods and biofuel production systems in North America. Insect Sci 17: 220–236.

15

Diversification versus Specialization in Complex Ecosystems

Riccardo Di Clemente[1,2]*, **Guido L. Chiarotti**[2], **Matthieu Cristelli**[2], **Andrea Tacchella**[2,3], **Luciano Pietronero**[2,3,4]

1 IMT Institute for Advanced Studies Lucca, Lucca, Italy, 2 Istituto dei Sistemi complessi ISC-CNR, UOS Sapienza, Roma, Italy, 3 Dipartimento di Fisica, Università di Roma "Sapienza", Roma, Italy, 4 London Institute for Mathematical Sciences, London, United Kingdom

Abstract

By analyzing the distribution of revenues across the production sectors of quoted firms we suggest a novel dimension that drives the firms diversification process at country level. Data show a non trivial macro regional clustering of the diversification process, which underlines the relevance of geopolitical environments in determining the microscopic dynamics of economic entities. These findings demonstrate the possibility of singling out in complex ecosystems those micro-features that emerge at macro-levels, which could be of particular relevance for decision-makers in selecting the appropriate parameters to be acted upon in order to achieve desirable results. The understanding of this micro-macro information exchange is further deepened through the introduction of a simplified dynamic model.

Editor: Helge Thorsten Lumbsch, Field Museum of Natural History, United States of America

Funding: M. C., A. T. and G. L. C. are funded by Italian PNR project "CRISIS-Lab"; R. D. C. is funded by EU Project nr. 611272 "GROWTHCOM". The funders had no role in study design, data collection and analysis, decision to publish, or preparation of the manuscript.

Competing Interests: The authors have declared that no competing interests exist.

* Email: riccardo.diclemente@isc.cnr.it

Introduction

Countries and firms are fundamental actors sharing complex economic and social ecosystems. Their evolutive paths lead to structurally different scenarios: firms are specialized entities while countries, as recently shown, are diversified [1,2]. This raises a question on the mechanisms driving specialized entities to organize themselves into diversified super-structures. Is diversification a matter of size, of time horizon, or both? Are there other hidden dimensions governing the diversification process?

A similar scenario holds in biological ecosystems [3]: species (firms) tend to be substantially specialized, while groups of species competing on the same ecosystem (countries), appear to be diversified. Inspired by this argument in this paper we investigate the key mechanisms this picture is grounded on. It has been recently shown that this kind of analogy between economic and biological systems could gives rise to fruitful insights on elementary mechanisms [4].

Identifying the diversification drivers at the various scales is a challenging task in all disciplines since diversification processes are ubiquitous in nature [5] and economic systems [6,7]. In our view economic ecosystems represent an ideal (paradigmatic) playground for an empirical investigation.

We therefore analyze the distribution of revenues across production sectors of quoted firms aggregated by country (Bloomberg database [8,9]). Not surprisingly the analysis confirms that country competitiveness is mainly driven by diversification of productive systems, while firms' competitiveness is mainly a matter of specialization. The macroscopic signature of these macro-micro level discrepancies is reflected by the nested triangular structure of the country-sector binary matrix contrasting the essential randomness of the firm-sector binary matrix (see Methods section).

We argue that this is a specific observation of a general feature of complex systems: the shift from the macro to the micro level generally entails the loss of those features characterizing the former level. As in biology [10], the emerging diversification at macro level cannot be properly addressed at the level of individual species/firms. However, the environment in which the micro level is embedded preserves a sort of a macro level *memory* which enables to identify those micro level features that could emerge at larger scales [11].

Guided by this idea we show that, in the specific case of economic ecosystems, the microscopic feature emerging at the macro scale is the firm's diversification barrier α (see fig. 1). Moreover the α's of different countries aggregate on macro-regional (multi-country) scale. This *zoom-in zoom-out* framework thus enables the identification of the proper micro-variable selecting the emerging (aggregated) macro-properties. This is of particular relevance in socio-economic systems, since it may help decision-makers to select the correct variable to be acted upon at the (micro) specialized level, in order to achieve desirable results at the (macro) diversified level.

In this respect the traditional economic literature has extensively studied the effect of institutions, policies and economic environments under which diversification has an impact on firm revenues

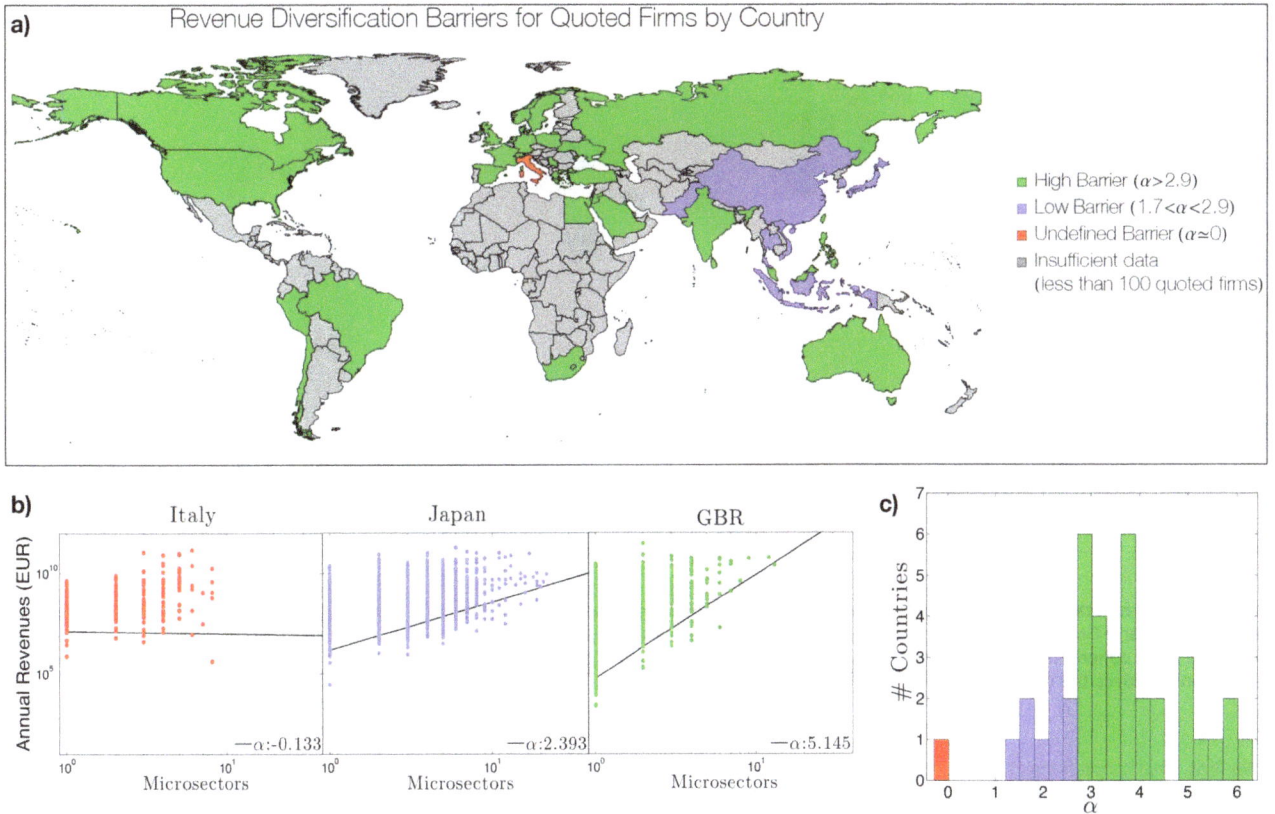

Figure 1. Revenue diversification barrier α. a. The worldwide distribution of the *revenue diversification barrier* α. The α tends to reflect geographical proximity and to cluster at the macro regional level. **b.** The scatter plot of firm revenues against firm diversification for thee paradigmatic countries. Except for Italy, the data draw a peculiar shape with a clear lower boundary. The angular coefficient of this linear boundary is what we define as the *revenue diversification barrier* α. **c.** The histogram of α. Colors are consistent with those used in panels a. and b.

[12–14]. However, the general picture which emerges from the standard approach is usually non conclusive as to whether diversification patterns affect firm revenues. Instead as mentioned, in the present work, we find that firm revenues are correlated to diversification, but the signature of this correlation appears in a highly non-trivial way as a selection rule which prevents firms from occupying a part of the diversification-revenues plane. We argue that the subtleness of this dependence - namely that high diversification implies high revenues while high revenues does not imply diversification. - is at the basis of the strongly debated economic literature about this field. We explored possible correlation between firms diversification and their size as measure by the number of employes without finding any significant signal.

We also propose a simple mathematical model mimicking the firms diversification dynamics in which firms evolve via a random walk in a random potential. Firm's survival rate depends on the values of the potential in the state reached by a particular firm environment in which firms compete. Surviving firms tend to diversify in time with a given probability. Such a minimal model is able to reproduce the main features observed in the data analysis.

Results

The dataset we use consists of annual revenues of quoted firms disaggregated into Bloomberg's sector code and downloaded in May 2013. The database contains about 38000 firms and about 2000 sectors.

We proceed similarly to the work of [2] where an archival export dataset is considered to measure intangible assets determining the competitiveness of countries. It is worth noticing that in both analyses the datasets were not collected with the purpose of the analyses in which they were subsequently used.

As previously mentioned, the identification of the diversification drivers at the various scales is a challenging task in all disciplines. In Economics, in particular, it is unclear, but crucial, how the dynamics at micro-level determines the one at the macro-level and *vice versa*. This paper aims to shed some light on this very relevant question which affects how the economy should support the concrete implementation of economic policy decisions with a more scientific grounding.

The analysis confirms the recent finding [2] that country competitiveness is mainly driven by diversification of productive systems.

Coherently with the evidence of a triangular structure of country-product matrix in [1,2,15,16], in the present analysis the same triangular feature is also found in the country-sector matrix obtained by aggregating firms on the basis of its legal address (see Information S1). The same matrix constructed at the firm level looses its nestedness and is similar to a random matrix with the same density (for further discussion see Methods section), reflecting firm specialization. This raises a rather fundamental question: what is the mechanism that organizes the information present into an almost random matrix, at the firm's level, in a nested matrix, at the country level?

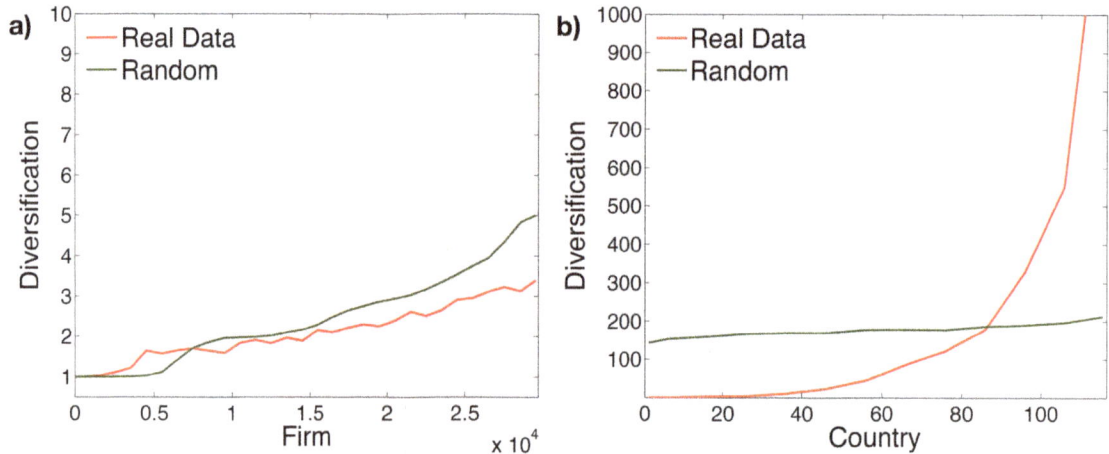

Figure 5. Comparison between the real data (red) and a random realization with same density (green). a. The firm-sector matrix exhibits a pattern similar to a random case emphasizing the firm' specialization. **b.** On the contrary aggregating the data on country level a non-random pattern emerges, corresponding to the presence of a nested structure.

Model

We propose an extremely simplified model that embodies in our view the minimal traits necessary to shed light on the meaning of the revenue diversification barrier α. Firms are mimicked as random walkers moving in a random potential, seeking local minima. The height of such minima is representative of a firm's performance (due to its simplicity the model does not distinguish between firm performance and firm revenues): the lower the value of the potential, the better the performance. Markets (countries) differ in their tolerance (τ) with respect to poor performances, i.e. in the probability for a firm to fail given its level of performance. Surviving firms, i.e. those with good performances, have the chance (P_{div}) to diversify, while failed firms are replaced with new ones with the lowest possible level of diversification.

The random potential is a realization of a simple gaussian discrete random walk, with 0 mean and unit variance. We generate 100 equally spaced discrete points of the potential. The potential $V(x)$ is then made periodic via a reflection, and is made continuous via a linear interpolation, the period being 200. Thus $V(x) = V(x+k*200)$ holds for any real x and for any integer k. Finally $V(x)$ is scaled to have maximum equal to 1 and minimum equal to 0.

Each firm starts at a random x_0 coordinate and is made to evolve as a brownian particle in the potential defined by $V(x)$. It seeks for local minima by evolving with the Metropolis-Hastings algorithm.

At each time step a proposal $x_t^{(p)}$ for a new value of x_t is drawn from a gaussian distribution $N(x_t,\sigma)$. The parameter σ needs to be chosen such that the typical jump distance for a firm will be inside a typical local minima. This typical width is of order 1, by construction, thus we have chosen $\sigma=0.1$. The proposal is then accepted with probability $P=e^{(V(x_{t-1})-V(x_t^{(p)}))/T}$. If the proposal is accepted we set $x_t=x_t^{(p)}$ else $x_t=x_{t-1}$.

We define the performance of a firm as $P(t)=1-V(x_t)$. Every 100 time-steps we compute the average performance $\bar{P}$ in such time window: the firm either survives with probability $1-\bar{P}^\tau$ or fails. If the firm survives it has the chance to increase its diversification of 1, with probability P_{div}. By making an analogy between the performance as defined in the present model and the revenues of a firm, we can observe in Fig. 3 how the model produces patterns very similar to those observed in the real

dataset. Interestingly there is still a linear lower bound in the doubly logarithmic diversification vs. performance scatter plot. Within this model the diversification is clearly proportional to the life span of a given firm. The similarity between real data scatter plot and the model produced data can thus be interpreted in view of the question raised in the introduction: diversification is a dynamic process that develops over time and the boundary in the diversification-performance relation is set by the competitiveness of the environment in which the economic entities are immersed. In other words what we observe in real data is compatible with diversification being a dynamic process that goes on as long as a firm is able to survive. How long it will survive given its profits depends on the tolerance of the ecosystem. The differences in tolerance generates the differences in the diversification boundaries that we observe across countries. The values of α have a clear dependence on τ and P_{div} as shown in the phase diagram in Fig. 4. In particular α decreases when the ecosystem tolerance increases. P_{div} acts as a simple multiplier of the life span of a firm in determining its diversification.

Discussion

The analysis of the distribution of firm revenues across production sectors aggregated by country manifests a peculiar triangular shape. This enables us to define a country dependent revenue diversification barrier "α", which represents a novel macroscopic dimension driving the microscopic diversification process.

We have shown that this new macro feature shows a non trivial geographical clustering, which points out the importance and implication of the geo-political environment in the diversification patterns. α can be interpreted as the microscopic signature responsible for micro-macro information exchange showing that though the economic complexity methods it is possible to single out the microscopic variables governing the macroscopic dynamic.

Within our finding the microscopic firms' differentiation dynamics can be interpreted as a "*Darwinan*" competitive process in which the firms survival to diversification depends on the characteristics of the macroscopical (country like) environment. To further confirm this picture, a time dependent analysis on similar data is called for. Moreover, to better understand the meaning of this newly introduced dimension α, a comparison with other

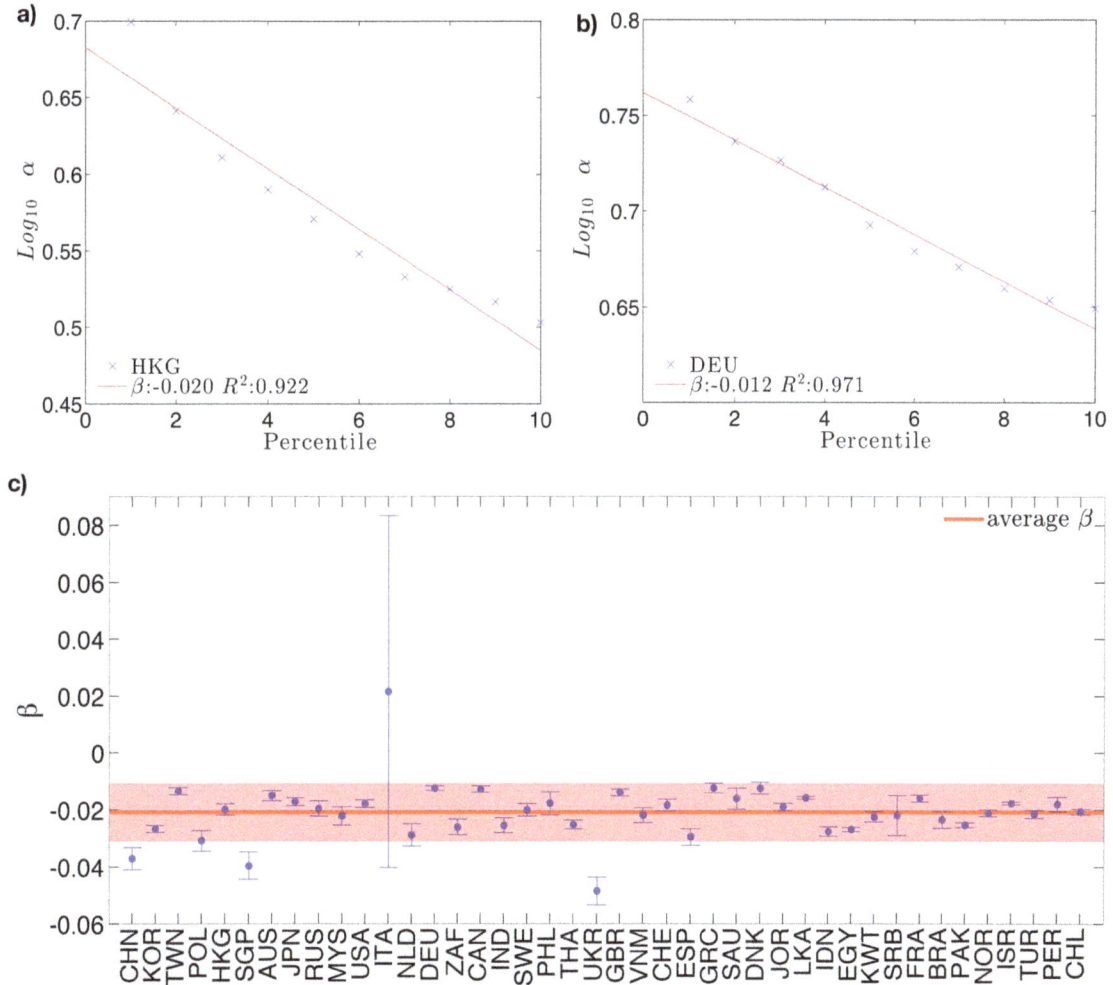

Figure 6. Dependency of α on different percentile cut-offs for two sample countries. a–b. The decay is well fitted by an exponential law $y = Ae^{\beta x}$ for all the countries examined. Values of β from regressions are shown in **c.** where each blue dot represents the coefficient β and its standard error for a specific country. The solid red line is the average value of β on all countries with more than 100 quoted firms. The shaded area in the plot marks one standard deviation. Most of the countries display a consistent decay of α with the percentile used thus making the particular choice of a percentile not relevant.

country dependent business environment indicators is called for and it will be implement in the future. These may include: Small and Medium Enterprises (SME) contributions to countries GDP, Global Competitiveness Index (GCI), and similar. We stres that the present analysis is restricted to quoted firms. It could be interesting to ask whether the influence of SMEs will affect the observed properties of α.

Methods

Triangularity vs. randomness

The firm diversification level is the number of sectors developed by the firm. The real binary firm-sector matrix has a density close to 0.05. We generate a random matrix with same size and density of the real one. In figure 5a we show a comparison of the firm diversification, sorted by fitness [2], between the real data (depicted in red) and the random case (green). The two diversification trends show a similar pattern. This outlines the firms' high specialization and the absence of triangular structure in the matrix. Instead, in Fig. 5b, the real country-sector matrix, generated aggregating firms at country level on the basis of the

legal address, exhibits a clearly nested (triangular) structure such as the country-product matrix [2].

Definition of the revenue diversification barrier and its robustness

The diversification barrier α is measured as the slope of the lower boundary of the scatter plot of diversification vs. revenues in logarithmic space. The lower boundary is defined as the lower 5th percentile of the distribution of revenues for a given diversification level.

We check the sensitivity of α with respect to a variation of the percentile used to define the lower bound.

In fig. 6a–b different values of α for different percentiles are shown, for each country with at least 100 quoted firms. The plot clearly shows a decay trend which is common to (almost) all the countries. We then study in detail this decay of α. In figure 6c we show the angular coefficient (β) of a linear regression between the logarithm of α and the percentile, together with the respective standard error, for each country. For the majority of the countries β lies within one standard deviation from the average (red solid

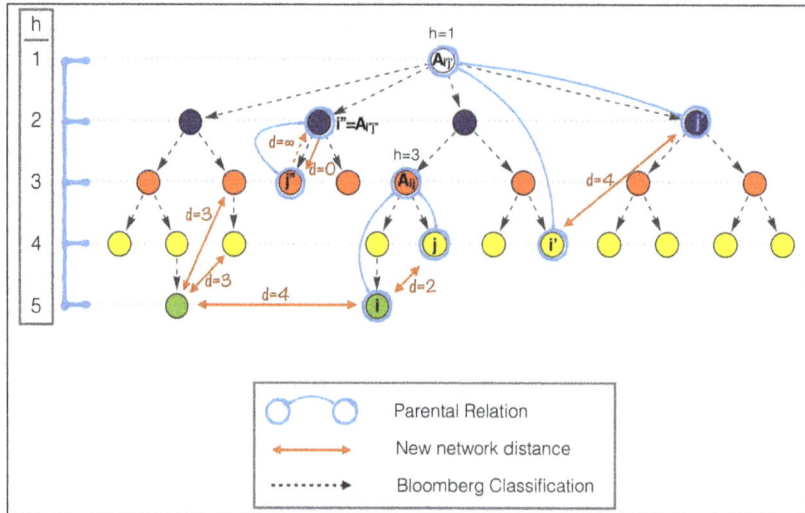

Figure 7. Examples of network distance as defined in Eq. 1.

line). This shows that the consistency of our analysis is not affected by a particular choice of the percentile. Italy shows an anomalous sensitivity dependence with respect to other countries. The χ^2 test over the β regressions in the fifth percentile accept the linear hypothesis at 95% for all the countries. The database we use it is available in Dataset S1.

Diversification coherence

As mentioned, the BICS classification itself defines a topological distance between the codes, more precisely a tree. Each node in the tree corresponds to a more fine specification of the parent element.

Relying on this information we want to develop a measure of how coherently a firm is diversified. In particular we want to be able to weight diversification by a distance among the BICS categories in which diversification occurs: a company diversified in

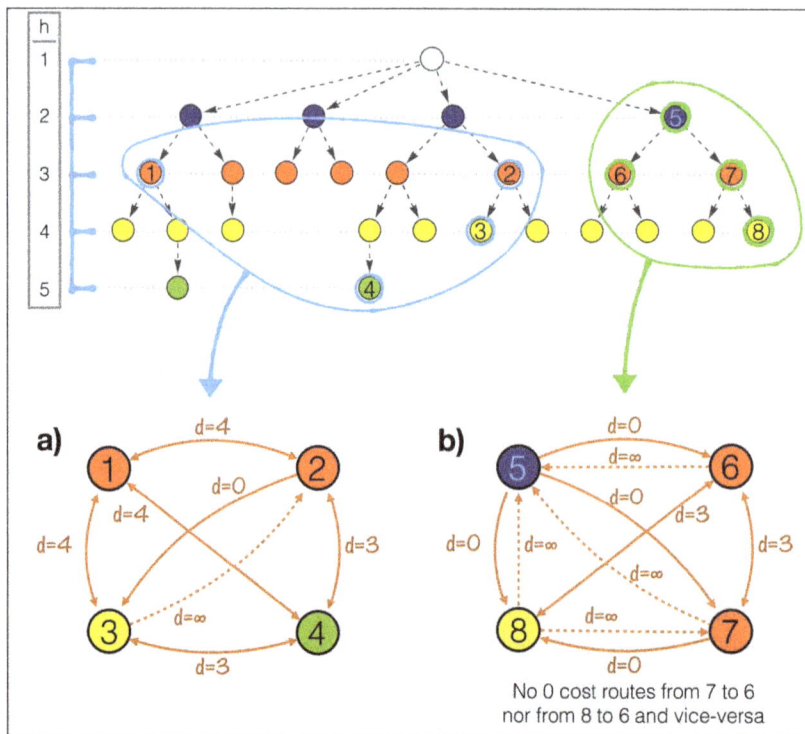

Figure 8. New resulting network. The resulting networks with link weights equal to $d_{i,j}$ for two hypothetical situations are shown in panels a and b. On these networks minimal spanning trees are determined via the Chu-Liu/Edmond's algorithm.

many very close subsectors might be considered less diversified than a company which has revenues only in two very distant sectors.

To this purpose we must take into account the fact that having revenues in a given sector and in one of its subsectors, at any level, does not add to the diversification. For this reason we cannot use the simple topological distance defined by the hierarchical tree implied by the BICS codes. Our approach is to define a new directed network, which is derived from the relations present in the BICS categorization, but with appropriate distances (or link weights). On such a network we use the total weight of the minimal (directed) spanning tree between all the nodes in which a company has revenues as a measure of its coherency.

To this end we need to define a distance (or link weights) that needs to have the following properties:

1. The distance between a sector and one of his subsectors must be 0 (producing pens and red pens does not add to diversification)

2. The distance between two subsectors of the same sector is proportional to the depth of the two subsectors (red pens and blue pens are more far apart than red pens with wooden body and red pens with plastic body)

3. As a consequence of the first property the distance between two sectors (A and B) and two of their respective subsectors (Aa and Bb) must be the same (pens are as distant from rulers as red pens are from metal rulers)

4. The distance between a subsector and its parent element sector must be infinite (to avoid 0 cost spanning trees between subsectors).

As depicted in Fig. 7 this translates in the fact that the distance between two nodes must be a function of depth of the nearest common parent element, except when one of the two nodes is a subsector of the other one, in which case the distance is asymmetric (0 or ∞). In formulae the distance is written as follows:

$$d_{i,j} = \begin{cases} H - h(A_{i,j}) & \text{if } A_{i,j} \neq i \wedge A_{i,j} \neq j \\ 0 & \text{if } A_{i,j} = i \\ \infty & \text{if } A_{i,j} = j \end{cases} \quad (1)$$

where $A_{i,j}$ is the nearest common janitor to the nodes i and j, $h(A_{i,j})$ is its depth in the tree and H is the total depth of the tree plus 1. The application of this definition is illustrated in Fig. 8 where the resulting networks, with link weights equal to $d_{i,j}$, for two hypothetical situations are shown in panels a and b. On these networks minimal spanning trees are determined via the Chu-Liu/Edmond's algorithm [20–22].

Supporting Information

Dataset S1 The dataset. The dataset to replicate the main findings of the article. In the first column there are the companies indicated simply by the country of domicile, in the second column the number of company micro sectors developed and in the third column its total amount of annual revenues in euros.

Information S1 Data specification. The description of the dataset and how the data sanitation was performed.

Acknowledgments

We thank Andrea Zaccaria for many useful discussions. We acknowledge Bloomberg platform for the dataset.

Author Contributions

Conceived and designed the experiments: RDC GLC MC AT LP. Performed the experiments: RDC GLC MC AT LP. Analyzed the data: RDC GLC MC AT LP. Wrote the paper: RDC GLC MC AT LP.

References

1. Hidalgo CA, Hausmann R (2009) The building blocks of economic complexity. Proceedings of the National Academy of Sciences 106: 10570–10575.
2. Tacchella A, Cristelli M, Caldarelli G, Gabrielli A, Pietronero L (2012) A new metrics for countries' fitness and products' complexity. Scientific reports 2.
3. Garlaschelli D, Caldarelli G, Pietronero L (2003) Universal scaling relations in food webs. Nature 423: 165–168.
4. Haldane AG, May RM (2011) Systemic risk in banking ecosystems. Nature 469: 351–355.
5. Knoll AH, Carroll SB (1999) Early animal evolution: emerging views from comparative biology and geology. Science 284: 2129–2137.
6. Gomez-Mejia LR (1992) Structure and process of diversification, compensation strategy, and firm performance. Strategic management journal 13: 381–397.
7. Ansoff HI (1957) Strategies for diversification. Harvard business review 35: 113–124.
8. (2013) Index METHODOLOGY Global Fixed Income Family. Bloomberg Finance L.P.
9. (may 2013) Data collected. Bloomberg Finance L.P.
10. Cracraft J (1985) Biological diversification and its causes. Annals of the Missouri Botanical Garden: 794–822.
11. Walker I (1972) Biological memory. Acta biotheoretica 21: 203–235.
12. Fauver L, Houston J, Naranjo A (2003) Capital market development, international integration, legal systems, and the value of corporate diversification: A cross-country analysis. Journal of Financial and Quantitative Analysis 38: 135–158.
13. Campa JM, Kedia S (2002) Explaining the diversification discount. The Journal of Finance 57: 1731–1762.
14. Khanna T, Palepu K (2000) Is group affiliation profitable in emerging markets? an analysis of diversified indian business groups. The Journal of Finance 55: 867–891.
15. Cristelli M, Gabrielli A, Tacchella A, Caldarelli G, Pietronero L (2013) Measuring the intangibles: A metrics for the economic complexity of countries and products. PloS one 8: e70726.
16. Tacchella A, Cristelli M, Caldarelli G, Gabrielli A, Pietronero L (2013) Economic complexity: conceptual grounding of a new metrics for global competitiveness. Journal of Economic Dynamics and Control 37: 1683–1691.
17. Faccio M, Lang LH (2002) The ultimate ownership of western european corporations. Journal of financial economics 65: 365–395.
18. Shleifer A, Vishny RW (1997) A survey of corporate governance. The journal of finance 52: 737–783.
19. Bianco M, Casavola P (1999) Italian corporate governance:: Effects on financial structure and firm performance. European Economic Review 43: 1057–1069.
20. Chu YJ, Liu TH (1965) On shortest arborescence of a directed graph. Scientia Sinica 14: 1396.
21. Edmonds J (1967) Optimum branchings. Journal of reserach of the national bureau of standards section B- Mathematical Science 4: 233.
22. Edmonds/Chu-Liu algorithm Repository. https://github.com/mlbright/edmonds. Accessed 2014 October 21.

16

Marine Communities on Oil Platforms in Gabon, West Africa: High Biodiversity Oases in a Low Biodiversity Environment

Alan M. Friedlander[1,2]*, **Enric Ballesteros**[3], **Michael Fay**[2,4,5], **Enric Sala**[2,3]

1 Fisheries Ecology Research Laboratory, Department of Biology, University of Hawaii, Honolulu, Hawaii, United States of America, **2** Pristine Seas, National Geographic Society, Washington, DC, United States of America, **3** Centre d'Estudis Avançats-CSIC, Blanes, Spain, **4** Wildlife Conservation Society, Bronx, New York, United States of America, **5** Special Advisor, Presidence de la République, Libreville, République Gabonaise

Abstract

The marine biodiversity of Gabon, West Africa has not been well studied and is largely unknown. Our examination of marine communities associated with oil platforms in Gabon is the first scientific investigation of these structures and highlights the unique ecosystems associated with them. A number of species previously unknown to Gabonese waters were recorded during our surveys on these platforms. Clear distinctions in benthic communities were observed between older, larger platforms in the north and newer platforms to the south or closer to shore. The former were dominated by a solitary cup coral, *Tubastraea* sp., whereas the latter were dominated by the barnacle *Megabalanus tintinnabulum*, but with more diverse benthic assemblages compared to the northerly platforms. Previous work documented the presence of limited zooxanthellated scleractinian corals on natural rocky substrate in Gabon but none were recorded on platforms. Total estimated fish biomass on these platforms exceeded one ton at some locations and was dominated by barracuda (*Sphyraena* spp.), jacks (Carangids), and rainbow runner (*Elagatis bipinnulata*). Thirty-four percent of fish species observed on these platforms are new records for Gabon and 6% are new to tropical West Africa. Fish assemblages closely associated with platforms had distinct amphi-Atlantic affinities and platforms likely extend the distribution of these species into coastal West Africa. At least one potential invasive species, the snowflake coral (*Carijoa riisei*), was observed on the platforms. Oil platforms may act as stepping stones, increasing regional biodiversity and production but they may also be vectors for invasive species. Gabon is a world leader in terrestrial conservation with a network of protected areas covering >10% of the country. Oil exploration and biodiversity conservation currently co-exist in terrestrial and freshwater ecosystems in Gabon. Efforts to increase marine protection in Gabon may benefit by including oil platforms in the marine protected area design process.

Editor: Christina A. Kellogg, U.S. Geological Survey, United States of America

Funding: This work was funded by the Waitt Foundation, Blancpain, Davidoff Cool Water, National Geographic Society, and the Wildlife Conservation Society. The funders had no role in study design, data collection and analysis, decision to publish, or preparation of the manuscript.

Competing Interests: The authors received funding from commercial sources (Blancpain and Davidoff Cool Water).

* Email: alan.friedlander@hawaii.edu

Introduction

Gabon, West Africa, is situated in the Guinea Current Large Marine Ecosystem (GCLME), one of the world's most productive ocean regions, which is rich in fisheries resources, petroleum production, and is crucial to the lives of the region's 300 million coastal residents [1]. Despite this importance, the marine biodiversity of Gabon is poorly known owing to limited financial resources, lack of regional expertise, and a greater emphasis on extractive resources. Previous research has focused on the deep-sea biodiversity of the continental margin associated with oil and gas exploration [2], or related to continental shelf and slope fisheries resources [3]. The continental margins possess small methane-rich cold-see sites and associated non-biogenic carbonate reefs [1,4,5]. However, most of Gabon's shelf area is sandy or muddy, with limited hard bottom habitat. The dozens of oil platforms along the shelf may provide the only hard substrate for thousands of km² [3]. The marine life on these structures has never been studied and here we report on the first scientific surveys of Gabon's oil platforms.

The sea bottom of the Gulf of Guinea is characterized by terrigenous and sandy muds with low habitat and seascape diversity [6]. The Congo River, the second most voluminous river in the world after the Amazon, supplies a large amount of particulate and dissolved organic matter to the ocean, both at the surface and at the seafloor through the Congo Canyon and into the offshore waters of Gabon [7]. In addition, the Ogooué River in central Gabon has an average discharge rate of 4,758 m³ s⁻¹, making it the third most important river in west equatorial Africa [8,9]. Although the water discharge is one order of magnitude lower compared to the Congo River, the particulate load of the two rivers is similar [10]. Collectively these river systems greatly influence the marine ecosystem of Gabon.

The shelf off Gabon is dominated by sandy, sand-shell and gravel bottoms, becoming muddy toward the shelf [3]. Factors contributing to the enrichment in nutrients are related to seasonal up-welling, the discharge from the Congo and Ogooué rivers, and shelf-break upwelling [11]. A strong primary productivity gradient exists from north to south with the south showing higher productivity [12]. Nearshore waters off Gabon are generally warmer, with lower salinity and higher oxygen concentrations compared with more offshore environments [3]. These gradients are likely important to the distribution and abundance of marine organisms along the continental shelf of Gabon.

Ocean productivity in Gabon is dominated by seasonal equatorial upwelling and is strongly influenced by freshwater and sediments from the Congo and Ogooué rivers, particularly in nearshore waters [13]. The Benguela Current is a strong biogeographic barrier that prevents species from the southwestern region of the African continent and the Indian Ocean from colonizing the Gulf of Guinea [14,15]. As a result of these oceanographic processes, the marine communities of the Gulf of Guinea are considered unique and a hotspot of marine biodiversity due primarily to the high proportion of endemic species [16]. Despite the high productivity of the region and its high degree of endemism, we know very little about Gabon's subtidal communities, with the exception of demersal fishes targeted by the large, industrial trawl fishery [3]. There are no true coral reefs in Gabon, and most of the largely unmapped reefs are rocky [17]. Except for a checklist of scleractinian corals for the region [17], there have been no studies of Gabon's subtidal reefs.

Extensive oil exploration and development began onshore in Gabon in the 1950s but offshore exploration did not begin until the 1960s [18]. There are currently >40 offshore oil platforms in Gabon yet virtually nothing is known about the marine biological communities associated with these platforms. Oil and gas platforms act as artificial reefs on continental shelves and provide hard substrate in open water that might otherwise be unavailable to marine organisms requiring such habitat [19,20]. As a result of the limited reefs and the high biodiversity potential of the region, these platforms may provide a unique and important habitat for marine communities in Gabon.

Offshore oil and gas platforms are among the largest artificial structures in the ocean [21]. Worldwide there are >7,500 oil platforms [22,23], and these structures are colonized by diverse ecological communities that have, in some instances, been shown to enhance biodiversity and fisheries production [24–26]. Approximately 40,000 oil and gas wells have been drilled in the northern Gulf of Mexico (GOM) since the 1940s [27]. There are currently about 3,600 oil and/or gas production platforms in the GOM and these platforms act as hard substratum upon which many reef organisms can settle and grow in a region where very little hard substratum now exists, nor has existed in recent geological time [27,28]. Oil and gas platforms in southern California harbor diverse fish and invertebrate populations that are considered *de facto* artificial reefs and serve as important habitat for a number of species [29–31]. Off the north-western Australia continental shelf, a diverse range of taxa were observed on oil industry structures, including reef-dependent species and transient pelagics [23]. These platforms provide habitat that potentially increases the growth and survival of individuals, affording shelter for protection from predation and spawning substrate, and acting as a visual attractant for organisms not otherwise dependent on hard bottom, such as pelagic species [32]. However, hazards to shipping, exclusion of certain fisheries (e.g., trawling), spread of alien species, and potential spills and leaks have all been cited as some of the negative consequences associated with oil platforms, particularly older ones that are no longer in production [33–35].

The objectives of this study were to describe the marine communities on oil platforms in Gabon, determine the biogeographic affinities of the associated species, and discuss the implications of these platforms for marine conservation.

Methods

The Government of Gabon granted all necessary permission to conduct this research. No vertebrate sampling was conducted and therefore no approval was required by the Institutional Animal Care and Use Committee.

Oil platforms

The oil platforms off Gabon investigated in this study sit on sandy and muddy bottoms at depths ranging from 26 to 102 m (Table 1, Fig. 1). The platforms had three to four vertical pillars linked by crossbeams at different depths. The shallowest crossbeams were at depths between 6–10 m, and the second shallowest ('deep' crossbeam hereafter) between 13–24 m (Table 1). There were a number of deeper crossbeams depending on the size and depth of the platform but these were beyond scuba depths and were not surveyed. We conducted quantitative surveys of benthic communities and fish assemblages along the two shallowest crossbeams at each of the 10 oil platforms surveyed. Since these crossbeams were not independent samples, values from benthic and fish surveys were averaged to derive density estimates for each platform.

Benthic cover

Benthic cover was quantified *in situ* on the crossbeams using 0.25 m×0.25 m quadrats divided into 25 sub-quadrats of 5×5 cm [36,37]. Eight quadrats were positioned haphazardly along every crossbeam, resulting in 24–32 quadrats for every depth and platform. Algae and other benthic cover (sponges, hydrozoans, soft and hard corals, barnacles, mollusks, echinoderms, and ascidians) within each quadrat were identified to the lowest possible taxonomic level. The percentage of sub-quadrats in which a species appeared was recorded and used as a measure of occurrence. A highly abundant taxon that appeared in all 25 subquadrats would produce a presence of 100%, whereas the total lack of a taxon would produce a presence of 0% [36]. The final abundance of a taxon within each depth and oil platform was then calculated as the mean of the percentage presence values of the quadrats sampled. The sum abundance of all quadrats exceeded 100% due to the combined presence of several taxa on the same sub-quadrat. Field visual survey methods did not allow identification of the smallest species (turf and filamentous algae, hydrozoans), which were sometimes pooled into groups during visual estimations. Vagile crustaceans (e.g. decapods) were not quantified in the quadrats, although the presence of lobsters during the survey was noted since they were conspicuous and commercially important.

Fish assemblages

We conducted underwater visual fish censuses on oil platforms using two methods. First, we estimated the numerical density and size of individuals of each species along 4-m wide × 4-m tall belt transects along the 'shallow' crossbeams and the 'deep' crossbeams on each oil platform. Depth and length of transects varied corresponding to the length of the crossbeam where the transects were located (Table 1). Crossbeam lengths were measured upon completion of each quantitative survey. Because many of the schooling pelagic species (e.g., jacks, barracudas) were loosely associated with the platform structure, we estimated the numbers

Table 1. Descriptions of oil platforms sampled along the coast of Gabon, West Africa.

Date	Platform no.	Company	Install year	Lat.	Long.	Bottom depth (m)	Deep* crossbeam			Shallow beam			Flare
							Depth (m)	Length (m)	Area (m⁻²)	Depth (m)	Length (m)	Area (m⁻²)	
9-Oct-12	1	Total Gabon	1973	−1.509	8.827	43	19.0	16.0	64.0	6.0	13.0	52.0	Yes
9-Oct-12	2	Total Gabon	1980	−1.738	8.738	45	17.0	19.6	78.4	6.0	16.6	66.4	No
10-Oct-12	3	Total Gabon	1975	−1.103	8.594	62	19.0	18.3	73.0	6.0	15.0	60.0	No
10-Oct-12	4	Total Gabon	1974	−1.155	8.475	102	20.2	18.1	72.5	7.1	12.4	49.5	No
13-Oct-12	5	Vaalco	2001	−3.717	10.436	45	13.3	13.0	52.0	-	-	-	No
13-Oct-12	6	Vaalco	2001	−3.877	10.590	45	13.2	13.0	52.0	-	-	-	No
15-Oct-12	7	Perenco	1975	−2.039	9.105	45	23.8	6.0	24.0	6.7	9.0	36.0	No
18-Oct-12	8	Perenco	1975	−1.357	8.892	26	15.0	10.0	40.0	-	-	-	No
18-Oct-12	9	Perenco	1975	−1.177	8.703	31	20.0	5.0	78.5	10.0	5.0	78.5	No
18-Oct-12	10	Perenco	1975	−1.033	8.592	50	19.0	20.0	80.0	8.0	18.0	72.0	Yes

* 'Deep' is the second shallowest crossbeam of each platform.

and sizes of these species that were visible during the dive to calculate their standing stock around each platform since they were underrepresented on quantitative transects. Although estimates associated with this method are constrained by schooling behavior, water clarity, and species identification, it complements the finer resolution belt transects by providing valuable information on highly vagile, and commercially important species that would have otherwise not been quantified.

Length estimates of fishes from both survey methods were converted to mass (M) using the following length–mass relationship: $M = aTL^b$, where the parameters a and b are constants for the allometric growth equation and TL is total length in cm [38]. From belt transects, all biomass estimates were converted to metric tons per hectare (t ha^{-1}) and numerical abundance estimates were converted to number of individuals m^{-2}. All fishes observed were categorized into four trophic groups (piscivores, carnivores, planktivores, and herbivores; [39,40]), and classified into province and regional affinities based on [41].

Statistical analysis

Non-metric multi-dimensional scaling (nMDS) analysis was used to examine differences in benthic communities and fish assemblages among oil platforms (PRIMER v.5 – [42]). Separate Bray–Curtis similarity matrices were created using abundance of benthic components at the lowest possible taxonomic level and fish numerical density (no. m^{-2}). Prior to conducting the nMDS, benthic data were arcsin square root transformed, while fish density was square root transformed. The primary taxa vectors driving the ordination (Pearson correlation Product movement correlations ≥0.9 for benthos, ≥for 0.7 for fishes) were overlaid on the nMDS plot to visualize the major taxa that explained the spatial distribution patterns observed

Benthic taxon and fish species diversity were calculated from the Shannon-Weaver diversity index [43]: $H' = - \sum (p_i \ln (p_i))$, where p_i is the proportion of all individuals counted that were of taxa i. Fish trophic biomass data did not conform to parametric statistical assumptions despite various transformations, therefore comparisons among trophic groups were conducted using a Kruskal-Wallis rank-sum test (H) with Dunn's test for unplanned multiple comparisons [44].

To describe the pattern of variation in community structure (patterns of distribution of biomass/abundance of high-level functional groups within the community) and their relationship to environmental-human gradients we used linear ordination methods. Linear models are appropriate for these data because a preliminary detrended correspondence analysis showed short gradient lengths (<2 SD) [45]. To explore the spatial distribution of community structure across oil platforms and its relationship with environmental variables (see below), we performed a direct gradient analysis (redundancy analysis: RDA) on the log-transformed community and environmental data matrices using the ordination program CANOCO for Windows version 4.0 [46]. The RDA introduces a series of explanatory (environmental) variables and resembles the model of multivariate multiple regression, allowing us to determine what linear combinations of these environmental variables determine the gradients. We pooled data from all taxa into the following community groups to facilitate the large-scale analysis: biomass of fish trophic groups (e.g., piscivores, carnivores, planktivores, and herbivores) and higher order benthic groups (e.g., Rhodophyta, Porifera, Cnidaria-Anthozoa, Cnidaria-Hydrozoa, Crustacea, Mollusca-Bivalvia).

The environmental data matrix included the following variables: latitude, longitude, depth, age of oil platform, platform area, distance offshore, and distance from the Ogooué and Congo river

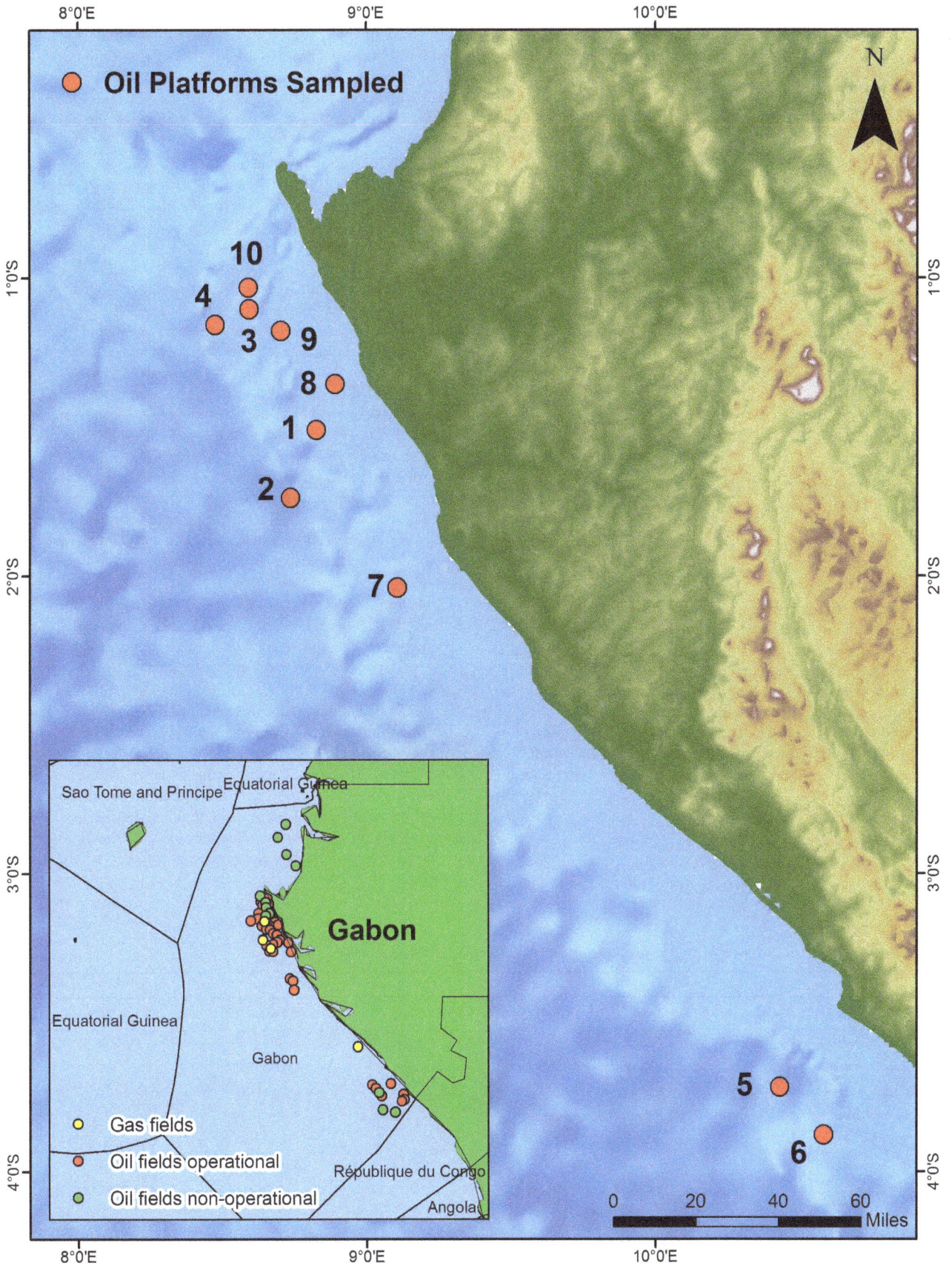

Figure 1. Oil platforms sampled along the coast of Gabon, West Africa. Platform numbers and descriptions are found in Table 1.

mouths. Correlations among environmental variables were examined using Spearman's rank-order correlation coefficients. Distance to the Congo and Ogooué rivers were significantly correlated with latitude and longitude (all p<0.05) and were excluded from the analysis. Longitude and depth, which were significantly correlated with latitude and platform area respectively (p<0.05), also were excluded from the analysis. To rank environmental variables in their importance for being associated with the structure of communities, we used a forward selection where the statistical significance of each variable was judged by a Monte-Carlo permutation test [47].

Results

Benthic composition

A total of 45 benthic taxa were recorded within quadrats on oil platforms sampled during our expedition (Table S1). Anthozoans made up 32.2% of the mean percent presence, followed by Cirriped crustaceans (19.9%), sponges (10.9%), bivalve mollusks (10.2%), and hydrozoans (8.3%). A solitary, undetermined cup coral (*Tubastraea* sp.) was the dominant species on many of the platforms, occurring on all platforms and accounting for 30.7% of the mean percent presence of all taxa (Table 2, Fig. 2). A barnacle, *Megabalanus tintinnabulum*, comprised 18.6% of the mean percent presence of all taxa and was present on 80% of the platforms. A sponge, *Haliclona* sp., accounted for 9.2% of the mean percent presence of all taxa, followed by an unidentified hydrozoan (9.2%), and two bivalve mollusks – *Crassostrea gasar* (7.6%) and *Dendostrea frons* (6.4%).

Based on benthic community composition, the platforms formed three distinct clusters in ordination space (Fig. 3). The clusters dominated by *Tubastraea* sp. were highly concordant and well separated from the platforms to the south (5 and 6) and the two inshore platforms (8 and 9). These southern and inshore platforms were dominated by the barnacle *Megabalanus tintinnabulum*, but had much more diverse assemblages compared to the ones dominated by *Tubastraea* sp. The average number of taxa was highest on the inshore platforms ($\overline{X}$ = 18.0±5.7) and lowest on the *Tubastraea* dominated ones ($\overline{X}$ = 14.8±2.4). Diversity was highest on the southern platforms ($\overline{X}$ = 2.0±0.1) and lowest on the *Tubastraea* dominated platforms ($\overline{X}$ = 1.5±0.4).

Although not present on quadrats, we commonly observed two commercial species of lobsters on the platforms: spiny lobsters (*Panulirus regius*) and slipper lobsters (*Scyllarides herklotsii*). Spiny lobsters were found in groups of 2–4 individuals and were of small size (10–15 cm total body length), while slipper lobsters were often found alone or in pairs and attained larger adult sizes (20–25 cm total length). No estimates of lobsters by platform are provided as we did not use a specific protocol for quantifying these species.

Fish assemblage characteristics

Based on all quantitative and qualitative surveys we observed a total of 65 fish species associated with oil platforms during this study (Table S2). Of these, 22 (34%) are new records for Gabon and 4 (6%) are new to the tropical West African coast. Assemblage richness consisted of 32% tropical eastern Atlantic (TEA) endemics and 31% amphi-Atlantic species (those found on both sides of the Atlantic), followed by circumtropical species (17%). Based on quantitative surveys, amphi-Atlantic species accounted for 73% of the numerical abundance, and 76% of the biomass observed on oil platforms along the coast of Gabon. Species endemic to the TEA only accounted for 21% of the numerical abundance and 16% of the biomass observed on quantitative surveys, while circumtropical species accounted for an additional 4% of the abundance and 2% of the biomass. Of the 65 species observed, only 38 (42%) were recorded on our quantitative surveys. Most of the species not observed on quantitative surveys were either pelagic or encountered at the bases of the platforms, associated with the seabed, and therefore not encountered on transects on the crossbeams.

The number of fish species on transects ranged from a low of 7 on platform 6 to a high of 22 on platform 1 ($\overline{X}$ = 14.6±4.1). Higher numbers of species were observed on the platforms to the north, which were also the oldest platforms with the highest abundance of *Tubastraea* sp. The number of individuals observed on transects ranged from 0.9 m^{-2} on platform 2 to 8.5 m^{-2} on platform 10 ($\overline{X}$ = 3.9±2.3), with no apparent pattern observed. Fish biomass among platforms also showed no clear pattern, with the highest biomass observed on platform 6 (10.8 t ha^{-1}) and the lowest on platform 8 (0.9 t ha^{-1}) with a mean of 4.3 (±3.3). Fish diversity varied without pattern and averaged 1.1 (±0.3) with the lowest diversity on platform 6 (H' = 0.5) and highest on platform 9 (H' = 1.6). Unlike the benthic communities, the fish assemblages associated with oil platforms did show spatial separation, with a small number of species accounting for much of the spatial variability (Fig. 3).

Table 2. Percent occurrence of major taxa found on oil platforms in Gabon.

Group	Taxa	Mean % (sd)	% freq.
Cnidaria-Anthozoa	*Tubastraea* sp.	47.34 (39.50)	100
Crustacea-Cirripedia	*Megabalanus tintinnabulum*	28.63 (33.33)	80
Sponge	*Haliclona* sp.	14.23 (15.61)	80
Cnidaria-Hydrozoa	*Hidrarians* unidentified	12.73 (16.29)	90
Mollusca-Bivalvia	*Crassostrea gasar*	7.61 (9.52)	100
Mollusca-Bivalvia	*Dendostrea frons*	6.36 (9.30)	60
Red alga	*Antithamnionella elegans*	6.15 (7.36)	60
Green alga	*Bryopsis plumosa*	5.25 (6.92)	50
Bryozoa	*Pentapora*-like	3.03 (5.10)	30
Red alga	Red algal turf	2.36 (5.15)	40

The mean is the percentage of the sub-quadrats were the taxa was observed. % freq. is the percentage of platforms upon which this taxon was observed (n = 10).

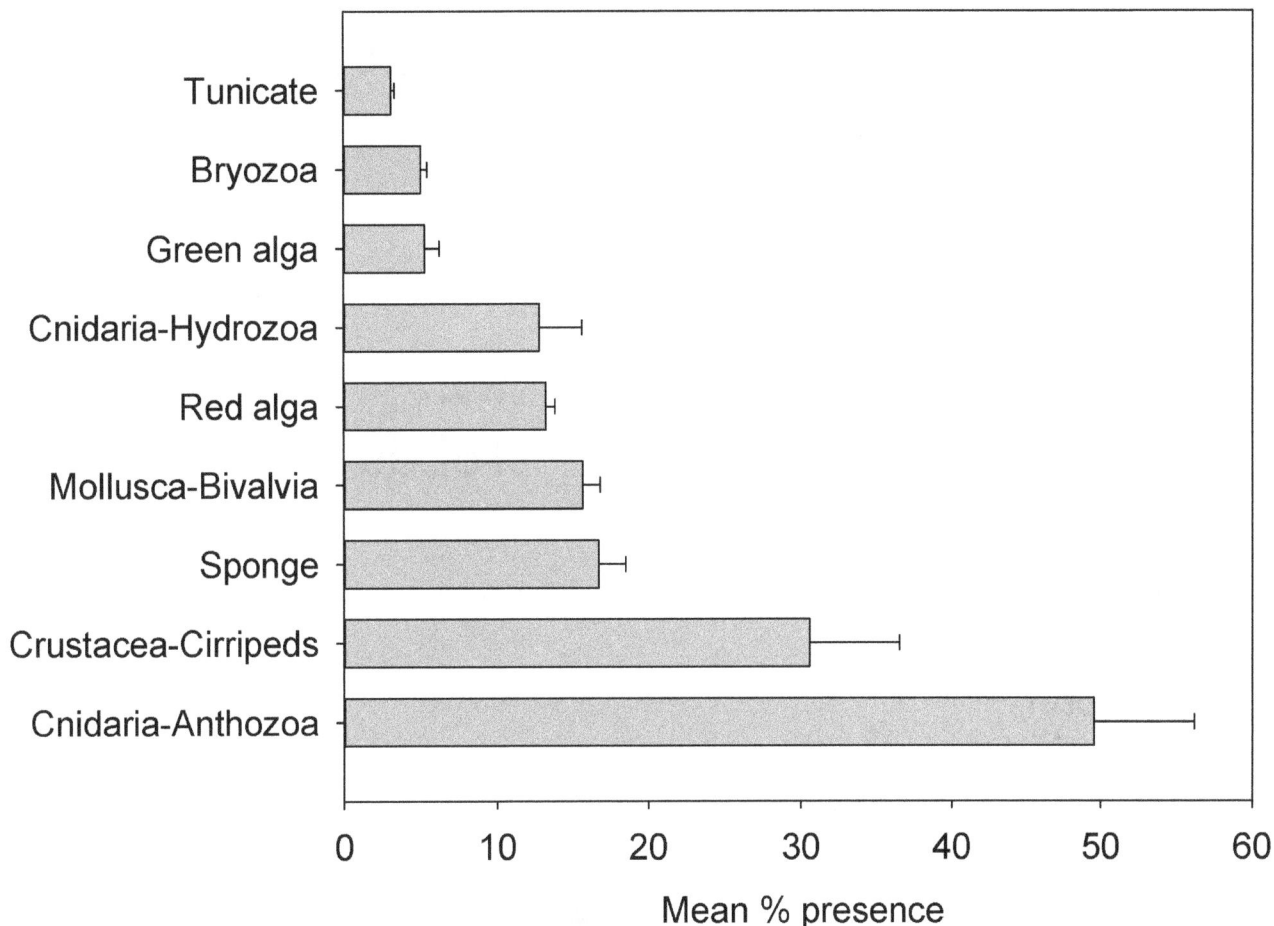

Figure 2. Benthic taxa on transects associated with oil platforms in Gabon, West Africa. Values are mean percent presence and standard error of the mean.

Fish trophic structure

Based on transect data, there were significant differences in the biomass and abundances among trophic groups observed on oil platforms (Fig. 4). Planktivores accounted for 74.0% of the numerical abundance and 36.8% of the overall biomass on platform transects. Planktivore abundance was highest at the two most southern platforms, which was also where the highest primary productivity occurs. Piscivores were the next most important trophic group by weight, accounting for 35.8% of the total biomass but 15.7% of the total abundance, with the highest biomass observed at the three north platforms (4, 9, and 10). Herbivores accounted for an additional 20.6% of the total biomass but only 4.3% of the numerical abundance. More than 80% of the total herbivore biomass was observed on the two most southern platforms. Invertivores accounted for 6.9% of the biomass and 5.9% of the abundance with no apparent pattern among platforms.

Fish species composition

The creolefish, *Paranthias furcifer*, was the dominant fish species both numerically (32.2%) and by weight (26.3%) on oil platform transects (Table 3 and 4). The brown chromis, *Chromis multilineata*, was the second most important species numerically (27.1%) but only accounted for 3.4% of the total biomass. This species was followed in abundance by the African sergeant,

Abudefduf hoefleri, comprising 13.4% of the total assemblage by numbers and 7.2% by weight. The second most important species by weight was the Bermuda chub, *Kyphosus sectator*, accounting for 23.5% of the total biomass, followed by blue runner, *Caranx crysos* (14.9%).

Associated fish assemblage

Since many of the vagile pelagic schooling species were not recorded on transects, we estimated total abundance and sizes of these species over the course of the dive that were associated with each oil platform. Nine species, primarily jacks and barracuda, made up the vast majority of schooling pelagic species associated with these platforms (Table 5). Estimates of standing stock were based on the estimated numbers in each school and the median size of individuals in the school. The standing stock of these pelagic species averaged 0.83 t ($\pm$0.24) per platform with a minimum of 0.57 tonnes on platform 8 and a maximum of 1.18 tonnes on platform 1. Great barracuda (*Sphyraena barracuda*) accounted for 33.2% of the average standing stock, followed by longfin jack crevalle (*Caranx fischeri* - 33.0%), and rainbow runner (*Elagatis bipinnulata* - 12.7%)

Fish spawning and courtship

On October 12, 2012, at 3 pm local time, we observed pair spawning of red snapper, *Lutjanus dentatus*, on platform no. 4 off

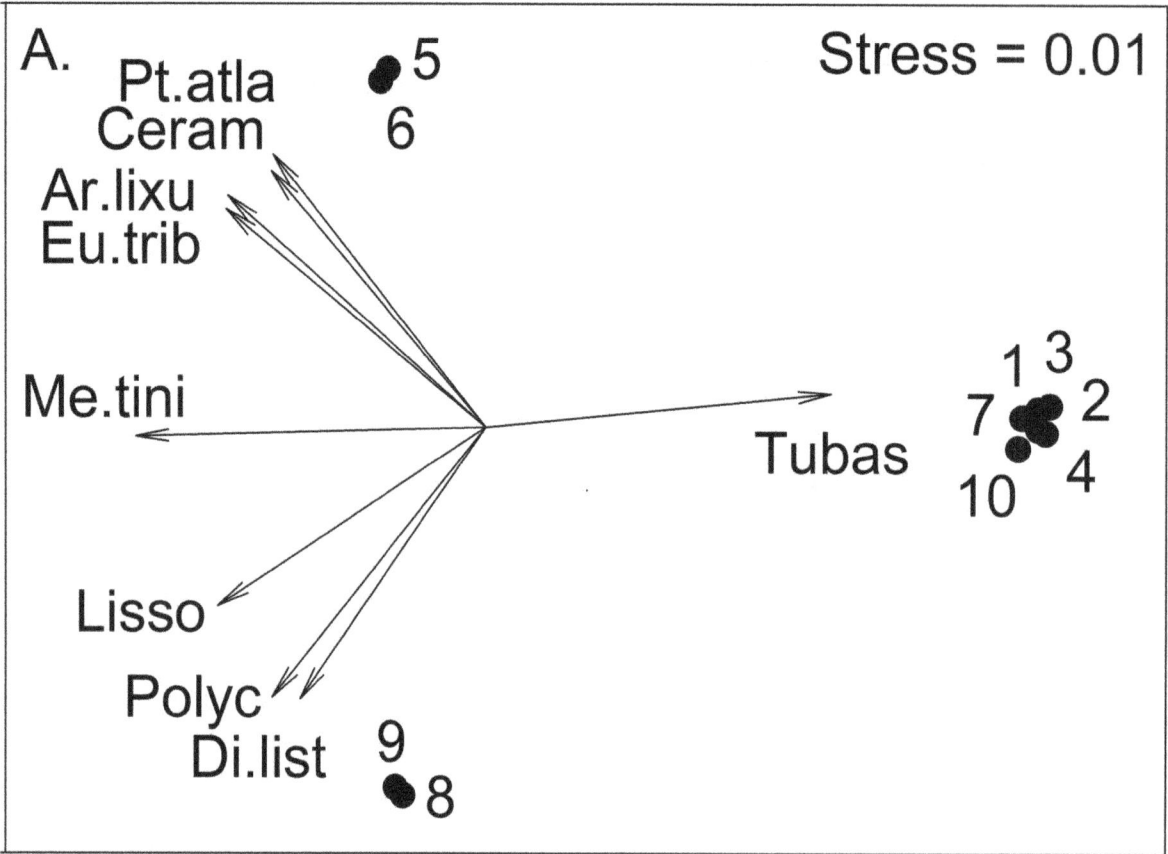

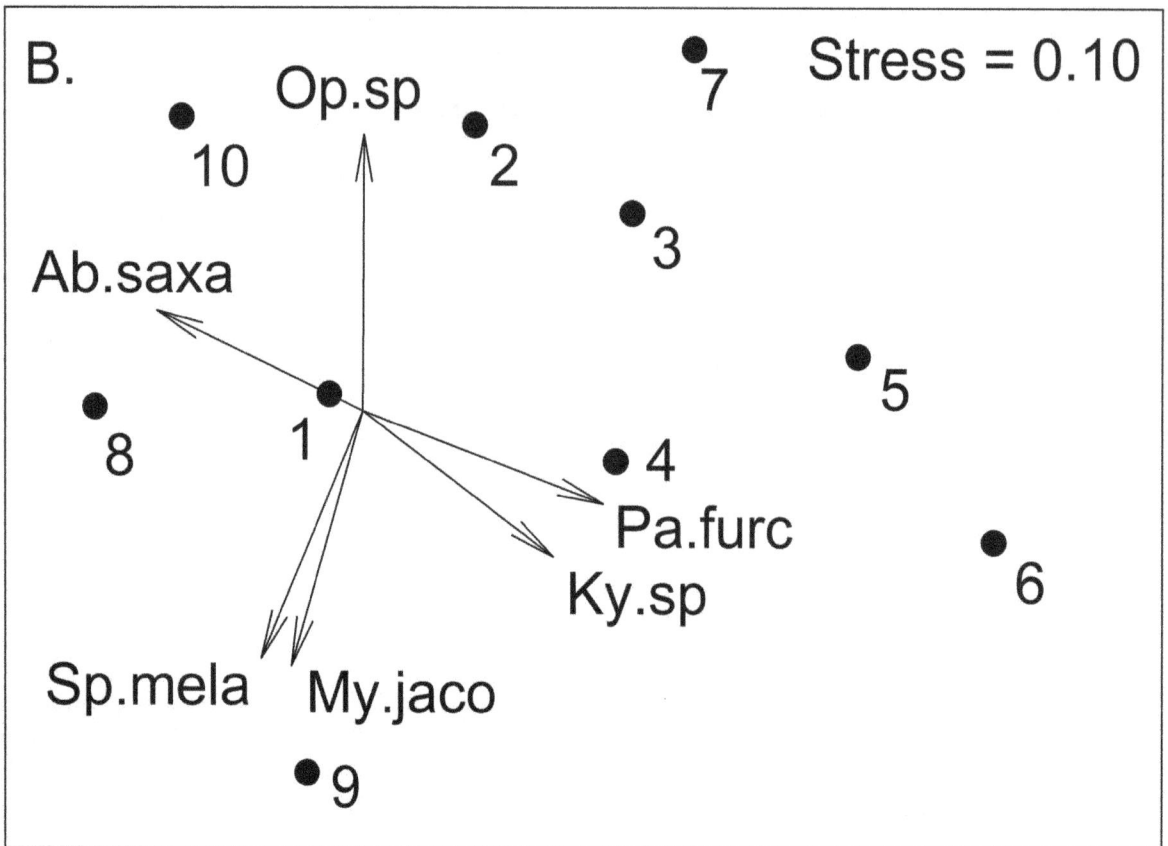

Figure 3. Non-metric multidimensional scaling plot of benthic communities and fish assemblages associated with oil platforms in Gabon. Vectors are the primary taxa driving the ordination (Pearson Product movement correlations ≥0.9). A. benthic communities species codes: Ar.lixu = *Arbacia lixula v. africana*, Ceram = *Ceramiaceae* unidentified, Di.list = *Diplosoma listerianum*, Eu.trib = *Eucidaris tribuloides v. africana*, Liss = *Lissoclinum* sp., Me.tini = *Megabalanus tintinnabulum*, Poly = Polyclinidae unidentified, Pt.atla = *Pteria atlantica*, Tubas = *Tubastraea* sp. B. Fish assemblage species codes: Ab.saxa = *Abudefduf saxatilis*, Ky.sp = *Kyphosus* sp., My.jaco = *Myripristis jacobus*, Op.sp = *Ophioblennius* sp., Pa.furc = *Paranthias furcifer* Sp.mela = *Spicara melanurus*.

Cape Lopez. A male (~70 cm) and a female (~50 cm) swam together from deeper water towards the surface, forming a spiral, and swimming apart while releasing eggs and sperm at 10 m depth. We observed two consecutive spawning events; the first outside of the platform structure, and the second within the structure. Immediately after spawning, dozens of creolefish swam towards the spawn cloud, to eat the eggs. We also observed the courtship behavior of the yellow jack, *Caranx bartholomaei*, at several platforms. Yellow jacks formed schools of up to several hundred individuals, most with typical silver body coloration. In the afternoon, a small number of individuals developed a dark coloration, and silver-dark pairs formed and swam together away from the school. Although we did not observe spawning, we believe this is courtship behavior that has been identified for other species of jacks (e.g., [48]).

Community structure

Ordination of biomass/abundance of functional groups among oil platform communities revealed three groups of sites mainly consisting of: (1) platforms dominated by anthozoans and red algae (2) those dominated by piscivorous feeding fishes and sponges, and (3) platforms dominated by planktivorous and herbivorous fishes along with hydroids and bivalves (Fig. 5). The first two axes of the RDA biplot explained 78% of the functional group variance and 97% of the functional group-environment relationship (Table 6). The main factors involved in this ordination are platform age and area, which are orthogonal to one another in ordination space (Fig. 5, Table 7). Platform age and latitude showed similar ordination trajectories, reflecting the history of expansion of oil exploration in Gabon starting in Port Gentil and moving south to the border with Congo. Platform area is correlated with water depth and distance offshore, with assemblages becoming more diverse with increasing platform area, water depth, and distance from shore.

Discussion

This work represents the first scientific surveys of the marine ecosystems on oil platforms off West Africa. We found an extraordinary diversity and biomass of fishes and rocky invertebrates on these platforms that are in sharp contrast to much of Gabon's marine environment, which is dominated by soft sediment communities. Quantitative transect surveys were most useful in describing the biodiversity and abundance of "resident" fishes and benthic communities on oil platforms but they underestimate the standing stock of pelagic species that are not as closely associated with the platforms as other species. The estimated fish biomass on oil platforms exceeded 1 ton and was dominated by pelagic species (barracuda, rainbow runner, jacks). Large snappers (primarily *L. dentatus*) were often deeper and not easily observed, so the actual standing stock of fish biomass on these platforms is likely much higher than our estimates suggest. Elsewhere, the potential production around platforms is high as indicated by estimates of biomass of cod (*Gadus morhua*) and saithe (*Pollachius virens*) around oil platforms in the North Sea, which range from 7–12 tonnes [49].

Owing to the known prevalence of resource species on these platforms, spear fishermen frequent them, particularly out-of-service ones near the population centers of Libreville and Port Gentil. Videos and anecdotal information suggest substantial hunting of goliath grouper (*Epinephelus itajara*), barracuda, cobia (*Rachycentron canadum*), Guinean pompano (*Trachinotus maxillosus*), cubera snapper (*Lutjanus agennes*), jacks (Carangidae) and several species of carcharhinid sharks. Despite this fishing pressure, fish biomass at some platforms is larger than is reported in the literature for most tropical reef fishes, and even higher than many pristine reefs surveyed in the Pacific [50–52]. This high standing stock of fishes is possible because of the high productivity of the region, and because many of the platforms are largely unfished since they are de facto MPAs owing to security restrictions. As on pristine reefs, top predators account for a large part of the fish biomass. Before the systematic spearfishing of the large groupers, snapper, jacks, and sharks, biomass and dominance of predators must have been even higher.

Most of the fish biomass on the platforms was composed of pelagic species with broad biogeographic distributions. However, much of the observed species richness consisted of demersal species, many of which had distinct and unique assemblages. Studies at nearby São Tomé and Príncipe found the archipelago fish fauna to consist of both western and eastern Atlantic species and noted that the easterly flowing Equatorial currents (the seasonal Equatorial Counter Currents and the subsurface Equatorial Undercurrent) link the western Atlantic and the eastern Atlantic at this latitude [53,54]. Likewise, the marine invertebrate fauna of São Tomé and Príncipe is known to consist of a mix of the two faunal regions [17,55–59]. Oil platforms off Gabon may therefore help extend this unique assemblage onto the West African shelf.

Despite its proximity to the African continent, these platforms harbor a unique fish fauna, with a large proportion of the assemblage composed of amphi-Atlantic species. Several amphi-atlantic species, which, in the Eastern Atlantic occur only around oceanic islands (e.g., *Epinephelus adscensionis, Paranthias furcifer, Mulloidychtis martinicus, Bodianus pulchellus, Chromis multilineata, Gnatholepis thomsoni, Melychthis niger*, [53]) were observed on the oil platforms off Gabon. The wrasse *Thalassoma newtoni* was considered endemic to the islands of São Tomé but has recently been recorded from Senegal [60]. We observed this species on all of the platforms surveyed, where it was very common. These facts highlight how these platforms act like small islands in an otherwise featureless seascape across the continental shelf of Gabon. We observed spawning and spawning behavior for two commercially important snappers on the platforms, and they are likely spawning sites for numerous other species as well. Therefore, these platforms are likely sources for the replenishment of other platforms and the scarce reefs in the region, acting like distinct populations in a metapopulation [61]. They could be of great value to fisheries conservation for commercially harvested fish species that associated with these platforms.

Some species of invertebrates from our surveys have not been successfully identified to the species level and taxonomists are

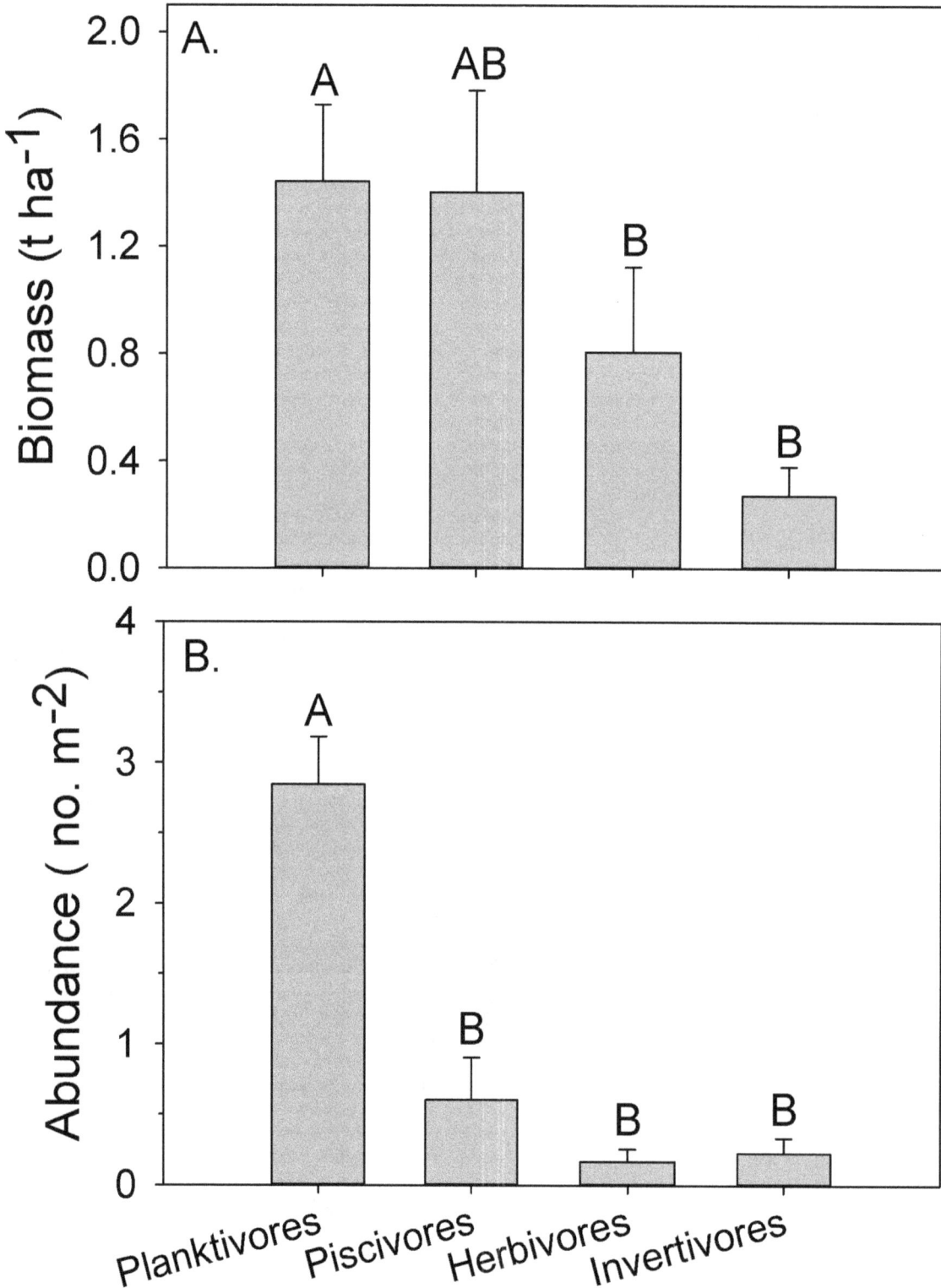

Figure 4. Comparisons of trophic (a) biomass (t ha⁻¹) and (b) abundance (no. m⁻²) at oil platforms in Gabon. Values are means and standard error. Kruskal-Wallis Rank Sum comparisons among trophic groups were statistically different for biomass (H = 12.3, p = 0.007) and abundance (H = 35.3, p < 0.001). Trophic groups with the same letter are not significantly different (Dunn's unplanned multiple comparisons procedures, $\alpha = 0.05$).

Table 3. Biomass (g m^{-2}) of the top ten species observed on quantitative transects on oil platforms in Gabon, West Africa.

Family	Species	Common name	Distribution	Trophic	Mean (sd)	%
Groupers	*Paranthias furcifer*	Creolefish	Atlantic	Plank	112.7 (149.3)	26.3
Chubs	*Kyphosus sp*	Bermuda Chub	Atlantic	Herb	101.0 (213.5)	23.5
Jacks	*Caranx crysos*	Blue Runner	Atlantic	Pisc	63.8 (127.7)	14.9
Damselfishes	*Abudefduf hoefleri*	African Sergeant	West Africa	Plank	30.8 (32.5)	7.2
Snappers	*Lutjanus fulgens*	Golden African Snapper	West Africa	Pisc	16.2 (48.6)	3.8
Damselfishes	*Chromis multilineata*	Brown Chromis	Atlantic	Plank	14.4 (14.8)	3.4
Surgeonfishes	*Acanthurus monroviae*	Monrovia Doctorfish	Atlantic	Herb	13.9 (14.9)	3.2
Barracuda	*Sphyraena barracuda*	Great Barracuda	Circumtropical	Pisc	11.1 (15.3)	2.6
Jacks	*Caranx fischeri*	Longfin Crevalle Jack	West Africa	Pisc	10.8 (12.4)	2.5
Hawkfishes	*Cirrhitus atlanticus*	West African Hawkfish	West Africa	Invert	7.4 (7.6)	1.7

Values are means and one standard deviation. Trophic groups – Plank = planktivores, Herb = herbivores, Pisc = piscivores, Invert = invertivores. Distributions derived from [82,83]. % is the percent contribution of each species to total fish biomass.

currently working on them. Although several species of scleractinian corals have been reported from Gabon [10], no true coral reefs are found. Despite the fact that oil platforms provide suitable substrate for the settlement of corals in other locations, no hermatypic corals were recorded on the platforms during our surveys. The ahermatypic dendrophyllid *Tubastraea* is circumtropical in distribution but likely a relatively recent invader to the Gulf of Guinea, possibly coming from the Indo-Pacific via the Panamenian region (Óscar Ocaña, pers. comm.). The taxonomy of the genus *Tubastraea* is currently under debate.

Conservation implications

Artificial reefs are increasingly being applied in mitigation efforts for natural systems – including their use as physical barriers to discourage illegal trawling in seagrass beds in Western Europe [62]. Oil platforms have been described as "de facto marine protected areas" [63], because they exclude trawl fishing and their large internal spaces offer shelter to fishes and other organisms. Platforms are complex structures, involving numerous crossbeams and large interstitial spaces, which increase overall habitat complexity and likely support high reef fish diversity and abundance.

The presence of oil platforms in the marine environment can create a number of environmental and social problems. The sinking of the Deepwater Horizon drilling rig created the largest marine oil spill in history (nearly 5 million barrels) and resulted in significant ecological and economic damage to the GOM region [64,65]. Smaller spills and leaks from offshore platforms commonly occur around the world, and as oil exploration and extraction moves into ever deeper water and into stormier and icier seas, potential risks will increase [66,67]. Invasive and exotic species have been reported on oil platforms around the world, where they compete with native species for space [20,28,68]. This is also an issue with decommissioned platforms that may be moved to a distant locations and enhance the spread of these exotic and invasive species [69]. Offshore platforms can also be a source of chronic stress that can lead to sub-lethal impacts on resident benthic organisms, resulting in loss of biodiversity [70].

Although often controversial, rigs-to-reef programs that allow decommissioned oil platforms to stay in place have gained some support by governmental and non-governmental organizations

Table 4. Numerical abundance (no. m-2) of the top ten species observed on quantitative transects on oil platforms in Gabon, West Africa.

Family	Species	Common Name	Distribution	Trophic	Mean (sd)	%
Groupers	*Paranthias furcifer*	Creolefish	Atlantic	Plank	1.25 (1.68)	32.2
Damselfishes	*Chromis multilineata*	Brown Chromis	Atlantic	Plank	1.05 (1.13)	27.1
Damselfishes	*Abudefduf hoefleri*	African Sergeant	West Africa	Plank	0.52 (0.55)	13.4
Jacks	*Caranx juveniles*	Jacks	Atlantic	Pisc	0.31 (0.99)	8.0
Damselfishes	*Abudefduf saxatilis*	Sergeant-Major	Atlantic	Plank	0.14 (0.18)	3.6
Snappers	*Lutjanus fulgens*	Golden African Snapper	West Africa	Pisc	0.11 (0.34)	2.8
Chubs	*Kyphosus sp*	Bermuda Chub	Atlantic	Herb	0.10 (0.18)	2.6
Jacks	*Caranx crysos*	Blue Runner	Atlantic	Pisc	0.07 (0.14)	1.8
Damselfishes	*Stegastes imbricatus*	Cape Verde Gregory	West Africa	Herb	0.06 (0.08)	1.5
Wrasses	*Thalassoma newtoni*	Newton Wrasse	West Africa	Invert	0.06 (0.09)	1.5

Values are means and one standard deviation. Trophic groups – Plank = planktivores, Herb = herbivores, Pisc = piscivores, Invert = invertivores. Distributions derived from [82,83]. % is the percent contribution of each species to total fish abundance.

Table 5. Estimates of abundance and sizes of large schooling resource species associated with oil rigs in Gabon.

Species	Platform 1	Platform 2	Platform 3	Platform 4	Platform 5	Platform 6	Platform 7	Platform 8	Platform 9	Platform 10
Caranx bartholomaei										100
Yellow Jack										(40–45)
Caranx crysos					100	100				300
Blue Runner					(40)	(40)				(40–45)
Caranx fischeri	200	200	100	100	75	75	100	75	75	
Longfin Crevalle Jack	(50–60)	(50–60)	(50–60)	(50–60)	(50–75)	(50–75)	(60–75)	(65–70)	(65–75)	
Caranx hippos			50	50			75			
Crevalle Jack			(50–65)	(50–65)			(60–70)			
Caranx latus			50							
Horse-eye Jack			(60–65)							
Elagatis bipinnulata	100	100	50	75	50	50	75			150
Rainbow Runner	(65–70)	(65–70)	(65–75)	(65–75)	(75)	(75)	(65–80)			(70–80)
Rachycentron canadum								2		
Cobia								(120–130)		
Sphyraena barracuda	200	200	100	100	50	50		100	100	100
Great Barracuda	(80–100)	(80–100)	(75–100)	(75–100)	(75–90)	(75–90)		(80–100)	(80–100)	(75–100)
Trachinotus ovatus	50		25		75	75				
Pompano	(45–50)		(45–50)		(50–65)	(50–65)				

Values are approximate number of individuals associated with the oil rig with minimum and maximum size (TL [total length] cm) in parentheses.

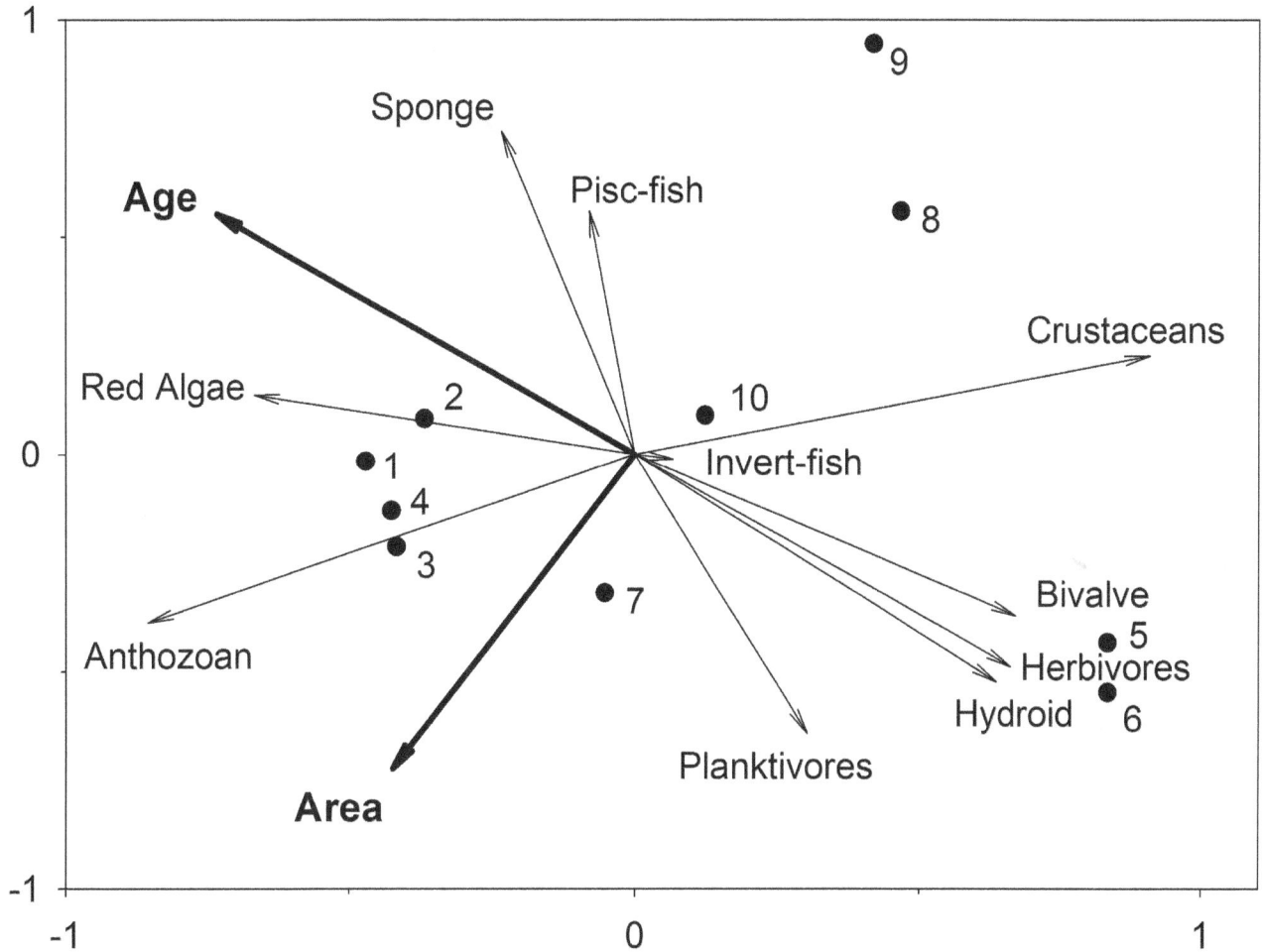

Figure 5. Biplot of results of redundancy analysis on biological (biomass of fish trophic groups and higher order benthic groups [e.g, Red alga, Sponge, Cnidaria-Anthozoa, Cnidaria-Hydrozoa, Crustacea-Cirripedia, Mollusca-Bivalve) and environmental data (e.g., age of platform, platform area).

[71]. They are particularly effective when they serve as no-take reserves and included in a marine protected area network [72,73]. Preliminary evidence indicates that decommissioned rigs in shallower waters can help rebuild declining fish stocks and therefore have high conservation value. Love and colleagues provided data that support the beneficial role of rigs for southern

California's bocaccio rockfish (*Sebastes paucispinis*) populations [74]. They estimated that, of the average number of juveniles that survive annually across the geographic range of the species, approximately 20% (equivalent to ~430,000 individuals) were supported by eight southern Californian rigs.

Table 6. Results of redundancy analysis (RDA) on log-transformed data on fish biomass of trophic groups and abundance of higher-level benthic groups with environmental variables (e.g., age of platform, distance offshore, latitude, platform area).

Axes	1	2	3	Total variance
Eigenvalues	0.59	0.19	0.01	1
FG-environment correlations	0.96	0.97	0.56	
Cumulative percentage variance				
of FG data	58.92	77.93	79.15	
of FG-environment relation	73.74	97.32	98.87	
Sum of all eigenvalues				1
Sum of all canonical eigenvalues				0.80

FG = functional groups.

Table 7. Conditional effects of Monte-Carlo permutation results on the redundancy analysis (RDA).

Variable	Lambda A	P	F
Age	0.40	**0.01**	5.4
Area	0.25	**0.01**	5.0
Offshore	0.09	0.13	1.9
Lattitude	0.06	0.17	1.6

Variables in bold are significantly correlated with the RDA axes.

Population persistence in the deep sea relies heavily on connectivity between deep-sea communities on isolated reefs [75]. Over large distances, small, isolated reefs may act as stepping stones within an inhospitable matrix of soft sediment [19]. Propagules that settle on to artificial reefs in locations relatively isolated from natural reefs would likely not have found suitable natural habitat before perishing. In this case, artificial reefs could potentially increase biomass production by increasing settlement [19].

The addition of artificial reefs in the deep sea is likely to increase ecological connectivity, which will have important biogeographical consequences [19]. In the Gulf of Mexico, where the amount of natural hard substratum is limited, oil platforms contribute substantially to local and regional abundance of reef habitat and the abundance of reef-associated fishes [76–80]. Platforms have been shown to facilitate the expansion of coral populations in the Gulf of Mexico and therefore they possess an intrinsic environmental value through the presence of coral populations [20]. However it has also been shown that these platforms can act as vectors for invasive species.

Rocky bottoms are extremely scarce on the continental shelf of Gabon [3]. The addition of artificial reefs may contribute substantially to local and regional abundance of hard substrate habitat and the abundance of reef-associated fishes [76–80] and is likely to increase ecological connectivity [75]. Moreover, if fish populations are limited by the amount of available habitat, then the addition of suitable artificial habitat increases the environmental carrying capacity, resulting in a sustained increase in population biomass (the "production" hypothesis; [17]). Polovina and Sakai [81] showed that regional catch per unit effort of the Pacific giant octopus (*Octopus dofleini*) increased after the addition of artificial reefs in northern Japanese waters, and therefore supporting fisheries production. However, fish observed on artificial reefs may simply have been attracted to those locations from surrounding habitat (the "attraction" hypothesis; [17]). Our results suggest that these platforms increase local production through enhanced settlement, increased reproductive output, and likely through reduced natural and fishing mortality.

The Government of Gabon made history in 2002 when they placed nearly 11% of their land area under permanent protection in 13 National Parks. To date three marine protected areas have been designated within this region (Akanda National Park, Mayumba National Park, and Pongara National Park) and are located in inshore territorial and coastal waters, covering ~1,745 km^2; of these only Mayumba National Park is entirely a no-take MPA, whereas Akanda and Pongara National Parks are terrestrial in nature and have marine buffer components. In February 2014, President Ali Bongo Ondimba announced the creation of a system of Marine National Parks covering 20% of Gabon's exclusive economic zone. Between oil exclusion zones and Marine National Parks all of the oil infrastructure as of 2014 will be in no-take zones. These platforms should become core areas for replenishment of the marine parks and some of Gabon's fisheries.

Supporting Information

Table S1   List of macroalgae and invertebrates observed on oil platforms in Gabon.

Table S2   List of fish species observed on oil platforms in Gabon. Species ordered phylogenetically based on Eschmeyer 2013. Provinces based on [51,66–67], NWA = Northwestern Atlantic, SWA = Southwestern Atlantic, MAR = Mid-Atlantic Ridge, NEA = Northeast Atlantic, TEA = Tropical Eastern Atlantic. New = new record for Gabon.

Acknowledgments

This expedition was made possible by the Waitt Foundation, National Geographic Society, and the Wildlife Conservation Society. Thanks to the crew of the Waitt Institute research vessel for making our work safe and easy. Thanks to President Ali Bongo Ondimba, White Lee, Koumba Koumbila, Parcs Gabon, Total Gabon, Perenco and Vaalco for making it possible for us to survey Gabon's waters. The following people provided assistance with species taxonomy: Serge Gofas (mollusks), Óscar Ocaña (corals), María Jesús Uriz (sponges), and Xavier Turon (ascidians).

Author Contributions

Conceived and designed the experiments: AMF EB MF ES. Performed the experiments: AMF EB. Analyzed the data: AMF EB MF ES. Contributed reagents/materials/analysis tools: AMF EB MF ES. Wrote the paper: AMF EB MF ES.

References

1. McGlade JM, Cury P, Koranteng KA, Hardman-Mountford NJ (2002) The Gulf of Guinea Large Marine Ecosystem: environmental forcing and sustainable development of marine resources. Amsterdam: Elsevier Science. 428 p.
2. Sibuet M, Vangrieshcim A (2009) Deep-sea environment and biodiversity of the West African Equatorial margin. Deep Sea Res. Part II Top Stud Oceanogr 56: 2156–2168.
3. Bianchi G (1992) Study of demersal assemblages of the continental shelf and upper slope of Congo and Gabon, based on the trawl surveys of the RV "Dr. Fridtjoff Nansen". Mar Ecol Prog Ser 85: 9–23.
4. Ondréas H, Olu K, Fouquet Y, Charlou JL, Gay A, et al. (2005) ROV study of a giant pockmark on the Gabon continental margin. Geo-Marine Letters 25: 281–292.

5. Roy OL, Caprais JC, Fifis A, Fabri MC, Galeron J, et al. (2007) Cold-seep assemblages on a giant pockmark off West Africa: spatial patterns and environmental control. Mar Ecol 28: 115–130.

6. Bentahila Y, Hebrard O, Ben Othman D, Luck JM, Seranne M, et al. (2006) Gulf of Guinea continental slope and Congo (Zaire) deep-sea fan: Sr–Pb isotopic constraints on sediments provenance from ZaiAngo cores. Mar Geol 226: 323–332.

7. Baudin F, Disnar JR, Martinez P, Dennielou B (2010) Distribution of the organic matter in the channel-levees systems of the Congo mud-rich deep-sea fan (West Africa). Implication for deep offshore petroleum source rocks and global carbon cycle. Mar Petrol Geol 27: 995–1010.

8. Vanden Bossche JP, Bernacsek GM (1990) Source book for the inland fishery resources of Africa: Vol. 2. CIFA Technical Paper, no. 18/1. Rome: FAO. 411 p.

9. Reid GM (1996) Ichthyogeography of the Guinea–Congo rain forest, West Africa. Proc Roy Soc Edinb B 104: 285–312.

10. Seranne M, Bruguier O, Moussavou M (2008) U–Pb single zircon grain dating of Present fluvial and Cenozoic aeolian sediments from Gabon: consequences on sediment provenance, reworking, and erosion processes on the equatorial West African margin. Bull Societé Géologique de France 179: 29–40.

11. Longhurst AR, Pauly D (1987) Ecology of tropical oceans. San Diego: Academic Press. 407 p.

12. Voituriez B, Herbland A (1982) Comparaison des systèmes productifs de l'Atlantique tropical Est: dômes thermiques, upwelling côtier et upwelling équatorial. Rapp P-V, Reun Cons Int Explor Mer 180: 114–130.

13. Lœuff PL, von Cosel R (1998) Biodiversity patterns of the marine benthic fauna on the Atlantic coast of tropical Africa in relation to hydroclimatic conditions and paleogeographic events. Acta Oecologica 19: 309–321.

14. Wauthy B (1983) Introduction à la climatologie du Golf de Guinée. Oceanogr Trop 18: 103–138.

15. Arnaud E, Arnaud PM, Intès A, Le Loeuff P (1976) Transport d'invertébrés benthiques entre l'Afrique du Sud et Sainte-Hélène par les laminaires (Phaeophyceae), Bull Mus nat Hist nat Paris sér 3ᵉ Ecol génér 30: 49–55.

16. Roberts CM, McClean CJ, Veron JEN, Hawkins JP, Allen GR, et al. (2002) Marine biodiversity hotspots and conservation priorities for tropical reefs. Science 295: 1280–1284.

17. Laborel J (1974) West African reef corals, and hypothesis on their origin. Proceedings of the Second International Symposium on Coral Reefs' Brisbane Vol. 1: 425–443.

18. Lee ME, Alonso A, Dallmeier F, Campbell P, Pauwels OSG (2006) The Gamba complex of protected areas: an illustration of Gabon's biodiversity. Bull Biol Soc Wash 12: 229–241.

19. Scarborough-Bull A (1989) Some comparisons between communities beneath the petroleum platforms off California and in the Gulf of Mexico. In: Reggio, V.C. (Ed.), Petroleum structures as artificial reefs: A compendium. Fourth Int. Conf. on Artificial Habitats for Fisheries, Rigs-to-Reefs Special Session, Miami, FL: US Dept. Interior, Minerals Mgt. Service, New Orleans, OCS Study MMS 89-0021. pp. 47–50.

20. Atchison AD, Sammarco PW, Brazeau DA (2008) Genetic connectivity in corals on the Flower Garden Banks and surrounding oil/gas platforms, Gulf of Mexico. J Exp Mar Biol Ecol 365: 1–12.

21. Hamzah BA (2003) International rules on decommissioning of offshore installations: some observations. Mar Policy 27: 339–348.

22. Parente V, Ferreira D, dos Santos EM, Luczynskie E (2006) Offshore decommissioning issues: deductibility and transferability. Energ Policy 34: 1992–2001.

23. Pradella N, Fowler AM, Booth DJ, Macreadie PI (2013) Fish assemblages associated with oil industry structures on the continental shelf of north-western Australia. J Fish Biol 84: 247–255.

24. Bohnsack JA (1989) Are high densities of fishes at artificial reefs the result of habitat limitation or behavioral preference? Bull Mar Sci 44: 631–645.

25. Jørgensen T, Løkkeborg S, Soldal AV (2002) Residence of fish in the vicinity of a decommissioned oil platform in the North Sea. ICES J Mar Sci 59: S288–S293.

26. Macreadie PI, Fowler AM, Booth DJ (2011) Rigs-to-reefs: will the deep sea benefit from artificial habitat? Front Ecol Environ 9: 455–461.

27. Sammarco PW, Atchison AD, Boland GS (2004) Expansion of coral communities within the Northern Gulf of Mexico via offshore oil and gas platforms. Mar Ecol Prog Ser 280: 129–143.

28. Sammarco PW, Porter SA, Cairns SD (2010) A new coral species introduced into the Atlantic Ocean - Tubastraea micranthus (Ehrenberg 1834) (Cnidaria, Anthozoa, Scleractinia): An invasive threat? Aquat Invasions 5: 131–140.

29. Wolfson A, van Blaricom G, Davis N, Lewbel GS (1979) The marine life of an offshore oil platform. Mar Ecol Prog Ser 1: 81–89.

30. Love MS, Caselle JE, Snook L (2000) Fish assemblages around seven oil platforms in the Santa Barbara Channel area. Fish Bull 98: 96–117.

31. Helvey M (2002) Are southern California oil and gas platforms essential fish habitat? ICES J Mar Sci: Journal du Conseil 59 (suppl): S266–S271.

32. Bohnsack JA, Johnson DL, Ambrose RF (1991) Ecology of artificial reef habitats. In: Seaman W Jr, Sprague LM, editors.Artificial habitats for marine and freshwater fisheries.New York: Academic Press. pp. 61–108.

33. Olsgard F, Gray JS (1995) A comprehensive analysis of the effects of offshore oil and gas exploration and production on the benthic communities of the Norwegian continental shelf. Mar Ecol Prog Ser 122: 277–306.

34. Schroeder DM, Love MS (2004) Ecological and political issues surrounding decommissioning of offshore oil facilities in the Southern California Bight. Ocean Coast Manage 47: 21–48.

35. Page HM, Dugan JE, Culver CS, Hoesterey JC (2006) Exotic invertebrate species on offshore oil platforms. Mar Ecol Prog Ser 325: 101–107.

36. Sala E, Ballesteros E (1997) Partitioning of space and food resources by three fishes of the genus Diplodus (Sparidae) in a Mediterranean rocky infralittoral ecosystem. Mar Ecol Prog Ser 152: 273–283.

37. Cebrian E, Ballesteros E, Canals M (2000) Shallow rocky bottom benthic assemblages as calcium carbonate producers in the Alboran Sea (Southwestern Mediterranean). Oceanol Acta 23: 311–322.

38. Froese R, Pauly D (2011) FishBase. World Wide Web electronic publication. Available: www.fishbase.org. Accessed 2014 May 5.

39. Floeter SR, Ferreira CEL, Dominici-Arosemena A, Zalmon IR (2004) Latitudinal gradients in Atlantic reef fish communities: trophic structure and spatial use patterns. J Fish Biol 64: 1680–1699.

40. Ferreira CEL, Floeter SR, Gasparini JL, Ferreira BP, Joyeux JC (2004) Trophic structure patterns of Brazilian reef fishes: a latitudinal comparison. J Biogeogr 31: 1093–1106.

41. Almada VC, Toledo JF, Brito A, Levy A, Floeter SR, et al. (2013) Complex origins of the Lusitania biogeographic province and northeastern Atlantic fishes. Front Biogeogr 5: 20–28.

42. Clarke KR, Gorley RN (2001) PRIMER version 5: user manual/tutorial. Plymouth, UK: Primer-E Ltd. 91 p.

43. Ludwig JA, Reynolds JF (1988) Statistical Ecology. New York, New York: John Wiley and Sons. 337 p.

44. Zar JH (1999) Biostatistical analysis, 4th edition. India: Pearson Education. 662 p.

45. ter Braak C, Šmilauer P (2002) CANOCO Reference Manual and CanoDraw for Windows User's Guide: Software for Canonical Community Ordination (version 4.5). Section on permutation methods. Microcomputer Power (Ithaca NY, USA). 500 p.

46. ter Braak C (1994) Canonical community ordination. Part I: Basic theory and linear methods. Ecoscience 1: 127–140.

47. ter Braak C, Verdonschot P (1995) Canonical correspondence analysis and related multivariate methods in aquatic ecology. Aquat Sci 55: 1–35.

48. Sala E, Aburto O, Paredes G, Thompson G (2003) Spawning aggregations and reproductive behavior of reef fishes in the Gulf of California. Bull Mar Sci 72: 103–121.

49. Soldal AV, Svellingen I, Jørgensen T, Løkkeborg S (2002) Rigs-to-reefs in the North Sea: hydroacoustic quantification of fish in the vicinity of a "semi-cold" platform. ICES J Mar Sci: Journal du Conseil 59 (suppl): S281–S287.

50. Sandin SA, Smith JE, DeMartini EE, Dinsdale EA, Donner SD, et al. (2008) Degradation of coral reef communities across a gradient of human disturbance. PLoS ONE 2008; 3(2):e1548.

51. Friedlander AM, Ballesteros E, Beets J, Berkenpas E, Gaymer CF, et al. (2013) Effects of isolation and fishing on the marine ecosystems of Easter Island and Salas y Gómez, Chile. Aquat Conserv 23: 515–531.

52. Friedlander AM, Zgliczynski BJ, Ballesteros E, Aburto-Oropeza O, Bolaños A, et al. (2012) The shallow-water fish assemblage of Parque Nacional Isla del Coco, Costa Rica: structure and patterns in an isolated, predator-dominated ecosystem. Rev Biol Trop (Suppl. 3): 321–338.

53. Wirtz P, Ferreira CEL, Floeter SR, Fricke R, Gasparini JL, et al. (2007) Coastal fishes of São Tomé and Príncipe islands, Gulf of Guinea (Eastern Atlantic Ocean) – an update. Zootaxa 1523: 1–48.

54. Afonso P, Porteiro FM, Santos RS, Barreiros JP, Worms J, et al. (1999) Coastal marine fishes of São Tomé Island (Gulf of Guinea). Arquipélago Life Mar Sci 17: 65–92.

55. Scheltema RS (1971) The dispersal of the larvae of shallow-water benthic invertebrate species over long distances by ocean currents. In: Crisp PJ, editor.Fourth European Marine Biology Symposium.Cambridge: Cambridge University Press. pp. 7–28.

56. Scheltema RS (1995) The relevance of passive dispersal for the biogeography of Caribbean molluscs. Am Malac Bull 11: 99–115.

57. Wirtz P (2001) New records of marine invertebrates from the Cape Verde Islands. Arquipélago Life Mar Sci 18A: 81–84.

58. Wirtz P (2003) New records of marine invertebrates from São Tomé Island. J Mar Biol Assoc UK 83: 735–736.

59. Wirtz P (2004) Four amphi-Atlantic shrimps new for São Tomé and Príncipe (eastern central Atlantic). Arquipélago Life Mar Sci 21: 83–85.

60. Wirtz P (2012) Seven new records of fish from Ngor Island, Senegal. Arquipélago, Life Mar Sci 29: 77–81.

61. Kritzer JP, Sale PF (2010) Marine metapopulations. San Diego, California: Elsevier Academic Press. 576 p.

62. González-Correa JM, Bayle JT, Sánchez-Lizaso JL, Valle C, Sánchez-Jerez P, et al. (2005) Recovery of deep Posidonia oceanica meadows degraded by trawling. J Exp Mar Biol Ecol 320: 65–76.

63. Schroeder DM, Love MS (2004) Ecological and political issues surrounding decommissioning of offshore oil facilities in the Southern California Bight. Ocean Coast Manage 47: 21–48.

64. White HK, Hsing PY, Cho W, Shank TM, Cordes EE, et al. (2012) Impact of the Deepwater Horizon oil spill on a deep-water coral community in the Gulf of Mexico. Proc Nat Acad Sci 109: 20303–20308.

65. Sumaila UR, Cisneros-Montemayor AM, Dyck A, Huang L, Cheung W, et al. (2012) Impact of the Deepwater Horizon well blowout on the economics of US Gulf fisheries. Can J Fish Aquat Sci 69: 499–510.

66. Jernelöv A (2010) The threats from oil spills: Now, then, and in the future. Ambio 39: 353–366.

67. Trevors JT, Saier MH (2010) The legacy of oil spills. Water Air Soil Poll 211: 1–3.

68. Rooker JR, Holt GJ, Pattengill CV, Dokken Q (1997) Fish assemblages on artificial and natural reefs in the Flower Garden Banks National Marine Sanctuary, USA. Coral Reefs 16: 83–92.

69. Yeo DC, Ahyong ST, Lodge DM, Ng PK, Naruse T, et al. (2010) Semisubmersible oil platforms: understudied and potentially major vectors of biofouling-mediated invasions. Biofouling 26: 179–186.

70. Street GT, Montagna PA (1996) Loss of genetic diversity in Harpacticoida near offshore platforms. Mar Biol 126: 271–282.

71. Macreadie PI, Fowler AM, Booth DJ (2012) Rigs-to-reefs policy: can science trump public sentiment? Front Ecol Environ 10: 179–180.

72. Schroeder DM, Love MS (2002) Recreational fishing and marine fish populations. CalCOFI Rep 43: 182–190.

73. Caselle JE, Love MS, Fusaro C, Schroeder D (2002) Trash or habitat? Fish assemblages on offshore oilfield seafloor debris in the Santa Barbara Channel, California. ICES J Mar Sci 59: S258–S265.

74. Love MS, Schroeder DM, Lenarz W, MacCall A, Bull AS, et al. (2006) Potential use of offshore marine structures in rebuilding an overfished rockfish species, bocaccio (*Sebastes paucispinis*). Fish Bull 104: 383–390.

75. Cowen RK, Sponaugle S (2009) Larval dispersal and marine population connectivity. Ann Rev Mar Sci 1: 443–66.

76. Gallaway BJ, Lewbel GS (1981) The ecology of petroleum platforms in the northwestern Gulf of Mexico: a community profile. Washington, DC: FWS 10BS-82/27, Open File Report 82–03. USFWS Office of Biology Services.

77. Stanley DR, Wilson CA (1989) Utilization of offshore platforms by recreational fishermen and SCUBA divers off the Louisiana coast. Bull Mar Sci 44: 767–775.

78. Stanley DR, Wilson CA (1990) A fishery-dependent based study of fish species composition and associated catch rates around oil and gas structures off Louisiana. Fish Bull 88: 719–730.

79. Scarborough-Bull A, Kendall JJ Jr (1994) An indication of the process: offshore platforms as artificial reefs in the Gulf of Mexico. Bull Mar Sci 55: 1086–1098.

80. Shipp RL, Bortone SA (2009) A perspective of the importance of artificial habitat on the management of red snapper in the Gulf of Mexico. Rev Fish Sci 17: 41–47.

81. Polovina JJ, Sakai I (1989) Impacts of artificial reefs on fishery production in Shimamaki, Japan. Bull Mar Sci 44: 997–1003.

82. Floeter SR, Rocha LA, Robertson DR, Joyeux JC, Smith-Vaniz WF, et al. (2008) Atlantic reef fish biogeography and evolution. J Biogeogr 35: 22–47.

83. Wirtz P, Brito A, Falcón JM, Freitas R, Fricke R, et al. (2013) The coastal fishes of the Cape Verde Islands–new records and an annotated check-list. Spixiana 36: 113–142.

17

Trait-Specific Responses of Wild Bee Communities to Landscape Composition, Configuration and Local Factors

Sebastian Hopfenmüller*, Ingolf Steffan-Dewenter, Andrea Holzschuh

Department of Animal Ecology and Tropical Biology, Biocenter, University of Würzburg, Würzburg, Germany

Abstract

Land-use intensification and loss of semi-natural habitats have induced a severe decline of bee diversity in agricultural landscapes. Semi-natural habitats like calcareous grasslands are among the most important bee habitats in central Europe, but they are threatened by decreasing habitat area and quality, and by homogenization of the surrounding landscape affecting both landscape composition and configuration. In this study we tested the importance of habitat area, quality and connectivity as well as landscape composition and configuration on wild bees in calcareous grasslands. We made detailed trait-specific analyses as bees with different traits might differ in their response to the tested factors. Species richness and abundance of wild bees were surveyed on 23 calcareous grassland patches in Southern Germany with independent gradients in local and landscape factors. Total wild bee richness was positively affected by complex landscape configuration, large habitat area and high habitat quality (i.e. steep slopes). Cuckoo bee richness was positively affected by complex landscape configuration and large habitat area whereas habitat specialists were only affected by the local factors habitat area and habitat quality. Small social generalists were positively influenced by habitat area whereas large social generalists (bumblebees) were positively affected by landscape composition (high percentage of semi-natural habitats). Our results emphasize a strong dependence of habitat specialists on local habitat characteristics, whereas cuckoo bees and bumblebees are more likely affected by the surrounding landscape. We conclude that a combination of large high-quality patches and heterogeneous landscapes maintains high bee species richness and communities with diverse trait composition. Such diverse communities might stabilize pollination services provided to crops and wild plants on local and landscape scales.

Editor: Francesco de Bello, Institute of Botany, Czech Academy of Sciences, Czech Republic

Funding: This research received funding from the European Community's Seventh Framework Programme (FP7/2007–2013) under grant agreement no 244090, STEP Project (Status and Trends of European Pollinators: www.step-project.net). This publication was funded by the German Research Foundation (DFG) and the University of Wuerzburg in the funding programme Open Access Publishing. The funders had no role in study design, data collection and analysis, decision to publish, or preparation of the manuscript.

Competing Interests: The authors have declared that no competing interests exist.

* Email: sebastian.hopfenmueller@uni-wuerzburg.de

Introduction

Global food security and stable ecosystem services like pollination are major challenges that the fast growing human population has to deal with in the next decades [1,2]. Agricultural landscapes where remaining natural and semi-natural habitats are often highly fragmented and degraded [3] suffer from loss of pollinators and increasing effects of global change pressures [4,5]. Therefore the conservation of pollinating insects should be a major issue for providing pollination services to agricultural and natural ecosystems [6,7]. Wild bees are one of the most important pollinator groups [8] and their diversity can influence pollination services [9,10]. In agroecosystems themselves pollinators can be negatively influenced by isolation from semi-natural habitats [11,12]. Still there is little knowledge how wild bee diversity can be enhanced in semi-natural habitats to provide a high and stable spillover of wild bees to agroecosystems and secure pollination of insect-pollinated plants. Land-use change is considered the major driver of global biodiversity change [13], and therefore understanding patterns and driving factors of wild bee diversity in agricultural landscapes is an essential precondition for maintaining stable ecosystems and crop pollination worldwide.

One of the most species rich but highly fragmented habitats in central Europe are calcareous grasslands, that are in severe decline since the middle of the 19th century [3]. This is due to the decrease of historical land-use such as shepherding, as well as forestation and fertilization. Through the severe loss of these habitats many of the remaining fragments are strongly isolated and many species specialized on these habitats are threatened [14]. This shows that the identification of factors that influence species diversity, especially of habitat specialists on these habitat patches is important for conservation and restoration. The relative importance of different factors influencing species richness and population viability, like local factors (e.g. habitat area and quality) and landscape factors (eg. landscape composition, landscape configuration and habitat connectivity) are still controversially discussed or unclear [15–18]. This might be because different factors could affect different life-history traits of bees and therefore trait-specific analysis are helpful to disentangle the importance of these factors. The different definition and use of traits has led to some confusion, especially in plant ecology [19]. Therefore new approaches have been developed and the use of trait based measures like community weighted mean and functional diversity have been proposed [20]. Nevertheless, in studies dealing with wild bee diversity the use of life-history traits usually based on categories is a widely used approach (e.g. [12,21,22]). Community shifts can be identified by using community weighted mean [23] but this might overlook species groups that share different traits and are smaller in numbers (e.g. small social bees [24]). As bee diversity effects on pollination can be driven by functional complementarity [10], knowledge about trait-specific performance of bees are also needed to preserve pollination services that should consequently only sufficiently provided by a combination of bee-traits.

Despite the often shown positive species-area relationship [25,26] and the widespread notion that size matters, several studies showed that habitat quality can be even more important for insect populations than habitat area [27–29]. In wild bees habitat quality includes both quality of food (pollen and nectar diversity) and of nesting resources (mainly sun exposed soil in central Europe), whereby quality of nesting resources has rarely been tested (but see [30]). Habitat area and quality thus seem to be important factors for insect diversity but it is still unclear how important the heterogeneity of the surrounding landscape is [18]. This question should be addressed by separating the effects of landscape composition and configuration that are both expected to influence species diversity in habitat patches [17]. Landscape composition - particularly the percentage of semi-natural habitats in the landscape - has been shown to affect insect diversity [31,32]. In contrast, effects of landscape configuration - for example edge or patch density - that are independent of effects of landscape composition have rarely been tested [17] because these are often confounding factors (landscapes with high amount of semi-natural habitat are often also highly structured, but see [33,34]). Landscape configuration can be measured in different ways and the most simple is patch density, i.e. the number of patches in the landscape [34]. The re-allocation of agricultural land causing lower patch densities is still a threat to edge habitats that provide resources and can promote dispersal [35,36]. A recent meta-analysis found only weak effects of landscape configuration on wild bee diversity in agroecosystems [12] but most of the studies included in this analysis were not explicitly designed to focus on landscape configuration. There are still no studies focusing on the effects of landscape configuration in semi-natural habitats, but as linear landscape structures and edge habitats are important foraging and nesting resources for wild pollinators [36], there is a strong need to understand how important the landscape configuration is for pollinator communities.

While landscape configuration describes the arrangement of patches in the landscape independently of habitat type, habitat connectivity describes the areas and arrangement of habitat patches (e.g. of calcareous grasslands) in the landscape around a focal habitat patch. High connectivity of calcareous grasslands has been shown to positively affect butterflies and plants ([37], but see [38]). Few studies have tested the effect of habitat connectivity on wild bees so far and none of them found an effect of habitat connectivity on wild bees [39,24]. This might be because these studies did no detailed trait analyses, although e.g. habitat specialists are expected to react stronger to reduced connectivity than habitat generalists [37]. Futheremore, habitat specialists should be influenced more by local habitat area than by landscape composition [40]. Another study also showed that solitary bees show stronger response to habitat area than social bees, but social Halictidae did show stronger response to habitat area than solitary Halictidae [24]. Therefore, traits like trophic rank, habitat specialization, sociality and size should be considered as they have been shown to disentangle factors affecting pollinator diversity [22,40–43].

In this study we aimed to test the relative importance of local factors (habitat area and quality) and landscape factors (habitat connectivity, landscape composition and configuration) of calcareous grasslands for wild bee diversity and different traits of wild bees. We selected 23 calcareous grassland with gradients of local and landscape factors and recorded their bee communities during one season. We expected bee diversity in calcareous grasslands to be influenced by both, local and landscape factors, but traits like habitat specialists or large social species (bumblebees) were expected to differ in their response to local and landscape effects.

Materials and Methods

Study region

The study was conducted in Upper Franconia, north-eastern Bavaria, Germany (see Fig. 1). The study region is characterized by its geology, which consists mainly of Jurassic limestone forming a hilly lowland plateau. The total extent of the study region was 45×50 km with altitude varying between 350 and 585 m a.s.l. Mean annual precipitation varies between 650 and 900 mm. The current land use in this area is predominantly characterized by a small-scaled mosaic of arable land, forest, meadows and semi-natural habitats. Important semi-natural habitats are calcareous grasslands that are characteristic for the region and mostly located on hillsides of small valleys.

Study sites and local habitat parameters

For this study 23 calcareous grasslands were chosen that ranged in size from 0.2 to 11.8 ha. All grasslands were characterized by high flower diversity to reduce for potential effects of flower resources. Flowering plant species excluding graminaceous and tree species were recorded during transect walks (description see section "bees") and flower cover was estimated. The effect of number of flowering plant species and flower cover (i.e. flower units/ha) on both abundance and species richness of wild bees was tested using simple linear regressions, and did not show significant results (p-values>0.1). Therefore, we focused on habitat quality in terms of nesting resources and choose habitat slope which is a factor that can influence nest density of bees [44,45] and which is easy to measure. As most bee species in central Europe are soil nesting and direct counting of nesting resources is difficult, we used habitat slope as a simplified factor that influences nesting resources [45]. All grasslands were regularly sheep-grazed (minimum once a year) but two were mown in the end of summer. The effect of historical land-use could not be investigated but might have an influence on current species composition. Habitat area was calculated in ArcGIS 9.3 [46] using orthorecticified digital aerial photos from 2008 with a resolution of 0.2 m (provided by Bayerische Vermessungsverwaltung). As the investigated grasslands were quite homogenous in microstructure, the slope angle did not differ much within the whole grassland. Therefor we used average habitat slope which was calculated in ArcGIS 9.3 [46] using digital contour line maps (provided by Bayerische Vermessungsverwaltung).

Landscape parameters

Landscape parameters were calculated for landscape sectors with 1 km radius around the patch edges of the grasslands. Based on aerial photos (provided by Bayerische Vermessungsverwaltung), land-use was mapped in the field in August–September 2010. Patches had to be larger than 100 m² to be included in the mapping. Digitizing of field mapping and calculation of landscape parameters were done in ArcGIS 9.3 [46]. We calculated three landscape parameters (see also Fig. 1): (1) the percentage of semi-

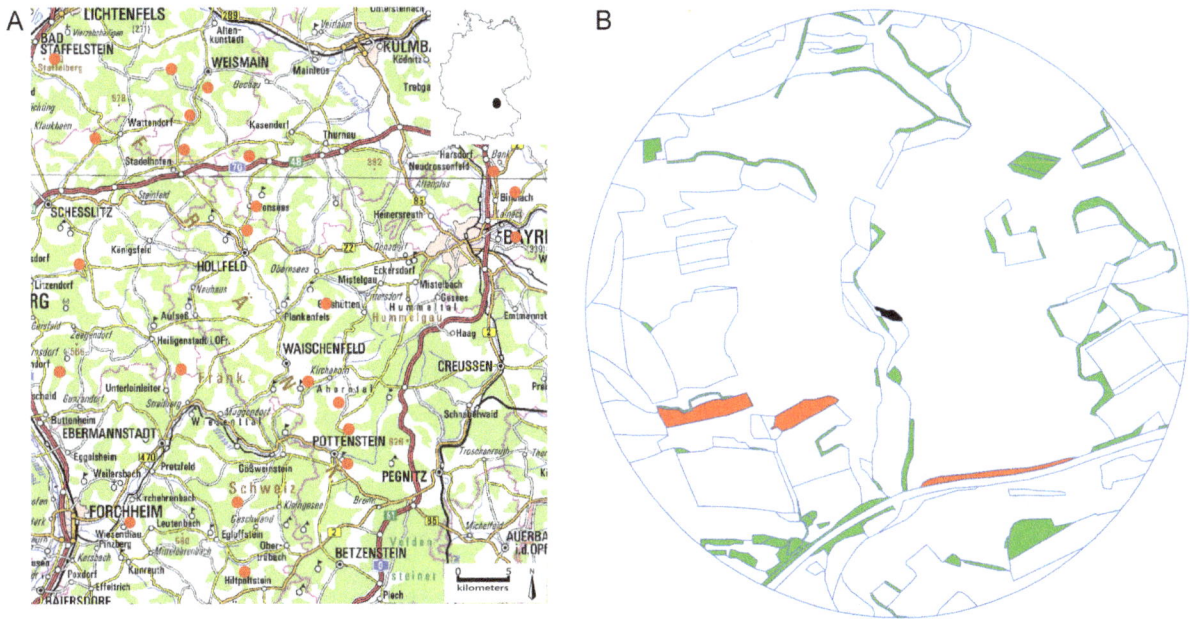

Figure 1. Overview of the study region and example site for illustration of the used landscape metrics. (A) Study region with all sampled sites (red dots). (B) Example site where the black patch in the middle is the sampled calcareous grassland, red patches are calcareous grasslands in the surrounding landscape, green patches are other semi-natural habitats and blue lines are borders between different land-use patches. The Connectivity Index takes area and distance of red patches to the black patch (sampled site) into account. Landscape composition is the percentage of semi-natural habitats (all green and red patches). Landscape configuration is the number of patches (blue lines) in the landscape (patch density).

natural habitats as a measure of landscape composition, where we focused on important bee habitats, i.e. calcareous grasslands, orchards, fallows, ruderal areas, plant species-rich margins and hedgerows. (2) The patch density of the landscape as a measure of landscape configuration, was calculated as the number of all land-use patches in the landscape divided by total landscape area (in km²). (3) The habitat connectivity of calcareous grasslands as a measure taking configuration and area of calcareous grasslands in the landscape into account. The habitat connectivity was calculated as Connectivity Index (CI) developed by Hanski [47]. This index has been shown to be a good predictor for species loss of habitat specialists [37], because it takes distance and area of surrounding habitat patches (in this study calcareous grasslands) into account and is described by the following equation:

$$CI_i = \sum_{i \neq j} \exp(-\alpha d_{ij}) A_j^b$$

where A_j is the size (in m²) of the surrounding habitat patch j and d_{ij} the distance (in km) from the patch to the central focus patch i. The parameter α sets the survival rate of migrants as $1/average$ *migration distance* in km, whereas b scales the size of the habitat patches. According to literature $\alpha = 1$ was chosen as 1 km appears to be an average dispersal range for wild bees [48–50]. For the parameter b a value of 0.5 was chosen as [51] suggested that with increasing patch size, the ratio of patch edge to patch area decreases following $A^{0.5}$. We also calculated the percentage of calcareous grasslands per landscape that was highly correlated with connectivity $(r = 0.82)$, but as it was also correlated with percentage of semi-natural habitats $(r = 0.42)$ we did not use percentage of calcareous grasslands in any analysis.

Bees

Wild bees (Hymenoptera: Apoidea) were sampled five times from April to August 2010 in "variable transect walks" [52] covering an area of approximately 0.1 ha per study site. Transects had no fixed direction, but were directed to attractive nesting and food resources for bees, whose position could change from month to month. Sampling was conducted from 10.30 to 17.00 h in April and May and from 9.30 to 17.30 in June, July and August. Sampling was only conducted when the temperature was at least 16°C with low wind and sunny weather. Within each transect walk all bees (except honey bees) were caught with a net during a 45 min. period with 9 subunits of 5 min. All individuals that could be identified in the field were recorded and released, otherwise they were stored in ethylacetate and brought to the lab for further identification. Sampling time was stopped during notations or handling of the caught bees. Permissions for sampling of bees and access to protected areas were given by the government of Upper Franconia.

All individuals were identified to species level. Species that were difficult to determine or very rare ones were sent to a specialist for identification. Number of individuals and number of species determined for each study site represented the sum of all five transect walks conducted on that site.

As wild bees are expected to show trait specific responses to habitat loss and fragmentation [40,42,53], species were grouped according to their life-history traits: we separated cuckoo bees, habitat specialists, solitary bees and small (Halictidae p.p.) and large (bumblebees) social bees according to Westrich [54] (see Fig. 2). Habitat specialization does not represent a real life-history trait [19] but as it is an important characteristic in wild bees we here use the term trait also for habitat specialization. The majority of cuckoo bees were solitary habitat generalists and habitat specialists were almost exclusively solitary species. The complete

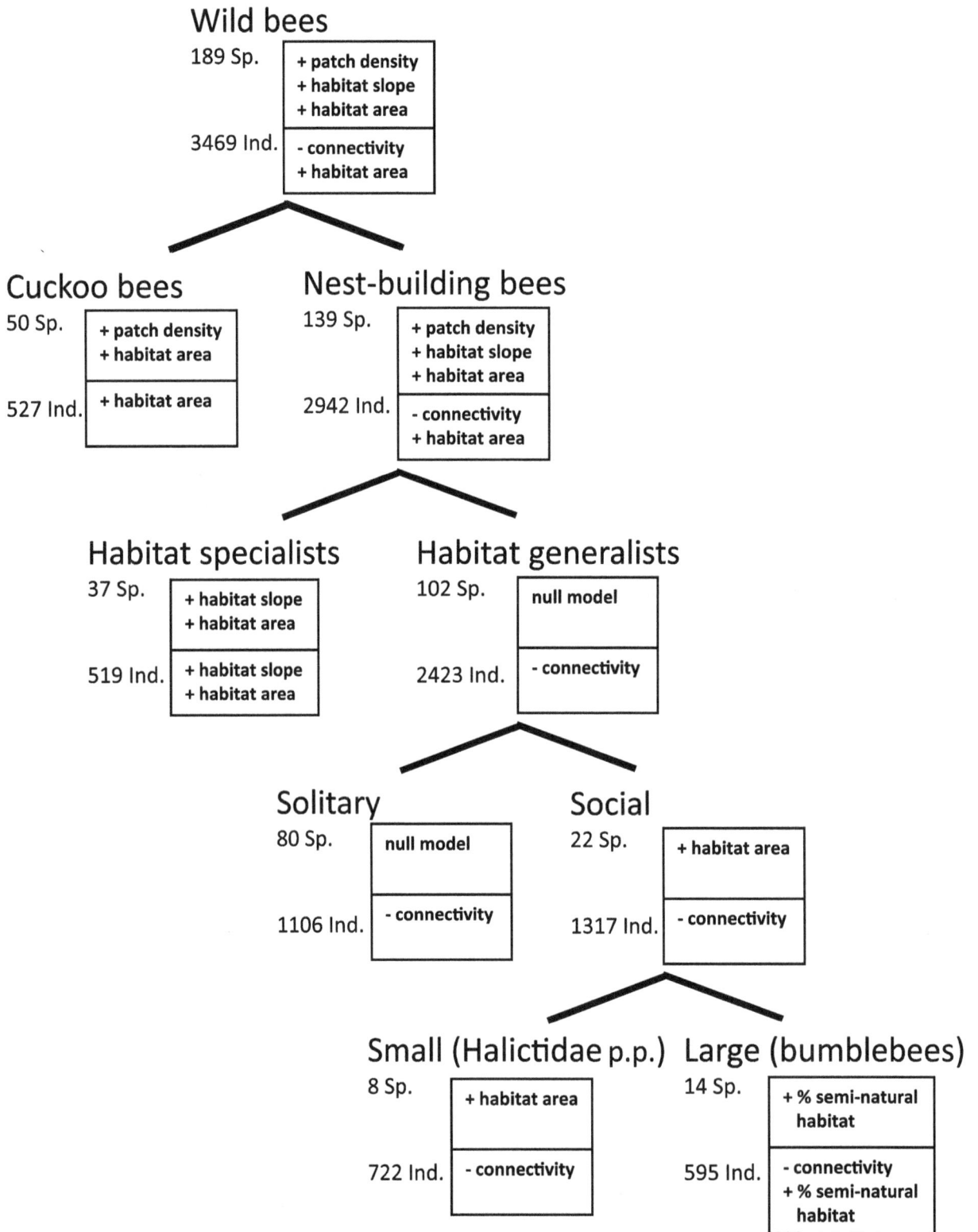

Figure 2. Overview of tested bee-traits. Number of species (Sp.) and of individuals (Ind.) are given for each group and factors that significantly affect them (+ and − indicate the relation of the effects).

bee species list with abundance, frequency and trait category can be found in Table S1.

Statistical analysis

Statistical analyses were performed in R version 2.13.0 [55]. Habitat and landscape effects on bees were tested with general linear models (GLM). The response variables were species richness and abundance and met the assumptions of normality and homoscedasticity [56]. The predictor variables were connectivity (CI), percentage of semi-natural habitats (arcsin$\sqrt{p}$-transformed to increase linearity in proportion data, see [57]), patch density, habitat slope and habitat area (log$_{10}$-transformed to increase linearity, see [58]). Predictor variables were not inter-correlated in Spearman rank correlation tests (all $|R| \leq 0.35$, see [59]). Model fit was checked following Zuur et al. [56]. Full models were manually simplified in backward steps beginning with the highest non-significant p-values from F-tests with type 1 sums of squares. This was done until only significant ($p<0.05$) variables were left in the model.

To check if two trait categories (e.g. cuckoo bees and nest-building bees) react different to a predictor variable we compared the slopes of regression lines using linear mixed effects models (lme). If the GLMs revealed a significant effect of a predictor variable (e.g. patch area) on both of the dichotomous trait categories, we tested in an lme whether the interaction between predictor and trait category was significant, i.e. whether the slope of regression lines differed for different trait categories. Fixed factors were the predictor of interest (patch area, patch density or connectivity), trait category and the interaction between the predictor of interest and trait category. Response variables were the richness or abundance of the bees within the dichotomous trait categories. Study site was included as random factor.

To check for sufficiency of sampling, first-order jack-knife estimates of species richness were calculated for every grassland patch (with pooled survey rounds) and for the total study region using EstimateS version 9 [60]. Mean observed bee species richness was 67% (range: 62–72%) of estimated species richness of the grassland patches and 83% of the total study region. Estimated sampling sufficiency was not related to habitat area ($F = 0.28$, $p = 0.599$) nor did estimated species richness change the results of analyses compared to observed species richness of the grassland patches.

As bees might have large dispersal distances we also checked for spatial autocorrelation.

Therefore we calculated Bray-Curtis similarity of bee communities between all site pairs in EstimateS [60] and spatial distance between all site pairs in ArcGIS 9.3 [46]. We tested the similarity matrix for spatial autocorrelation using a Mantel test (package vegan in R [55]) and found no significant spatial autocorrelation ($p = 0.142$).

Results

In total, 3469 wild bee individuals of 189 species belonging to 25 genera were collected on the 23 calcareous grasslands, representing 55% of the wild bee species occurring in Upper Franconia [61]. In Figure 2 the number of individuals and of species are given for all tested groups of wild bees. A total of 35 species were endangered according to the Red List of Bavaria [62] and most of those were habitat specialists.

Local factors

Habitat area of the calcareous grasslands strongly affected wild bees: species richness and abundance of total wild bees, of cuckoo

bees and of nest-building bees increased with increasing habitat area (Table 1, Fig. 2 and 3). Within the group of nest-building bees, species richness and abundance of habitat specialists, but not of total habitat generalists, increased with increasing habitat area. Within the group of total habitat generalists, only the abundance of small social generalists (Halictids) increased with increasing habitat area (resulting also in an increase of total social generalists), while large social generalists alone (bumblebees) and solitary generalists were not affected by habitat area. Cuckoo bees did not show a stronger response to habitat area than nest-building bees as the slopes of the regression lines were not significantly different (interaction term in lme with $p>0.1$).

Steep slopes of the grasslands positively affected species richness and abundance of habitat specialists (resulting also in a positive effect on the species richness of nest-building species and of total wild bees), but did neither effect cuckoo bees nor any of the habitat generalist bee groups (Table 1, Fig. 3).

Landscape Factors

Landscape configuration affected total wild bees, cuckoo bees and nest-building bees: the richness of these groups was higher in landscapes with higher patch density (number of patches in 1 km radius around the grasslands) (Table 1, Fig. 2–4). Cuckoo bees and nest-building bees did not differ in the strength of their response to landscape configuration (interaction term in lme with $p>0.1$).

Landscape composition only affected the species richness and abundance of large social habitat generalists (bumblebees), which increased with increasing percentage of semi-natural habitat (Table 1). The gradient in percentage of semi-natural habitats showed one very low and two high values that clearly separate in the graphical plot of bumblebee richness and percentage of semi-natural habitats (Fig. 4e). Therefore we calculated additional models, but removing any of the three data points did not remove the significant effect of percentage of semi-natural habitats ($F_{1,20} = 5.1$, $p = 0.035$; $F_{1,20} = 6.0$, $p = 0.024$; $F_{1,20} = 12.3$, $p = 0.002$) and removing all three data points still resulted in a marginally not significant effect ($F_{1,18} = 3.6$, $p = 0.074$).

High connectivity of the calcareous grasslands did not show the expected positive effect on habitat specialists, but had a negative effect on the abundance of all groups except on cuckoo bees and habitat specialists (Table 1, Fig. 5). Solitary and social as well as small and large social bees did not differ in the strength of their response to connectivity (interaction term in lme with $p>0.1$).

Discussion

In our study, we investigated the effects of local and landscape factors on bee communities in calcareous grasslands. We could show that the relative importance of these factors differs among bee groups with different combinations of life-history traits: habitat specialists were affected by local factors only, small social habitat generalists and cuckoo bees were affected by both local and landscape factors, and solitary habitat generalists and large social habitat generalists were affected by landscape factors only.

Local Factors

Habitat specialists, cuckoo bees and small social generalists showed a positive species-area relationship. The positive relationship between species richness and habitat area is a common and often shown pattern in ecology [26,63]. However, different species groups may react differentially, for example specialized bees and higher trophic levels are expected to be more sensitive to habitat loss than generalists [41,42]. The positive species-area relationship

Table 1. Results of general linear models.

Response variables	Predictors	d.f.	F value	P
Total wild bee richness	Patch density (+)	1,19	14.3	0.001
	Habitat slope (+)	1,19	11.9	0.003
	Habitat area (+)	1,19	42.0	<0.001
Total wild bee abundance	Connectivity (−)	1,20	13.0	0.002
	Habitat area (+)	1,20	9.8	0.005
Nest-building bee richness	Patch density (+)	1,19	5.6	0.028
	Habitat slope (+)	1,19	13.2	0.002
	Habitat area (+)	1,19	17.3	<0.001
Nest-building bee abundance	Connectivity (−)	1,20	11.2	0.003
	Habitat area (+)	1,20	7.1	0.015
Cuckoo bee richness	Patch density (+)	1,20	6.5	0.020
	Habitat area (+)	1,20	18.4	<0.001
Cuckoo bee abundance	Habitat area (+)	1,21	5.1	0.034
Habitat generalist richness	Null model	1,22	-	-
Habitat generalist abundance	Connectivity (−)	1,21	16.5	<0.001
Habitat specialist richness	Habitat slope (+)	1,20	16.5	<0.001
	Habitat area (+)	1,20	19.8	<0.001
Habitat specialist abundance	Habitat slope (+)	1,20	6.5	0.020
	Habitat area (+)	1,20	7.3	0.014
Social generalist richness	Habitat area (+)	1,21	6.5	0.019
Social generalist abundance	Connectivity (−)	1,21	13.5	0.001
Solitary generalist richness	Null model	1,22	-	-
Solitary generalist abundance	Connectivity (−)	1,21	7.0	0.016
Small social generalist richness	Habitat area (+)	1,21	8.4	0.009
Small social generalist abundance	Connectivity (−)	1,21	6.5	0.019
Large social generalist richness	% semi-natural habitat (+)	1,21	8.5	0.008
Large social generalist abundance	Connectivity (−)	1,20	7.7	0.014
	% semi-natural habitat (+)	1,20	23.3	<0.001

Effects of connectivity, patch density, percentage semi-natural habitats, patch slope and patch area on abundance and richness of different groups of wild bees. (+) and (−) indicate the relation of the effects.

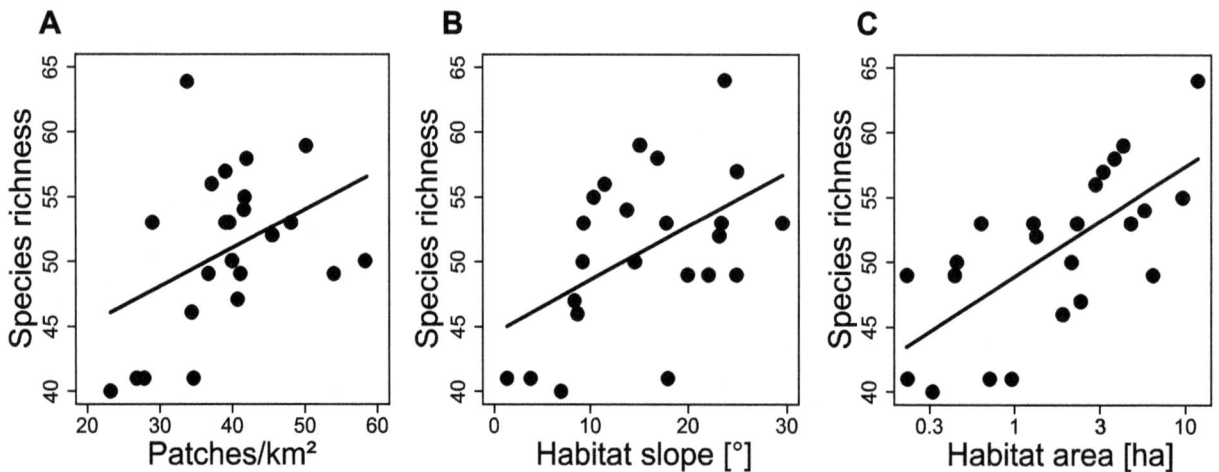

Figure 3. Factors affecting total wild bee richness. Relationship between patch density (A), habitat slope (B) and habitat area (C) and total wild bee richness. Regression lines: (A) y = 0.29x+39.20, (B) y = 0.41x+44.47, (C) y = 3.69x+48.89.

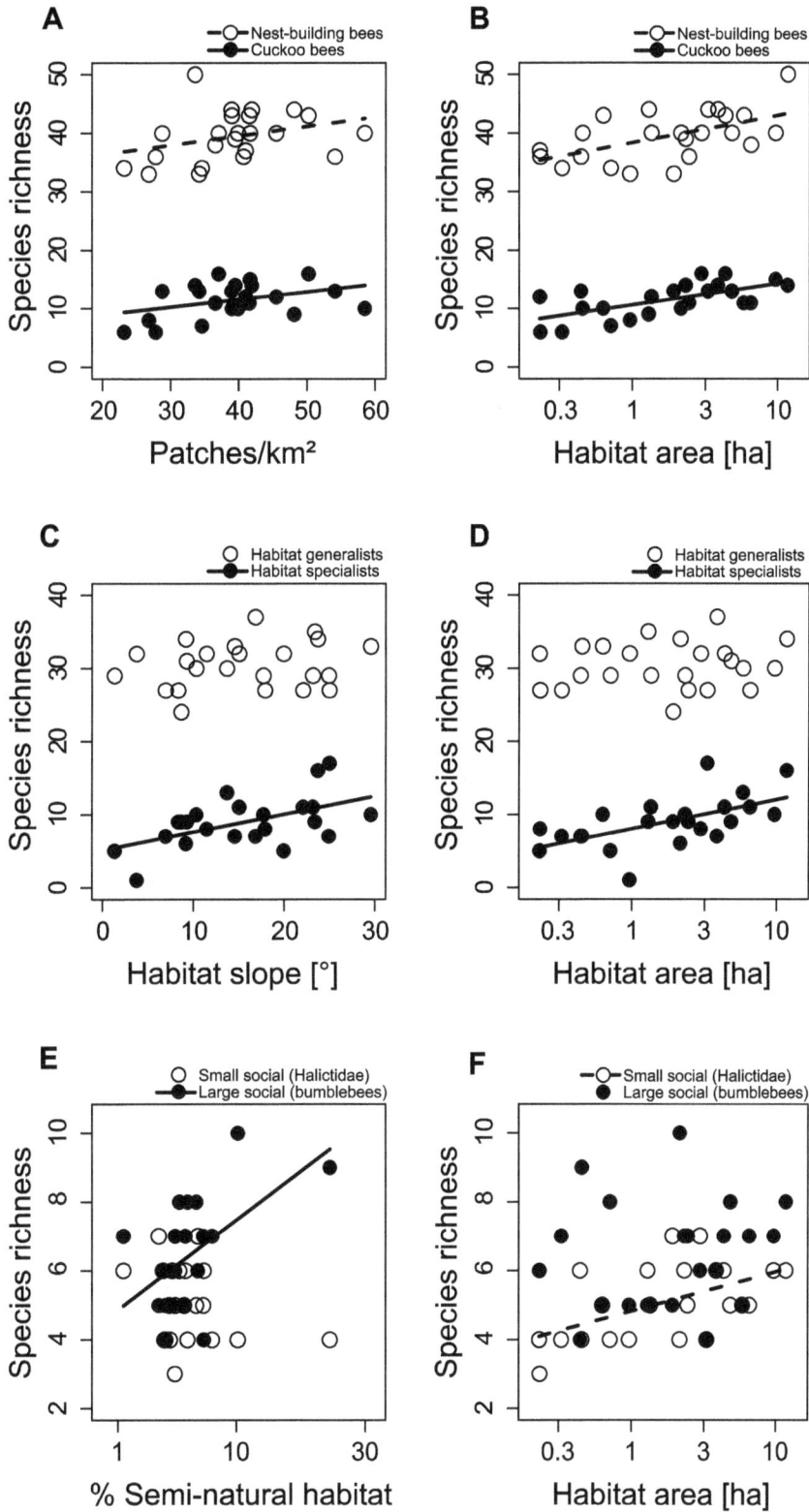

Figure 4. Local and landscape factors affecting different bee traits. Effects of patch density (A) and habitat area (B) on nest-building and cuckoo bees. Effects of habitat slope (C) and habitat area (D) on habitat specialists and generalists. Effects of percentage semi-natural habitat (E) and habitat area (F) on small and large social generalists. Regression lines of significant relationships: (A) Nest-building bees: y = 0.16x+33.15; Cuckoo bees: y = 0.13x+6.37; (B) Nest-building bees: y = 4.66x+38.35; Cuckoo bees: y = 3.70x+10.62; (C) Habitat specialists: y = 0.25x+5.11; (D) Habitat specialists: y = 4.03x+8.02; (E) Large social: y = 11.32x+3.73; (F) Small social: y = 1,14x+4,82.

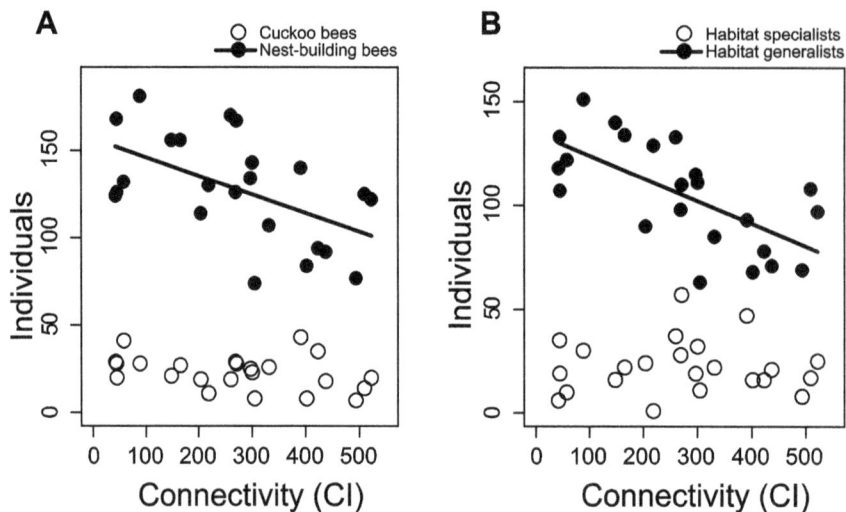

Figure 5. Effects of connectivity on wild bee abundance. Effects of connectivity on nest-building bees and cuckoo bees (A) and on habitat specialists and generalists (B). Regression lines: (A) y = −0,11x+156,67; (B) y = −0,11x+134,84.

for small social generalists but not for solitary and large social generalists is in accordance with results of Öckinger et al. [40]. Social species need large amounts of resources within their foraging range to provide food to a large number of larvae and as all social Halictidae have a quite short foraging distance compared to the large social bumblebees [64] they are expected to be more dependent on local than on landscape factors. Resource concentration as described by Root [65] might also influence species richness and abundance as large calcareous grasslands provide diverse food and more abundant and nesting resources especially for habitat specialists and a variety of host species for cuckoo bees.

Species richness and abundance of habitat specialists increased with increasing habitat slope. This coincides with our expectations that habitat specialists highly depend on the quality of their habitat. Habitat slope influences habitat quality of calcareous grasslands as steep slopes promote erosion and bare soil and increase solar radiation on south exposed slopes compared to flat areas. Potts and Willmer [44] showed that the nest density of the ground-nesting bee *Halictus rubicundus* was positively related with slope angle of the nesting sites. As most of the central European bees are ground nesting [14], bare soil is an essential factor for diverse bee communities [30]. High soil temperature and low soil humidity might be a more important factor for nesting sites of calcareous grassland specialist than for generalists. Nevertheless, there is still a huge lack of knowledge if and how nesting site availability can regulate bee communities [63].

Landscape Factors

We found landscape configuration i.e. patch density as an important factor for total wild bee richness, and both cuckoo bee and nest-building bee richness. Patch density increases the amount of edges and corridors that can act as food and nesting resource like hedgerows, field margins and ditches [36,66,67] and promote dispersal [35]. There is still a lack of knowledge how important landscape composition and configuration are for animal diversity [17]. A recent meta-analysis pointed out the importance of landscape composition but not configuration for wild bees in agroecosystems [12]. This seems contrasting to our findings, but Kennedy et al. [12] analyzed wild bee diversity in crop systems that provide foraging resources but mostly no nesting resources

like calcareous grasslands do. This might explain the different results because crop systems need certain amounts of semi-natural habitats in their surroundings to be visited by a variety of pollinators [12,68] but high quality bee habitats should profit of a highly structured landscape with a variety of other habitats especially linear habitats like ditches and forest edges, which host additional species [54,66]. We found that cuckoo bees showed a positive response to patch density like nest-building bees but not a stronger one. Higher trophic levels like cuckoo bees are expected to react stronger to habitat fragmentation than their hosts [38] but there are also studies showing the opposite [24,58]. There is almost no knowledge if nest site fidelity and dispersal distances of cuckoo bees differ from their hosts, but as they have no nesting site they should be more mobile. As most cuckoo bee species have a wide host range [54] they might be not strongly dependent on local habitat quality, but might profit of dispersal corridors provided by structurally rich landscape.

We found no relationship between the percentage of semi-natural habitat in the surrounding landscape and either total species richness or abundance. The large social bumblebees were the only group showing a relationship in which richness as well as abundance increased with increasing percentage of semi-natural habitats. Bumblebees that have a high dispersal capacity [49] seemed to be promoted by large amounts of food resources at a landscape scale [69]. Williams et al. [70] showed that colony growth of bumblebees was driven by flower resources on a landscape scale and thus stable flower resources like semi-natural habitats should enhance abundance of bumblebees. As semi-natural habitats like orchards or calcareous grasslands also provide a variety of nesting cavities [54] the species richness of bumblebees should also be enhanced by the amount of these habitats.

In contrast to our expectations, habitat specialists did not benefit from high habitat connectivity. We found decreasing abundances of all groups except cuckoo bees and habitat specialists, with increasing connectivity. This effect might be the result of a concentration of bees on strongly fragmented grasslands as described in "the landscape-moderated concentration and dilution hypothesis" [71]. Bees that nest in the landscape (like habitat generalists and bumblebees) should also be found on calcareous grasslands as these provide a stable food resource. Our results suggest that in landscapes with high connectivity between the grasslands, foraging bees dilute and disperse on several grasslands whereas in landscapes with low connectivity bees concentrate on a few grassland patches. This implies that bees, especially generalists profit from nearby calcareous grasslands and highly connected landscapes could have positive effects on the fitness of wild bees due to shorter foraging distances. Effects and importance of habitat connectivity differ among recent studies [28,37,40,72] and are therefore controversially discussed [15,16,73]. In contrast to butterflies or plants [37,71] the species richness of bees has not yet been shown to be influenced by habitat connectivity in studies that tested this factor [39,40,74]. The Connectivity Index was developed to explain butterfly movement and dispersal in fragmented landscapes [47] and was shown to be good predictor of habitat specialized butterflies [37]. For wild bees connectivity seemed to influence the foraging behavior leading to a dilution in highly connected landscapes.

The portion of local and landscape effects explaining species richness and abundance of the different species groups varied in this study. Habitat specialists were only affected by local factors whereas cuckoo bees and bumble bees were predominantly influenced by landscape configuration and composition, respectively. Structurally rich landscapes with low land-use intensity can accommodate a diverse bee fauna and are therefore important target areas for conservation. But such landscapes are more and more altered to homogeneous landscapes and hence important factors influencing bee communities have to be considered for providing stable ecosystems in the future. According to our results we conclude that large and high-quality habitats are important for diverse bee communities. However, landscape configuration enhanced total wild bee richness and landscape composition at least bumblebee richness and abundance. This implies that structure and quality of agricultural landscapes are also of importance. Decision makers in ecosystem service planning and conservation should therefore strongly promote and restore areas that include both large high-quality habitats and landscapes with high configurational complexity.

Acknowledgments

We would like to thank the Bayerische Vermessungsverwaltung for providing us with geospatial data, Regierung von Oberfranken for permissions and Dr. Klaus Mandery for identification of critical bee species. We are grateful for the comments of two anonymous reviewers, who highly improved the manuscript.

Author Contributions

Conceived and designed the experiments: SH ISD AH. Performed the experiments: SH. Analyzed the data: SH. Wrote the paper: SH ISD AH.

References

1. Godfray HCJ, Beddington JR, Crute IR, Haddad L, Lawrence D, et al. (2010) Food Security: The Challenge of Feeding 9 Billion People. Science 327 (5967): 812–818.
2. Ehrlich PR, Ehrlich AH (2013) Can a collapse of global civilization be avoided. Proceedings of the Royal Society B: Biological Sciences 280 (1754): 20122845.
3. Poschlod P, WallisDeVries MF (2002) The historical and socioeconomic perspective of calcareous grasslands—lessons from the distant and recent past. Biological Conservation 104 (3): 361–376.
4. Garibaldi LA, Aizen MA, Klein AM, Cunningham SA, Harder LD (2011) Global growth and stability of agricultural yield decrease with pollinator dependence. Proceedings of the National Academy of Sciences 108 (14): 5909–5914.
5. González-Varo JP, Biesmeijer JC, Bommarco R, Potts SG, Schweiger O, et al. (2013) Combined effects of global change pressures on animal-mediated pollination. Trends in Ecology & Evolution (28): 524–530.
6. Ollerton J, Winfree R, Tarrant S (2011) How many flowering plants are pollinated by animals. Oikos 120 (3): 321–326.
7. Garibaldi LA, Steffan-Dewenter I, Winfree R, Aizen MA, Bommarco R, et al. (2013) Wild Pollinators Enhance Fruit Set of Crops Regardless of Honey Bee Abundance. Science 339 (6127): 1608–1611.
8. Klein A, Vaissiere BE, Cane JH, Steffan-Dewenter I, Cunningham SA, et al. (2007) Importance of pollinators in changing landscapes for world crops. Proceedings of the Royal Society B: Biological Sciences 274 (1608): 303–313.
9. Blüthgen N, Klein A (2011) Functional complementarity and specialisation: The role of biodiversity in plant–pollinator interactions. Basic and Applied Ecology 12 (4): 282–291.
10. Fründ J, Dormann CF, Holzschuh A, Tscharntke T (2013) Bee diversity effects on pollination depend on functional complementarity and niche shifts. Ecology 94 (9): 2042–2054.
11. Garibaldi LA, Steffan-Dewenter I, Kremen C, Morales JM, Bommarco R, et al. (2011) Stability of pollination services decreases with isolation from natural areas despite honey bee visits. Ecol Letters 14 (10): 1062–1072.
12. Kennedy CM, Lonsdorf E, Neel MC, Williams NM, Ricketts TH, et al. (2013) A global quantitative synthesis of local and landscape effects on wild bee pollinators in agroecosystems. Ecol Lett 16 (5): 584–599.
13. Sala OE (2000) Global Biodiversity Scenarios for the Year 2100. Science 287 (5459): 1770–1774.
14. Westrich P (1996) Habitat requirements of central European bees and the problems of partial habitats. In: Matheson A, editor. The conservation of bees. London: Academic Press.
15. Doerr VAJ, Barrett T, Doerr ED (2011) Connectivity, dispersal behaviour and conservation under climate change: a response to Hodgson et al. Journal of Applied Ecology 48 (1): 143–147.
16. Hodgson JA, Moilanen A, Wintle BA, Thomas CD (2011) Habitat area, quality and connectivity: striking the balance for efficient conservation. Journal of Applied Ecology 48 (1): 148–152.
17. Fahrig L, Baudry J, Brotons L, Burel FG, Crist TO, et al. (2011) Functional landscape heterogeneity and animal biodiversity in agricultural landscapes. Ecology Letters 14 (2): 101–112.
18. Hadley AS, Betts MG (2012) The effects of landscape fragmentation on pollination dynamics: absence of evidence not evidence of absence. Biological Reviews 87 (3): 526–544.
19. Violle C, Navas M, Vile D, Kazakou E, Fortunel C, et al. (2007) Let the concept of trait be functional! Oikos 116: 882–892.
20. Dias ATC, Berg MP, Bello F de, van Oosten AR, Bílá K, et al. (2013) An experimental framework to identify community functional components driving ecosystem processes and services delivery. Journal of Ecology 101 (1): 29–37.
21. Ekroos J, Rundlöf M, Smith HG (2013) Trait-dependent responses of flower-visiting insects to distance to semi-natural grasslands and landscape heterogeneity. Landscape Ecology 28 (7): 1283–1292.
22. Hoiss B, Krauss J, Potts SG, Roberts S, Steffan-Dewenter I (2012) Altitude acts as an environmental filter on phylogenetic composition, traits and diversity in bee communities. Proceedings of the Royal Society B: Biological Sciences 279 (1746): 4447–4456.
23. Ricotta C, Moretti M (2011) CWM and Rao's quadratic diversity: a unified framework for functional ecology. Oecologia 167 (1): 181–188.
24. Jauker B, Krauss J, Jauker F, Steffan-Dewenter I (2013) Linking life history traits to pollinator loss in fragmented calcareous grasslands. Landscape Ecol 28 (1): 107–120.
25. MacArthur RH, Wilson EO (1967) The theory of Island biogeography. Princeton, NJ: Princeton Univ. Press. 203 p.
26. Rosenzweig ML (1995) Species diversity in space and time. New York: Cambridge University Press. 436 p.
27. Thomas JA, Bourn NAD, Clarke RT, Stewart KE, Simcox DJ, et al. (2001) The quality and isolation of habitat patches both determine where butterflies persist in fragmented landscapes. Proceedings of the Royal Society B: Biological Sciences 268 (1478): 1791–1796.
28. Franzen M, Nilsson SG (2010) Both population size and patch quality affect local extinctions and colonizations. Proceedings of the Royal Society B: Biological Sciences 277 (1678): 79–85.
29. Thomas JA, Simcox DJ, Hovestadt T (2011) Evidence based conservation of butterflies. J Insect Conserv 15 (1–2): 241–258.
30. Potts SG, Vulliamy B, Roberts S, O'Toole C, Dafni A, et al. (2005) Role of nesting resources in organising diverse bee communities in a Mediterranean landscape. Ecological Entomology 30 (1): 78–85.
31. Steffan-Dewenter I, Münzenberg U, Bürger C, Thies C, Tscharntke T (2002) Scale-Dependent Effects of Landscape Context on Three Pollinator Guilds. Ecology 83 (5): 1421–1432.

32. Le Féon V, Schermann-Legionnet A, Delettre Y, Aviron S, Billeter R, et al. (2010) Intensification of agriculture, landscape composition and wild bee communities: A large scale study in four European countries. Agriculture, Ecosystems & Environment 137 (1–2): 143–150.

33. Holzschuh A, Steffan-Dewenter I, Tscharntke T (2010) How do landscape composition and configuration, organic farming and fallow strips affect the diversity of bees, wasps and their parasitoids. J Anim Ecology 79 (2): 491–500.

34. Flick T, Feagan S, Fahrig L (2012) Effects of landscape structure on butterfly species richness and abundance in agricultural landscapes in eastern Ontario, Canada. Agriculture, Ecosystems & Environment 156: 123–133.

35. Holzschuh A, Steffan-Dewenter I, Tscharntke T (2009) Grass strip corridors in agricultural landscapes enhance nest-site colonization by solitary wasps. Ecological Applications 19 (1): 123–132.

36. Rands SA, Whitney HM (2011) Field Margins, Foraging Distances and Their Impacts on Nesting Pollinator Success. Plos One 6(10): e25971.

37. Brückmann SV, Krauss J, Steffan-Dewenter I (2010) Butterfly and plant specialists suffer from reduced connectivity in fragmented landscapes. Journal of Applied Ecology 47 (4): 799–809.

38. Steffan-Dewenter I, Tscharntke T (2000) Butterfly community structure in fragmented habitats. Ecol Letters 3 (5): 449.

39. Meneses Calvillo L, Meléndez Ramírez V, Parra-Tabla V, Navarro J (2010) Bee diversity in a fragmented landscape of the Mexican neotropic. Journal of Insect Conservation 14 (4): 323–334.

40. Öckinger E, Lindborg R, Sjödin NE, Bommarco R (2012) Landscape matrix modifies richness of plants and insects in grassland fragments. Ecography 35 (3): 259–267.

41. Ewers RM, Didham RK (2006) Confounding factors in the detection of species responses to habitat fragmentation. Biol Rev 81 (01): 117.

42. Bommarco R, Biesmeijer JC, Meyer B, Potts SG, Poyry J, et al. (2010) Dispersal capacity and diet breadth modify the response of wild bees to habitat loss. Proceedings of the Royal Society B: Biological Sciences 277 (1690): 2075–2082.

43. Biesmeijer JC (2006) Parallel Declines in Pollinators and Insect-Pollinated Plants in Britain and the Netherlands. Science 313 (5785): 351–354.

44. Potts SG, Willmer P (1997) Abiotic and biotic factors influencing nest-site selection by Halictus rubicundus, a ground-nesting halictine bee. Ecological Entomology 22 (3): 319–328.

45. Sardiñas HS, Kremen C (2014) Evaluating nesting microhabitat for ground-nesting bees using emergence traps. Basic and Applied Ecology 15 (2):161–168.

46. ESRI (2008) ArcGIS 9.3. Redlands, CA, USA.

47. Hanski I (1994) A Practical Model of Metapopulation Dynamics. Journal of Animal Ecology 63 (1): 151.

48. Franzén M, Larsson M, Nilsson SG (2009) Small local population sizes and high habitat patch fidelity in a specialised solitary bee. Journal of Insect Conservation 13 (1): 89–95.

49. Lepais O, Darvill B, O'Connor S, Osborne J, Sanderson R, et al. (2010) Estimation of bumblebee queen dispersal distances using sibship reconstruction method. Molecular Ecology 19 (4): 819–831.

50. Krewenka KM, Holzschuh A, Tscharntke T, Dormann CF (2011) Landscape elements as potential barriers and corridors for bees, wasps and parasitoids. Biological Conservation 144 (6): 1816–1825.

51. Moilanen A, Nieminen M (2002) Simple connectivity measures in spatial ecology. Ecology. Ecology 83 (4): 1131–1145.

52. Westphal C, Bommarco R, Carré G, Lamborn E, Morison N, et al. (2008) Measuring bee diversity in different european habitats and biogeographical regions. Ecological Monographs 78 (4): 653–671.

53. Goulson D, Lye G, Darvill B (2008) Decline and Conservation of Bumble Bees. Annual Review of Entomology 53 (1): 191–208.

54. Westrich P (1990) Die Wildbienen Baden-Württembergs. Stuttgart: Ulmer.

55. R Development Core Team (2011) R.

56. Zuur AF, Ieno EN, Elphick CS (2010) A protocol for data exploration to avoid common statistical problems. Methods in Ecology and Evolution 1:3–14.

57. Sokal RR, Rohlf F (2003) Biometry. New York: Freeman.

58. Krauss J, Alfert T, Steffan-Dewenter I (2009) Habitat area but not habitat age determines wild bee richness in limestone quarries. Journal of Applied Ecology 46 (1): 194–202.

59. Dormann CF, Elith J, Bacher S, Buchmann C, Carl G, et al. (2013) Collinearity: a review of methods to deal with it and a simulation study evaluating their performance. Ecography 36 (1): 27–46.

60. Colwell RK (2013) EstimateS: Statistical estimation of species richness and shared species from samples. Version 9.

61. Mandery K (2001) Die Bienen und Wespen Frankens. Ein historischer Vergleich über neue Erhebungen und alte Sammlungen: (Hymenoptera: Aculeata). Nürnberg: Bund Naturschutz in Bayern.

62. BayLfU (2003) Rote Liste gefährdeter Tiere Bayerns. Augsburg: Bayerisches Landesamt für Umweltschutz.

63. Roulston TH, Goodell K (2011) The Role of Resources and Risks in Regulating Wild Bee Populations. Annual Review of Entomology 56 (1): 293–312.

64. Greenleaf S, Williams N, Winfree R, Kremen C (2007) Bee foraging ranges and their relationship to body size. Oecologia 153 (3): 589–596.

65. Root RB (1973) Organization of a Plant-Arthropod Association in Simple and Diverse Habitats: The Fauna of Collards (Brassica Oleracea). Ecological Monographs 43 (1): 95.

66. Diekötter T, Walther-Hellwig K, Conradi M, Suter M, Frankl R (2006) Effects of Landscape Elements on the Distribution of the Rare Bumblebee Species Bombus muscorum in an Agricultural Landscape. Biodivers Conserv 15 (1): 57–68.

67. Marshall E, West T, Kleijn D (2006) Impacts of an agri-environment field margin prescription on the flora and fauna of arable farmland in different landscapes. Agriculture, Ecosystems & Environment 113 (1–4): 36–44.

68. Holzschuh A, Dudenhöffer J, Tscharntke T (2012) Landscapes with wild bee habitats enhance pollination, fruit set and yield of sweet cherry. Biological Conservation 153: 101–107.

69. Westphal C, Steffan-Dewenter I, Tscharntke T (2003) Mass flowering crops enhance pollinator densities at a landscape scale. Ecol Letters 6 (11): 961–965.

70. Williams NM, Regetz J, Kremen C (2012) Landscape-scale resources promote colony growth but not reproductive performance of bumble bees. Ecology (93(5)): 1049–1058.

71. Tscharntke T, Tylianakis JM, Rand TA, Didham RK, Fahrig L, et al. (2012) Landscape moderation of biodiversity patterns and processes - eight hypotheses. Biological Reviews 87 (3): 661–685.

72. Raatikainen KM, Heikkinen RK, Luoto M (2009) Relative importance of habitat area, connectivity, management and local factors for vascular plants: spring ephemerals in boreal semi-natural grasslands. Biodiversity and Conservation 18 (4): 1067–1085.

73. Hodgson JA, Thomas CD, Wintle BA, Moilanen A (2009) Climate change, connectivity and conservation decision making: back to basics. Journal of Applied Ecology 46 (5): 964–969.

74. Steffan-Dewenter I (2003) Importance of Habitat Area and Landscape Context for Species Richness of Bees and Wasps in Fragmented Orchard Meadows. Conservation Biology 17 (4): 1036–1044.

18

Retention of Habitat Complexity Minimizes Disassembly of Reef Fish Communities following Disturbance: A Large-Scale Natural Experiment

Michael J. Emslie*, Alistair J. Cheal, Kerryn A. Johns

Australian Institute of Marine Science, Townsville, Queensland, Australia

Abstract

High biodiversity ecosystems are commonly associated with complex habitats. Coral reefs are highly diverse ecosystems, but are under increasing pressure from numerous stressors, many of which reduce live coral cover and habitat complexity with concomitant effects on other organisms such as reef fishes. While previous studies have highlighted the importance of habitat complexity in structuring reef fish communities, they employed gradient or meta-analyses which lacked a controlled experimental design over broad spatial scales to explicitly separate the influence of live coral cover from overall habitat complexity. Here a natural experiment using a long term (20 year), spatially extensive ($\sim$115,000 kms^2) dataset from the Great Barrier Reef revealed the fundamental importance of overall habitat complexity for reef fishes. Reductions of both live coral cover and habitat complexity had substantial impacts on fish communities compared to relatively minor impacts after major reductions in coral cover but not habitat complexity. Where habitat complexity was substantially reduced, species abundances broadly declined and a far greater number of fish species were locally extirpated, including economically important fishes. This resulted in decreased species richness and a loss of diversity within functional groups. Our results suggest that the retention of habitat complexity following disturbances can ameliorate the impacts of coral declines on reef fishes, so preserving their capacity to perform important functional roles essential to reef resilience. These results add to a growing body of evidence about the importance of habitat complexity for reef fishes, and represent the first large-scale examination of this question on the Great Barrier Reef.

Editor: Maura (Gee) Geraldine Chapman, University of Sydney, Australia

Funding: The study was supported by Australian Institute of Marine Science and the Australian Government's Marine and Tropical Sciences Research Facility and National Environment Research Program (Tropical Ecosystems Hub). The funders had no role in study design, data collection and analysis, decision to publish, or preparation of the manuscript.

Competing Interests: The authors have declared that no competing interests exist.

* Email: m.emslie@aims.gov.au

Introduction

Habitat complexity is fundamentally important for the maintenance of high biodiversity across a range of ecosystems [1–5]. Coral reef ecosystems are among the most diverse on the planet with reefs with higher habitat complexity often housing more species than less complex reefs due to the greater variety of niches and shelter [6–8]. Habitat complexity on coral reefs has two major components; the underlying substrate rugosity and the skeletal structure provided by live and dead hard corals. Coral reefs are subject to many types of disturbance that can have negligible to severe impacts on coral cover and habitat complexity. For example, disturbances such as *Acanthaster planci* (crown-of-thorns starfish) outbreaks and coral bleaching cause coral mortality but leave skeletons intact [9–11], so habitat complexity remains largely unchanged in the short term. Subsequently, coral skeletons may erode due to natural processes causing longer term declines in habitat complexity. Conversely, waves from storms can obliterate entire coral colonies removing the habitat complexity previously afforded by their skeletons [11,12]. However, loss of coral structures due to storms or skeletal erosion will not necessarily lead to low habitat complexity if substrate rugosity is high. Indeed, reefs with high substrate rugosity should maintain a greater diversity of organisms than reefs with low substrate rugosity once hard corals are removed, with the exception of those organisms fundamentally dependent on intact coral skeletons or living coral tissue for survival.

Disturbances on coral reefs can dramatically impact the diversity, abundance and community structure of reef fishes, because many fish species are closely associated with live corals and their structures [6–8,13–15]. To date, many studies have attributed changes in fish communities to loss of hard coral cover [9,13,16–20]. Numerous reef fishes rely on hard corals for food and/or shelter and many of these species decline in abundance following hard coral decline [9,16–22]. However, numerous fish species with seemingly limited reliance on hard corals *per se* (e.g. non-corallivorous butterflyfishes, large predators, some herbivorous fishes) have also declined in abundance following disturbances, and in these cases the role of habitat complexity has been implicated [8,9,11,13,20,23]. Declines in abundance and diversity of reef fishes following disturbances can be detrimental to ecosystem functioning and reef resilience due to a reduction in

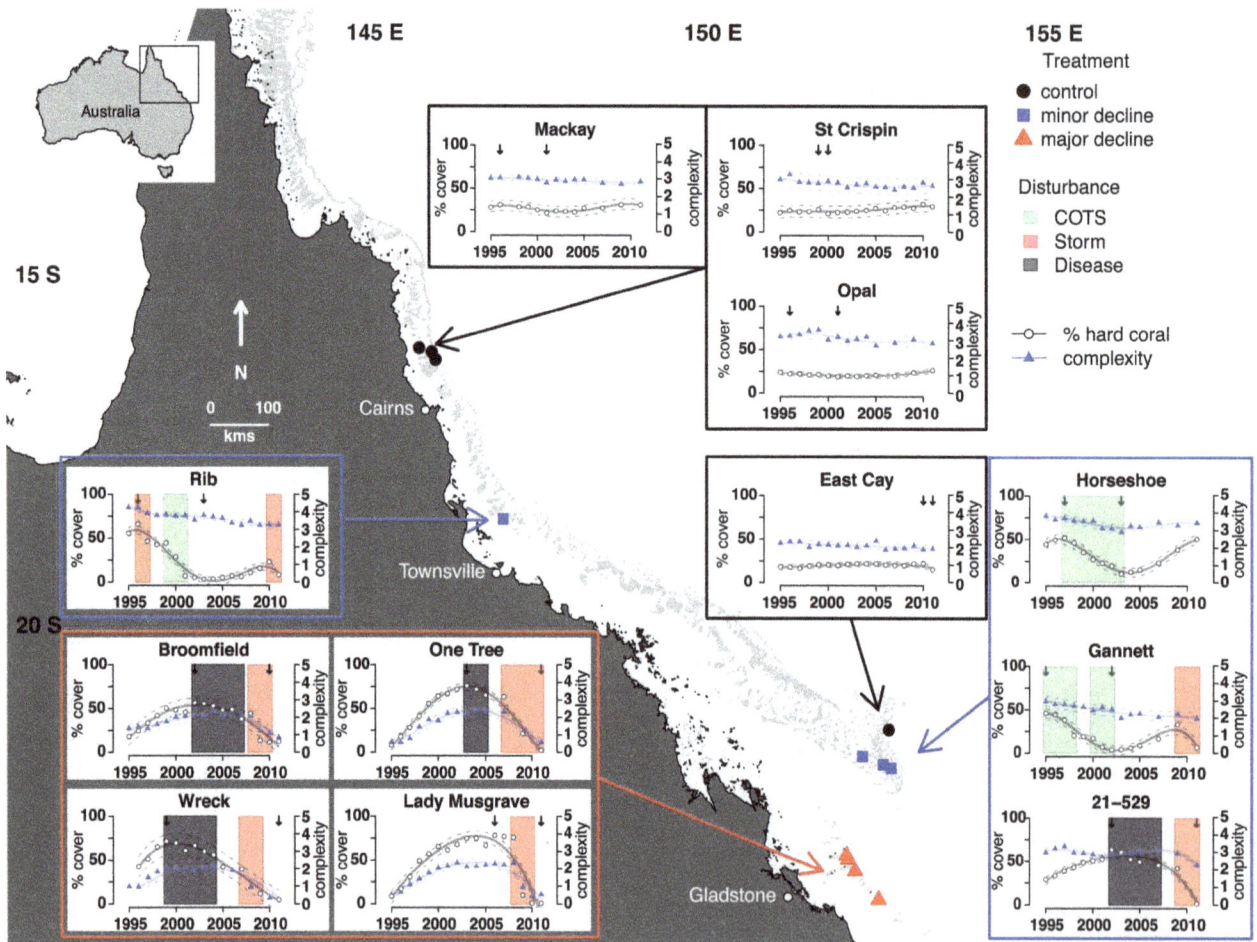

Figure 1. Location of the study reefs in each of the three treatments (Major Decline, Minor Decline and Control). Small panels display trends in hard coral cover and habitat complexity, along with shaded periods of time when disturbances (COTS = *Acanthaster planci* outbreaks, storms & coral disease) occurred. Points are raw data means, while solid lines indicate modelled average trends and dotted lines show 2 x standard errors from a linear mixed effects model fitted separately to hard coral cover and habitat complexity. Arrows mark the years of greatest and least hard coral cover.

the capacity of reef fishes to perform trophic functions. For example, a reduction in the number and diversity of herbivorous fishes decreases their capacity to prevent proliferation of macro-algae that may limit recovery of corals following disturbances [24–32]. Clearly, declines in both live corals and habitat complexity must be important to reef fishes, and disentangling the relative influence of each will provide clues to the relative threat to reef fishes of disturbances which do and do not alter habitat complexity.

It has previously been demonstrated through experimentation [7,33,34] and longer term datasets [12–14,16–22] that reductions in habitat complexity and live coral cover adversely affect reef fish communities. Manipulative experiments have generally been conducted at restricted spatio- temporal scales, typically small (~10 s of m^2) patch reefs surveyed over several months [7,33,34], and results are difficult to scale up to ecosystem levels. Projects conducted over larger spatio-temporal scales have generally employed gradient/regression type analyses (e.g. [13]) or meta-analyses (e.g. [11]), which are useful approaches for highlighting relationships among variables, changes in variables along a gradient and for integrating many disparate datasets, but lack rigorous experimental designs with which to definitively attribute

causation. Here we use data collected from reefs spread over 115,000 km^2 of the Great Barrier Reef (GBR), gathered over 20 years and employ a natural experiment to formally test how the loss of live coral versus loss of habitat complexity influences reef fish community structure, the diversity of reef fish families and functional groups, and the abundance of individual species.

Methods

Sampling

Data were gathered as part of the Long Term Monitoring Program at the Australian Institute of Marine Science (GBRMPA permit number G13/36390.1); in which fish and benthic communities have been surveyed on 47 reefs of the GBR since 1995. Large-scale disturbances, such as storms and *A. planci* outbreaks that have occurred over the last two decades on the GBR [22,35–36], facilitate opportunities to test macro-ecological hypotheses that due to their scope, require manipulations of a scale (100 s kilometres) that are logistically impossible for researchers to attempt using traditional experimental frameworks [37].We were able to perform a natural experiment to investigate the effects of reductions in live coral cover versus habitat complexity on reef fish communities, by retrospectively assigning replicate reefs into three treatments based on the effects of disturbances. Eight reefs were chosen based on comparable levels of live coral cover (>50%) and subsequent similar and very large relative declines in cover (~90%) due to disturbances. These reefs were separated into two equal treatments based on relative reductions in habitat complexity: 1. a major decline in habitat complexity from high/moderate to very low levels (hereafter "Major Decline"), and 2. a minor decline in habitat complexity from high to moderate levels (hereafter "Minor Decline"). A further four reefs had minimal declines in hard coral cover and no change in habitat complexity (hereafter "Control"; Fig. 1). Even though reefs in each treatment were unevenly distributed geographically (Fig. 1), 77% of fish species were common to all reefs in the study thus enabling valid comparisons of changes to fish communities. Furthermore, our analysis determined the magnitude of change in individual species abundance and community structure, plus the proportion of the community affected (irrespective of identity) before and after disturbances. Thus species identity *per se* was not important but rather the magnitude of changes and the proportion of the community affected.

Three sites of five permanently marked 50 m transects were situated in comparable reef slope habitats (n = 15 transects per reef) and were surveyed on SCUBA annually from 1995 until 2006 and then biennially thereafter. From 1995 until 2005, the benthic community was described using a 30-cm video swathe along the transects. Forty frames from each video transect were sampled and the benthic organisms beneath five points projected on to each frame in a quincunx pattern were identified to the finest taxonomic resolution possible, yielding 200 samples per transect. After 2006, a digital still image was taken every metre along each transect, and forty images were selected and analysed as before [38]. These data were then converted to percent cover of total hard coral for use in univariate analyses. For multivariate analyses, data were converted to percent cover of finer taxonomic groupings that included different growth forms of the most abundant coral family Acroporidae and other hard corals (including all other non-Acroporidae hard coral families), fire coral (genus *Millepora*), soft corals, coralline, turf and macro-algae, rubble, dead coral, sand, abiotic, sponges and other (rare benthic organisms of very low abundance e.g., ascidians, anemones). Fish communities were surveyed concurrently on the same transects using underwater visual census. The abundance and number of species of fishes recorded during surveys were taken from a list of 215 mobile, diurnally active species (including the families Acanthuridae, Chaetodontidae, Labridae, Lethrinidae, Lutjanidae, Pomacentridae, Scaridae, Siganidae, Zanclidae and the commercially important *Plectropomus* spp., hereafter "coral trout"). While parrotfishes are now considered as a tribe Scarinae within the family Labridae, we use the term "Scaridae" to distinguish this group of fishes from other Labridae. We define "species richness" as the number of species recorded and use this term hereafter. Cryptic species such as gobies and blennies were not included. Two transect widths were used: 50×1 m belts for the Pomacentridae and 50×5 m belts for the remaining families [39]. Habitat complexity was independently estimated retrospectively by two observers using a scale of zero (least complex - minimal vertical relief, few holes, crevices and overhangs) to five (most complex - high vertical relief, many holes, crevices and overhangs) from 360° video panoramas taken at the start of each transect. This 0 to 5 scale correlates strongly with a range of other rugosity metrics and has been found to be a good predictor of reef fish diversity and abundance [40].

Analyses

To provide the clearest picture of absolute changes in fish communities under varying degrees of change in habitat complexity, we compared metrics of reef fish communities at times of greatest (hereafter "Before") and least (hereafter "After") percent coral cover (indicated by arrows in Fig 1). All analyses were conducted in R [41]. To visualise the changes in fish and benthic communities before and after disturbances, we performed a non-metric Multi-Dimensional Scaling (nMDS) based on the Bray-Curtis similarity co-efficient using the iso-MDS package. To reduce the influence of highly abundant taxa, benthic cover data were row centred and square-root transformed. Similarly, to visualise changes to the whole community rather than a few highly abundant species, fish abundances were row centred and fourth root transformed prior to analysis. To examine the magnitude of change in fish and benthic communities before and after disturbances, we conducted a permutational multivariate analysis of variance using distance matrices and assessed the sums of squares for each Treatment and used the ADONIS function from the VEGAN package in R [41]. As the Treatment by Time interaction was significant, we re-ran the analysis separately for each Treatment (Major Decline, Minor Decline, Control).

Changes in fish and benthic communities were further investigated using Bayesian hierarchical models [42], fitted separately for hard coral cover, habitat complexity, total fish species richness and the species richness of eight reef fish families surveyed (Acanthuridae, Chaetodontidae, Labridae, Lethrinidae, Lutjanidae, Pomacentridae, Scaridae, Siganidae), plus the commercially important coral trout (*Plectropomus* spp.). In order to assess the effects of loss of habitat complexity and live coral on functional roles performed by reef fishes, we examined changes to the species richness of broad functional groups including corallivorous and generalist butterflyfishes, herbivores, planktivores and predators. Models had the fixed factors of Time (Before or After) and Treatment (Major Decline, Minor Decline, Control), and random factors of reef, site and transect. Most variables were modelled against a gaussian distribution in the MCMCglmm package [43]; however some were modelled against negative binomial distributions (log link) to account for zero-inflation and over-dispersion inherent in ecological count data [44] (Table S1). Negative-binomial models were fitted through Just Another Gibbs Sampler (JAGS) via the R2JAGS package in R and used non-informative, flat gaussian priors and the posterior distributions were derived from three Markov chain Monte Carlo (MCMC) (see Table S1 for further model details including number of iterations, burn in and thinning). Model convergence and mixing of Markov chains was assessed visually from trace plots and autocorrelation of the chains was always less than 0.2. Inferences about temporal changes were based on 95% Bayesian Higher Posterior Density (HPD) intervals of cell means predicted from posterior distributions of model parameters. Specific post-hoc contrasts were examined including differences in Time (before and after disturbance) among Treatments and differences among Treatments.

We assessed changes in the abundance of individual reef fish species by plotting a comparable metric to account for differences in initial coral cover [45], calculated as the percent change in abundance from before to after disturbance;

$$\%\text{difference} = \ln[(A_{a,i} - A_{b,i})/A_{b,i}] \times 100$$

Where A_b and A_a were mean values at before and after disturbance respectively. Fish species were only included in these analyses if their summed abundance was ≥10 per reef (= 15 transects) in one of the two years. Changes in individual species abundance were then averaged across the four reefs within each Treatment.

Results

Benthic and fish community structure changed from times of greatest to least coral cover, but the magnitude of change varied among habitat complexity treatments (Fig 2). On reefs with a major decline in complexity, there were substantial shifts in the structure of both fish communities (ADONIS Time: $F = 19.134$, d. f. $= 1$, $Pr(>F) = 0.001$) and benthic communities (ADONIS Time: $F = 85.902$, d. f. $= 1$, $Pr(>F) = 0.001$) (Fig 2). Similarly, a large shift occurred in the benthic communities on reefs with minor declines in habitat complexity, (ADONIS Time: $F = 32.429$, d. f. $= 1$, $Pr(>F) = 0.001$), but a much smaller shift was evident for the fish communities (ADONIS Time: $F = 2.1751$, d. f. $= 1$, $Pr(>F) = 0.059$) on these reefs compared to those in the Major Decline treatment (Fig 2). Very little change occurred in either the fish communities (ADONIS Time: $F = 0.3885$, d. f. $= 1$, $Pr(>F) = 0.909$) or benthic communities (ADONIS Time: $F = 1.0507$, d. f. $= 1$, $Pr(>F) = 0.304$) on Control reefs (Fig 2).

Hard coral cover declined in all treatments but the decline was negligible on Control reefs. Habitat complexity only declined substantially on Major Decline reefs; reductions were minimal on reefs in the Minor Decline treatment and were similar to changes at Control reefs (Fig 3). Reductions in fish total species richness and the species richness of the Chaetodontidae and Labridae occurred on reefs in both complexity decline treatments, though the loss was greatest on in the Major Decline reefs (Fig 3). Also, species richness of Acanthuridae, Lutjanidae, Pomacentridae, Scaridae and coral trout declined on reefs in the Major Decline treatment, but not on those in the Minor Decline or Control treatments (Fig 3). There were large declines of species richness of all functional groups of fishes on Major Decline reefs (Fig 3). However, the species richness of only two functional groups, corallivorous butterflyfishes and predators, declined on Minor Decline reefs and these reductions were substantially smaller than those on reefs in the Major Decline treatment. There was no substantial decline in species richness of any functional group on Control reefs (Fig. 3).

Changes in the abundance of individual species varied substantially among the three habitat complexity treatments (Fig 4), with major declines in habitat complexity impacting a greater number of species than minor declines. On Major Decline reefs, 75% of species declined in abundance, 56% of species lost half their abundance and 18% were locally extirpated (declined to zero) (Fig 4). In comparison, the abundance of less than half (48%) of the fish species declined on reefs in the Minor Decline treatment, 24% declined in abundance by half and only 3% of species were locally extirpated (Fig 4). Fish species on Control reefs were far less affected; 22% of species declined in abundance, with only 3% declining by half and no species being locally extirpated (Fig 4).

The major loss of habitat complexity also greatly reduced the capacity of reef fishes to perform their functional roles. Among the functionally important herbivorous fishes, fourteen species declined in abundance by 50% or more on reefs that underwent major declines in habitat complexity, compared to four species on reefs with a minor decline and only one species on Control reefs. Additionally, abundances of some commercially important fishery species such as coral trout, were reduced to zero on Major Decline reefs, but declined by less than 5% on Minor Decline reefs. In addition, obligate corallivores accounted for a large proportion of the species that declined in abundance in the Minor Decline treatment, but accounted for a much smaller proportion of the substantially greater number of species that declined on reefs with major declines in complexity.

Discussion

Using long-term data at ecologically meaningful scales on the GBR, this study has demonstrated the fundamental importance of habitat complexity for the maintenance of diverse fish communities, which is critical for maintaining healthy ecosystem function. Among reefs which underwent large declines in live coral cover, it was only on those reefs where habitat complexity also declined markedly that reef fish communities underwent wholesale reductions in diversity, species abundances and functional capacity. Previously small scale manipulative experiments [7,33,34], gradient/regression type analyses [13,20], or meta-analyses [11] had proposed the importance of habitat complexity for reef fishes, but whether these results reflected a broad-scale truth had not been rigorously tested. Our large-scale, natural experiment was able to demonstrate the generality of habitat complexity as a fundamental driver of reef fish community structure on the GBR, supporting findings in other regions [11,13,20,46,47]. We showed that major loss of habitat complexity affected a broad array of reef fishes from all trophic/functional groups. Additionally, although major loss of hard coral but not habitat complexity caused declines in some fish species, mostly those intimately associated with hard corals, the role of corals was not as important if overall habitat complexity remained moderate to high. Such results suggest that reefs which undergo major reductions in overall habitat complexity following disturbances will support depauperate reef fish communities, with a reduced ability to perform critical functional roles that contribute to the resilience of coral reefs.

While decreases in abundance of coral dependent species following loss of live coral were expected irrespective of changes in habitat complexity [20,22], the sweeping reductions in abundance of most reef fish species following major reduction in habitat complexity was more surprising (but see [11,13]). Large predatory fishes, planktivorous damselfishes and various herbivores were included in these decreases despite most having no obvious dependence on corals, implying that these fishes are dependent on habitat complexity for their survival, most likely through the provision of shelter and food sources. Clearly, habitat complexity affords shelter not only through live corals, but also through dead coral skeletons and by caves, cracks and fissures in the substrate. Where fish abundance declined due to lack of shelter, it was uncertain whether this resulted from migration to more suitable habitat, either around the reef or into deeper water, or from increased mortality resulting from the lack of refugia from predation. Whatever the mechanism of these declines, such dramatic shifts in reef fish community structure have implications for the ecological functioning of coral reef communities.

The extirpation of numerous species of fishes following major declines in habitat complexity contributed to a major reduction in fish diversity, with species from a range of trophic affiliations lost. High fish diversity usually equates to increased functional diversity (the number of functional groups at a site) and functional redundancy (the number of species within a functional group), both key components of reef resilience [48–53]. Higher functional diversity should enhance the capacity of a reef to deal with disturbances while functional redundancy provides a form of ecological insurance for the maintenance of a functional role despite losses of some species due to disturbances. Thus it seems

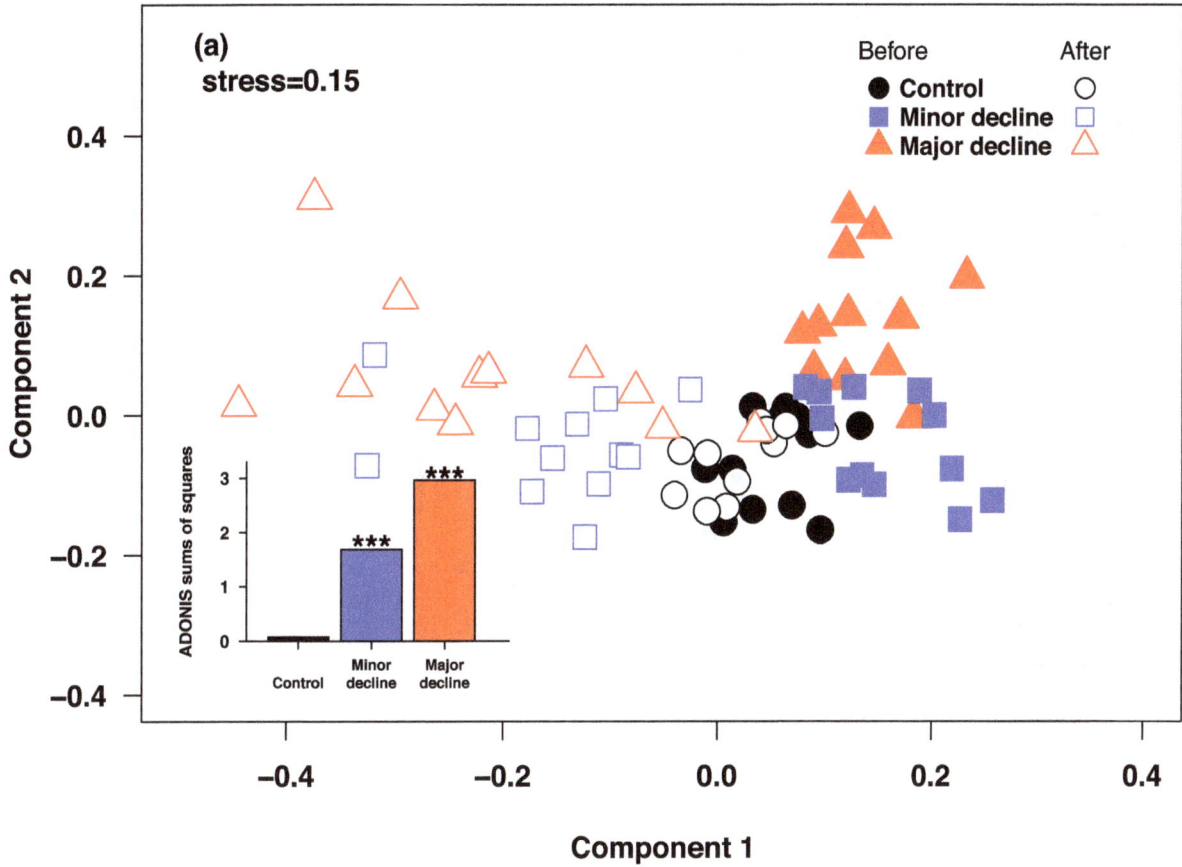

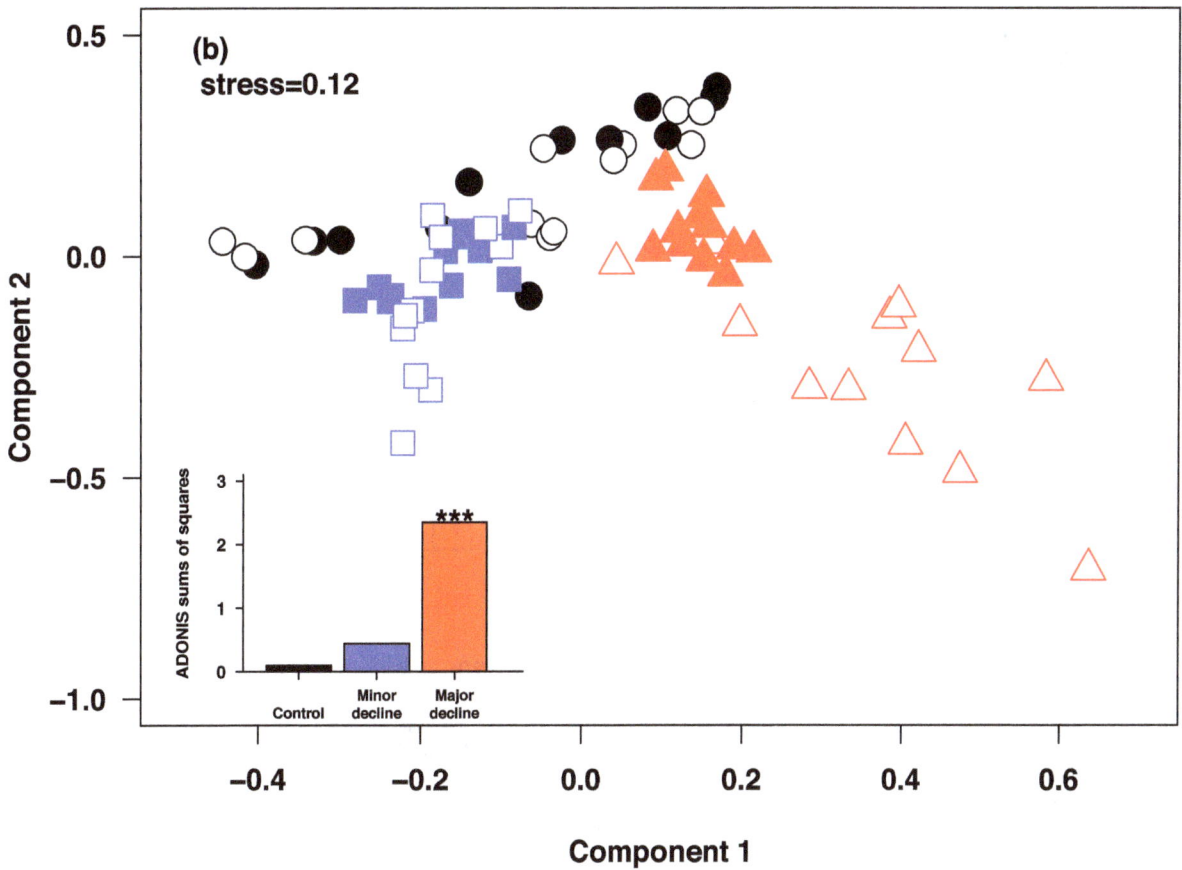

Figure 2. Multi-dimensional plot based on Bray-Curtis similarity coefficients of (a) square-root transformed percent benthic cover and (b) fourth-root transformed fish species abundances. Each panel presents changes to communities following disturbances for the three treatments (Major Decline, Minor Decline and Control). A full model ADONIS analysis revealed a significant interaction for both benthic communities (ADONIS Treatment*Time: F = 14.293, d. f. = 2, Pr(>F) = 0.001) and fish communities (ADONIS Treatment*Time: F = 4.9225, d. f. = 2, Pr(>F) = 0.001). Changes from times of greatest to least coral cover were further examined by separate ADONIS for each individual Treatment (Major Decline, Minor Decline and Control), and the small inset bar graphs display the effect sizes (Sums of Squares) from these individual analyses. ***: Pr(>F) = <0.001

highly likely that resilience will be diminished following major losses of habitat complexity. For example, the functional contribution of herbivorous fishes to reef resilience has been well established. Many species of herbivorous reef fishes have the capacity to prevent algal overgrowth and aid coral recovery through their grazing activities, thereby preventing undesirable shifts to a macro-algal dominated state [24,32,54]. In this study, the disappearance of fourteen species of herbivorous fishes on reefs where there were major declines of habitat complexity is likely to result in increased vulnerability to such phase shifts (but see [55]).

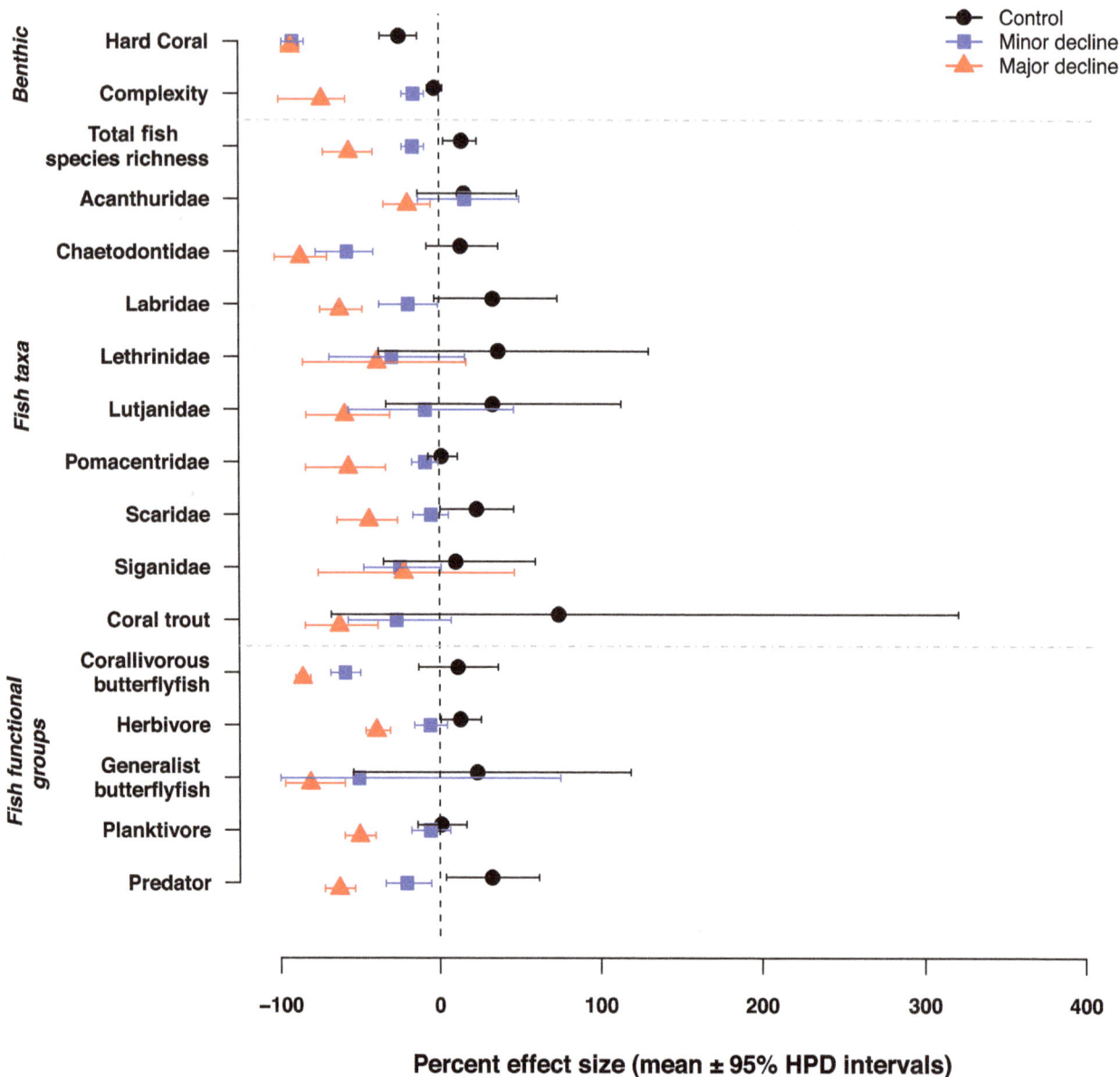

Figure 3. Differences in hard coral cover, habitat complexity, total species richness of fishes and species richness of eight fish families and five broad functional groups for each of the three treatments (Major Decline, Minor Decline and Control). Data are average effect sizes from generalized linear mixed effects model expressed as a per cent change from the time of greatest to least coral cover. Inferences about temporal changes were based on 95% Bayesian Highest Posterior Density (HPD) intervals of cell means predicted from posterior distributions of model parameters derived via Markov-chain Monte Carlo (MCMC) sampling. Effects are considered significant if the HPD intervals do not intersect zero.

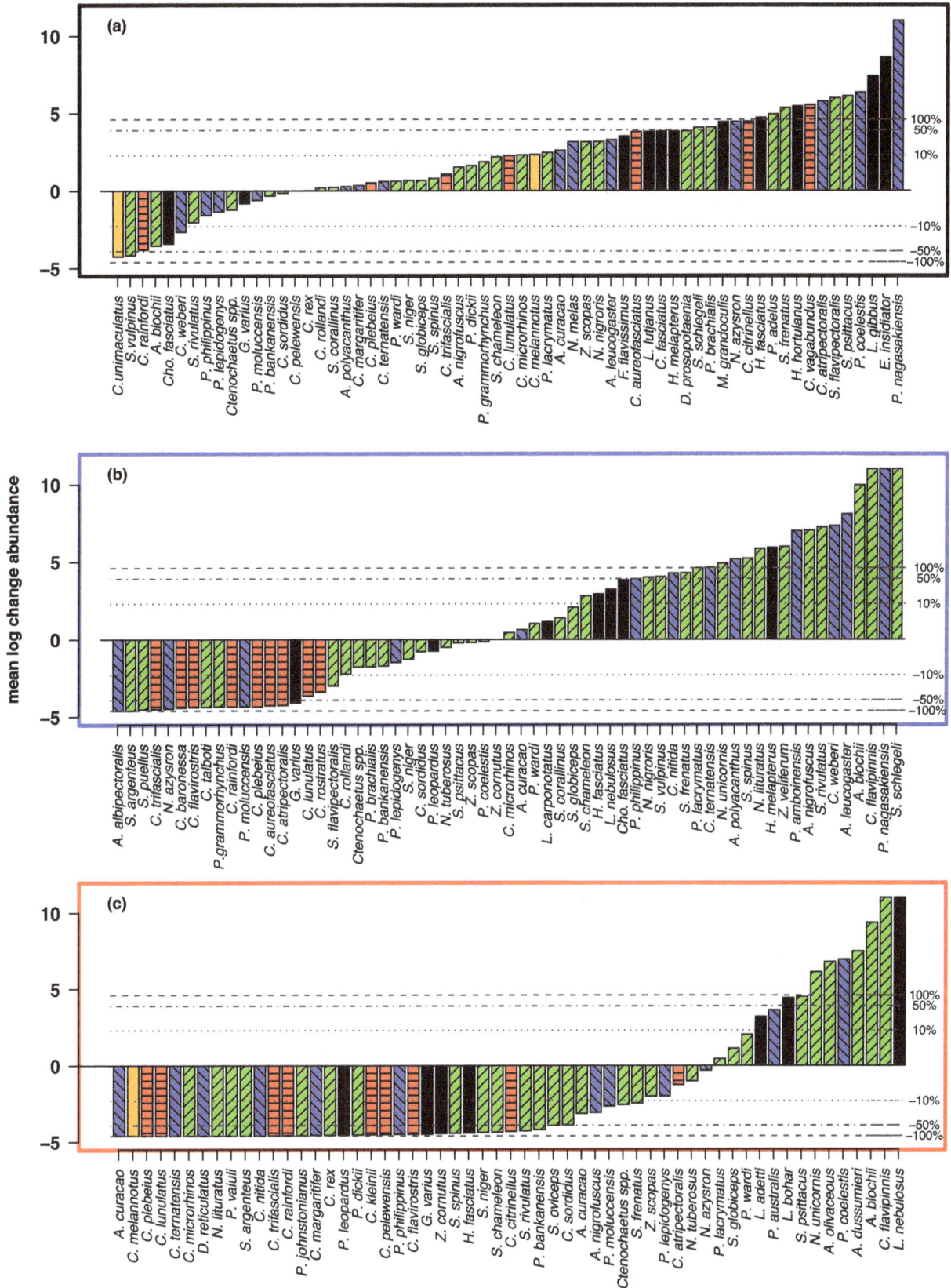

Figure 4. Average percentage change in abundance of individual fish species between times of greatest and least hard coral cover for (a) control reefs (b) reefs that underwent minor declines in complexity (c) reefs that underwent major decline in complexity. Fish species were only included in analyses if their reef wide abundance was ≥10 in one of the two years. Changes in individual species abundance at each reef were then averaged across the four reefs in each Treatment (Major Decline, Minor Decline and Control). Note that the y axis scale is in

natural log units and dotted horizontal lines represent 10, 50 and 100% changes in abundance and that error bars were not included to improve clarity. Coloured bars represent trophic affiliations: green with right diagonal hatching = herbivores, blue with left diagonal hatching = planktivores, red with horizontal hatching = corallivorous butterflyfishes, orange solid bars = generalist (non-coral feeding) butterflyfish, black solid bars = predators. A list of species abbreviations on the x-axis and their corresponding species names are found in Table S2.

While the role of herbivorous fishes in reef resilience has been well established, the contributions of many other reef fishes to reef resilience and healthy ecosystem functioning is less clear. However, what is certain is that the loss of a range of coral reef species performing many functional roles will likely have unknown consequences for ecosystem functioning. For example, reductions in the diversity and abundance of corallivorous fishes (e.g. butterflyfishes) will lower coral mortality [56], because corallivorous butterflyfishes can consume between 9 and 13% of the available tissue biomass of coral, representing 50 to 80% of the total annual productivity [57]. The loss of corallivorous fishes following disturbances will therefore remove substantial predation pressures from newly recruited corals and may ultimately aid recovery. Conversely, the loss of corallivorous fishes may deleteriously affect recovery as high diversity and abundance of corallivorous butterflyfishes has been demonstrated to slow or halt the transmission of coral disease [58]. Future research focused on the role played by corallivorous butterflyfishes in coral dynamics shortly following disturbances could aid our understanding of what impact, if any, the loss of corallivorous fishes plays in reef resilience and ecosystem functioning.

It appears that the short term loss following disturbances of adult fishes not directly dependent on live coral relates more closely to the lack of available shelter rather than to loss of living corals *per se*. Similarly, findings of diverse coral reef fish assemblages on artificial structures largely devoid of corals supports the idea that shelter provided by habitat complexity is fundamentally important to coral reef fish communities [59–61]. However, many reef fishes use live coral as a cue for settlement, including taxa that do not utilise live coral as adults [18]. Although fish communities may be relatively unaffected by coral mortality when habitat complexity is retained, shifts in community structure may lag behind disturbances if fish recruitment is suppressed by limited availability of living coral, while natural mortality of surviving fishes continues. Furthermore, the erosion of coral skeletons after some disturbances such as *A. planci* outbreaks, coral bleaching and coral disease slowly decreases habitat complexity, and may also produce lagged declines in fishes [20,62]. However, in this study adult fish populations were not depleted while habitat complexity remained, providing a buffer to fish population declines while coral is recovering in those cases. Thus in normal circumstances, lagged effects are likely to be balanced by coral recovery and new fish recruitment as long as complexity remains following disturbance. Nevertheless, lagged effects in reef fishes may potentially become more important in future decades, especially if predictions of increased coral bleaching and ocean acidification are correct [63]. In summary, while the retention of habitat complexity reduces the short term impact of disturbances on fish communities, the regeneration of live coral is essential for the maintenance of complex habitats and therefore, to the recovery and long term persistence of diverse reef fish communities.

While previous studies have identified the link between habitat complexity and reef fishes, many of these studies have focused on subsets of the fish community (e.g. [20,21,64], but see [46]). We were able to tease apart the roles of reductions in coral cover versus habitat complexity on a large proportion of diurnally active and conspicuous reef fish communities over ecologically meaningful scales. To our knowledge, this is the first large-scale natural experiment conducted on the GBR to investigate the fundamental contribution of habitat complexity in driving reef fish community change. These results illustrated that reef fish communities are more adversely affected by disturbances which degrade both live coral cover and habitat complexity (i.e. storms), than those which reduce cover of live corals only (i.e. coral bleaching and outbreaks of *A. planci*). Such results should be of interest to reef managers, particularly given our finding that the major fishery target species, the coral trout (*Plectropomus* spp.) disappeared from sites of major complexity decline, with socio-economic ramifications for fishers utilising this resource. In addition, the impact of storms on reef fish communities at sites where coral skeletons account for most of the habitat complexity will be equally devastating irrespective of any zoning to protect target species from fishing. In effect, the benefits afforded by reserve zoning can be reversed almost instantaneously. Conversely, protection of fish communities at sites where complexity of the underlying substrate is high would better preserve important functional processes performed by reef fishes, encouraging rapid recovery in the event that coral cover is removed. Given the prospect of increases in storm intensity with climate change [65] which may lead to the architectural collapse of coral reefs [66], protecting sites with high underlying substrate complexity should be considered to alleviate vulnerability to disassembly of reef fish communities, reductions in the functional roles they perform and much diminished reef resilience.

Acknowledgments

We thank all colleagues, past and present, who have contributed to gathering the data used in this study. The insightful comments of Nick Graham, Aaron MacNeil and Murray Logan are also appreciated.

Author Contributions

Conceived and designed the experiments: MJE AJC KAJ. Performed the experiments: MJE AJC KAJ. Analyzed the data: MJE KAJ. Contributed to the writing of the manuscript: MJE AJC KAJ.

References

1. MacArthur RH, MacArthur JW (1961) On bird species diversity. Ecology 42: 594–598.
2. Heck KL Jr, Wetstone GS (1977) Habitat complexity and invertebrate species and abundance in tropical seagrass. J Biogeogr 4: 135–142.
3. Russell BC (1977) Population and standing crop estimates for rocky reef fishes of North-East New Zealand, New Zealand. J Mar Freshw Res 11: 23–36.
4. Crowder LB, Cooper WE (1982) Habitat structural complexity and the interactions between bluegills and their prey. Ecology 63: 1802–1813.

5. Spies TA (1998) Forest structure: a key to the ecosystem. Northwest Sci 72: 34–39.

6. Risk MJ (1972) Fish diversity on a coral reef in the Virgin Islands. Atoll Res Bull 153: 1–7.

7. Syms C, Jones GP (2000) Disturbance, habitat structure and the dynamics of a coral reef fish community. Ecology 81: 2714–2729.

8. Gratwicke B, Speight MR (2005) Effects of habitat complexity on Caribbean marine fish assemblages. Mar Ecol Prog Ser 292: 301–310.

9. Sano M, Shimizu M, Nose Y (1987) Long-term effects of destruction of hermatypic corals by *Acanthaster planci* infestation on reef fish communities at Iriomote Island, Japan. Mar Ecol Prog Ser 37: 191–199.

10. Sano M (2004) Short-term effects of a mass coral bleaching event on a reef fish assemblage at Iriomote Island, Japan. Fish Sci 70: 41–46.

11. Wilson SK, Graham NAJ, Pratchett MS, Jones GP, Polunin NVC (2006) Multiple disturbances and the global degradation of coral reefs: are reef fishes at risk or resilient? Glob Change Biol 12: 2220–2234. (doi: 10.1111/j.1365-2486.2006.01252.x)

12. Halford AR, Cheal AJ, Ryan D, Williams DMcB (2004) Resilience to large-scale disturbance in coral and fish assemblages on the Great Barrier Reef. Ecology 85: 1892–1905.

13. Graham NAJ, Wilson SK, Jennings S, Polunin NVC, Bijoux JP, et al. (2006) Dynamic fragility of oceanic coral reef ecosystems. Proc Natl Acad Sci USA 103: 8425–8429.

14. Emslie MJ, Cheal AJ, Sweatman H, Delean S (2008) Recovery from disturbance of coral and reef fish communities on the Great Barrier Reef, Australia. Mar Ecol Prog Ser 371: 177–190.

15. Pratchett MS, Munday PL, Wilson SK, Graham NAJ, Cinner JE, et al. (2008) Effects of climate induced coral bleaching on coral-reef fishes – ecological and economic consequences. Ocean Mar Biol Ann Rev 46: 251–296.

16. Woodley JD, Chornesky EA, Clifford PA, Jackson JBC (1981) Hurricane Allen's impact on Jamaican coral reefs. Science 214: 749–755.

17. Bell JD, Galzin R (1984) Influence of live coral cover on coral reef fish communities. Mar Ecol Prog Ser 15: 265–274.

18. Jones GP, McCormick MI, Srinivasan M, Eagle J (2004) Coral declines threaten fish biodiversity in marine reserves. Proc Natl Acad Sci USA 101: 8251–8253.

19. Cheal AJ, Wilson SK, Emslie MJ, Dolman AM, Sweatman H (2008) Response of reef fish communities to coral declines on the Great Barrier Reef. Mar Ecol Prog Ser 372: 211–223.

20. Graham NAJ, Wilson SK, Pratchett MS, Polunin NVC, Spalding MD (2009) Coral mortality versus structural collapse as drivers of corallivorous butterflyfish decline. Biodivers Conserv 18: 3325–3336.

21. Luckhurst BE, Luckhurst K (1978) Analysis of influence of substrate variables on coral-reef fish communities. Mar Biol 49: 317–323.

22. Emslie MJ, Pratchett MS, Cheal AJ (2011) Effects of different disturbance types on butterflyfish communities of Australia's Great Barrier Reef. Coral Reefs 30: 461–471. (doi 10.1007/s00338-011-0730-x)

23. Wilson SK, Dolman AM, Cheal AJ, Emslie MJ, Pratchett MS, et al. (2009) Maintenance of fish diversity on disturbed coral reefs. Coral Reefs 28: 3–14.

24. Hughes TP (1994) Catastrophes, phase shifts and large-scale degradation of a Caribbean coral reef. Science 265: 1547–1551.

25. Hughes TP, Rodrigues MJ, Bellwood DR, Ceccarelli D, Hoegh-Guldberg O, et al. (2007) Phase shifts, herbivory and the resilience of coral reefs to climate change. Curr Biol 17: 360–365.

26. Lewis SM (1986) The role of herbivorous fishes in the organisation of a Caribbean reef community. Ecol Monogr 56:183–200.

27. Mantyka CS, Bellwood DR (2007) Direct evaluation of macroalgal removal by herbivorous coral reef fishes. Coral Reefs 26:435–442.

28. Burkepile DE, Hay ME (2008) Herbivore species richness and feeding complementarity affect community structure and function on a coral reef. Proc Natl Acad Sci USA 105:16201–16206.

29. Burkepile DE, Hay ME (2010) Impact of herbivore identity on algal succession and coral growth on a Caribbean reef. PLoS ONE 5(1):e8963.

30. Mumby PJ (2006) The impact of exploiting grazers (Scaridae) on the dynamics of Caribbean coral reefs. Ecol Appl 16:747–769.

31. Mumby PJ, Hastings A, Edwards HJ (2007) Thresholds and resilience of Caribbean coral reefs. Nature 450:98–101.

32. Cheal AJ, MacNeil MA, Cripps E, Emslie MJ, Jonker M, et al. (2010) Coral-macroalgal phase shifts or reef resilience: links with diversity and functional roles of herbivorous fishes on the Great Barrier Reef. Coral Reefs 29:1005–1015.

33. Lewis AR (1997) Effects of experimental coral disturbance on the structure of reef fish communities on large patch reefs. Mar Ecol Prog Ser 161: 37–50.

34. Coker DJ, Graham NAJ, Pratchett MS (2012) Interactive effects of live coral and structural complexity on the recruitment of reef fishes. Coral Reefs 31: 919–927. (doi 10.1007/s00338-012-0920-1)

35. Osborne K, Dolman AM, Burgess SC, Johns KA (2011) Disturbance and the dynamics of coral cover on the Great Barrier Reef (1995-2009). PLoS ONE 6(3): e17516. doi:10.1371/journal.pone.0017516

36. De'ath G, Fabricius KE, Sweatman H, Puotinen M (2012) The 27-year decline of coral cover on the Great Barrier Reef and its causes. Proc Natl Acad Sci USA 109 (44): 17995–17999. (doi/10.1073/pnas.1208909109)

37. Wellington GM, Victor BC (1985) El Nino mass coral mortality: a test of resource limitation in a coral reef damselfish population. Oecologia 68: 15–19.

38. Jonker M, Johns K, Osborne K (2013) Surveys of benthic reef communities using underwater digital photography and counts of juvenile corals. Long-term Monitoring of the Great Barrier Reef Standard Operational Procedure Number 10. Australian Institute of Marine Science, Townsville, Queensland, Australia.

39. Halford AR, Thompson AA (1994) Visual census surveys of reef fish. Long-term Monitoring of the Great Barrier Reef Standard Operational Procedure Number 3. Australian Institute of Marine Science, Townsville, Queensland, Australia.

40. Wilson SK, Graham NAJ, Polunin NVC (2007) Appraisal of visual assessments of habitat complexity and benthic composition on coral reefs. Mar Biol 151: 1069–1076. (doi 10.1007/s00227-006-0538-3)

41. R Core Team (2013) R: A language and environment for statistical computing. R Foundation for Statistical Computing, Vienna, Austria. ISBN 3-900051-07-0. Available: http://www.R-project.org. Accessed 2014 Jul 20.

42. Gelman A, Hill J (2007) Data Analysis Using Regression and Multilevel/Hierarchical Models. Cambridge University Press: New York.

43. Hadfield JD (2010) MCMC methods for multi-response generalized linear mixed models: the MCMCglmm R package. J Stat Softw 33:1–22.

44. Zuur AF, Saveliev AA, Ieno EN (2012) Zero inflated models and generalized linear mixed models with R. Highland Statistics Limited, Newburgh, UK.

45. Graham NAJ, McClannahan TR, MacNeil MA, Wilson SK, Polunin NVC, et al. (2008) Climate warming, Marine Protected areas and the ocean-scale integrity of coral reef ecosystems. PLoS ONE 3(8): e3039. (doi:10.1371/journal.pone.0003039)

46. Friedlander AM, Parrish JD (1998) Habitat characteristics affecting reef fish assemblages on a Hawaiian coral reef. J Exp Mar Biol Ecol 224: 1–30.

47. Graham NAJ, Nash KL (2012) The importance of structural complexity in coral reef ecosystems. Coral Reefs 32: 315–326.

48. Elmqvist T, Folke C, Nyström M, Peterson G, Bengtsson J, et al. (2003) Response diversity, ecosystem change, and resilience. Front Ecol Environ 1: 488–494.

49. Folke C, Carpenter S, Walker B, Scheffer M, Elmqvist T, et al. (2004) Regime shifts, resilience and biodiversity in ecosystem management. Annu Rev Ecol Evol Syst 35: 557–581.

50. Walker B (1992) Biodiversity and ecological redundancy. Conserv Biol 6: 18–23.

51. Petersen GC, Allen CR, Holling CS (1998) Ecological resilience, biodiversity and scale. Ecosystems 1: 6–18.

52. Hooper DU, Chapin FS, Ewel JJ, Hector A, Inchausti P, et al. (2005) Effects of biodiversity on ecosystem functioning: a consensus of current knowledge. Ecol Monogr 75: 3–35.

53. Folke C (2006) The re-emergence of a perspective for social-ecological systems analyses. Glob Environ Change 16: 253–267.

54. Bellwood DR, Hughes TP, Folke C, Nyström M (2004) Confronting the coral reef crisis. Nature 429: 827–833.

55. Cheal AJ, Emslie MJ, MacNeil MA, Miller I, Sweatman H (2013) Spatial variation in the functional characteristics of herbivorous fish communities and the resilience of coral reefs. Ecol Appl 23: 174–188.

56. Rotjan RD, Lewis SM (2008) Impact of coral predators on tropical reefs. Mar Ecol Prog Ser 367: 73–91.

57. Cole AJ, Lawton RJ, Wilson SK, Pratchett MS (2012) Consumption of tabular acroporid corals by reef fishes: a comparison with plant-herbivore interactions. Funct Ecol 26(2): 307–316.

58. Cole AJ, Chong-Seng KM, Pratchett MS, Jones GP (2009) Coral feeding fishes slow progression of black band disease. Coral Reefs 28: 965.

59. Alevizon WS, Gorham JC (1989) Effects of artificial reef deployment on nearby resident fishes. Bull Mar Sci 44:646–661.

60. Burt JA, Feary DA, Cavalcante G, Bauman AG, Usseglio P (2013) Urban breakwaters as reef fish habitat in the Persian Gulf. Mar Poll Bull 72: 342–350.

61. Pradella N, Fowler AM, Booth DJ, Macreadie PI (2014) Fish assemblages associated with oil industry structures on the continental shelf of north-western Australia. J Fish Biol 84: 247–255. (doi:10.1111/jfb.12274)

62. Graham NA, Wilson SK, Jennings S, Polunin NVC, Robinson J, et al. (2007) Lag effects in the impacts of mass coral bleaching on coral reef fish, fisheries and ecosystems. Cons Biol 21:1291–1300. (doi 10.1111/j.1523-1739.2007.00754.x)

63. Hoegh-Guldberg O, Mumby PJ, Hooten AJ, Steneck RS, Greenfield P, et al. (2007) Coral reefs under rapid climate change and ocean acidification. Science 318:1737–1742. (doi 10.1126/science.1152509)

64. Noonan SHC, Jones GP, Pratchett MS (2012) Coral size, health and structural complexity: effects on the ecology of a coral reef damselfish. Mar Ecol Prog Ser 456: 127–137.

65. Webster PJ, Holland GJ, Curry JA, Chang HR (2005) Changes in tropical cyclone number, duration and intensity in a warming environment. Science 309: 1844–1846.

66. Alvarez-Filip L, Dulvy NK, Gill JA, Cote IM, Watkinson AR (2009) Flattening of Caribbean coral reefs: region-wide declines in architectural complexity. Proc R Soc Lond B Biol Sci 276: 3019–3025.

19

Impact of Climate Change on Potential Distribution of Chinese Caterpillar Fungus (*Ophiocordyceps sinensis*) in Nepal Himalaya

Uttam Babu Shrestha[1]*, **Kamaljit S. Bawa**[2,3]

1 Institute for Agriculture and the Environment, University of Southern Queensland, Toowoomba, Queensland, Australia, **2** Department of Biology, University of Massachusetts, Boston, Massachusetts, United States of America, **3** Ashoka Trust for Research in Ecology and Environment (ATREE), Bangalore, India

Abstract

Climate change has already impacted ecosystems and species and substantial impacts of climate change in the future are expected. Species distribution modeling is widely used to map the current potential distribution of species as well as to model the impact of future climate change on distribution of species. Mapping current distribution is useful for conservation planning and understanding the change in distribution impacted by climate change is important for mitigation of future biodiversity losses. However, the current distribution of Chinese caterpillar fungus, a flagship species of the Himalaya with very high economic value, is unknown. Nor do we know the potential changes in suitable habitat of Chinese caterpillar fungus caused by future climate change. We used MaxEnt modeling to predict current distribution and changes in the future distributions of Chinese caterpillar fungus in three future climate change trajectories based on representative concentration pathways (RCPs: RCP 2.6, RCP 4.5, and RCP 6.0) in three different time periods (2030, 2050, and 2070) using species occurrence points, bioclimatic variables, and altitude. About 6.02% (8,989 km^2) area of the Nepal Himalaya is suitable for Chinese caterpillar fungus habitat. Our model showed that across all future climate change trajectories over three different time periods, the area of predicted suitable habitat of Chinese caterpillar fungus would expand, with 0.11–4.87% expansion over current suitable habitat. Depending upon the representative concentration pathways, we observed both increase and decrease in average elevation of the suitable habitat range of the species.

Editor: Helge Thorsten Lumbsch, Field Museum of Natural History, United States of America

Funding: Rufford Small Grants for Nature Conservation (RSGs), UK and Conservation Trust Grant of National Geographic Society, USA supported for the field data collection of this work. The funders had no role in study design, data collection and analysis, decision to publish, or preparation of the manuscript.

* Email: ubshrestha@yahoo.com

Introduction

The climate of our planet is changing at an unprecedented rate. Global average temperature has increased by 0.85°C from 1880 to 2012 and it is likely to increase further by a minimum of 0.3°C–1.7°C (RCP 2.6) to a maximum of 2.6°C–4.8°C (RCP 8.5) by the end of this century relative to 1986–2005 temperature [1]. Ecosystems and species have already responded to global climate change [2]. Changes in community structure, composition, and dynamics have been observed at ecosystem levels [3–4] whereas alteration in phenology, modification of physiology, and shifts in distribution have been documented at the level of species [5]. Shifts in plant species distribution may increase vulnerability to extinction [6]. Therefore, understanding the potential impacts of climate change on the distribution of species is important for mitigation of future biodiversity losses [7].

The Himalaya, a region of unique biodiversity, rich cultural and ethnic diversity, and varied topography has warmed up by about 1.5°C, three times more than the global average, during the last 25 years from 1982 to 2006 [8]. Climate change in the Himalaya has already impacted hydrology, agriculture, ecosystems, and species [9]. Similarly, changes in phenology of vegetation [8] and altitudinal shifts in vegetation communities [10] and species such

as *Rheum nobile*, *Saussurea stella*, *Rhodiola bupleuroides*, *Ponerorchis chusua*, *Microgynaecium tibeticum*, *Meconopsis simplicifolia*, and *Pedicularis trichoglossa* [11] have been noted. Comparatively higher resolution (50×50 km) regional climate models show that temperature and precipitation in the Himalayan region will continue to increase in future [12], and these changes are further likely to impact the distribution of biodiversity, as for example, predicted for Rhododendrons [13]. However, potential shift of many other plant species caused by past and future climate changes in this biologically diverse region of the world are largely unknown [11,13].

Historical records have been largely used in documenting potential shifts in the distribution of species due to the past climate. However, in many cases, historical data on species distribution are lacking in the Himalaya. Moreover, such approaches might not be sufficient to predict the changes in current predicted distribution due to future changes in climate. Species distribution modeling (SDM), based on the quantitative relationship between environmental variables and species occurrence points, has made it possible to recognize species' niche requirements [14] and to understand the impacts of climate change on species distribution [15]. Various SDM methods, statistical, machine learning, and classification and distance are in current use to model species distribution [16]. Among them, a general purpose machine learning method—MaxEnt, developed by Phillips et al. (2006) [17]—is one of the most popular methods [18] and used to model a wide range of plant and animal species such as Malayan Sun Bear [19], Snow leopard [20] Lantana [21], Yew [22], Rhododendrons [13] and habitats globally [18–23]. It performs very well to estimate current as well as future distributions of species due to climate change [24].

Here we use the MaxEnt model to map the current potential distribution of Chinese caterpillar fungus (Ophiocordyceps sinensis) and to predict changes in the potential distribution of the species under future climate change trajectories in Nepal Himalaya. We also use the model to predict changes in the current potential elevation range of Chinese caterpillar fungus due to future climate change. We assume that both the extent of suitable habitat and range of Chinese caterpillar fungus will change with the change in climate in the area.

Materials and Methods

Study Area

Nepal, a mountainous country is situated on the southern slope of the central Himalaya. The country covers an area of 147,181 km^2 and has the widest elevational gradient of any country in the world, ranging from 60 m.a.s.l (meters above sea level) to Mount Everest, the highest point of the world at 8,848 m.a.s.l.. This vast altitudinal gradient creates a wide variation in physiographic, climatic, topographic, and edaphic conditions resulting in rich biodiversity. The Chinese caterpillar fungus, which is reported from 27 mountain districts of Nepal, is one of the most important biological and socio-economic components of Nepal's rich biodiversity.

Nepal Himalaya has been experiencing climate change and its impact; mean annual maximum temperature increased by 0.6°C per decade during 1977–2000 [25]. Similarly, climate model projections for Nepal show temperature increases of 0.5–2.0°C (mean 1.4°C) by the 2030s; 1.7–4.1°C (mean 2.8°C) by the 2060s; and 3.0–6.3°C (mean 4.7°C) by the 2090s [26]. Although there is no clear trend in precipitation, different models show an increase in monsoon precipitation and a decrease in winter precipitation in the future [26].

Species

Chinese caterpillar fungus is an endemic species of the Himalayan countries: Bhutan, China, India and Nepal. It a parasitic complex formed by a parasitic relationship between the fungus, Ophiocordyceps sinensis and the caterpillar of 'ghost' moth species belonging to the genus Thitarodes [27]. In Nepal, this caterpillar-fungus complex occurs in the open grasslands at elevations between 3,500–5,000 m.a.s.l. [28–29]. Although it has various therapeutic usages, the major trade of Chinese caterpillar fungus, with the popular name "Himalayan Viagra", is due to its presumed effects as an aphrodisiac and powerful tonic [27]. It is one of the most expensive natural medical resources of the world [30–31]. In September 2012, the highest price of Chinese caterpillar fungus in China—the major trade destination—was $140,000/kg (for the best quality product), two times more expensive than gold [32]. Although the fungus has an extraordinarily high price, it is collected extensively by the poorest mountain communities to support their livelihood. In recent years, the availability of the Chinese caterpillar fungus in its natural habitat has been dwindling and may threaten the livelihoods of the poorest mountain people [29]. The decline has been attributed, in part, to climate change [27,29,33].

Species location data

The species occurrence data used for this study were collected from our own field surveys conducted in six districts (Dolpa, Manang, Dolakha, Gorkha, Darchula, Bajhang) of Nepal between 2011 and 2013. Altogether, 37 species occurrence points distributed in central, western and eastern Nepal were used to model the potential distribution (Figure 1). We secured the necessary permits from the National Trust for Nature Conservation (NTNC) to conduct field studies in protected areas of Manang and Gorkha.

Environmental and bioclimatic data

To model potential distribution of Chinese caterpillar fungus across our study area, we first collected 19 bioclimatic layers, and one topographic (altitude) layer. The 19 grid-based bioclimatic variables (Table 1) were downloaded from the worldclim datasets (www.worldclim.com). Bioclimatic variables calculated from monthly temperature and precipitation values were generated through interpolation of average monthly climate data from weather stations at 2.5 arc minute spatial resolution [34]. They are ecologically meaningful variables that describe annual trends, seasonality, and extremes of temperature and precipitation. Altitude was derived from digital elevation data from the Shuttle Radar Topographic Mission (SRTM); this was then resampled into 2.5 arc minute (~5 km) spatial resolution by using the nearest neighbor resampling technique in ArcGIS.

We calculated pairwise correlations and removed highly correlated variables (r$^2 \leq$0.80) to minimize the impact of multi-collinearity and over-fitting of the model. The remaining seven (Bio 3, Bio 7, Bio 11, Bio 14, Bio 15, Bio 18, and Bio 19) bioclimatic variables (Table 2) were used to model the distribution of Chinese caterpillar fungus in current and future climate conditions. All datasets were changed into ASCII files in ArcGIS as required by the MaxEnt software. We repeated the same process to generate predicted maps in three different future climate scenarios for 2030, 2050 and 2070. We did not take account of land use or land cover change, human disturbances, species dispersal or changes in biotic interactions in the model.

To determine the future distribution of the species under different climate trajectories, we used datasets of future climate from the International Center for Tropical Agriculture (www.ccafs-climate.org), a global agricultural research institution. We selected three future GHG (greenhouse gas) concentration trajectories, also known as representative carbon pathways (RCP 2.6, RCP 4.5 and RCP 6.0), for three different time periods (2030, 2050 and 2070) as adopted by the IPCC in its fifth Assessment Report (AR5) [1]. RCP 2.6 is the lowest GHG concentration pathway in which radioactive forcing (global energy imbalances) levels reach 3.1 W/m^2 by mid-century and drops 2.6 W/m^2 by 2100 [35]. RCP 4.5 is a stabilization scenario in which the total radiative forcing reaches to 4.5 W/m^2 by 2100 and stabilizes due to the employment of a range of technologies and strategies for reducing GHG emissions [36]. Likewise, RCP 6.0 also represents stabilization by 2100, this time at 6.0 W/m^2 by 2100 [37]. We selected a global circulation model, HadGEM2-CC (Hadley Global Environment Model 2 Carbon Cycle) developed by the Hadley Center, United Kingdom [38]. HadGEM2 models have been used to perform all the CMIP5 (Coupled Model Intercomparison Project Phase 5) centennial experiments including ensembles of simulations of the RCPs. HadGEM2-CC is one of the models used by the international governmental panel on climate change (IPCC) in its fifth Assessment Report (AR5). For consistency, we used the same seven bioclimatic variables (Bio 3, Bio 7, Bio 11, Bio 14, Bio 15, Bio 18, and Bio 19) we used for

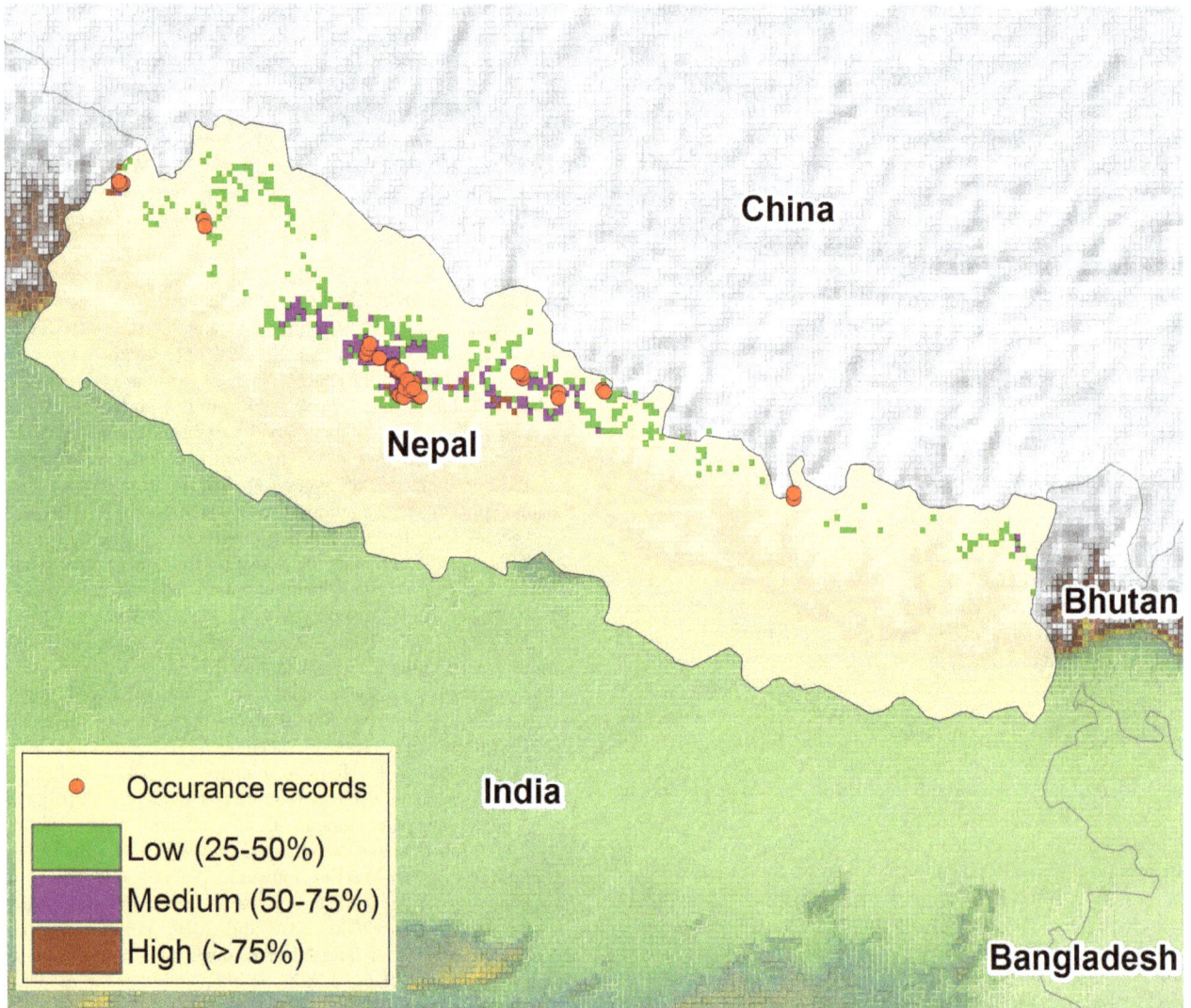

Figure 1. Predicted potential distribution of Chinese caterpillar fungus under current bioclimatic conditions and location of occurrence used for modeling.

modeling current potential distribution of the species to predict future distribution. Those data are statistically downscaled from a Global Circulation Model (GCM) based on the sum of interpolated anomalies to high resolution monthly climate surfaces from Worldclim [34].

Modeling

We used freely available MaxEnt software to model the current and future distributions of Chinese caterpillar fungus. MaxEnt is a general purpose machine learning method that estimates the probability distribution of a species occurrence based on environmental conditions of a location in which the species is found by calculating the distribution of maximum entropy i.e. the most spread out distribution in space for a given set of constraints [17]. It is the most popular species distribution modeling method with more than 1000 published usages since 2005 [18,23]. MaxEnt has also outperformed other methods and has shown higher predictive accuracy than other methods [6,39]. MaxEnt performs well to estimate potential range shifts of species due to

climate change [24]. The method is easy to use and has the functionality to use presence only data.

We used the following parameter values in the MaxEnt model: random test percentage $= 25\%$, regularization multiplier $= 1$, maximum number of backgrounds points $= 10,000$, maximum iterations $= 1,000$ and convergence threshold $= 0.00001$. This means that the model uses 25% of the original presence data against 10,000 random background points (pseudo-absences) for modeling the distribution. Model robustness is commonly evaluated by AUC values that range from 0 to 1; AUC values between 0.5–0.7 are considered low, 0.7–0.9 moderate and >0.9 high [16,35]. We also used AUC values to determine the model accuracy. The output of MaxEnt is continuous data with values ranging from 0 (lowest) to 1 (highest) probability of distribution. We imported the MaxEnt output data and reclassified into three classes of habitat suitability: low suitability (25–50% probability of occurrence), medium suitability (50–75% probability of occurrence) and high suitability (>75% probability of occurrence) by omitting the values below 25% as non-suitable habitat based on the logistic threshold, with suitable conditions predicted above the

Table 1. Bioclimatic variables used for modeling habitat of Chinese caterpillar fungus.

Variables Name	Code	Data Source	Resolution
Altitude	Alt	STRM	90 m
Annual Mean Temperature	BIO 1	Worldclim	~5 km (2.5 arc min)
Mean Diurnal Range [Mean of monthly (Max Temperature - Min Temperature)]	BIO 2	Worldclim	~5 km (2.5 arc min)
Isothermality (BIO 2/BIO 7) (*100)	BIO 3	Worldclim	~5 km (2.5 arc min)
Temperature Seasonality (Standard Deviation*100)	BIO 4	Worldclim	~5 km (2.5 arc min)
Max Temperature of Warmest Month	BIO 5	Worldclim	~5 km (2.5 arc min)
Min Temperature of Coldest Month	BIO 6	Worldclim	~5 km (2.5 arc min)
Temperature Annual Range (BIO 5-BIO 6)	BIO 7	Worldclim	~5 km (2.5 arc min)
Mean Temperature of Wettest Quarter	BIO 8	Worldclim	~5 km (2.5 arc min)
Mean Temperature of Driest Quarter	BIO 9	Worldclim	~5 km (2.5 arc min)
Mean Temperature of Warmest Quarter	BIO 10	Worldclim	~5 km (2.5 arc min)
Mean Temperature of Coldest Quarter	BIO 11	Worldclim	~5 km (2.5 arc min)
Annual Precipitation	BIO 12	Worldclim	~5 km (2.5 arc min)
Precipitation of Wettest Month	BIO 13	Worldclim	~5 km (2.5 arc min)
Precipitation of Driest Month	BIO 14	Worldclim	~5 km (2.5 arc min)
Precipitation Seasonality (Coefficient of Variation)	BIO 15	Worldclim	~5 km (2.5 arc min)
Precipitation of Wettest Quarter	BIO 16	Worldclim	~5 km (2.5 arc min)
Precipitation of Driest Quarter	BIO 17	Worldclim	~5 km (2.5 arc min)
Precipitation of Warmest Quarter	BIO 18	Worldclim	~5 km (2.5 arc min)
Precipitation of Coldest Quarter	BIO 19	Worldclim	~5 km (2.5 arc min)

threshold and unsuitable below. This reclassification of suitable conditions allowed us to compare the change in classes over time and space.

We also estimated the total area of predicted habitat of Chinese caterpillar fungus under current and future climatic conditions by calculating the number of 'presence' grid cells multiplied by their spatial resolution. To observe potential changes in the altitudinal distribution of Chinese caterpillar fungus, we extracted elevation values of the pixels of predicted presence from the predicted and current distribution maps generated by MaxEnt models. We compared the mean elevation values of the areas of predicted presence for current and future climate scenarios using independent sample T-test.

Results

Our study demonstrates for the first time the potential distribution of Chinese caterpillar fungus habitat in Nepal Himalaya. Out of seven predictor bioclimatic variables used for this study, the relative contribution of two variables (Bio 11 and Bio 15) to the model was more than 89%. Mean temperature of the coldest quarter (Bio 11) had the highest contribution (74.5%) to the model followed by precipitation seasonality (Bio 15), with 15.3% contribution. The bioclimatic variable, isothermality (Bio 3) had the lowest contribution (0.3%). The MaxEnt model's jackknife test of variable importance showed that the mean temperature of the coldest quarter (Bio 11) was the variable with the highest gain when used in isolation and the gain decreased when it was omitted (Figure 2). The validity of the model for current distribution of Chinese caterpillar fungus was high with AUC = 0.98 indicating that the selected variables described the distribution of Chinese caterpillar fungus very well.

Responses of each environmental variable that influenced the predicted suitable distribution of Chinese caterpillar fungus are shown in response curves (Figure 3). These response curves show changes in the logistic prediction when each environmental variable changes by keeping all other environmental variables at their average sample value. Models showed that the distribution of Chinese caterpillar fungus is highly controlled by temperature. It prefers areas where the mean annual temperature remains below 15°C and the mean temperature of the coldest quarters stays slightly lower than zero but not less than minus 10°C. Precipitation in the driest month and precipitation seasonality from 75–150 mm might also be useful predictors but excessive precipitation above 200 mm was not found to be very apposite.

The potential distribution of Chinese caterpillar fungus under current bioclimatic conditions and occurrence locations is shown in Figure 1. Based on our model, about 6.02% (8,989 km²) of the country is suitable for Chinese caterpillar fungus habitat, and the fungus occurs in 26 mountainous districts of Nepal. The greatest concentration of pixels (potential habitat or predicted presence) is observed in Dolpa, Rukum, Manang, Myagdi, and Jumla districts when the suitable habitat map is overlaid with the district map of Nepal.

To estimate the potential impact of climate change on the distribution of Chinese caterpillar fungus, the predicted distribution of Chinese caterpillar fungus in future climate scenarios is shown in Figure 4. Across all scenarios in three different periods, the predicted area of Chinese caterpillar fungus distribution was predicted to increase. The estimated area of the predicted distribution is given in Figure 5. The maximum expansion (a 4.87% addition to the current potential suitable area) would occur under RCP 2.6 by 2070 whereas the minimum expansion (0.11% addition in current potential suitable area) was predicted for the year 2030 for the same pathway. Much of the expansion

Table 2. Correlation matrix of altitude and bioclimatic variables.

Code	Alt	BIO1	BIO2	BIO3	BIO4	BIO5	BIO6	BIO7	BIO8	BIO9	BIO10	BIO11	BIO12	BIO13	BIO14	BIO15	BIO16	BIO17	BIO18	BIO19	LULC
Alt	-1.00	-1.00	-0.06	**0.13**	0.10	-0.99	-0.99	**-0.14**	-1.00	-0.99	-1.00	**-1.00**	-0.76	-0.81	**0.00**	-0.74	-0.79	0.31	**-0.58**	**0.46**	**0.12**
BIO1		1.00	0.05	**-0.14**	-0.10	0.99	0.99	**0.13**	1.00	0.99	1.00	**1.00**	0.76	0.80	**0.01**	0.73	0.78	-0.28	**0.57**	**-0.44**	**-0.13**
BIO2			1.00	**0.08**	0.73	0.15	-0.08	**0.88**	0.08	0.00	0.09	**0.01**	-0.20	-0.07	**-0.51**	0.40	-0.10	-0.44	**-0.27**	**-0.35**	**0.18**
BIO3				**1.00**	**-0.54**	**-0.23**	**-0.13**	**-0.40**	**-0.15**	**-0.15**	**-0.17**	**-0.11**	**0.05**	**0.01**	**-0.07**	**-0.01**	**0.04**	**-0.49**	**0.30**	**-0.49**	**0.01**
BIO4					1.00	0.03	-0.21	**0.93**	-0.07	-0.14	-0.05	**-0.16**	-0.35	-0.24	**-0.32**	0.15	-0.27	0.04	**-0.52**	**0.14**	**0.17**
BIO5						1.00	0.97	**0.26**	0.99	0.98	0.99	**0.98**	0.70	0.76	**-0.02**	0.73	0.74	-0.24	**0.48**	**-0.38**	**-0.11**
BIO6							1.00	**0.01**	0.98	0.99	0.98	**0.99**	0.78	0.80	**0.09**	0.65	0.79	-0.20	**0.60**	**-0.37**	**-0.16**
BIO7								**1.00**	**0.16**	**0.09**	**0.18**	**0.08**	**-0.19**	**-0.05**	**-0.43**	**0.38**	**-0.09**	**-0.16**	**-0.37**	**-0.09**	**0.15**
BIO8									1.00	0.99	1.00	**1.00**	0.75	0.80	**0.00**	0.74	0.78	-0.29	**0.57**	**-0.45**	**-0.12**
BIO9										1.00	0.99	**0.99**	0.76	0.80	**0.02**	0.69	0.78	-0.26	**0.58**	**-0.41**	**-0.13**
BIO10											1.00	**0.99**	0.74	0.79	**-0.01**	0.74	0.77	-0.28	**0.55**	**-0.43**	**-0.12**
BIO11												**1.00**	**0.77**	**0.80**	**0.03**	**0.71**	**0.79**	**-0.28**	**0.60**	**-0.44**	**-0.13**
BIO12													1.00	0.98	**0.18**	0.63	0.99	-0.22	**0.92**	**-0.38**	**-0.16**
BIO13														1.00	**0.06**	0.74	0.99	-0.32	**0.88**	**-0.47**	**-0.13**
BIO14															**1.00**	**-0.33**	**0.07**	**0.62**	**0.14**	**0.46**	**-0.14**
BIO15																1.00	0.72	-0.64	**0.53**	**-0.73**	**-0.06**
BIO16																	1.00	-0.31	**0.90**	**-0.46**	**-0.14**
BIO17																		1.00	**-0.32**	**0.96**	**-0.13**
BIO18																			**1.00**	**-0.46**	**-0.14**
BIO19																				**1.00**	**-0.08**

Unrelated variables (Correlation ≤0.80) used for the study BIO3, BIO7, BIO 11, BIO 15, BIO 18, BIO 19, LULC.

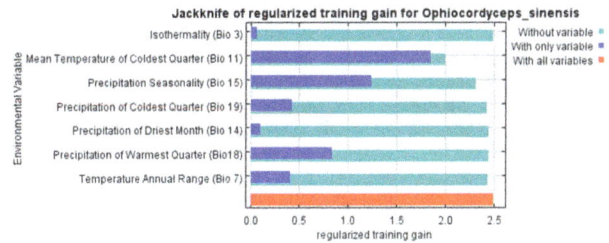

Figure 2. Results of jackknife test of relative importance of predictor variables for Chinese caterpillar fungus.

occurs in western Nepal whereas there are nominal changes only in the eastern parts of the country. To be precise, maximum increases in the area of suitable habitat occurs in Dolpa, Mustang, and Mugu districts under RCP 2.5 in 2050 whereas the greatest reduction in habitat area occurs in the Taplejung district of Eastern Nepal. Taplejung might lose its entire area of suitable habitat under RCP 4.5 by the year 2050. Similar loss of the entire area of Chinese caterpillar fungus habitat was predicted in Humla district under RCP 6.0 by the year 2070. However, expansion of potential habitat might occur in two new districts, Dolpa and Pachthar under RCP 6.0 by 2070. Under both the RCP 4.5 and RCP 6.0 trajectories, the area of predicted distribution increases continuously up to 2050 then decreases by 2070 whereas in the RCP 2.6 trajectory, the distribution continuously increases from 2030 to 2070 and reaching a maximum area in 2070 (Figure 5).

Expansion in both lower and upper altitudinal limits was observed under all future climate change pathways (Figure 6). The average altitude of current distribution and that of future distributions increased under four pathways: RCP 2.6 (2070), RCP 4.5 (2050), RCP 6.0 (2050) and RCP 6.0 (2070) whereas a decrease was found in all other pathways (Table 3). However, none of the changes were statistically significant (p = 0.01) except RCP 4.5 (2070) based on the independent sample T-test between the altitude means of current potential distribution range and that of future potential distribution range.

Discussion

Our work represents the first attempt to model the impact of future climate change on the distribution of Chinese caterpillar fungus—a flagship species of the Himalaya. Climate change projections for the Nepal Himalaya predict that average annual temperatures will increase by 3.0–6.3°C (mean 4.7°C) by the 2090s and that the warming would be greater in higher elevation regions—habitats of the Chinese caterpillar fungus [26]. Since climate change has already impacted species' habitats including those of several species of fungi worldwide [40–42], it will likely affect the distribution of this species in Nepal as well. We here provide a model that could be used to predict how the distribution of species might be affected by changes in future climate depending upon the different representative concentration pathways.

It is widely believed that temperature and humidity play important roles in yields [27–28], abundance [43], and the probability of infection and sporulation [44] of Chinese caterpillar fungus. Furthermore, Chinese caterpillar fungus distribution is also assumed to be affected by winter and summer temperatures and the seasonality of precipitation [45–46]. Our observations that the mean temperature of the coldest quarter and the seasonality of precipitation affect the distribution of Chinese caterpillar fungus is

in line with observations that temperature and humidity play important roles in the ecology and physiology of the fungus. Furthermore, bioclimatic response variables that affect the predicted distribution of the species concur with the knowledge of local communities about the habitat needs of the species [46].

Although the potential distribution predicted by the model is not the observed distribution, there is close congruence between the two distributions. The districts of Dolpa, Rukum, Manang, Myagdi, and Jumla showed a greater area of potential distribution (maximum concentration of the pixels). These districts are the major producers of Chinese caterpillar fungus in Nepal, contributing approximately 60% of the total volume of Chinese caterpillar fungus traded in Nepal [47]. Spatial overlap of the potential occurrence with districts where Chinese caterpillar fungus is reported is 96.29% (26 districts in the model vs. 27 districts in actual). However, the model slightly overestimates the potential distribution for the Kaski district and underestimates it for Dolakha District. Nonetheless, both the maximum spatial overlapping and high AUC value suggests that the model has a high level of accuracy.

Our models show that the potential distribution of Chinese caterpillar fungus will extend under all future climate scenarios, suggesting that additional new suitable habitats will be created for Chinese caterpillar fungus with the future climate change. However, this expansion may not guarantee an increase in the production of Chinese caterpillar fungus in future. Studies in other parts of the world have shown both expansion and shrinkage of potential habitat in response to future climate change [48–50]. Both the reductions and increase in range size were predicted for the endemic flora of California under future climate change scenarios [48] as well as higher plants in Europe [49]. Range expansion of lemur parasites is predicted for Madagascar [50] whereas shrinkage of Rhododendrons has been predicted in the Himalaya [13]. Likewise a 60% reduction in the range of Ngwayir (*Pseudocheirus occidentalis*) is predicted in Australia [51].

It has been unofficially reported that the altitudinal limit of Chinese caterpillar fungus seems to have shifted upwards by 200–500 m in China due to past climate change [27,33], which may have also caused local extinction of the fungus from Mount Emai in the Chinese province of Sichuan, located on the edge of the Tibetan plateau [52–53]. However, we found no noteworthy changes in the altitudinal ranges of the modeled potential distributions of Chinese caterpillar fungus under predicted future climate change trajectories in this study. Furthermore, it is unclear what methodology Yang (2008) had used to observe the shifts in the altitudinal range of Chinese caterpillar fungus [33]. Due to the unavailability of historical data on the distribution of Chinese caterpillar fungus in Nepal Himalaya, monitoring the impact of past climate change on the distribution of Chinese caterpillar fungus is impossible.

One of the major current drivers of decline of Chinese caterpillar fungus is extensive harvesting [27,29]. Every year about 100 thousand people are involved in harvesting Chinese caterpillar fungus in Nepal [54]. The abundance of Chinese caterpillar fungus in the future will largely depend on human use. Although the area of suitable habitat for Chinese caterpillar fungus is predicted to expand with future climate change, the availability of fungus is likely to be largely governed by human impacts on populations. Future distribution models will have to take into account actual and potential harvesting intensities at different locations. Our model does not account for harvesting pressure or future transformation of land. Furthermore, since the Chinese caterpillar fungus complex involves an array of linked biotic interactions of diverse organisms with different life cycles and

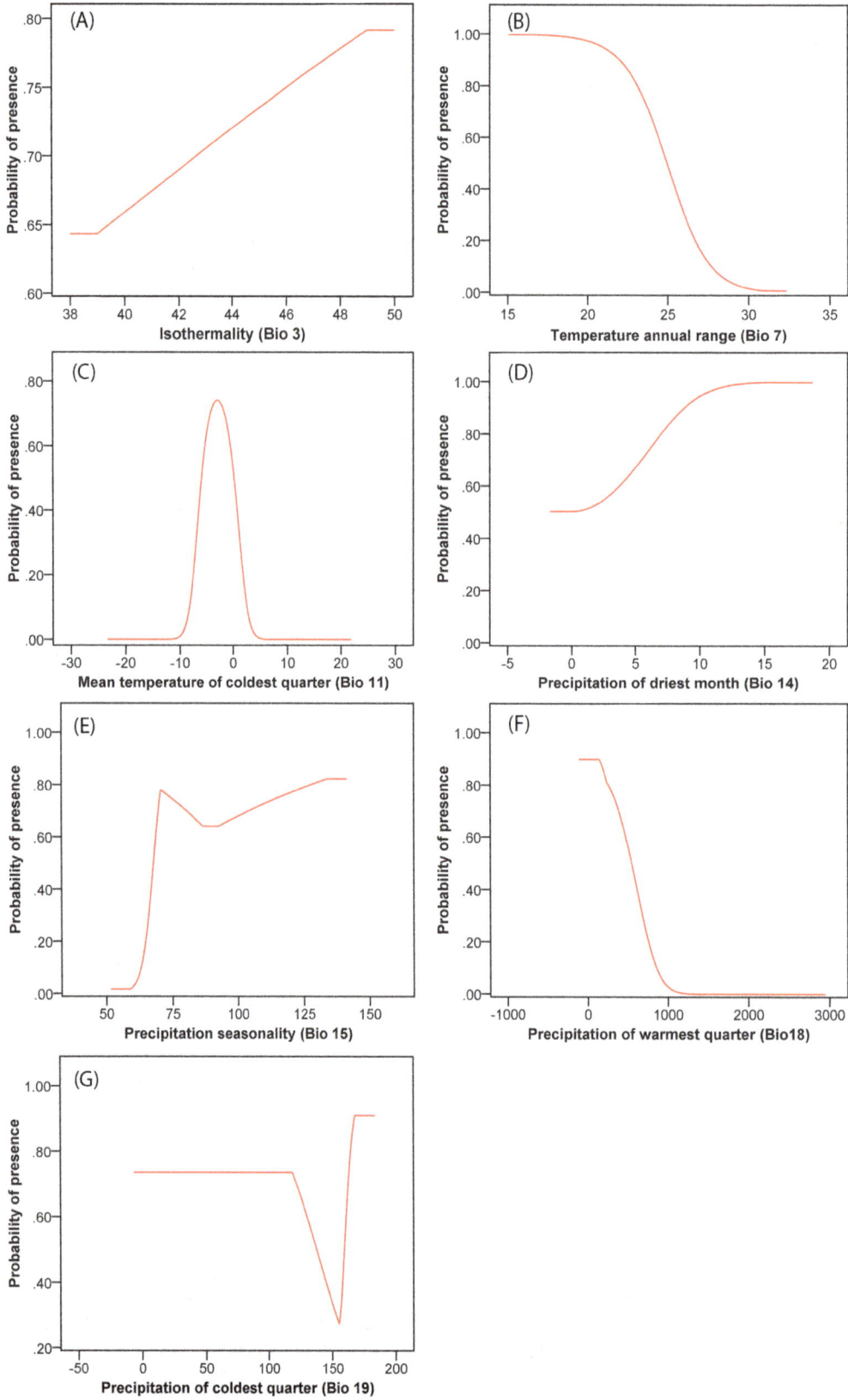

Figure 3. Response curves for the predictors of the MaxEnt Model.

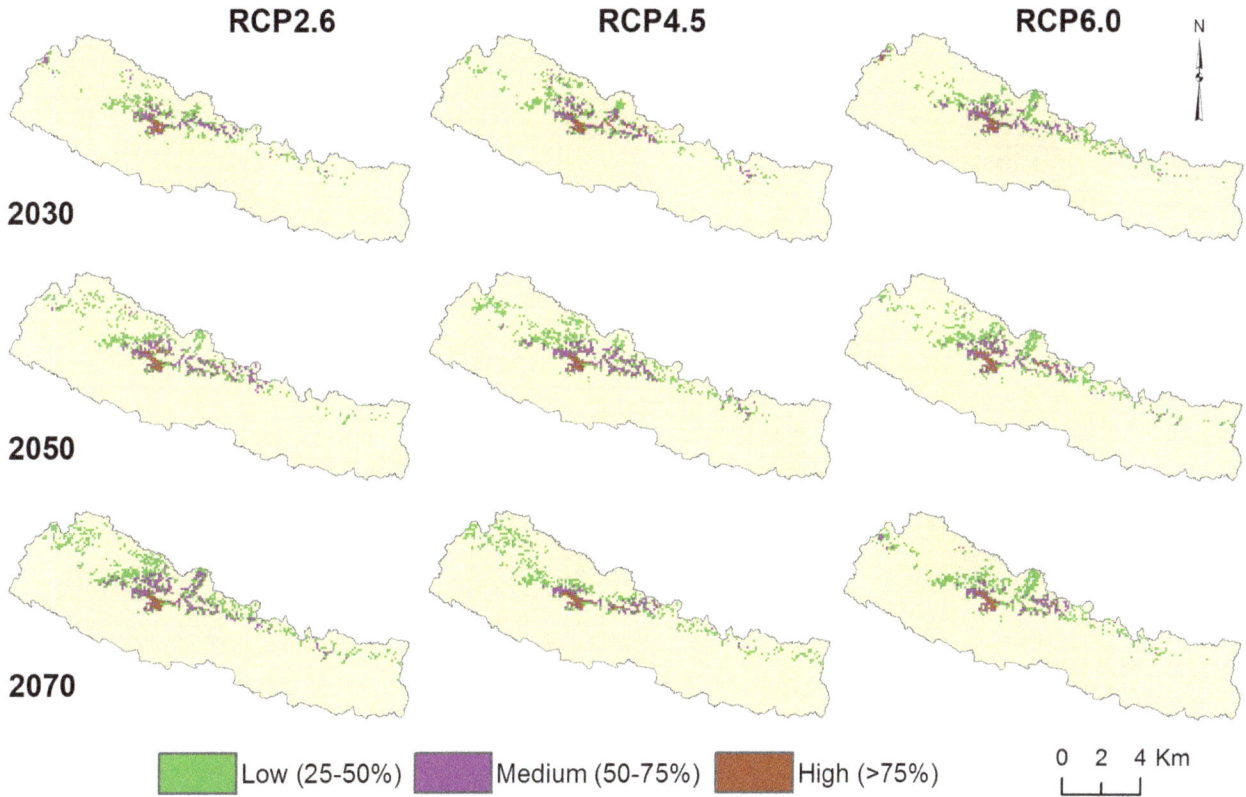

Figure 4. Predicted future distribution of Chinese caterpillar fungus in future climate scenarios.

generation spans, future climate change may affect the interacting species differentially. Future research would therefore need to incorporate harvesting, future land use and land cover change, and biotic interactions in distribution models of not only this species but other species as well.

Although this study lays down the foundation for studying the dynamics of Chinese caterpillar fungus distribution through time and space, it has some limitations. Bioclimatic variables used in this study have a 5×5 km spatial resolution. This resolution may be acceptable for species with broad range sizes including Chinese

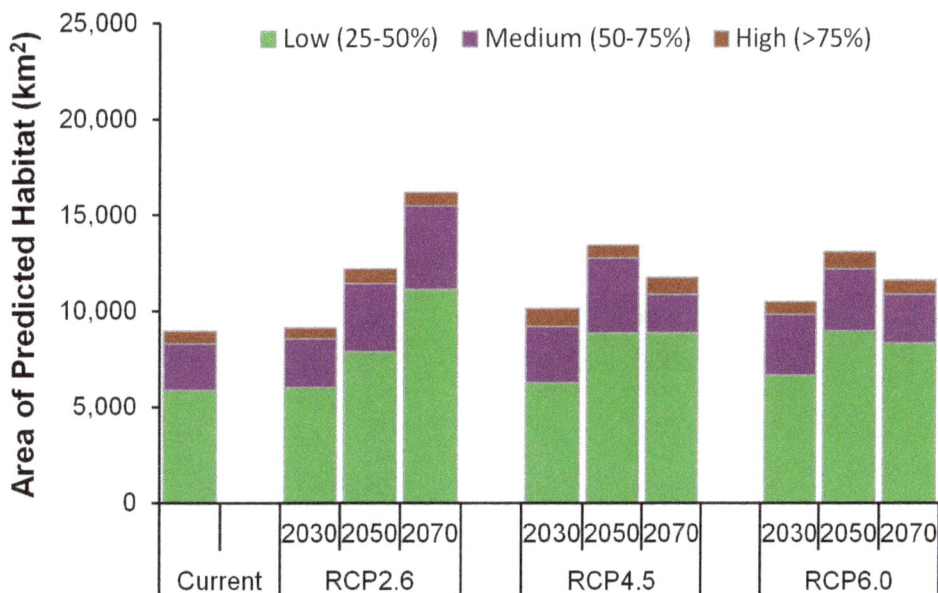

Figure 5. The estimated areas of the predicted distribution of Chinese caterpillar fungus.

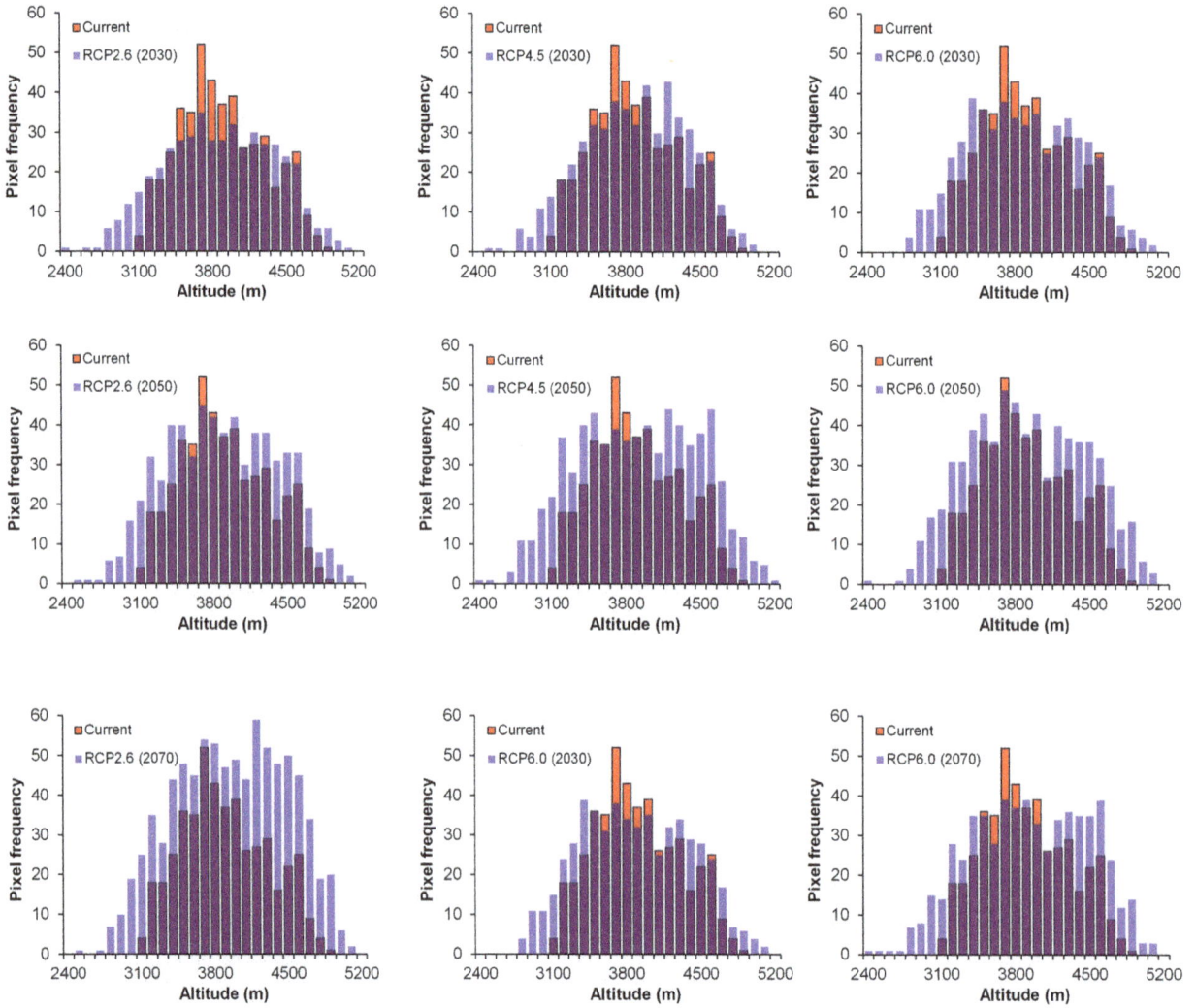

Figure 6. Change in average elevation in distribution of Chinese caterpillar fungus in three future climate scenarios.

caterpillar fungus [55], however, given the diverse topography of the Himalayan region, finer scale resolution of the environmental data might be needed to produce results at scales relevant to species with more restricted habitat requirements. As we used a limited number of samples from only six districts while this species is reported from 27 mountain districts of Nepal, this resolution was used as a compromise between fine scale resolution and the precision of the coordinates for the species' occurrences. More

Table 3. Independent sample T-test between average altitude of current predicted distribution and future predicted distributions.

Climate trajectories	Year	Mean	SD	P value of independent sample T-test
RCP2.6	2030	3921.68	523.03	0.302
RCP2.6	2050	3933.59	523.11	0.500
RCP2.6	2070	3998.86	529.83	0.104
RCP4.5	2030	3948.13	486.09	0.861
RCP4.5	2050	3955.49	566.19	0.938
RCP4.5	2070	3878.31	525.48	0.011
RCP6.0	2030	3929.22	514.55	0.416
RCP6.0	2050	3959.01	538.16	0.841
RCP6.0	2070	3982.11	548.55	0.337

occurrence records spread across the entire country of Nepal and finer resolution of environmental data that also capture microclimates, edaphic conditions, vegetation dynamics, and landscape heterogeneity might enable more sophisticated models to be developed in future.

Despite these limitations, the study provides information on the suitable climate space or bioclimatic envelope responsible for the potential distribution of Chinese caterpillar fungus. Furthermore, understanding the potential distribution of one of the most exploited species of this region is a necessary to design an efficient conservation plan. This study provides evidence of the distribution of suitable habitat for the Chinese caterpillar fungus in Nepal Himalaya using a robust bioclimatic model. It can be further useful in assessing threats and evaluating gaps in protected areas in order to conserve this economically important species. Further-

more, our evaluation of the potential impacts of future climate change on the potential suitable habitat of the species is important in devising adaptive responses and precautionary measures for the sustainable management of the habitat of the species in future.

Acknowledgments

We thank Kamal Nepali, Shivish Bhandari, Bikram Pandey, Chhabi Thapa and Rabindra Parajuli for their support in the field. We thank Kathryn Reardon-Smith and anonymous reviewer for valuable comments for improving the manuscript.

Author Contributions

Conceived and designed the experiments: UBS KB. Performed the experiments: UBS. Analyzed the data: UBS. Wrote the paper: KB UBS.

References

1. IPCC (2013) Climate Change 2013: The physical science basis. Contribution of working group I to the fifth assessment report of the Intergovernmental Panel on Climate Change. In: Stocker TF, Qin D, Plattner GK, Tignor M, Allen SK, Boschung J, Nauels A, Xia Y, Bex V, Midgley PM, editors. Cambridge University Press, Cambridge, United Kingdom and New York, NY, USA. pp. 1535.

2. Hansen AJ, Neilson RP, Dale VH, Flather CH, Iverson LR, et al. (2001) Global change in forests: responses of species, communities, and biomes. Bioscience 51: 765–779.

3. Walther GR, Post E, Convey P, Menzel A, Parmesan C, et al. (2002) Ecological responses to recent climate change. Nature 416: 389–395.

4. Parmesan C (2006) Ecological and evolutionary responses to recent climate change. Ann Rev Ecol Evol Syst 37: 637–669.

5. Bellard C, Bertelsmeier C, Leadley P, Thuiller W, Courchamp F (2012) Impacts of climate change on the future of biodiversity. Ecol Lett 15: 365–377.

6. Summers DM, Bryan BA, Crossman ND, Meyer WS (2012) Species vulnerability to climate change: impacts on spatial conservation priorities and species representation. Glob Change Biol 18(7): 2335–2348.

7. Pressey RL, Cabeza M, Watts ME, Cowling RM, Wilson KA (2007) Conservation planning in a changing world. Trends Ecol Evol 22: 583–592.

8. Shrestha UB, Gautam S, Bawa KS (2012) Widespread climate change in the Himalayas and associated changes in local ecosystems. PLoS ONE 7(5): e36741.

9. Xu J, Grumbine RE, Shrestha A, Eriksson M, Yang X, et al. (2009) The melting Himalayas: Cascading effects of climate change on water, biodiversity, and livelihoods. Conserv Biol 23(3): 520–530.

10. Joshi PK, Rawat A, Narula S, Sinha V (2012) Assessing impact of climate change on forest cover type shifts in Western Himalayan Eco-region. J For Res 23(1): 75–80.

11. Telwala Y, Brook BW, Manish K, Pandit MK (2013) Climate-induced elevational range shifts and increase in plant species richness in a Himalayan biodiversity epicentre. PLoS ONE 8(2): e57103.

12. Kulkarni A, Patwardhan SKKK, Ashok K, Krishnan R (2013) Projected climate change in the Hindu Kush–Himalayan region by using the high-resolution regional climate model PRECIS. Mt Res Dev 33(2): 142–151.

13. Kumar P (2012) Assessment of impact of climate change on Rhododendrons in Sikkim Himalayas using Maxent modelling: limitations and challenges. Biod Conserv 21: 1251–1266.

14. Hirzel AH, Lay GL, Helfer V, Randin C, Guisan A (2006) Evaluating the ability of habitat suitability models to predict species presences. Ecol Model 199(2): 142–152.

15. Beaumont IJ, Pitman AJ, Poulsen M, Hughes L (2007). Where will species go? Incorporating new advances in climate modelling into projections of species distributions. Glob Change Biol 13(7): 1368–1385.

16. Franklin J (2010) Mapping Species Distributions – Spatial inference and prediction, ecology. biodiversity and conservation, Cambridge University Press, Cambridge. 336 p.

17. Phillips SJ, Anderson RP, Schapire RE (2006) Maximum entropy modeling of species geographic distributions. Ecol Model 190(3–4): 231–259.

18. Fourcade Y, Engler JO, Rödder D, Secondi J (2014) Mapping species distributions with MAXENT using a geographically biased sample of presence data: A performance assessment of methods for correcting sampling bias. PLoS ONE 9(5): e97122.

19. Nazeri M, Jusoff K, Madani N, Mahmud AR, Bahman AR, et al. (2012) Predictive modeling and mapping of Malayan Sun Bear (Helarctos malayanus) distribution using maximum entropy. PLoS ONE 7(10): e48104.

20. Li J, Wang D, Yin H, Zhaxi D, Jiagong Z, et al. (2014) Role of Tibetan Buddhist monasteries in Snow Leopard conservation. Conserv Biol 28(1): 87–94.

21. Taylor S, Kumar L, Reid N, Kriticos DJ (2012) Climate Change and the Potential Distribution of an Invasive Shrub, Lantana camara L.. PLoS ONE 7(4): e35565.

22. Poudel RC, Möller M, Gao L-M, Ahrends A, Baral SR, et al. (2012) Using morphological, molecular and climatic data to delimitate yews along the Hindu Kush-Himalaya and adjacent regions. PLoS ONE 7(10): e46873.

23. Merow C, Smith MJ, Silander Jr. JA (2013) A practical guide to MaxEnt for modeling species' distributions: what it does, and why inputs and settings matter. Ecography 36(10): 1058–1069.

24. Hijmans RJ, Graham CH (2006) The ability of climate envelope models to predict the effect of climate change on species distributions. Glob Chang Biol 12(12): 2272–2281.

25. Shrestha AB, Wake CP, Mayewski PA, Dibb JE (1999) Maximum temperature trends in the Himalaya and its vicinity: an analysis based on temperature records from Nepal for the period 1971–94. J Climate 12: 2775–2786.

26. NCVST (2009) Vulnerability through the eyes of vulnerable. Climate change induced Uncertainties and Nepal's Development Predicaments. Institute for Social and Environmental Transition-Nepal Normal (ISET-N). 95 p.

27. Winkler D (2010) Caterpillar fungus (Ophiocordyceps sinensis) production and sustainability on the Tibetan plateau and in the Himalayas. Asian Medicine 5: 291–316.

28. Devkota S (2010) Ophicordyceps sinensis (Yarsagumba) from Nepal Himalaya: status, threats and management strategies. In: Zhang PH, editor.Cordyceps Resources and Environment. Grassland Supervision Center by the Ministry of Agriculture, People's Republic of China. pp. 91–108.

29. Shrestha UB, Bawa KS (2013) Trade, harvest, and conservation of caterpillar fungus (Ophiocordyceps sinensis) in the Himalayas. Biol Conserv 159: 514–520.

30. Stone R (2008) Last stand for the body snatcher of the Himalayas? Science 322: 1182.

31. Shrestha UB (2012) Asian medicine: A fungus in decline. Nature 482: 35.

32. Xuan C (2012) How long can the Caterpillar fungus craze last? China Dialogue. Available: http://www.chinadialogue.net/article/show/single/en/5143-How-long-can-the-caterpillar-fungus-craze-last. Accessed 24 December 2012.

33. Yang DR (2008) Recommendations on strengthening the protection of the resources of the rare Tibet Plateau fungus—Cordyceps sinensis and its ecological environment' (Jia qiang qing zang gao yuan zhen xi zhen jun—Dong cong xia cao zi yuan he sheng tai huan jing bao hu de jian yi) Unpublished report to Chinese Academy of Science (in Chinese).

34. Hijmans RJ, Cameron SE, Parra JL, Jones PG, Jarvis A (2005) Very high resolution interpolated climate surfaces for global land areas. Int J Climatol 25(15): 1965–1978.

35. van Vuuren D, den Elzen M, Lucas P, Eickhout B, Strengers B, et al. (2007) Stabilizing greenhouse gas concentrations at low levels: an assessment of reduction strategies and costs. Climat Chang 81: 119–159.

36. Clarke L, Edmonds J, Jacoby H, Pitcher H, Reilly J, Richels R (2007) Scenarios of Greenhouse Gas Emissions and Atmospheric Concentrations. Sub-report 2.1A of Synthesis and Assessment Product 2.1 by the U.S. Climate Change Science Program and the Subcommittee on Global Change Research. Department of Energy, Office of Biological & Environmental Research, Washington DC, USA. pp. 154.

37. Fujino J, Nair R, Kainuma M, Masui T, Matsuoka Y (2006) Multi-gas mitigation analysis on stabilization scenarios using AIM global model. The Energy Journal Multi-Greenhouse Gas Mitigation and Climate Policy, Special Issue # 3: 343–354.

38. Collins WJ, Bellouin N, Doutriaux-Boucher M, Gedney N, Halloran P, et al. (2011) Development and evaluation of an Earth-System model—HadGEM2. Geosci Model Dev Discuss 4: 997–1062.

39. Elith J, Graham CH, Anderson P, Dudik M, Ferrier S, et al. (2006) Novel methods improve prediction of species distributions from occurrence data. Ecography 29: 129–151.

40. Gange AC, Gange EG, Sparks TH, Boddy L (2007) Rapid and recent changes in fungal fruiting patterns. Science 316: 71.

41. Kauserud H, Heegaard E, Semenov MA, Boddy L, Halvorsen R, et al. (2010) Climate change and spring-fruiting fungi. P Roy Soc B-Biol Sci 277: 1169–1177.

42. Gange AC, Gange EG, Mohammad AB, Boddy L (2011) Host shifts in fungi caused by climate change? Fungal Ecol 4(2): 184–190.

43. Weckerle CS, Yang Y, Huber FK, Li Q (2010) People, money, and protected areas: the collection of the caterpillar mushroom *Ophiocordyceps sinensis* in the Baima Xueshan Nature Reserve, Southwest China. Biod Conserv 19: 2685–2698.

44. Zhang YJ, Sun BD, Zhang S, Wang M, Liu XZ, et al. (2010) Mycobiotal investigation of natural *Ophiocordyceps sinensis* based on culture-dependent investigation. Mycosystema 29: 518–527.

45. Cannon PF, Hywel-Jones NL, Maczey N, Norbu L, Samdup T, et al. (2009) Steps towards sustainable harvest of *Ophiocordyceps sinensis* in Bhutan. Biodivers Conserv 18: 2263–2281.

46. Shrestha UB, Bawa KS (2014) Harvesters' perceptions of population status and conservation of Chinese caterpillar fungus in the Dolpa region of Nepal (under review).

47. Thapa YB, Chettri R (2011) Handbook for sustainable collection of Yarsagumba, Department of Forest Resource, Government of Nepal, Kathmandu, Nepal (Yarsagumba Digo Sankalan Prabidi Sahayogi Pustika) (in Nepali).

48. Loarie SR, Carter BE, Hayhoe K, McMahon S, Moe R, et al. (2008) Climate change and the future of California's endemic flora. PLoS ONE 3(6): e2502.

49. Bakkenes M, Alkemade JRM, Ihle F, Leemans R, Latour JB (2002) Assessing effects of forecasted climate change on the diversity and distribution of European higher plants for 2050. Global Change Biol 8: 390–407.

50. Barrett MA, Brown JL, Junge RE, Yoder AD (2013) Climate change, predictive modeling and lemur health: Assessing impacts of changing climate on health and conservation in Madagascar. Biol Cons 157: 409–422.

51. Molloy SW, Davis RA, Van Etten JB (2013) Species distribution modelling using bioclimatic variables to determine the impacts of a changing climate on the western ringtail possum (*Pseudocheirus occidentals*; Pseudocheiridae). Environm Conserv 1–11.

52. Li QS, Li L, Yin DH, Fu SQ, Huang TF, et al. (1991) Biological characteristics of *Cordyceps sinensis*. Special Wild Econ. Anim Plant Res 1: 42–45 (in Chinese).

53. Li Y, Wang XL, Jiao L, Jiang Y, Li H, et al. (2011). A survey of the geographic distribution of *Ophiocordyceps sinensis*. J Microbiol 49: 913–919.

54. Shrestha UB, Shrestha BB, Shrestha S, Ghimire S, Nepali K (2014). Chasing Chinese caterpillar fungus (*Ophiocordyceps sinensis*) harvesters in the Himalayas: Harvesting practice and its conservation implications in western Nepal. Soc Nat Res In press.

55. Seo C, Thorne JH, Hannah L, Thuiller W (2009). Scale effects in species distribution models: implications for conservation planning under climate change. Biol lett 5: 39–43.

20

Multi-Scales Analysis of Primate Diversity and Protected Areas at a Megadiverse Region

Míriam Plaza Pinto[1*¤], **José de Sousa e Silva-Júnior**[2], **Adriana Almeida de Lima**[3],
Carlos Eduardo Viveiros Grelle[4,5]

1 Programa de Pós-Graduação em Ecologia, Universidade Federal do Rio de Janeiro, Rio de Janeiro, Rio de Janeiro, Brazil, 2 Coordenação de Zoologia, Museu Paraense Emílio Goeldi, Belém, Pará, Brazil, 3 Departamento de Ecologia, Universidade Federal do Rio Grande do Norte, Natal, Rio Grande do Norte, Brazil, 4 Departamento de Ecologia, Universidade Federal do Rio de Janeiro, Rio de Janeiro, Rio de Janeiro, Brazil, 5 Laboratório de Vertebrados, Universidade Federal do Rio de Janeiro, Rio de Janeiro, Rio de Janeiro, Brazil

Abstract

In this paper, we address the question of what proportion of biodiversity is represented within protected areas. We assessed the effectiveness of different protected area types at multiple scales in representing primate biodiversity in the Brazilian Legal Amazon. We used point locality data and distribution data for primate species within 1°, 0.5°, and 0.25° spatial resolution grids, and computed the area of reserves within each cell. Four different approaches were used – no reserves (A), exclusively strict use reserves (B), strict and sustainable use reserves (C), and strict and sustainable use reserves and indigenous lands (D). We used the complementarity concept to select reserve networks. The proportions of cells that were classified as reserves at a grid resolution of 1° were 37%, 64%, and 88% for approaches B, C and D, respectively. Our comparison of these approaches clearly showed the effect of an increase in area on species representation. Representation was consistently higher at coarser resolutions, indicating the effect of grain size. The high number of irreplaceable cells for selected networks identified based on approach A could be attributed to the use of point locality occurrence data. Although the limited number of point occurrences for some species may have been due to a Wallacean shortfall, in some cases it may also be the result of an actual restricted geographic distribution. The existing reserve system cannot be ignored, as it has an established structure, legal protection status, and societal recognition, and undoubtedly represents important elements of biodiversity. However, we found that strict use reserves (which are exclusively dedicated to biodiversity conservation) did not effectively represent primate species. This finding may be related to historical criteria for selecting reserves based on political, economic, or social motives.

Editor: Cédric Sueur, Institut Pluridisciplinaire Hubert Curien, France

Funding: Míriam Plaza Pinto was supported by FAPERJ doctoral scholarship (http://www.faperj.br/) and CNPq pos-doctoral scholarship (PDJ 150721/2010-2; http://www.cnpq.br/) and Carlos Eduardo Viveiros Grelle by CNPq productivity fellowship. This study was supported by grants from CNPq and FAPERJ (Auxílio Instalação – proc. E-26/111.829/2010 and Jovem Cientista do Estado). The funders had no role in study design, data collection and analysis, decision to publish, or preparation of the manuscript.

Competing Interests: The authors have declared that no competing interests exist.

* Email: miriamplazapinto@yahoo.com.br

¤ Current address: Departamento de Ecologia, Universidade Federal do Rio Grande do Norte, Natal, Rio Grande do Norte, Brazil

Introduction

The Amazon is the largest Brazilian biome and has been subjected to several destructive changes during the last decade [1]. It contains over half of all global tropical forest and provides habitat for a wide range of biodiversity [2]. Considering the magnitude of loss of habitat and general ecological relationships (for example species area relationships), some studies predict that an extinction debt will have to be paid in the future [3,4].

Deforestation itself may not be the only cause of threats to species (see [5] for an example relating to Amazonian mammals). It has consequences other than habitat area reduction, such as fragmentation [6] and selective logging [7]. Selective logging is spatially diffuse and difficult to monitor [8]. Other indirect effects include hunting [9] and fire [7,10,11], since deforestation benefits hunters by increasing an area's accessibility [12], while border effects, heat, and desiccation are conducive to the occurrence of fire [13,14]. Other factors such as invasive exotic species [15] are a

further threat to the persistence of native species. Some threats, such as agriculture, hunting, and rural or urban expansion, act synergistically [16] in affecting tropical mammal species [14]. When there are several competing demands on land use, it is important to prioritize areas with the specific purpose of conserving biological diversity.

Protected areas cover 12.9% of the global terrestrial area, of which 5.8% is in reserves with strict use regulation (IUCN categories I–IV) [17]. Latin America contains more sustainable use reserves (20%; categories V and VI) as well as non-categorized reserves than any other continent [18]. A total of 11.1% of the entire area of the Amazon is strictly protected [19]. Protected areas are heterogeneously distributed among the Brazilian biomes, of which the Amazon has the greatest proportion of protected areas. Regarding categories of protection, 5.7% of its area is under strict use regulation, 1.9% is allocated to sustainable use reserves, and 17.7% to indigenous land [20].

Protected areas are spatially correlated with low deforestation and selective logging rates in humid tropical forests [2]. For example, deforestation and selective logging in protected areas are lower compared to adjacent lands in a region with rapid agro-industrialization rates south of the Brazilian Amazon [7]. Natural and indigenous reserves constitute refuges for tropical biodiversity, featuring fewer incidences of deforestation and fire [21]. These areas are also generally protected against selective logging [8].

It is therefore assumed that protected areas secure biodiversity against several threats. However, there remains the question of the proportion of biodiversity that is represented within protected areas. An assessment of the extent to which conservation targets have already been achieved in existing conservation areas is of critical importance in this regard [22]. Gap analysis is a process for spatially comparing data on species or other biodiversity elements (especially conservation-related data) under various types of land use, to identify gaps in protection [23–25]. Several systematic conservation-planning studies have assessed how effective existing reserves are at representing specific biodiversity elements [26–29]. These include recent studies on South American mammals [30] and endemic Atlantic Forest primates [31].

Threats to the Amazon affect primate species of this biome in different ways [32]. The Amazon is the region with the greatest primate diversity in South America. There are 92 primate species occurring in the Brazilian Amazon [33], of which 26 are on the Brazilian list of threatened species [34]. Several species have narrow ranges [35]. Rivers play an important role in restricting dispersion of Amazonian primates, while interspecific differences in the size of geographic distribution areas may be associated with the capacity of different species to transcend these barriers [36]. Primates are a well-studied group relative to other mammals, likely because most of them are visible, diurnal, have a large body size, and are arboreal, noisy and colorful [37].

Our aim was to assess the effectiveness of current protected areas in representing primate diversity in the Brazilian Amazon using grids of different resolutions. This evaluation considered strict and sustainable use reserves and indigenous lands, as well as different combinations thereof. Strict use reserves are areas dedicated exclusively to nature conservation, whereas sustainable reserves aim to reconcile nature conservation and sustainable use of natural resources. Indigenous lands are socio-cultural areas that are not explicitly created with the purpose of conserving biological diversity, but nevertheless seem to contribute to this end [8,21]. For this study, we constructed a database of point occurrence data and we also used the IUCN distribution database [38]. A minimum area criterion was used to classify a grid cell as a reserve. We also mapped irreplaceability patterns indicating priority regions needed to guarantee representation of all species.

Materials and Methods

We constructed a database with locality data for all primate species occurring within Legal Amazon in Brazil (Figure 1). Legal Amazon is a politically categorized area used by the Brazilian government for planning, coordinating, controlling, and executing actions for regional development, which encompasses this biome (see http://www.sudam.gov.br/amazonia-legal). Locality data were initially compiled from the Museu Paraense Emílio Goeldi (MPEG) mammal collection. Using the Web of Science (http://apps.isiknowledge.com) and Scielo (http://www.scielo.br/) databases, we searched for articles published between 2000 and 2013 (last update) using each genus name as a separate keyword. We also searched the *Neotropical Primates* and *Checklist* journals that were not indexed in these databases. Locality data for primate

species recorded by experts (primate specialists) were extracted from these articles. Consulted articles are listed in Text S1, including specific reviews for each genus. Some localities were georeferenced using Global Gazetteer Version 2.1 (http://www.fallingrain.com/world, last accessed in December 2009), National Geospatial-Intelligence Agency (http://www.nga.mil/portal/site/nga01/, last accessed in December 2009), and the speciesLink project (http://splink.cria.org.br/, last accessed in December 2009). Specific details for each genus can be found in Text S2. Primate species distribution data, outlining estimated ranges for each species, were obtained from IUCN [38].

We conducted our analysis at multiple scales using three grids. The first was composed of 433 cells at a spatial resolution of $1°$ latitude/longitude, the second of 1,687 cells at a resolution of $0.5°$, and the third of 6,667 cells at a resolution of $0.25°$. These grids were superimposed on the Legal Amazon map (IBGE <http://downloads.ibge.gov.br/downloads_geociencias.htm>, accessed in October 2009). To be included in the analysis, cells had to overlap with Legal Amazon by at least 25%. This ensured that at least 25% of each grid cell was part of the Amazon biome. All databases were incorporated in this grid, and the cells were considered the units for data analysis.

Maps of strict use reserves (IUCN categories I–IV) and sustainable use reserves (IUCN categories V and VI) were obtained from the Ministry of Environment (Ministério do Meio Ambiente - MMA) database (http://mapas.mma.gov.br/i3geo/datadownload.htm, accessed in January 2014). Indigenous lands maps were obtained from the National Indian Foundation (Fundação Nacional do Índio - FUNAI) database (http://mapas.funai.gov.br/). We computed the area of each protected area type (strict use, sustainable use, and indigenous land) within each cell. Cells that included 11,570 ha or more of protected areas were categorized as reserves (see below).

Our decision about the minimum area used to designate a cell as a reserve was based on a population viability analysis (PVA) previously conducted for *Brachyteles* [39], an Atlantic Forest genus. However, no PVA analysis has to date been carried out for a large Amazonian primate species. We assumed that the area needed to maintain viable populations was positively correlated to body size [40]. The species belonging to the *Brachyteles* genus have individual body weights of 9.4–12.1 kg [40]. The largest Amazonian primate genera include *Alouatta*, *Ateles*, and *Lagothrix* with weight ranges of 3.8–9.0 kg, 7.0–9.0 kg, and 7.0–12.0 kg, respectively [41].

Selection of priority areas was performed using the following four approaches:

A)   For this approach, the existence of actual reserves in the Legal Amazon was ignored.

B)   All cells with at least 11,570 ha of existing strict use protected areas were classified as reserves. These cells were automatically included in the network, and the remaining priority areas were selected taking these pre-existing reserves into account.

C)   As B), including sustainable use protected areas.

D)   As C), including indigenous lands.

We used gap analysis to ascertain whether the cells in which each species occurred coincided with existing reserves (for approaches B, C, and D). If there were no matches, the species was considered not to be represented in the existing reserve system. Gap analysis was also done using distribution data.

The reserve selection procedure was carried out using MARXAN software to implement a simulated annealing algo-

Figure 1. Legal Amazon region, South America. The Legal Amazon covers a greater part of the biome, and is located inside Brazilian borders. It is a politically categorized area used by the Brazilian government for planning, coordinating, controlling and executing actions for regional development.

rithm [42–45]. For each of the four approaches we performed 200 runs with 10^6 iterations. Unless each species occurs in only one cell, multiple solutions exist for achieving the goal of representing all species within the minimum number of cells. A high number of interactions increases the likelihood of finding good solutions (a minimum number of cells representing each of the species at least once) using the simulated annealing algorithm. We used an irreplaceability index for the top 100 best solutions. This index was calculated by summing the number of times each cell appeared in a solution, using a minimum value of 0 for cells that did not appear in any solution and a maximum value of 100 for cells that appeared in all solutions.

To test the effectiveness of the existing network of reserves relative to species representation, we randomly selected 10^4 networks composed of the same number of cells considered as reserves (for each approach B, C, and D). The frequency distribution of the number of primate species represented in the random networks was compared with the actual number of species represented in the reserve cells. Random networks were selected using a script written in R [46].

Results

The occurrence database yielded 1,690 localities for 87 primate species grouped within 16 genera in the Legal Amazon area. The following species each occurred in a single 0.25° grid cells: *Aotus vociferans* (Spix, 1823), *Callicebus regulus* Thomas, 1927, *Mico manicorensis* Van Roosmalen, Van Roosmalen, Mittermeier & Rylands, 2000, and *Mico marcai* (Alperin, 1993). The species with the highest number of occurrences in different cells was *Sapajus apella* (Linnaeus, 1758). For several species the number of point occurrences did not correspond to the number of cells of the grid in which they occurred, as some points were spatially grouped. Numbers of cells containing occurrence data for primate species at 0.25°, 0.5°, and 1° were 782 (11.73%), 510 (30.23%), and 266 (61.43%), respectively (Figure 2). Cells with high richness values were well dispersed across the grids.

Numbers of cells containing at least 11,570 ha of strict use reserves (approach B), strict and sustainable use reserves (approach C), or strict and sustainable use reserves and indigenous lands (approach D) were 161 (37.18%), 275 (63.51%), and 383 (88.45%), respectively, at a grid resolution of 1° (Tables 1 and 2). Results for other grid resolutions are also shown. Cells classified as reserves were often adjacent to one another.

The proportion of species represented in the current network of reserves considering occurrence data varied from 60% at a grid resolution of 0.25°, considering only strict use reserves, to 99% at a grid resolution of 1°, considering strict and sustainable use reserves and indigenous lands (Table 1). When using distribution data, the proportion of species represented varied from 91% at a grid resolution of 0.25°, considering only strict use reserves, to 100% at a grid resolution of 1° considering strict and sustainable use reserves (including the indigenous lands made no difference) (Table 2). Regardless of grid resolution, the number of species represented was higher when all types of reserves were included (Table 1, Figure 3). When using distribution data, the number of species represented was exactly the same for approaches C and D, i.e. including indigenous lands made no difference (Table 2). The number of species represented was not higher than expected by chance alone (Table 1). At a grid resolution of 1°, the number of species represented in protected areas was higher than the upper quartile of random networks when considering strict use and sustainable use reserves. When all types of protected areas or only

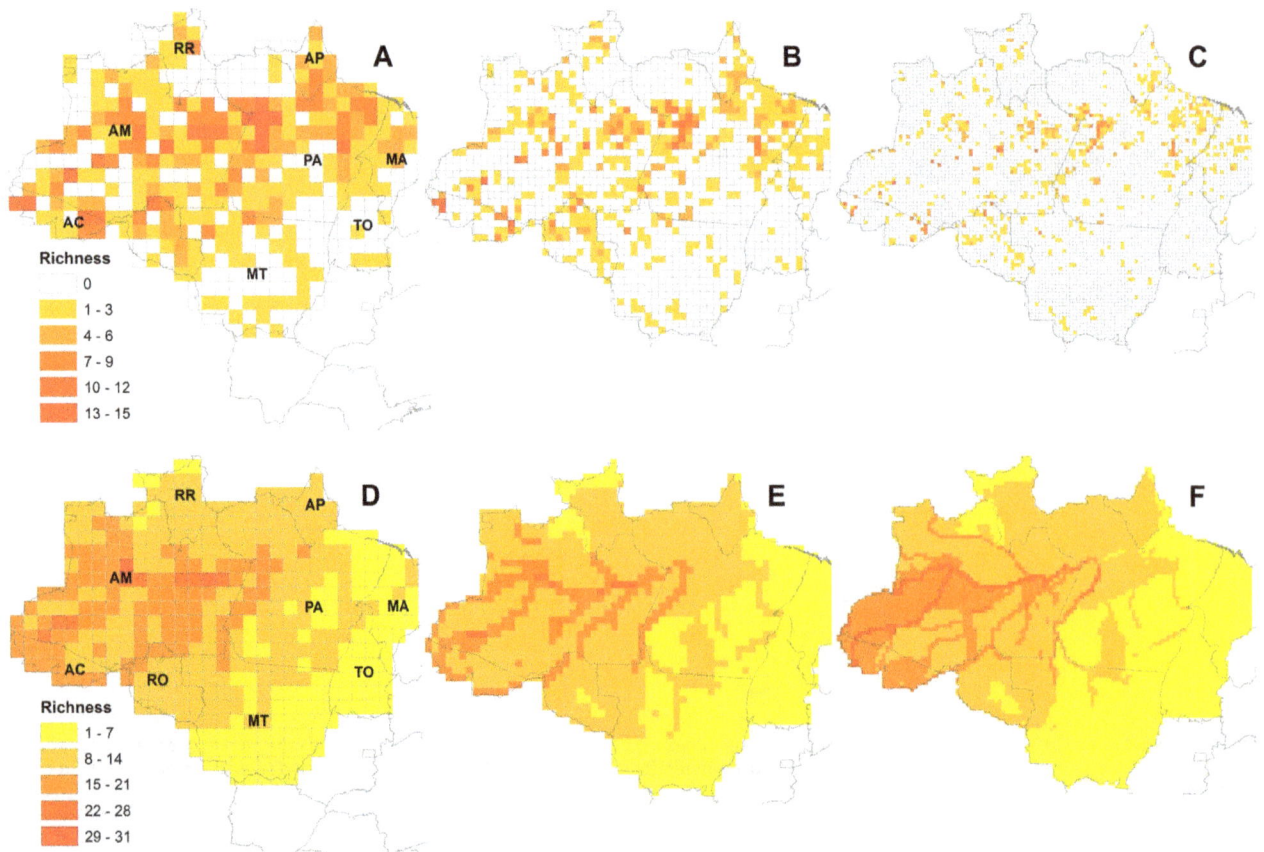

Figure 2. Spatial richness pattern of Amazonian primate species. Number of primate species occurring in each grid cell in the 1° grid, with a total of 433 cells (A and D), the 0.5° grid, with a total of 1687 cells (B and E), and the 0.25° grid, with a total of 6,667 cells (C and F) in Legal Amazon, based on point locality occurrence data (A, B and C) and distribution data (D, E and F). States are indicated in A and D as follows. AC – Acre, AM – Amazonas, AP – Amapá, MA – Maranhão, MT – Mato Grosso, PA – Pará, RO –Rondônia, RR – Roraima, TO – Tocantins.

Table 1. Selection of priority areas in Legal Amazon for conservation at three resolutions, based on point locality occurrence data for primates.

Resolution	Approach	Existing Protected Areas		Primate species		Random networks	
		N	P	N	P	N	P
1°	B	161	37.18%	75	86.21%	3859	0.39
1°	C	275	63.51%	84	96.55%	1608	0.16
1°	D	383	88.45%	86	98.85%	3948	0.39
0.5°	B	348	20.63%	66	75.86%	5560	0.56
0.5°	C	736	43.63%	78	89.66%	5463	0.55
0.5°	D	1183	70.12%	82	94.25%	8670	0.87
0.25°	B	840	12.60%	52	59.77%	9820	0.98
0.25°	C	2039	30.58%	74	85.06%	6895	0.69
0.25°	D	3575	53.62%	79	90.80%	9359	0.94

Resolution gives the size of the grid cells. **Approach** gives the approach to classifying grid cells as reserves (B takes only strict use reserves into account, C uses strict use and sustainable use and indigenous lands). Under **Existing Protected Areas, N** gives the number of grid cells classified as existing reserves on the basis that they contain at least 11,570 ha of existing reserves, and **P** gives the proportion that these represent of the total grid cells in Legal Amazon. Under **Primate Species, N** gives the number of primate species occurring in these grid cells, and **P** gives the proportion that these represent of all Amazonian primate species. Under **Random Networks, N** gives the number of randomly selected networks that protected more species than the existing reserve network, and **P** gives the proportion relative to 10.000 (which is the total number of random networks selected).

Table 2. Selection of priority areas in Legal Amazon for conservation at three resolutions, based on distribution data for primates.

Resolution	Approach	Existing Protected Areas		Primate species		Random networks	
		N	P	N	P	N	P
1°	B	161	37.18%	88	97.78%	9590	0.96
1°	C	275	63.51%	90	100.00%	0	<0.01
1°	D	383	88.45%	90	100.00%	0	<0.01
0.5°	B	348	20.63%	85	94.44%	9332	0.93
0.5°	C	736	43.63%	89	98.89%	5427	0.54
0.5°	D	1183	70.12%	89	98.89%	9030	0.90
0.25°	B	840	12.60%	82	91.11%	10000	1.00
0.25°	C	2039	30.58%	89	98.89%	6654	0.67
0.25°	D	3575	53.62%	89	98.89%	9433	0.94

Resolution gives the size of the grid cells. **Approach** gives the approach to classifying grid cells as reserves (B takes only strict use reserves into account, C uses strict use and sustainable use and indigenous lands). Under **Existing Protected Areas, N** gives the number of grid cells classified as existing reserves on the basis that they contain at least 11,570 ha of existing reserves, and **P** gives the proportion that these represent of the total grid cells in Legal Amazon. Under **Primate Species, N** gives the number of primate species occurring in these grid cells, and **P** gives the proportion that these represent of all Amazonian primate species. Under **Random Networks, N** gives the number of randomly selected networks that protected more species than the existing reserve network, and **P** gives the proportion relative to 10.000 (which is the total number of random networks selected).

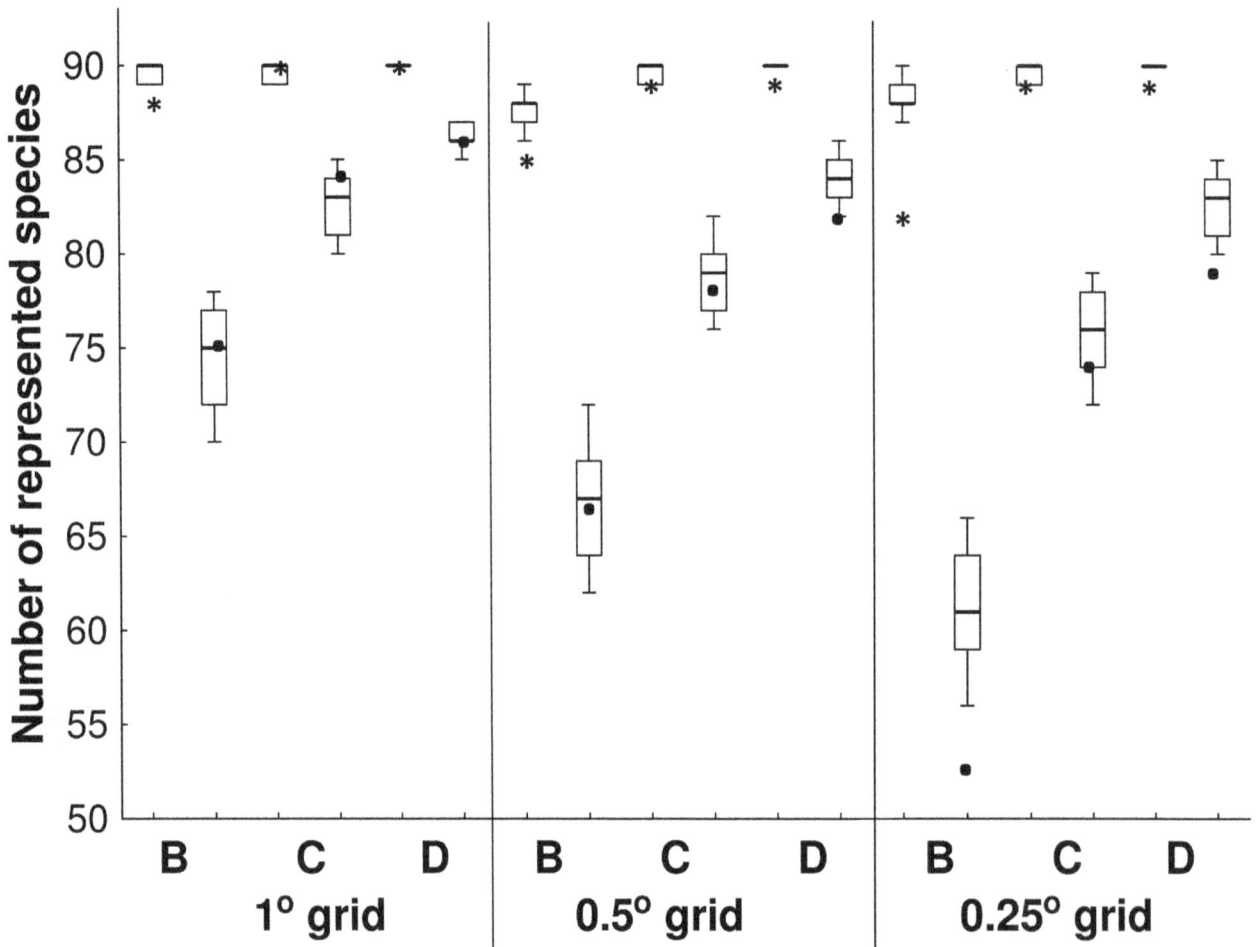

Figure 3. Number of primate species represented in actual and randomly selected reserve networks in Legal Amazon. Number of species represented in the existing reserve network (closed circles – point locality occurrence data; asterisks – distribution data) in relation to number represented in randomly selected networks (box and whiskers: median, quartiles and deciles). This analysis was conducted at three resolutions, using a 1°, 0.5°, and 0.25° grid. In approach B, only strict use reserves were included in the existing reserve network, in C both strict and sustainable use reserves were included, and in D, strict and sustainable use reserves and indigenous lands were included.

strict use reserves were considered, the number of species represented was lower than the lower decile of species in random networks using a grid resolution of 0.25° (Figure 3).

Networks selected in approach A that disregarded the existence of actual reserves in Legal Amazon contained 24, 31, and 32 cells at grid resolutions of 1°, 0.5°, 0.25°, respectively (Figure 4). At these resolutions, 14, 7, and 7 cells were respectively irreplaceable. Most of these irreplaceable cells were located in Amazonas, and one was at the border of the states of Amazonas and Rondônia. Some irreplaceable cells also occurred in Acre, Mato Grosso and Pará at a grid resolution of 1° (Figure 4).

When previously considering only existing strict use reserves, we selected cells to complement the actual reserve system throughout the entire biome. However, the cells with higher irreplaceability values were located in Amazonas (Figure 5 A, D and G). There were no additional areas selected in states other than Amazonas when we considered other types of existing reserves at grid resolutions of 1° and 0.5° (Figure 5 B, C, E, F, H and I).

Discussion

The existing reserve network in Legal Amazon does not completely represent the primate biodiversity of this region. Moreover, representation would probably be even lower if intraspecific diversity was considered. When we used occurrence data, representation was lower compared to random networks. The influence of an increase in area on species representation was clearly evident when we sequentially compared approaches B, C, and D. Considering all types of reserves together consistently enhanced species representation.

There was an effect of grain size. Considering existing reserves, representation of primate species was consistently higher when we used coarse resolution grids. The generalization of one point occurrence datum to an entire cell in a large grid (1°) compared with a small one (0.25°) may have inflated species representation. This effect is not so strong when using distribution data. The proportion of cells with data was positively related to grid cell size. Beyond that, a species can be considered represented in a reserve using a coarse grid resolution, but the occurrence data may not be coincident with reserves inside a large grid cell. Differences in grid resolutions affect characterization of richness patterns [47]. At

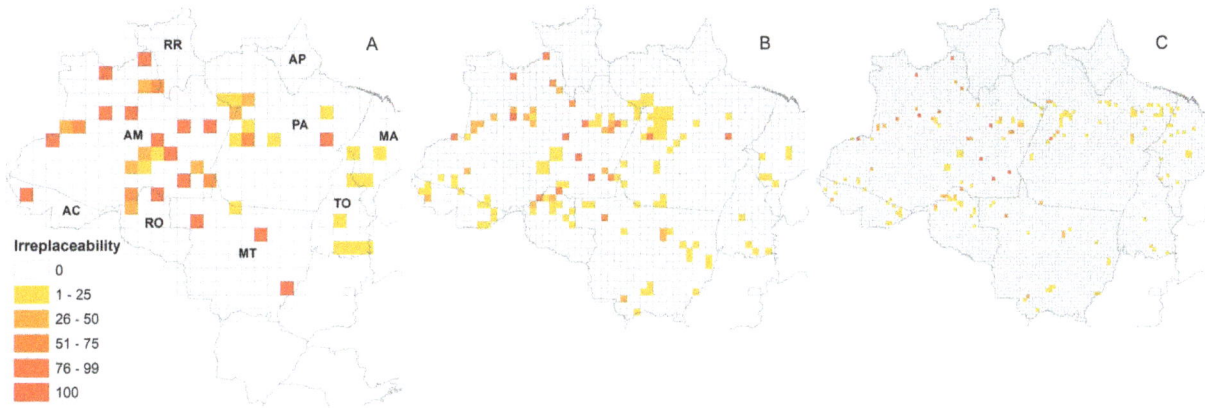

Figure 4. Identification of important grid cells for Amazonian primate conservation, ignoring existing reserves. Irreplaceability pattern was obtained using the 100 best network solutions for primate conservation in Legal Amazon. Grid cells were identified based on point locality occurrence data for primate species, ignoring existing reserves. Grid cells are colour coded according to the irreplaceability pattern, which is the number of solutions in which they were identified as important for primate conservation. At a grid resolution of 1° (A), 24 grid cells were identified in each solution. At a resolution of 0.5° (B), 31 grid cells were identified, and at 0.25° (C), 32 grid cells were identified. States are indicated in A as follows: AC –Acre, AM – Amazonas, AP – Amapá, MA – Maranhão, MT – Mato Grosso, PA – Pará, RO – Rondônia, RR – Roraima, TO – Tocantins.

coarser spatial resolutions data are less prone to false absences [47], but at gap analysis species representation may be overestimated by false presences. The combined effect of grain size and reserve types was sometimes considerable (when including all types of reserves, 54% of 0.25° cells, 70% of 0.5° cells and 88% of 1° cells were classified as reserves). Nevertheless, the number of species represented was mostly low considering the high proportion of the biome that was protected, especially at smaller grid sizes.

Small portions of some grid cells consisted of reserves. An important question is the minimum area required to classify a cell as a reserve. In another study [31], we also considered 11,570 ha as a baseline criterion of minimum area. This value was chosen based on a population viability study for larger primate species in the Atlantic Forest biome [39]. However, as previously mentioned there is no comparative study available for any large Amazon primate species.

Other studies have also used gap analysis to evaluate the representation of several species [29]. In Brazil, such studies were recently conducted for bird species from Cerrado using modeled geographic distribution [48], and for endemic primates in the Atlantic Forest [31]. Within South America, other important primate studies have focused on the identification of hotspots [49], areas of endemism [35,50], and areas of greater rarity [37]. The present study is therefore the first study conducted on primates in the entire Legal Amazon area using gap analysis and reserve selection.

We used both species point locality occurrence data and geographic distribution data in this study, and compared the results. The use of point locality data reduces type II or commission errors that falsely indicate the occurrence of a species at a particular place. However, it also disregards places where no sampling has been done but where the species may still occur. Thus, point locality occurrence data may be biased toward areas where more studies were carried out [51]. Compared with extrapolated geographic distribution, this type of data also makes it harder to visualize, understand, and relate diversity patterns to historic, environmental, or biological characteristics. But commission errors may be inflating species representation estimates when using distribution data. Our approach in this study was to apply spatial information at a wide scale. However, this is not a panacea

for conservation planning [23], and should be used in conjunction with fine-scale studies. We achieved our purpose, which was to evaluate the effectiveness of the existing Amazonian protected areas network, focusing primarily on strict use reserves, in representing primate species biodiversity. Based on the demonstrated multi-scale effect on species representation, we strongly recommend the use of a smaller grain size to reduce commission errors if the intention is to use point locality data.

Networks selected in approach A, which ignored the current reserve system, had low flexibility. Flexibility was assessed according to the number of irreplaceable cells. These cells were imperative to the network for the purpose of representing all species. The high number of irreplaceable cells could be explained by the use of point locality occurrence data. While the limited number of point occurrences of some species may be due to a lack of spatially dispersed studies and consequently of geographic distribution knowledge (Wallacean shortfall) [52,53], it may also be caused by actual restricted geographic species distribution. Species that restricted the flexibility of networks were mainly those occurring in a small number or just one cell (*Aotus vociferans*, *Callicebus regulus*, *Mico manicorensis*, and *Mico marcai*). A lack of sampling was evident since some of these species, e.g. *Aotus trivirgatus* (Humbolt, 1811), are expected to occur within larger areas.

A point worth emphasizing is that while a smaller network could in fact be effective in representing primate species, this does not imply that we can ignore the existing reserve system [54]. Rather, additional and complementary areas to those contained within the established system should be identified. Existing reserves already enjoy a structure, legal protection status, and recognition by society [27]. More importantly, they clearly represent important elements of biodiversity other than those that our study focused on (primates). Also, we used the criterion of just one occurrence to consider a primate species as being represented. Our findings would certainly have indicated a less effective reserve network if we had considered the occurrence of each species at more than one reserve as a criterion. This is a preliminary evaluation of primate conservation in Amazonian reserves from a multispecies perspective, and it can be considered the minimum to guarantee that each species is represented in at least one reserve. However, it is evident that an assessment of individual species' population

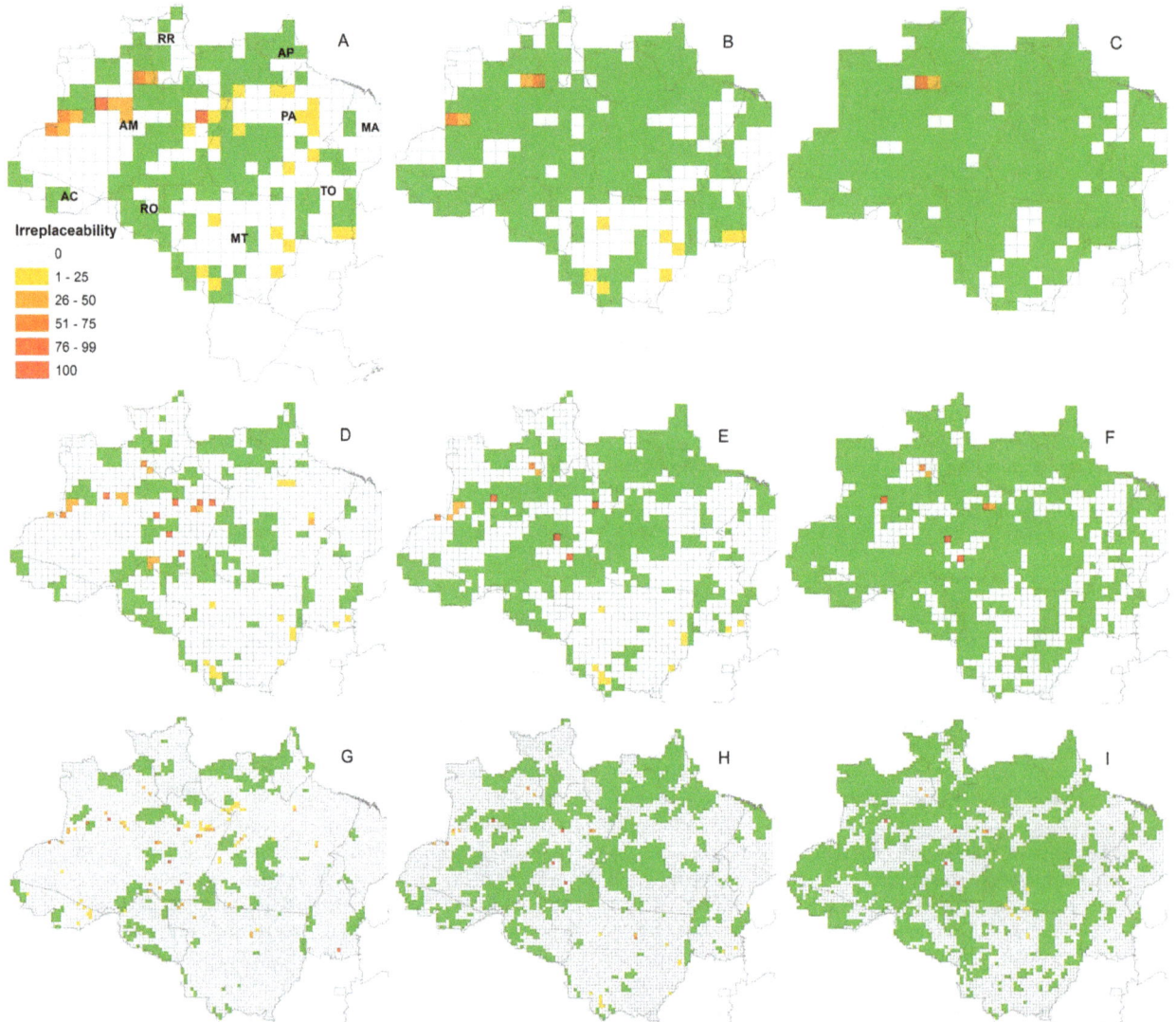

Figure 5. Identification of important grid cells for Amazonian primate conservation, taking existing protected areas into account.
Irreplaceability pattern was obtained using the 100 best network solutions for primate conservation in Legal Amazon. Grid cells were identified based on point locality occurrence data for primate species, taking into account already existing protected areas. This process was done at three resolutions: 1° (A, B and C), 0.5° (D, E and F) and 0.25° (G, H and I). Three categories of protected areas were used: strict use reserves (A, D and G), strict and sustainable use reserves (B, E and H), and strict and sustainable use reserves and indigenous lands (C, F and I). Grid cells are colour coded green for existing reserves, and yellow to red according to the irreplaceability pattern, which is the number of solutions in which each grid cell was identified as important, as indicated in the legend. States are indicated in A as follows: AC – Acre, AM – Amazonas, AP – Amapá, MA – Maranhão, MT – Mato Grosso, PA – Pará, RO – Rondônia, RR – Roraima, TO – Tocantins.

viability requires further exploration. It is also necessary to study each species individually and to include characteristics that relate to its persistence. Moreover, it is important to guarantee reserves with low anthropogenic disturbance, as these areas may be associated with higher primate density [55]. We found that strict use reserves alone did not represent primate species more effectively than random networks. These reserves are the only type of protected area that is exclusively dedicated to biodiversity conservation.

The existing Amazonian reserve system does not effectively represent primate species. We found this evaluation may be influenced by grid resolution, with an overestimation of species representation at coarser resolutions, and by type of data, with an overestimation of species representation using distribution data.

The historical selection of national reserves within the Brazilian Amazon was based on criteria other than primate conservation. For example, some focused on one or a limited number of species or vegetation types, while others were selected in an *ad hoc* manner [47], based on political, economic, or social motives, without using explicit criteria [56]. In Brazil, protected areas are biased toward low altitude areas, elevated terrains, and areas that are far from roads, urban aggregations, and agriculture-dominated areas [57]. We should continue to improve distribution knowledge and database quality for a variety of taxonomic groups [24]. A limited focus on a specific taxonomic group results in biased conclusions relating to that group, although there do exist coincidences in richness patterns of, e.g., orders of Amazonian mammals [58].

Beyond inventories, other studies are needed to improve knowledge about factors that influence population viability, differences in habitat quality, interrelations among species, environmental disturbances, and other important issues in ecology and conservation biology [23]. Field studies and planning may use concepts and tools from conservation biology applicable to fine scales to design local conservation strategies [23].

Acknowledgments

We thank Sally Hofmeyr and two anonymous reviewers, Fabiano Melo, Helena Bergallo, José Alexandre Diniz-Filho, Rafael Loyola and Rafael Laia for reading a previous version of this manuscript. We also thank Renato Gregorin for the first review of data relative to *Alouatta*.

Author Contributions

Conceived and designed the experiments: MPP CEVG. Performed the experiments: MPP AAL. Analyzed the data: MPP. Contributed reagents/materials/analysis tools: MPP CEVG JSSJ AAL. Wrote the paper: MPP CEVG JSSJ.

References

1. Achard F, Eva HD, Stibig HJ, Mayaux P, Gallego J, et al. (2002) Determination of deforestation rates of the world's humid tropical forests. Science 297: 999–1002. doi: 10.1126/science.1070656
2. Bird JP, Buchanan GM, Lees AC, Clay RP, Develey PF, et al. (2011) Integrating spatially explicit habitat projections into extinction risk assessments: a reassessment of Amazonian avifauna incorporating projected deforestation. Divers Distrib 18: 273–281. doi: 10.1111/j.1472-4642.2011.00843.x
3. Grelle CEV (2005) Predicting extinction of mammals in the Brazilian Amazon. Oryx 39: 347–350. doi: 10.1017/S0030605305000700
4. Wearn OR, Reuman DC, Ewers RM (2012) Extinction debt and windows of conservation opportunity in the Brazilian Amazon. Science 337: 228–232. doi: 10.1126/science.1219013
5. Grelle CEV, Fonseca GAB, Fonseca MT, Costa LP (1999) The question of scale in threat analysis: a case study with Brazilian mammals. Anim Conserv 2: 149–152.
6. Laurance WF, Vasconcelos HL (2009) Consequências ecológicas da fragmentação florestal na Amazônia. Oecol Bras 13: 434–451. doi: 10.4257/oeco.2009.1303.03
7. Asner GP, Rudel TK, Aide TM, Defries R, Emerson R (2009) A contemporary assessment of change in humid tropical forests. Conserv Biol 23: 1386–1395. doi: 10.1111/j.1523-1739.2009.01333.x
8. Asner GP, Knapp DE, Broadbent EN, Oliveira PJC, Keller M et al. (2005) Selective logging in the Brazilian Amazon. Science 310: 480–482. doi: 10.1126/science.1118051
9. Fa JE, Peres CA, Meeuwig J (2002) Bushmeat exploitation in tropical forests: an intercontinental comparison. Conserv Biol 16: 232–237. doi: 10.1046/j.1523-1739.2002.00275.x
10. Nepstad D, Moreira A, Verissimo A, Lefebvre P, Schlesinger P, et al. (1998) Forest fire prediction and prevention in the Brazilian Amazon. Conserv Biol 12: 951–953. doi: 10.1046/j.1523-1739.1998.012005951.x
11. Nepstad DC, Verissimo A, Alencar A, Nobre C, Lima E, et al. (1999) Large-scale impoverishment of Amazonian forests by logging and fire. Nature 398: 505–508. doi: 10.1038/19066
12. Peres CA, Nascimento HS (2006) Impact of game hunting by the Kayapo of south-eastern Amazonia: implications for wildlife conservation in tropical forest indigenous reserves. Biodivers Conserv 15: 2627–2653. doi: 10.1007/s10531-005-5406-9
13. Cochrane MA (2001) In the line of fire - Understanding the impacts of tropical forest fires. Environment 43: 28–38. doi: 10.1080/00139150109604505
14. Laurance WF, Useche DC (2009) Environmental synergisms and extinctions of tropical species. Conserv Biol 23: 1427–1437. doi: 10.1111/j.1523-1739.2009.01336.x
15. Strayer DL, Eviner VT, Jeschke JM, Pace ML (2006) Understanding the long-term effects of species invasions. Trends Ecol Evol 21: 645–651. doi: 10.1016/j.tree.2006.07.007
16. Brook BW, Sodhi NS, Bradshaw CJA (2008) Synergies among extinction drivers under global change. Trends Ecol Evol 23: 453–460. doi: 10.1016/j.tree.2008.03.011
17. Jenkins CN, Joppa L (2009) Expansion of the global terrestrial protected area system. Biol Conserv 142: 2166–2174. doi: 10.1016/j.biocon.2009.04.016
18. Brooks TM, Wright SJ, Sheil D (2009) Evaluating the success of conservation actions in safeguarding tropical forest biodiversity. Conserv Biol 23: 1448–1457. doi: 10.1111/j.1523-1739.2009.01334.x
19. Schmitt CB, Burgess ND, Coad L, Belokurov A, Besancon C, et al. (2009) Global analysis of the protection status of the world's forests. Biol Conserv 142: 2122–2130. doi: 10.1016/j.biocon.2009.04.012
20. Klink CA, Machado RB (2005) Conservation of the Brazilian Cerrado. Conserv Biol 19: 707–713. doi: 10.1111/j.1523-1739.2005.00702.x
21. Nepstad D, Schwartzman S, Bamberger B, Santilli M, Ray D, et al. (2006) Inhibition of Amazon deforestation and fire by parks and indigenous lands. Conserv Biol 20: 65–73. doi: 10.1111/j.1523-1739.2006.00351.x
22. Margules CR, Pressey RL (2000) Systematic conservation planning. Nature 405: 243–253. doi: 10.1038/35012251
23. Scott JM, Davis F, Csuti B, Noss R, Butterfield B, et al. (1993) Gap analysis: A geographic approach to protection of biological diversity. Wildlife Monogr 123: 3–41.
24. Jennings MD (2000) Gap analysis: concepts, methods, and recent results. Landscape Ecol 15: 5–20. doi: 10.1023/A:1008184408300
25. Pressey RL, Whish GL, Barrett TW, Watts ME (2002) Effectiveness of protected areas in north-eastern New South Wales: recent trends in six measures. Biol Conserv 106: 57–69. doi: 10.1016/S0006-3207(01)00229-4
26. Rodrigues ASL, Andelman SJ, Bakarr MI, Boitani L, Brooks TM, et al. (2004) Effectiveness of the global protected area network in representing species diversity. Nature 428: 640–643. doi: 10.1038/nature02422
27. O'Dea N, Araujo MB, Whittaker RJ (2006) How well do important bird areas represent species and minimize conservation conflict in the tropical Andes? Divers Distrib 12: 205–214. doi: 10.1111/j.1366-9516.2006.00235.x
28. Araujo MB, Lobo JM, Moreno JC (2007) The effectiveness of Iberian protected areas in conserving terrestrial biodiversity. Conserv Biol 21: 1423–1432. doi: 10.1111/j.1523-1739.2007.00827.x
29. Jantke K, Schleupner C, Schneider UA (2011) Gap analysis of European wetland species: priority regions for expanding the Natura 2000 network. Biodivers Conserv 20: 581–605. doi: 10.1007/s10531-010-9968-9
30. Tognelli MF, Arellano PIR, Marquet PA (2008) How well do the existing and proposed reserve networks represent vertebrate species in Chile? Divers Distrib 14: 148–158. doi: 10.1111/j.1472-4642.2007.00437.x
31. Pinto MP, Grelle CEV (2009) Reserve selection and persistence: complementing the existing Atlantic Forest reserve system. Biodivers Conserv 18: 957–968. doi: 10.1007/s10531-008-9513-2
32. Chapman CA, Peres CA (2001) Primate conservation in the new millennium: The role of scientists. Evol Anthropol 10: 16–33. doi: 10.1002/1520-6505(2001)10:1<16::AID-EVAN1010>3.0.CO;2-O
33. Paglia AP, Fonseca GAB, Rylands AB, Herrmann G, Aguiar LMS, et al. (2012) Annotated checklist of Brazilian mammals. 2nd ed. Occasional Papers in Conservation Biology, No 6. Arlington: Conservation International. 76p.
34. Machado ABM, Drummond GM, Paglia AP (2008) Livro vermelho da fauna brasileira ameaçada de extinção. Brasília: MMA, Belo Horizonte: Fundação Biodiversitas. 907p.
35. Silva JMC, Rylands AB, Silva-Júnior JS, Gascon C, Fonseca GAB (2005) Primate diversity patterns and their conservation in Amazonia. In: Purvis A, Gittleman L, Brooks T, editors. Phylogeny and conservation.Cambridge: Cambridge University Press. pp. 337–364.
36. Ayres JM, Cluttonbrock TH (1992) River boundaries and species range size in Amazonian primates. Am Nat 140: 531–537.
37. Harcourt AH (2006) Rarity in the tropics: biogeography and macroecology of the primates. J Biogeogr 33: 2077–2087. doi: 10.1111/j.1365-2699.2006.01557.x
38. IUCN (2012) IUCN Red List of Threatened Species. Version 2012.1. http://www.iucnredlist.org. Downloaded on November/2014.
39. Brito D, Grelle CEV (2006) Estimating minimum area of suitable habitat and viable population size for the northern muriqui (*Brachyteles hypoxanthus*). Biodivers Conserv 15: 4197–4210. doi: 10.1007/s10531-005-3575-1
40. Soulé ME (1987) Viable populations for conservation. Cambridge: Cambridge University Press. pp. 101–148.
41. Bicca-Marques JC, Silva VM, Gomes DF (2006) Ordem Primates. In: Reis NR, Peracchi AL, Pedro WA, Lima IP, editors. Mamíferos do Brasil. Londrina.

42. Game ET, Grantham HS (2008) Marxan user manual: for Marxan version 1.8.10. Australia: University of Queensland. Canada: Pacific Marine Analysis and Research Association.

43. Possingham H, Ball I, Andelman S (2000) Mathematical methods for identifying representative reserve networks. In: Ferson S, Burgman M, editors. Quantitative methods for conservation biology. New York: Springer. pp. 291–306.

44. Ball I, Possingham HP (2000) Marxan (v1.8.2). User manual.

45. Ball IR, Possingham HP, Watts ME (2009) Marxan and relatives: software for spatial conservation prioritization. In: Moilanen A, Wilson KA, Possingham HP, editors. Spatial conservation prioritization: quantitative methods & computational tools. Oxford: Oxford University Press. pp. 185–195.

46. R Development Core Team (2012) R: A language and environment for statistical computing. Vienna: R Foundation for Statistical Computing. http://www.R-project.org/.

47. Hurlbert AH, Jetz W (2007) Species richness, hotspots, and the scale dependence of range maps in ecology and conservation. PNAS 104: 13384–13389. doi: 10.1073/pnas.0704469104

48. Marini MA, Barbet-Massin M, Lopes LE, Jiguet F (2009) Major current and future gaps of Brazilian reserves to protect Neotropical savanna birds. Biol Conserv 142: 3039–3050. doi: 10.1016/j.biocon.2009.08.002

49. Harcourt AH (2000) Coincidence and mismatch of biodiversity hotspots: a global survey for the order, primates. Biol Conserv 93: 163–175. doi: 10.1016/S0006-3207(99)00145-7

50. Goldani A, Carvalho GS, Bicca-Marques JC (2006) Distribution patterns of neotropical primates (Platyrrhini) based on parsimony analysis of endemicity. Braz J Biol 66: 61–74. doi: 10.1590/S1519-69842006000100009

51. Rondinini C, Wilson KA, Boitani L, Grantham H, Possingham HP (2006) Tradeoffs of different types of species occurrence data for use in systematic conservation planning. Ecol Lett 9: 1136–1145. doi: 10.1111/j.1461-0248.2006.00970.x

52. Bini LM, Diniz JAF, Rangel T, Bastos RP, Pinto MP (2006) Challenging Wallacean and Linnean shortfalls: knowledge gradients and conservation planning in a biodiversity hotspot. Divers Distrib 12: 475–482. doi: 10.1111/j.1366-9516.2006.00286.x

53. Whittaker RJ, Araujo MB, Paul J, Ladle RJ, Watson JEM, et al. (2005) Conservation Biogeography: assessment and prospect. Divers Distrib 11: 3–23. doi: 10.1111/j.1366-9516.2005.00143.x

54. Pressey RL, Cowling RM (2001) Reserve selection algorithms and the real world. Conserv Biol 15: 275–277. doi: 10.1111/j.1523-1739.2001.99541.x

55. Fuller HL, Harcourt AH, Parks SA (2009) Does the population density of primate species decline from centre to edge of their geographic ranges? J Trop Ecol 25: 387–392. doi: 10.1017/S0266467409006063

56. Schulman L, Ruokolainen K, Junikka L, Saaksjarvi IE, Salo M, et al. (2007) Amazonian biodiversity and protected areas: Do they meet? Biodivers Conserv 16: 3011–3051. doi: 10.1007/s10531-007-9158-6

57. Joppa LN, Pfaff A (2009) High and Far: Biases in the Location of Protected Areas. Plos One, 4: e8273. doi: 10.1371/journal.pone.0008273

58. Sebastiao H, Grelle CEV (2009) Taxon surrogates among Amazonian mammals: Can total species richness be predicted by single orders? Ecol Indic 9: 160–166. doi: 10.1016/j.ecolind.2008.03.002

PERMISSIONS

LIST OF CONTRIBUTORS

Niels Jobstvogt
Oceanlab, University of Aberdeen, Aberdeen, United Kingdom
Aberdeen Centre for Environmental Sustainability (ACES), University of Aberdeen, Aberdeen, United Kingdom

Michael Townsend
National Institute of Water and Atmospheric Research (NIWA), Hamilton, New Zealand

Ursula Witte
Oceanlab, University of Aberdeen, Aberdeen, United Kingdom

Nick Hanley
Department of Geography and Sustainable Development, University of St Andrews, St Andrews, United Kingdom

Ruidong Wu
Institute of International Rivers and Eco-security, Yunnan University, Kunming, Yunnan, China
Yunnan Key Laboratory of International Rivers and Transboundary Ecosecurity, Yunnan University, Kunming, Yunnan, China

Yongcheng Long, Shuang Zhang3 Peng Zhao and Longzhu Wang
The Nature Conservancy China Program, Kunming, Yunnan, China

George P. Malanson
Department of Geography, The University of Iowa, Iowa City, Iowa, United States of America

Paul A. Garber
Department of Anthropology, University of Illinois, Urbana, Illinois, United States of America

Diqiang Li
Institute of Forest Ecology, Environment, and Protection, Chinese Academy of Forestry, Beijing, China

Hairui Duo
School of Nature Reserve, Beijing Forestry University, Beijing, China

Rafael de Fraga
Instituto Nacional de Pesquisas da Amazônia – Programa de Pós-graduação em Ecologia, Manaus, Amazonas, Brazil

Adam J. Stow
Department of Biological Sciences, Macquarie University, Sydney, New South Wales, Australia

William E. Magnusson and Albertina P. Lima
Instituto Nacional de Pesquisas da Amazônia – Coordenação de Biodiversidade, Manaus, Amazonas, Brazil

Jing Luo, Emily Walsh and Abhishek Naik
Department of Plant Biology & Pathology, Rutgers University, New Brunswick, New Jersey, United States of America

Wenying Zhuang and Lei Cai
State Key Laboratory of Mycology, Institute of Microbiology, Chinese Academy of Sciences, Beijing, China

Keqin Zhang
Laboratory for Conservation and Utilization of Bio-Resources and Key Laboratory for Microbial Resources of the Ministry of Education, Yunnan University, Kunming, China

Ning Zhang
Department of Plant Biology & Pathology, Rutgers University, New Brunswick, New Jersey, United States of America
Department of Biochemistry & Microbiology, Rutgers University, New Brunswick, New Jersey, United States of America

Viviane G. Ferro, Adriano S. Melo and Rafael Loyola
Departamento de Ecologia, Universidade Federal de Goiás, Goiânia, Goiás, Brazil

Priscila Lemes
Programa de Pós-Graduação em Ecologia e Evolução, Universidade Federal de Goiás, Goiânia, Goiás, Brazil

Danielle Fraser and Natalia Rybczynski
Department of Biology, Carleton University, Ottawa, Ontario, Canada
Palaeobiology, Canadian Museum of Nature, Ottawa, Ontario, Canada

Christopher Hassall
School of Biology, University of Leeds, Leeds, United Kingdom
Department of Biology, Carleton University, Ottawa, Ontario, Canada

Root Gorelick
Department of Biology, Carleton University, Ottawa, Ontario, Canada
Department of Mathematics and Statistics, Carleton University, Ottawa, Ontario, Canada
Institute of Interdisciplinary Studies, Carleton University, Ottawa, Ontario Canada

Annamaria Bevivino, Patrizia Paganin, Maite Sampedro Pellicer and Claudia Dalmastri
ENEA (Italian National Agency for New Technologies, Energy and Sustainable Economic Development) Casaccia Research Center, Technical Unit for Sustainable Development and Innovation of Agro-Industrial System, Rome, Italy

Giovanni Bacci
Consiglio per la Ricerca e la Sperimentazione in Agricoltura - Research Centre for the Soil-Plant System, Rome, Italy
Laboratory of Microbial and Molecular Evolution, Department of Biology, University of Florence, Florence, Italy

Alessandro Florio and Anna Benedetti
Consiglio per la Ricerca e la Sperimentazione in Agricoltura - Research Centre for the Soil-Plant System, Rome, Italy

Maria Cristiana Papaleo, Alessio Mengoni and Renato Fani
Laboratory of Microbial and Molecular Evolution, Department of Biology, University of Florence, Florence, Italy

Luigi Ledda
Dipartimento di Agraria, University of Sassari, Sassari, Italy

Catharina J. E. Schulp, Jasper Van Vliet and Peter H. Verburg
Faculty of Earth and Life Sciences, VU University Amsterdam, Amsterdam, the Netherlands

Benjamin Burkhard
Institute for Natural Resource Conservation, Kiel University, Kiel, Germany
Leibniz Centre for Agricultural Landscape Research ZALF, Müncheberg, Germany

Joachim Maes
European Commission - Joint Research Centre, Institute for Environment and Sustainability, Ispra, Varese, Italy

Stalin Nithaniyal
Department of Genetic Engineering, Center for DNA Barcoding, SRM University, Chennai, India
Interdisciplinary School of Indian System of Medicine, SRM University, Chennai, India

Steven G. Newmaster and Subramanyam Ragupathy
Centre for Biodiversity Genomics, University of Guelph, Guelph, Ontario, Canada

Devanathan Krishnamoorthy, Sophie Lorraine Vassou and Madasamy Parani
Department of Genetic Engineering, Center for DNA Barcoding, SRM University, Chennai, India

Xoaquín Moreira
Institute of Biology, Laboratory of Evolutive Entomology, University of Neuchâtel, Neuchâtel, Switzerland
Department of Ecology and Evolutionary Biology, University of California Irvine, Irvine, California, United States of America

Luis Abdala-Roberts and Kailen A. Mooney
Department of Ecology and Evolutionary Biology, University of California Irvine, Irvine, California, United States of America

Víctor Parra-Tabla
Departamento de Ecología Tropical, Campus de Ciencias Biológicas y Agropecuarias, Universidad Autónoma de Yucatán, Mérida, Yucatán, México

Brett Favaro
Department of Biology, University of Victoria, Victoria, Canada

Centre for Sustainable Aquatic Resources, Fisheries and Marine Institute of Memorial University of Newfoundland, St John's, Canada
Department of Ocean Sciences, Memorial University of Newfoundland, St. John's, Canada

Danielle C. Claar, Cameron Freshwater and Jessica J. Holden
Department of Biology, University of Victoria, Victoria, Canada

Caroline H. Fox
Department of Geography, University of Victoria, Victoria, Canada
Raincoast Conservation Foundation, Sidney, Canada

Allan Roberts
Bamfield Marine Sciences Centre, Bamfield East, Canada

UVic Research Derby
Department of Biology, University of Victoria, Victoria, Canada
Department of Geography, University of Victoria, Victoria, Canada
School of Environmental Studies, University of Victoria, Victoria, Canada

Yuhua Zhang, Yongfan Wang and Shixiao Yu
Department of Ecology, School of Life Sciences/State Key Laboratory of Biocontrol, Sun Yat-sen University, Guangzhou, China

Ashley B. Bennett and Rufus Isaacs
Department of Entomology and Great Lakes Bioenergy Research Center, Michigan State University, East Lansing, Michigan, United States of America

Timothy D. Meehan and Claudio Gratton
Department of Entomology and Great Lakes Bioenergy Research Center, University of Wisconsin - Madison, Madison, Wisconsin, United States of America

Riccardo Di Clemente
IMT Institute for Advanced Studies Lucca, Lucca, Italy
Istituto dei Sistemi complessi ISC-CNR, UOS Sapienza, Roma, Italy

Guido L. Chiarotti and Matthieu Cristelli
Istituto dei Sistemi complessi ISC-CNR, UOS Sapienza, Roma, Italy

Andrea Tacchella
Istituto dei Sistemi complessi ISC-CNR, UOS Sapienza, Roma, Italy
Dipartimento di Fisica, Universitàdi Roma "Sapienza", Roma, Italy

Luciano Pietronero
Istituto dei Sistemi complessi ISC-CNR, UOS Sapienza, Roma, Italy
Dipartimento di Fisica, Universitàdi Roma "Sapienza", Roma, Italy
London Institute for Mathematical Sciences, London, United Kingdom

Alan M. Friedlander
Fisheries Ecology Research Laboratory, Department of Biology, University of Hawaii, Honolulu, Hawaii, United States of America
Pristine Seas, National Geographic Society, Washington, DC, United States of America

Enric Ballesteros
Centre d'Estudis Avançats-CSIC, Blanes, Spain

Michael Fay
Pristine Seas, National Geographic Society, Washington, DC, United States of America
Wildlife Conservation Society, Bronx, New York, United States of America
Special Advisor, Presidence de la République, Libreville, République Gabonaise

Enric Sala
Pristine Seas, National Geographic Society, Washington, DC, United States of America
Centre d'Estudis Avançats-CSIC, Blanes, Spain

Sebastian Hopfenmüller, Ingolf Steffan-Dewenter and Andrea Holzschuh
Department of Animal Ecology and Tropical Biology, Biocenter, University of Würzburg, Würzburg, Germany

Michael J. Emslie, Alistair J. Cheal and Kerryn A. Johns
Australian Institute of Marine Science, Townsville, Queensland, Australia

Uttam Babu Shrestha
Institute for Agriculture and the Environment, University of Southern Queensland, Toowoomba, Queensland, Australia

Kamaljit S. Bawa
Department of Biology, University of Massachusetts, Boston, Massachusetts, United States of America
Ashoka Trust for Research in Ecology and Environment (ATREE), Bangalore, India

Míriam Plaza Pinto
Programa de Pós-Graduação em Ecologia, Universidade Federal do Rio de Janeiro, Rio de Janeiro, Rio de Janeiro, Brazil
Departamento de Ecologia, Universidade Federal do Rio Grande do Norte, Natal, Rio Grande do Norte, Brazil

José de Sousa e Silva-Júnior
Coordenação de Zoologia, Museu Paraense Emilio Goeldi, Belém, Pará, Brazil

Adriana Almeida de Lima
Departamento de Ecologia, Universidade Federal do Rio Grande do Norte, Natal, Rio Grande do Norte, Brazil

Carlos Eduardo Viveiros Grelle
Departamento de Ecologia, Universidade Federal do Rio de Janeiro, Rio de Janeiro, Rio de Janeiro, Brazil
Laboratório de Vertebrados, Universidade Federal do Rio de Janeiro, Rio de Janeiro, Rio de Janeiro, Brazil

Index